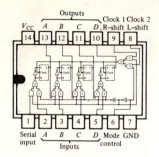

4-bit right-shift left-shift register
7495 (TTL)

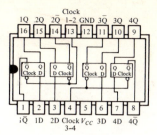

4-bit bistable latch
7475 (TTL)

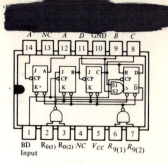

Decade ripple counter
7490 (TTL)

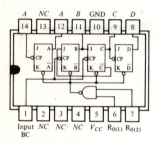

Divide-by-twelve ripple counter
7492 (TTL)

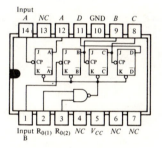

4-bit binary ripple counter
7493 (TTL)

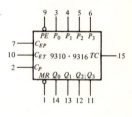

Synchronous up counters (TTL)
decade-9310
hexadecimal-9316

**INTRODUCTION
TO INTEGRATED
CIRCUITS**

**McGRAW-HILL SERIES ON THE
FUNDAMENTALS OF ELECTRONIC SCIENCE**

Consulting Editor: JAMES F. GIBBONS, Stanford University

GRINICH AND JACKSON: Introduction to Integrated Circuits
SIEGMAN: Introduction to Lasers and Masers

**McGRAW-HILL
BOOK COMPANY**

New York
St. Louis
San Francisco
Auckland
Düsseldorf
Johannesburg
Kuala Lumpur
London
Mexico
Montreal
New Delhi
Panama
Paris
São Paulo
Singapore
Sydney
Tokyo
Toronto

VICTOR H. GRINICH

Department of Electrical Engineering
Stanford University

HORACE G. JACKSON

Department of Electrical Engineering
and Computer Sciences
University of California, Berkeley

Introduction to Integrated Circuits

This book was set in Times New Roman.
The editors were Kenneth J. Bowman and Madelaine Eichberg;
the cover was designed by Joseph Gillians;
the production supervisor was Sam Ratkewitch.
The drawings were done by Vantage Art, Inc.
Kingsport Press, Inc., was printer and binder.

Library of Congress Cataloging in Publication Data

Grinich, Victor H.
    Introduction to integrated circuits.

    (McGraw-Hill series on the fundamentals of electronic science)
    Includes bibliographical references and index.
    1. Integrated circuits. 2. Electronic circuit design. I. Jackson, Horace G., joint author.
II. Title.
TK7874.G76   621.381'73   74-23920
ISBN 0-07-024875-3

**INTRODUCTION
TO INTEGRATED
CIRCUITS**

1 2 3 4 5 6 7 8 9 0 K P K P 7 9 8 7 6 5

CONTENTS

PREFACE

The process of electronic design has undergone a revolution in recent years. Until the mid-1960s a large number of electrical engineers were engaged in circuit design. Using discrete parts (such active elements as transistors and diodes, and passive components like resistors, capacitors, and inductors), they designed amplifiers, oscillators, switching networks, and a variety of other circuits that were then combined to build electronic systems such as voltmeters, TV receivers, computers, and so on. Because each circuit was specially designed for a particular application, its performance could be optimized to match that of the other basic circuits in the system. Furthermore, since there are usually several arrangements of such essential parts as transistors and resistors that will accomplish a specific circuit purpose, there was ample room for creativity in the design of each circuit. Thus there was always a critical need for those rare individuals called "clever circuit designers." Under these conditions, basic textbooks in electronic design could properly concentrate on such fundamental circuits as the single-stage amplifier and the flip-flop, leaving more sophisticated circuits for later courses and/or design experience.

While some design of this type, using discrete parts, is still done (and there is always a need for ingenious circuit designers), the advent of integrated circuits

(ICs) has reduced it considerably. Many of the circuits which previously were specially designed for each application are now available in a standard integrated form. As a consequence, design is increasingly a matter of selecting the proper combination of IC operational amplifiers, logic gates, counters, etc., and specifying their interconnections and operating conditions so that they will together perform a specified task. Integrated circuits have thus tended to standardize the design process, in some cases to such an extent that the circuit design is the implementing of a complete system block diagram. As a corollary, it follows that ICs have brought the process of system design within the grasp of a much broader range of users and have accordingly multiplied the applications of integrated electronics enormously.

The question then arises as to how first courses and basic textbooks should be organized to account for IC technology. Clearly they cannot deal exclusively with the design of single-stage amplifiers and flip-flops. It would be nearer the mark to simply treat the available range of ICs as circuit elements having specified terminal characteristics, organizing the first course in such a way that students begin immediately to gain practical experience in using ICs to build electronic systems for particular applications. Indeed, teaching this type of design forms one of the major objectives of this book.

At the same time, however, one must face the fact that circuit standardization could reduce the incentive for creativity; a designer is unlikely to use ICs in an unconventional way if he simply treats them as functional black boxes. Hence an important objective of this book is to show how the most important ICs operate internally. With additional attention to these features of the design and performance of ICs, a knowledgeable designer can obtain additional unconventional functions. We hope in this way to encourage creative circuit design with ICs.

Organization   To accomplish these objectives we make use of the following topical organization. We open with an Introduction, where we discuss the difference between an analog system and a digital system, and also describe three small electronic systems which illustrate the use of ICs in (1) an all-analog, (2) an all-digital, and (3) a combined analog-digital application.

The contents of subsequent chapters is as follows:

Chapter 1 discusses the basic circuit models for transistors and diodes, and we show how they are used to analyze some simple but important circuits common to many ICs.

Chapter 2 introduces the basic elements of logic analysis that will be used in the following chapters on digital circuits. We discuss some simple Boolean algebra, and logic reduction by means of Karnaugh maps. Diode gates are introduced for purposes of circuit reinforcement.

Chapter 3 provides a detailed study of the major types of IC logic gates and shows how they are characterized. The dc operation of RTL, DTL, TTL, ECL, and MOS circuits is covered.

Chapter 4 then discusses the use of these circuits in forming more complex functions than the single logic gate. Examples are the EXCLUSIVE-OR, digital comparator, adder, and decoder circuits. Circuit reinforcement is accomplished through a discussion of unused inputs, collector logic, and totem-pole output circuits.

Chapter 5 is on the fundamentals of the design and operation of latches and flip-flops. Circuit reinforcement is achieved from detailed designs with latches. An introduction to state tables is made by conversion among T, D, and J-K flip-flops. The Schmitt trigger and monostable multivibrator circuits are also briefly discussed.

Chapter 6 discusses the design of a variety of standard and nonstandard counters and shift registers that are made from flip-flops. Features needed for designing a digital clock are covered here.

Chapter 7 begins the second portion of the book. Using a basic hybrid-π model, we study the common-emitter amplifier, the emitter follower, and the common-base circuit. Single-stage and multistage amplifier circuits are considered.

Chapter 8 studies the differential amplifier (emitter-coupled pair). Half-circuits are used for analysis. Conversion to single-ended output and an IC constant current source are also studied. A detailed circuit analysis is made of two popular IC operational amplifiers.

Chapter 9 deals with applications of operational amplifiers that employ resistance feedback. Inverting, noninverting, and voltage-follower applications are considered. The generalized treatment of all feedback cases is not presented, but enough is done to provide an understanding of the advantages and pitfalls in using feedback. Frequency-compensation techniques are introduced.

Chapter 10 is an applications chapter concerned with the use of both digital and linear ICs in electronic instrumentation. A detailed analysis and design of a dual-slope digital voltmeter is included, as well as a digital-to-analog converter and a phase-locked-loop circuit.

Approach   Our approach to the material is similar to that of other books in this series. We attempt to present throughout a simple but still adequately quantitative description of circuit design techniques, using examples liberally to illustrate our methods and relying on a number of demonstrations to (1) show how ICs actually work and (2) assess the quality of agreement between design theory and experimental reality. We are confident that the reader's grasp of all the basic ideas will be improved enormously from the demonstrations. The best philosophy to adopt is that theory alone is only half a loaf.

Within each chapter we have liberally placed exercise problems (with answers at the back of the book) at the conclusion of main sections. Thus the student is able to assure himself of his comprehension of the section before proceeding to the next. At the end of each chapter there are 30 to 40 problems, of increased complexity, covering the subject matter of the whole chapter. The solutions to these problems are included in the Instructor's Manual. Problems indicated with a star (*) are the more advanced "challenge" problems.

In the way of background preparation, we assume little more than maturity on the part of the student; though previous familiarity with Kirchhoff's laws (as they apply in both dc and ac circuits), and some circuit experience with diodes and transistors as found in a basic electricity course is certainly helpful. Our intention here is to be as practical in our approach to the subject as we can, consistent with providing a firm basis for quantitative design.

Lesson Outline In general it is expected that most students using this text will already have had a basic electronic course. The following outline is then suggested for a one-quarter course, with three hours of lecture and a three-hour laboratory period each week:

Week 1 Chapter 1: Much of this material would only need to be reviewed. The section on MOS transistors should be left out.

Week 2 Chapter 2: We find that students readily assimilate the contents of this chapter in one week.

Week 3 {Chapter 3:} Covering the characteristics of the inverter and RTL the
Week 4 {   (Part I)} first week, and DTL and TTL the second week.

Week 5 Chapter 4: Taking *either* the simple adder *or* the decoder circuit as an example of combinational logic.

Week 6 Chapter 5: Covering the analysis and design of flip-flops (especially the *J-K*). The section on the Schmitt trigger and monostable multivibrator should be left out.

Week 7 Chapter 6: Concentrating on the design of synchronous counters.

Week 8 Chapter 7: Again, much of this material need only be a review for most students.

Week 9 Chapter 8: A full week, even with the analysis of the 741 omitted.

Week 10 Chapter 9: The analysis with an ideal op amp can readily be accomplished. Time will not allow the coverage of *both* the more formal feedback analysis *and* frequency compensation.

For a semester course, the three additional weeks allow for:

1 Complete coverage of Chapter 3 (except the MOS circuits) in three weeks.

2 Complete coverage of Chapter 9 in two weeks; that is, frequency compensation is now included.

3 The design of the DVM (in Chapter 10) makes for a satisfying conclusion to the course.

More complete details of these lesson outlines, as well as a two-quarter sequence, are given in the Instructor's Manual.

ACKNOWLEDGMENTS

As noted earlier the advent of the IC has necessitated changes in the teaching of the basic electronic circuit courses. This book is a result of these changes that were made at the University of California at Berkeley. It is therefore a pleasure to acknowledge that the new course outline at Berkeley was prepared under the capable leadership of Professor Donald O. Pederson, with able assistance also coming from Gary Baldwin and Bruce Wooley. We would also like to thank Peter Appel, who prepared many of the problems.

Most of the typing was done by Mrs. V. Donelson and Mrs. E. Ferriter; their skilled help is gratefully acknowledged.

We also acknowledge our gratitude to the many reviewers for their helpful suggestions.

To our editor, Professor J. F. Gibbons, we are indebted for his considerable aid and unbounded enthusiasm in the preparation of this manuscript.

Finally we wish to thank our wives and families for their enduring patience.

VICTOR H. GRINICH
HORACE G. JACKSON

INTRODUCTION

In this text we intend to introduce the main concepts that are needed to understand the operation of existing integrated circuits and their applications in basic information-processing circuits. Our concentration on integrated circuits (ICs) is justified by the fact that they are used wherever it is feasible to do so in modern electronic systems. Our concentration on information-processing applications arises from the "user's interest" that students from a variety of fields have in this subject.

I.1 ANALOG AND DIGITAL SYSTEMS

To set our work in a proper perspective, it is useful to begin with an examination of the two basic ways in which information is represented in an electronic system. In an *analog* system, a continuous property of an electrical signal, such as its amplitude or frequency, is made to be an exact replica of the original signal that is being processed. One example of an analog system is a public-address system, where the amplitude of the electrical signal throughout the amplifier is directly proportional to the variations in air pressure that are produced when a person speaks into a microphone. ICs that process signals for such systems are called *linear* or *analog* ICs.

In a *digital* system, on the other hand, electrical signals are used to represent *numbers* or the occurrence of *events*. In a traffic-control system, for example, cars that are to cross a major thoroughfare have to drive over a pressure plate in the street to actuate a system that will, after a suitable delay, turn the light GREEN in their direction. Such a system is designed to carry out a *predetermined* set of operations when a specific input condition (car passing over the pressure plate) has been satisfied. The designer of such a system can choose *whatever* electrical signals he pleases to represent events within the system, so long as the proper system output is obtained (light turns GREEN) after the proper system input is applied (car passes over pressure plate).

A variety of different signals, including dc levels, pulses, or appropriately modulated sine waves, could be used within the traffic-control system to achieve the desired system response. However, with ICs it is customary to choose *two* predetermined dc voltage levels (typically 0 V, called LOW, and $+3$ V, called HIGH) to represent events in such a system. The system is then designed to perform the necessary operations on a number of logic signals, each of which can take on one of these two voltage levels. ICs that process signals for such systems, including those that either *store* the logic values or perform the *logic operations*, are called *digital ICs*.

Of course, there are many electronic systems which must perform both analog and digital operations. In fact, most systems of practical importance require both analog and digital processing, and in these cases both analog and digital ICs are used in the system construction. The following examples will help to suggest the types of systems in which all analog, all digital, or a mixture of analog *and* digital circuits are used to obtain the required system performance.

I.1.1   A Purely Analog System: The Audio Amplifier

It is frequently useful to visualize the signal processing that occurs in an electronic system in terms of a block diagram. The block diagram consists of a series of interconnected boxes, each box representing a particular signal-processing operation that must be performed in the system under consideration. A block diagram of a low-power public-address system is shown in Fig. I.1a. This is a very simple example of a purely analog system. It consists of a carbon microphone, an audio amplifier, a speaker, and a battery (or a power supply) which is not shown. The function of the microphone is to faithfully convert the air-pressure variations that arrive at its surface into variations of the electrical resistance of the microphone. To obtain an electrical signal, the resistance fluctuations are then converted to a voltage by passing a constant current through the microphone. With a constant current of 11 mA, a common carbon microphone would typically develop an open-circuit peak-signal voltage of 220 mV and have an internal ac resistance of 250 Ω.

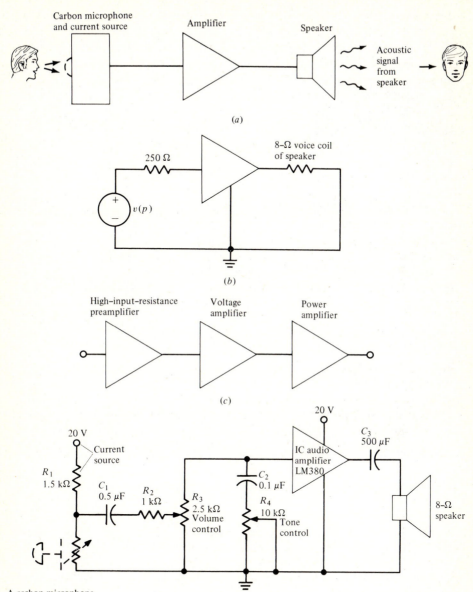

(a)

(b)

(c)

(d)

FIGURE I.1
An example of a small system—a public address system. (a) Block diagram of a public address system. (b) Electronic schematic for (a). (c) Subfunctions of the amplifier in (b). (d) Complete schematic for a public address system using one IC as the amplifier. (e) An IC audio amplifier mounted on a printed circuit board. (*Courtesy National Semiconductor Corporation.*)

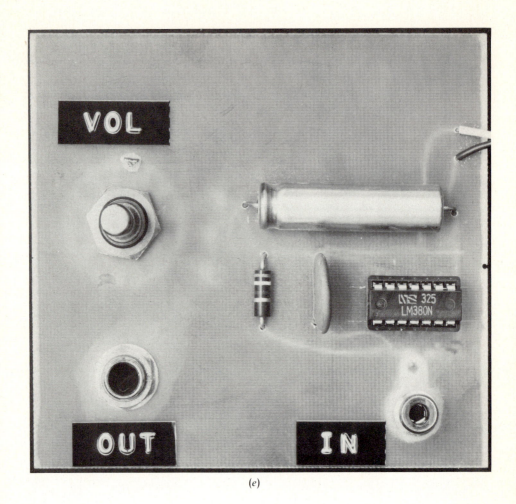

(e)

From an electrical viewpoint, the output of the first box in Fig. I.1a is therefore a series combination of a pressure-dependent voltage source $v(p)$ and a 250-Ω resistor. This combination provides the electrical input signal for the audio amplifier, as shown in Fig. I.1b. To this point the analog character of the system rests on the assumption that the instantaneous output voltage of the microphone will indeed be accurately proportional to the original variations in air pressure arriving at the surface of the microphone. To the extent that this is true, the first box has simply converted the information from one analog form to another.

The function of the audio amplifier is to increase the signal power to the point where we can obtain a usefully loud output from the speaker. It does this by converting dc power (taken from the battery) into ac power delivered to the speaker. If we want to have 5 W of peak power into an 8-Ω speaker, we need an amplifier that will deliver a peak output voltage of $V = (5 \text{ W} \times 8 \text{ }\Omega)^{1/2} = 6.3 \text{ V}$

across an 8-Ω load. This follows from $P = I^2R = V^2/R$, where P is the power into the resistor R, which has a voltage V across it when a circuit I flows through it. If the 220-mV signal is impressed directly on the amplifier input, the *voltage gain* of the amplifier must be at least 28.7 (i.e., 6.3/0.22). To retain the analog character of the system, we need to make sure that the output voltage of the amplifier is exactly (or at least reasonably) proportional to its input at every instant. One specification that is a translation of this requirement is the *frequency response* which the amplifier must have. In this case, since we want the system to reproduce essentially any sound that a human ear can hear, we require that the audio amplifier be capable of delivering its 6.3-V output at any signal frequency† from 40 Hz up to perhaps 20 kHz, providing only that the microphone could supply the necessary 220-mV input at the signal frequency in question. Then the amplifier should be designed to meet this condition if the second box is to preserve completely the analog character of the signal.

From thinking of this type we can make considerable progress toward setting out the specifications from which a circuit designer must work to build each box. If we were not using ICs but *discrete* parts to build the audio system, the next step for the circuit designer would be to expand the audio amplifier block into its own series of interconnected boxes. The first one might be specially designed to couple the signal to the amplifier effectively, intermediate boxes might be designed to increase the signal voltage level, and the final box might contain a special configuration of parts that would provide the required output while simultaneously meeting other design requirements. These subfunctions are represented in the boxes shown in Fig. I.1c and would form the specifications from which a circuit designer could begin to develop an audio amplifier that was to be made from discrete parts.

The advent of ICs has, however, nearly eliminated this last step. Fully integrated audio amplifiers are available which perform *all* the subfunctions indicated in Fig. I.1c. As shown in Fig. I.1d, it is only necessary to add discrete components that (1) couple the output signal of the amplifier to the speaker (C_3, a 500-μF capacitor) so that the desired ac signal can be transmitted to the load but with no dc current permitted to flow from the IC into the speaker, and (2) couple the microphone to the amplifier input through a combination of resistors and capacitors that can provide dc blocking (C_1) and controls for volume (R_3) and tone (R_2, C_2, R_4). For our modest power requirements systems can be constructed using one IC, as shown in Fig. I.1e. For a slightly more complicated system, such as a 250-W airport-paging system, we would need to design special output stages that would use discrete *power* transistors, with ICs then being used to provide all the low-power functions.

It follows from these latter remarks that, as we expand our view to more complicated analog systems, the block diagrams will become more complicated, and it is unlikely that we will be able to find a single IC that will perform all the

---

† The unit of frequency is Hertz, which is abbreviated as Hz.

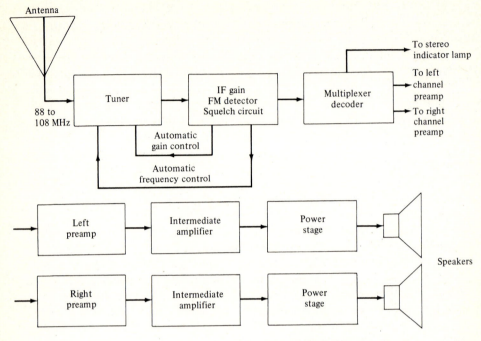

FIGURE I.2
Block diagram of a FM stereo receiver.

required analog signal-processing operations. However, available ICs will nearly always reduce the design of an analog system to the selection of a few analog ICs and some associated discrete parts. For example, the signal-processing part of an FM stereo receiver can be built using just four ICs, three discrete transistors for the very-high-frequency signal processing, and a few tuning components to select a desired station. The block diagram for such a receiver and a possible realization of it are shown in Fig. I.2. The two output amplifiers shown could be those used in Fig. I.1d. Again the system is an analog system because each box has an output that is exactly (or nearly) proportional to that part of its input that contains the desired signal. It turns out, however, that *digital* signal processing takes place inside the multiplex decoder, even though we do not see this at the block-diagram level.

I.1.2 A Purely Digital System: The Digital Clock

To elaborate on how a digital system and digital information processing are different from the analog systems just described, let us consider the basic operation of a digital clock. An oversimplified block diagram for such a clock is shown in Fig. I.3 and consists of a *time base*, *pulse counters*, and *display devices*. The

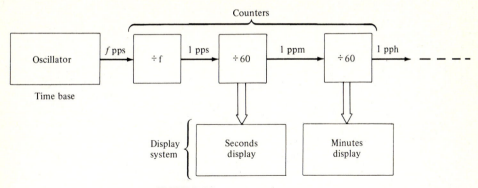

FIGURE I.3
Block diagram for the seconds and minutes portion of a digital clock.

manner in which these basic components function to indicate time is as follows. First, the block labeled *time base* is an electronic *oscillator* which is assumed to deliver *exactly f* pulses per second. Elapsed time can then be determined by simply counting output pulses from the oscillator and arranging to display the count in seconds, minutes, and hours. The accuracy of the clock is determined by the degree to which the time base delivers exactly f pulses each second.

To generate the display, the time base is followed by a tandem connection of counters that are, in the simplest case, arranged to give outputs each time the time base has delivered f, $60f$, and $3600f$ pulses. The outputs of the counters then correspond to SECONDS, MINUTES, and HOURS. To be more specific, the first counter in the chain is internally wired so as to give (at one of its output pins) *one* pulse for every f input pulses. It then returns to its zero state and begins to count input pulses once again. We will refer to this as a *divide-by-f* or $\div f$ counter. The pin that provides the single-pulse output after f input pulses is called the *terminal-count* pin.

Since the terminal output of the $\div f$ counter consists of 1 pulse per second, the $\div f$ counter may be followed by a $\div 60$ counter, which will then provide a terminal output pulse once every minute. And since the internal state of this $\div 60$ counter records the passage of seconds in each minute, a SECONDS display can be obtained by the use of other outputs of the counter to drive the appropriate display, as shown in Fig. I.3. Another $\div 60$ counter may then be connected in the chain to give a terminal count in HOURS. The internal state of this counter records the passage of MINUTES.

For practical digital clocks that can be powered from the ac power line, we note that the ac-power-line frequency is very accurately maintained at 60 Hz (that is, 60 cycles per second), and it can therefore serve as a very convenient time base for a digital clock. Counters, decoders, and display devices may then be arranged in the manner shown in detail in Fig. I.4 to produce a highly accurate and very practical digital clock.

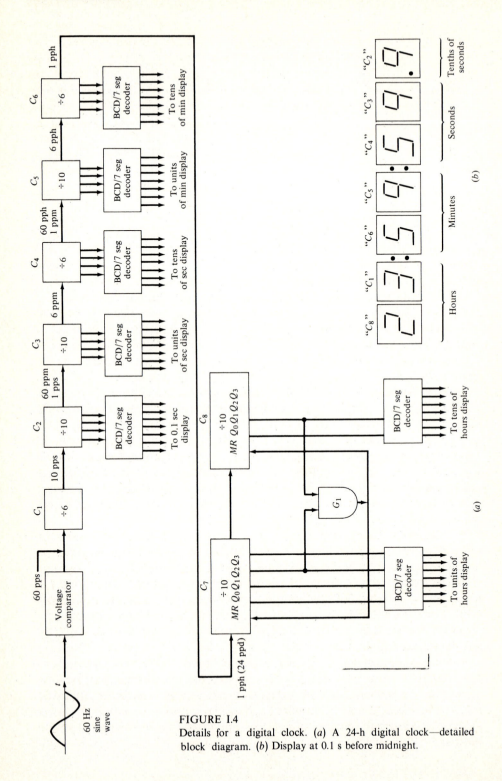

FIGURE I.4
Details for a digital clock. (*a*) A 24-h digital clock—detailed block diagram. (*b*) Display at 0.1 s before midnight.

The basic signal-processing operations that are performed in this clock may be described as follows. First, the 60-Hz pure sine wave is fed to a circuit that produces a standard output pulse (for example, 3 V in amplitude) that starts each time the 60-Hz sine wave passes through zero in the positive-going direction and lasts for the duration the sine wave is positive (8.33 ms). The IC which produces these pulses is quite similar to an IC amplifier and is called a *voltage comparator.*

The output of the voltage comparator is a sequence of 60 pulses per second (60 pps), each pulse having a standard size that can be used to drive an IC counter. As suggested in Fig. I.4, the first counter is wired to divide by 6, so its output is 10 pps, or 1 pulse every tenth of a second. Such ÷6 counters are available in IC form.

The output of the ÷6 counter, labeled $C1$ in Fig. I.4, is fed on to a ÷10 counter labeled $C2$, whose terminal output is then 1 pps or 60 pulses per minute (60 ppm). Divide-by-ten counters (*decade* counters) are available in a variety of forms in ICs. ICs of this level of complexity are referred to as medium-scale integration (MSI) circuits. In addition to the terminal-count output, the counter labeled $C2$ has four *status outputs* shown coming out from the bottom of $C2$ in Fig. I.4*a*. These status outputs can be used to determine the internal state of the counter at any time. Thus, if the counter is first set to read ZERO and then five input pulses are fed into it, a combination of LOW and HIGH voltages will appear on the four status outputs to indicate that exactly five pulses have entered the counter.

Four status-output *pins* (rather than 10) are used because the IC counter in fact does its counting in the *binary* or base 2 system, rather than the base 10 system that we are more accustomed to. Thus, the five input pulses will be represented internally by the *binary* number 0101. These digits represent the sum of 0 eight's, 1 four, 0 two's, and 1 one; for elaboration of the binary counting system, see Appendix C. A LOW level on an output terminal corresponds to a binary 0, while a HIGH level corresponds to a binary 1. The need for four status outputs follows directly from the fact that any decimal number from 0 to 9 can be represented as a four-digit binary number.

Because it is convenient to make IC counters that operate in the binary number system, but we want our indicated output to be in the base 10 system, we have to connect a special *decoding circuit* between the status outputs of the counter and the device that we are using to display the decimal state of the counter. Let us suppose we wish to use for each decimal number a seven-segment numerical display of the kind shown in Fig. I.5. The special circuit is then called a *BCD-to-seven-segment decoder* where BCD stands for binary-coded decimal. BCD-to-seven-segment decoders are also available in IC form and can be used directly to drive display devices of the form shown in Fig. I.5. The display device (also available as an integrated "circuit") has seven segments that can be independently lighted to give any number from 0 to 9. In summary, then, the second counter in Fig. I.4*a*, labeled $C2$, is a ÷10 counter that does its counting in the binary number system with internal connections that cause it to produce one terminal output pulse after 10 input pulses. Status outputs are provided to "read" the state of the

FIGURE I.5
Decimal display made from seven independent segments.

counter $C2$ in the binary number system, this state serving as the input to a BCD-to-seven-segment decoder. The BCD-to-seven-segment decoder in turn drives a seven-segment display device to indicate the appropriate integer in a base 10 system (decimal numbers 0 through 9).

The remainder of the clock in Fig. I.4a can now be seen to consist simply of a set of counters with appropriately connected decoder and display units attached. For example, $C3$ is a $\div 10$ counter with an output in SECONDS. It accumulates the information needed to construct the first digit (or UNITS digit) of the SECONDS display, and its output drives a $\div 6$ counter ($C4$) which accumulates the information necessary to construct the TENS OF SECONDS display. Additional counters, decoders, and seven-segment display devices are connected to read MINUTES, TENS OF MINUTES, HOURS, and TENS OF HOURS. For this 24-h clock, the display with its maximum value is shown in Fig. I.4b. Some logic operations are performed on the HOURS readout to reset the HOURS display to zero at the time it would read 24 HOURS. In particular, the *logic circuit* labeled $G1$ in Fig. I.4 functions so that when the decimal content of $C7$ is 4 *and* that of $C8$

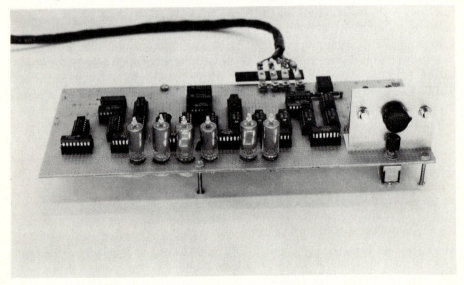

FIGURE I.6
Digital clock circuit. (*a*) Student built clock using SSI circuits. (*b*) A photo of the top surface of the chip for a digital clock (MOS-LSI MM5311). (*Courtesy National Semiconductor Corporation.*) (*c*) Photo of the package containing the chip in (*b*). (*Courtesy National Semiconductor Corporation.*)

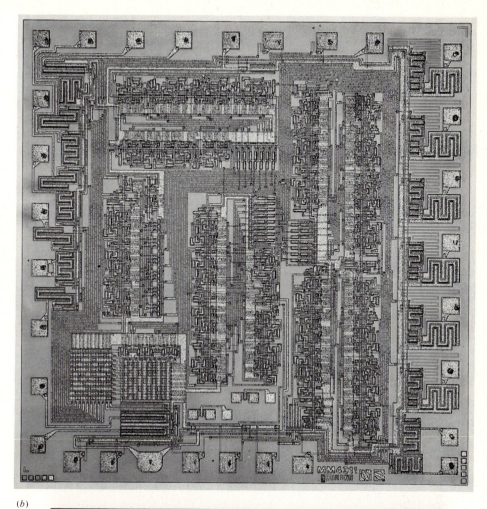

(b)

(c)

is 2 (i.e., midnight on a 24-h clock), the output of $G1$ drives the input terminals labeled MR which will *reset* counters $C7$ and $C8$ to the ZERO state.

The reader may well feel at this point that a large number of circuits are used to make the digital clock. While this is certainly true, it is also clear that really only four basic circuit types are used repeatedly: We need three $\div 6$ counters, four $\div 10$ counters, and seven BCD-to-seven-segment decoders which will drive the seven-segment display devices. This is, in fact, a characteristic of digital circuits in general: a few basic circuits blocks are used repeatedly to develop a given circuit function. Because this is so, manufacturers of digital ICs can concentrate on high-volume production of basic circuits, which tends to make the cost per circuit very low.

Figure I.6a shows a digital clock built by a student as a laboratory project. For this particular project the student elected to make counters using IC *flip-flops* and gates rather than MSI counter circuits. How to make flip-flops and then the counter using them will be discussed in Chaps. 5 and 6.

Today digital clocks are in such common use that IC manufacturers have made *all* the clock circuitry within *one* chip. ICs of this complexity are referred to as large-scale integration (LSI). A photograph of the top surface of an LSI chip for a digital clock is shown in Fig. I.6b. The packaged LSI circuit is shown in Fig. I.6c.

I.1.3   The Digital Voltmeter: An Analog-Digital Hybrid

To return to our original theme of analog and digital signal processing, we now consider as our final example the digital voltmeter (DVM). The basic purpose of this instrument is to give a digital readout of the value of a dc voltage applied to its input. The instrument cannot perform this function exactly, since the input can have any value (for example, π V), whereas the digital readout will be limited to the complexity of the system design we are willing to tolerate and the number of display devices we are willing to drive. For example, if we only display three digits as the output of the instrument, an actual input of π V will be displayed as 3.14 V. In effect, errors are introduced in the process of "rounding off" a number, and the displayed output is therefore not precisely proportional to the input. However, the overall system does present the correct answer if we are only interested in three-place accuracy in the output. And, of course, from a practical standpoint, we often prefer a purely electronic voltmeter over a moving-coil type of instrument, both for its speed of response and its digital readout.

To proceed to a discussion of the operating characteristics of a basic DVM, we show in Fig. I.7a a block diagram of a common way that a DVM is made. The voltage to be measured, V_x, is applied to the input terminal at the left. Then, in a fraction of a second the decimal reading of a counter, which is directly proportional to the input voltage, is displayed on the output display devices. The manner in which this is accomplished is as follows.

Initially the electronic switch S_1 grounds the *ramp generator*, and its output is set to zero. When we want to get a meter reading, the manual switch S_R is momentarily closed at time $t = t_0$, and the *control* block puts out signals that cause the following events:

1 The counter is *cleared* so that the output read in the display is zero.
2 The counter is then *enabled* so that it can begin to accummulate the count of clock pulses, as shown in Fig. I.7b.
3 Electronic switch S_1 connects V_x to the input of the ramp generator so that its output starts down with a slope proportional to V_x, as shown in Fig. 1.7c.

This state of affairs continues until the counter fills up to its capacity of N pulses, at which time $(t = t_1)$ the counter sends an *overflow* signal to the *control* block. This actuates switch S_1 so that V_{REF} is now the input to the ramp generator. V_{REF} is a stable fixed voltage source that is part of the DVM. The polarity of V_{REF} is selected opposite to that of V_x so the output of the ramp generator is now a positive-going ramp whose slope is proportional to V_{REF}, as shown in Fig. I.7c. The counter continues to keep on counting clock pulses, with its value having gone to zero at the time it "overflowed." At time t_2 the ramp-generator output crosses 0 V. This causes the *voltage-comparator* output to go HIGH for a brief period. This acts as an input signal causing *control* to:

1 Shut off the clock pulses going into the counter so that the counter stays with a count n in it.
2 Switches S_1 so that the ramp-generator input is grounded.

As we shall show in Chap. 10, the result of all this is that we obtain the following relationship between the unknown input voltage V_x and n the number of pulses that the counter counted between times t_1 and t_2:

$$n = \frac{N}{V_{REF}} V_x$$

Thus, if the capacity N of the counter is 1000 and the reference voltage V_{REF} was exactly 1 V, the display reading the count n is of the value

$$n = 1000 V_x$$

Thus the display value is the value of the unknown voltage V_x expressed in millivolts.

The design of the "hardware" that goes into the blocks in Fig. I.7a will be done in Chap. 10. A hand-held digital multimeter that measures dc and ac voltages and currents is shown in Fig. I.7d.

By way of comparison with other systems that we have described, it is interesting to note that the DVM employs six basic types of blocks, four of which we have been exposed to already. The ramp generator is made using an *operational amplifier*, a resistor, and a capacitor. An operational amplifier is a linear IC that is in many ways quite similar to the linear IC used in the construction of the simple audio system described previously. The voltage comparator is similar to

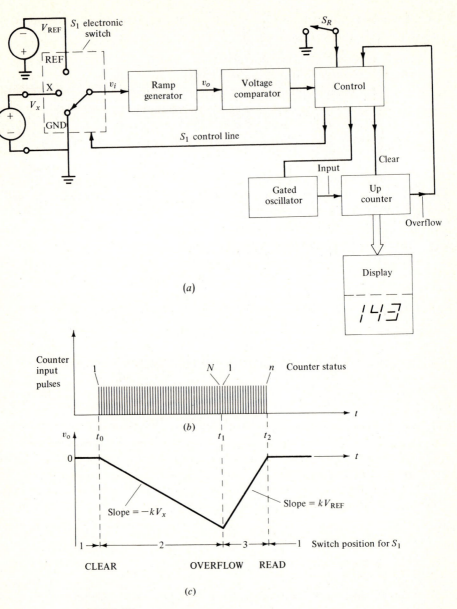

(a)

(b)

(c)

FIGURE I.7
Digital voltmeter. (a) Block diagram of a dual slope DVM. (b) Input pulses of the counter. (c) Waveform v_0 at the operational amplifier output. (d) A hand-held digital multimeter that measures dc and ac voltages and currents. It also functions as an ohmmeter. (*Courtesy Weston Instruments.*)

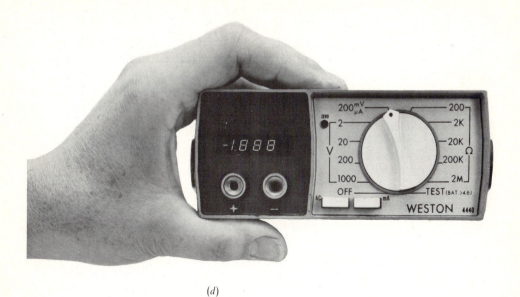

(*d*)

the one used in the digital clock. All other circuits in the DVM are digital ICs,†
including the control logic, the gated oscillator (made from one IC and a capacitor), and our old friend, the counter-decoder-display unit. The only really new
system component required for the DVM is the analog switch S_1. This switch
utilizes digital signals as inputs to enable it to switch the input analog signals
(voltages) to the ramp-generator input at the proper times. In effect, the switch S_1
is a simple form of a *digital-to-analog converter.* The voltage comparator is also a
common component in systems that perform mixed analog and digital operations.
Its inputs are analog and its output is digital, and it performs an *analog-to-digital
conversion.* Analog-to-digital conversion and/or digital-to-analog conversion are
essential functions in any system which employs both analog and digital signal-
processing techniques.

I.2   SUMMARY

In the preceding three system examples we found that amplifiers, counters, and
digital decoders showed up frequently, either as complete ICs or as parts inside an
IC. If we had examined other systems, such as digital computers, pocket electronic
slide rules, the automatic controls used in a steel mill, the mail-sorting equipment
in a Post Office, or even the control circuitry in a home washing machine, we

---

† All digital circuits are now available in one LSI circuit.

would again have found a need for these basic "components." The recurring need for these (and many other) common components makes it possible for an IC manufacturer to concentrate on producing standardized circuitry; it also makes it possible for us to gain a considerable insight into ICs in general by concentrating our attention primarily on a few basic types of circuits. Our intention is not to discuss all the various types of ICs but rather to provide a solid foundation on which the reader can build and extend his knowledge.

I.3 PREPARATION

One final point, regarding the level of preparation assumed, is in order before we begin our work. Since our main purpose is to show readers (with a variety of backgrounds and objectives) how ICs operate and how systems can be built using them, we have tried to keep the level of presentation as basic and simple as possible. However, we cannot realistically start in "square 1," so we do assume that the reader has a basic familiarity with Kirchhoff's laws and some related circuit theorems (Thévenin, Norton, and superposition). For readers who may wonder what is included in the term *basic familiarity*, we provide in Appendix B a resume of these basic circuit laws and theorems, with some exercises for those who may need to improve their grasp of these concepts.

We also do not propose to dwell at length on how ICs are *fabricated*, nor on the basic physics of the diodes and transistors, the *active devices* that form the heart of the IC business. We do, however, attempt to provide a general idea of internal structure, *chip* complexity, and related matters so that the reader will understand the terms that are used in the field to describe devices and circuits. Most of what we have to say on this topic is contained in Appendix A. There are also footnotes throughout the text to help the reader identify commercially manufactured components. To help the reader cope with all the new words used in describing ICs, we have included a glossary at the end of the text.

1

DEVICE MODELS AND INVERTER ANALYSIS

The analysis, design, and use of ICs rests on two major foundations. One of these is a set of elementary but fundamental circuit laws. The other is a set of simple but realistic circuit *models* for the diodes and transistors in the IC chip. The circuit laws that we use are Kirchhoff's current and voltage laws, Thévenin's and Norton's equivalent circuits, and the superposition theorem. For those not familiar with these laws, we have appended a brief review with some exercise problems in Appendix B.

In this chapter we present the basic device models that we will use to study ICs and apply them to analyze some simple but important circuits. In this discussion we will focus our attention on how to (1) select appropriate circuit models for the devices in a circuit, (2) verify *during the execution of the analysis* that the circuit models are valid, and (3) determine when we have achieved sufficient accuracy so we can terminate calculations.

One of the basic circuits that we wish to analyze is the transistor inverter, shown in Fig. 1.1. As we shall see later, this circuit performs an essential function in digital logic circuits. For our purposes, the basic properties of the circuit are contained in its *transfer characteristic*, i.e., the relationship between its output voltage V_{out} and its input voltage V_{in}.

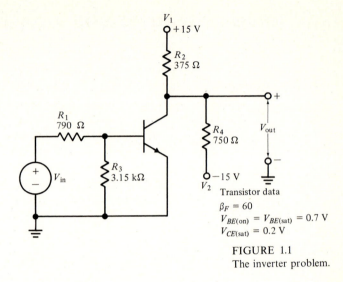

FIGURE 1.1
The inverter problem.

The transfer characteristic of the inverter shown in Fig. 1.1 may be *exper-imentally determined* by simply measuring V_{out} for a selected set of values of V_{in} and then plotting these results. Such an experimental transfer characteristic is shown in Fig. 1.2, where the input voltage has been limited to the range 0 to $+3V$.

The transfer characteristic can also be *computed* if we are given an appropriate *circuit model* for the transistor. Our major purpose in this chapter is to introduce some simple circuit models for the transistor which we can use to compute the circuit characteristics of inverters and other basic digital circuits. We will then be able to calculate the transfer function shown in Fig. 1.2 with a degree of accuracy that is adequate for most engineering designs. (The actual calculation

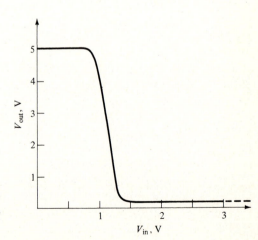

FIGURE 1.2
The transfer function of the inverter of
Fig. 1.1.

of the transfer characteristic and its comparison with experiment will be presented in Sec. 1.5.)

Our development of the subject of models for transistors will be organized as follows. First, we introduce the voltage-current law for a junction diode and show how circuit models for it may be constructed. To gain experience in the application of circuit models, we then use the diode models to analyze some simple but important diode circuits. Next we show how a circuit model for a bipolar transistor can be obtained as an appropriate combination of current sources and junction-diode models. We then give a careful discussion of how the parameters that describe the transistor model can be obtained from measurements or manufacturers' data sheets; finally, we return to the problem of computing the transfer characteristic of the inverter shown in Fig. 1.1. We conclude with a brief discussion of MOS field-effect transistors and their use in an inverter circuit.

PART I

1.1 JUNCTION-DIODE CHARACTERISTICS AND MODELS

A knowledge of voltage-current (VI) characteristics of a junction diode is important for two reasons. First, diodes are commonly used components in ICs; second, we will see that the circuit operation of bipolar transistors can also be described by the use of diodes as circuit elements. A knowledge of diode VI relationships is therefore essential to an understanding of the VI characteristics for a transistor. In this section we will present the current law for the junction diode and show how it leads to some simple diode-circuit models.

1.1.1 *p-n*–Junction-Diode Characteristics

It is convenient to begin our discussion of *p-n* junction diodes directly with a consideration of the VI characteristics which they exhibit, deferring a discussion of how these characteristics arise to Sec. 1.1.4. For this purpose we show in Fig. 1.3*a* an experimentally determined VI characteristic for a silicon diode. The diode symbol and reference directions are defined in Fig. 1.3*b*. As indicated on the figure, the terminal connection to the broad part of the arrow is called the *anode*, and the terminal located at the point of the arrow is called the *cathode*, these terms being carried over from an era when nearly all diodes were thermionic (vacuum tube) devices.

In Fig. 1.3*c* is shown the cross section of a *p-n* junction diode that corresponds to the diode symbol of Fig. 1.3*b*. The definition of the *p* and *n* layers is given in Sec. 1.1.4.

When a voltage having the polarity shown in Fig. 1.3*b* is applied to the diode, current will flow in the direction of the arrow. The diode is then said to be

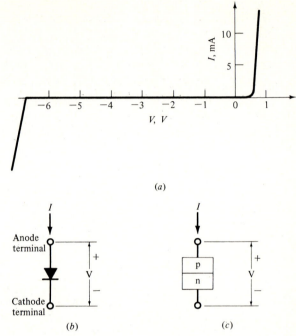

(a)

(b)   (c)

FIGURE 1.3
Diode characteristics and symbol. (a) The VI characteristic of a silicon diode; (b) general symbol for a diode with "easy" flow in the direction of the arrowhead; (c) cross section of a p-n diode with polarities as shown in part (b).

in the *forward-bias* or forward-conduction state. If the polarity of the applied voltage is reversed, a much smaller current will flow in a direction opposite to the direction of the arrow, and the diode is said to be in the *reverse-bias* state.

The theoretical VI characteristic From the physical theory of semiconductors it is possible to show that, for both forward and reverse biases, the VI law for an idealized p-n junction diode has the theoretical form

$$I = I_S(e^{V/V_T} - 1) \qquad (1.1)$$

where I_S is referred to as the *saturation current* of the diode.

The exponent in Eq. (1.1) is the ratio of V (the applied *diode* voltage) to V_T, which is commonly referred to as the *thermal* voltage. The thermal voltage is a quantity that depends only on the diode-junction temperature and is defined by

$$V_T = \frac{kT}{q} \qquad (1.2)$$

where k = Boltzmann's constant = 1.38×10^{-23} J/K
  T = absolute temperature, K
   = 273 + temperature in degrees Celsius
  q = charge of an electron = 1.602×10^{-19} C

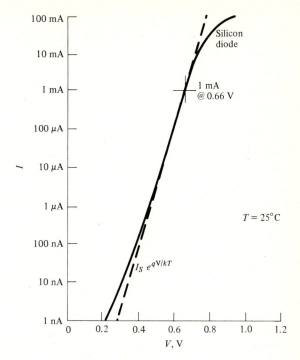

FIGURE 1.4
VI plot (linear V, log I) for a forward-biased silicon diode (*From A. S. Grove,* "*Physics and Technology of Semiconductor Devices,*" *Wiley, New York, 1967.*)

Upon substituting the listed values, we find that the thermal voltage is about 26 mV when the temperature is 300 K (this corresponds to 27°C, which is approximately "room" temperature).

Forward bias　The derivation of Eq. (1.1) is a topic that we can properly leave to books on semiconductor-device theory. We are, however, concerned with how accurately the theoretical law describes real diodes. To study this question let us compute the current that flows under a forward bias such that $V > 5V_T$. For this condition $e^{V/V_T} \gg 1$. Hence I and V are related simply by

$$I \approx I_S e^{V/V_T} \qquad (1.3a)$$

or equivalently, taking the natural logarithm of both sides of Eq. (1.3a),

$$\ln \frac{I}{I_S} = \frac{V}{V_T} \qquad (1.3b)$$

To compare Eq. (1.3b) with experimental data, we show in Fig. 1.4 a plot of log I versus V taken on a forward-biased silicon diode. In the same figure we show by dashed lines the "best-fit" plot of Eq. (1.3b). For this example we see that Eq. (1.3) does indeed give a good approximation to the experimental data over the range of currents from 10 μA to about 10 mA. In fact, by including a fixed "ohmic" resistor† of 1.4 Ω in series with the model represented by Eq. (1.3), the new model characteristic matches experimental data at the high-current end for another factor of 10 in the diode current (up to 100 mA). Hence the theoretical law gives a valid description of real-diode characteristics over a very wide current range.

A departure of the actual current from that predicted by Eq. (1.3) can be seen to occur at low currents. This effect can be accounted for by adding additional terms in Eq. (1.1) to account for special effects that occur in the immediate vicinity of any p-n junction. However, since we will not be concerned with accurate modeling at extremely low currents and low temperatures, we will only need to use the relationship given in Eq. (1.1) or Eq. (1.3).

Reverse bias　Turning now to the case of *reverse bias*, we see that when the applied voltage V is negative in Eq. (1.1) and its absolute magnitude is large compared to V_T, the exponential term will be small compared to 1. Equation (1.1) then indicates that the current will be constant:

$$I = -I_S \qquad \text{when } V < (-5V_T) \qquad (1.4)$$

We sometimes express this by saying the diode current is *saturated* with respect to voltage changes.

When this prediction is compared with experiment, it is found that, for silicon diodes at room temperature, the actual currents that flow under a reverse-bias condition are much larger‡ than the saturated values predicted by Eq. (1.1). Nevertheless, the reverse currents are still sufficiently small so that they can normally be neglected in any circuit that we will discuss.

We also mention in passing that the theoretical saturation current I_S is a strong function of temperature for all semiconductor materials. For silicon at a temperature of 27°C, I_S increases by 17 percent for a 1°C rise in temperature. Thus a diode with an I_S of 10^{-14} at 27°C will have at 28°C an I_S of 1.17×10^{-14}. Again, while this exact behavior is not always observed, the saturation current is nevertheless a sensitive function of temperature in all cases.

1.1.2　A Different View of the VI Relationship

Our discussion of Eq. (1.1) so far treats I as a *dependent* variable whose value is determined by the applied voltage V. However, in many practical situations a known forward current is established in the diode, and we want to find the diode voltage drop V. For forward-bias conditions where $I \gg I_S$, we can rewrite Eq. (1.3b) as

$$V = V_T \ln \frac{I}{I_S} = V_T \left(2.303 \log \frac{I}{I_S} \right) \qquad (1.5a)$$

---

† This "ohmic" resistance arises from the electrical resistance in the semiconductor material between the junction and the metal contacts at the surface of the chip.

‡ Furthermore, this "leakage" current increases nearly linearly with an increasing reverse-bias voltage until the reverse bias approaches a critical value called the *breakdown voltage*.

Substituting for V_T ($= 26$ mV at 300 K) we obtain,

$$V = (60 \text{ mV}) \log \frac{I}{I_S} \qquad (1.5b)$$

With the aid of Eq. (1.5b) we can answer the following question that will be of considerable use in predicting diode (and transistor base-emitter) voltages under forward-bias conditions. Suppose that a given diode has a voltage V_1 across it when it is passing a forward current I_1. What will the new voltage V_2 be if the forward-bias current is changed to I_2? To answer this we note that

$$V_1 = (60 \text{ mV}) \log \frac{I_1}{I_S} \qquad (1.6a)$$

and

$$V_2 = (60 \text{ mV}) \log \frac{I_2}{I_S} \qquad (1.6b)$$

Hence

$$V_2 - V_1 = (60 \text{ mV})\left(\log \frac{I_2}{I_S} - \log \frac{I_1}{I_S}\right) = (60 \text{ mV}) \log \frac{I_2}{I_1} \qquad (1.6c)$$

Therefore

$$V_2 = V_1 + (60 \text{ mV}) \log \frac{I_2}{I_1} \qquad (1.6d)$$

EXAMPLE 1.1   Consider a diode similar to that in Fig. 1.4 which is operating at a temperature of 27°C (300 K) and has a forward voltage of 660 mV when the diode current is 1 mA. If the forward current is increased to 10 mA, then we can substitute these values into Eqs. (1.6a) and (1.6c) to find the *increase* in diode voltage drop:

$$V_2 - V_1 = (60 \text{ mV}) \log \frac{10 \text{ mA}}{1 \text{ mA}} = 60 \text{ mV}$$

or, equivalently,

$$V_2 = V_1 + 60 \text{ mV} = 720 \text{ mV}$$

EXAMPLE 1.2   For further practice with diode VI characteristics, let us study some additional properties of the diode that conducts 1 mA at 660-mV forward bias at 300 K. First we compute I_S and plot the forward VI characteristic, assuming that Eq. (1.3) holds. Thus we rewrite Eq. (1.5b) as

$$I = I_S \, 10^{V/60}$$

where V is to be in millivolts. We then find the value of I_S that is required to satisfy the conditions $I = 1$ mA, $V = 660$ mV. The result is

$$I_S = I(10^{-V/60}) = (10^{-3})(10^{-660/60}) = 10^{-14} \text{ A}$$

Using this value, we can now replot the calculated value for I in amperes on a log scale versus V on a linear volts scale, as is done in Fig. 1.5. (This plot is also the dashed line in Fig. 1.4.) Figure 1.5 shows there is a tenfold (one decade) increase in I for every 60-mV increase in V. This is a very useful fact to remember in the analysis and design of circuits using junction diodes and transistors.

EXERCISES

These exercises use the diode relationship of Eq. (1.6d).

E1.1   Suppose you are given two different diodes (numbered 1 and 2), with their data at forward bias as follows:

$$V_{D(on)1} = 0.7 \text{ V} \qquad \text{at 10 mA}$$

$$V_{D(on)2} = 0.6 \text{ V} \qquad \text{at 0.1 mA}$$

If these diodes are placed in a series circuit and a current of 1 mA is made to flow in them (forward bias), what will the sum of the voltage drops across the diode be?

E1.2   A series circuit consists of a diode and a 1-kΩ resistor. With 1.66 V applied across the network, the diode passes a forward current of 1 mA. What must the applied voltage be changed to if the diode forward current is to be reduced to 0.1 mA?

1.1.3   Diode Regions of Operation

We have seen in the previous sections that the theoretical VI characteristic of a junction diode has an exponential form and that experimental data corroborate the theory over a wide range in current. To demonstrate this effectively, we plotted experimental and theoretical VI relations on *semilogarithmic coordinates* (log I versus V in Fig. 1.4) to show the range of currents over which the fit between theory and experiment was reasonable. Such a plot demonstrates the fit well, but it gives a misleading impression of how a diode VI curve actually appears on a normal curve-tracer presentation. To illustrate this difference, we show in Figs. 1.5 and 1.6 the forward-biased region of a silicon diode plotted on both semilogarithmic and linear coordinates, respectively. The *linear* plot is taken directly from a commercial curve tracer.

For the range of currents selected (0 to 10 mA), the linear plot appears to indicate that *zero* current flows below 600 mV. Of course this is not true; we know from Eq. (1.1) that zero current flows only at zero voltage. However, it is characteristic of a semiconductor diode that very little current flows until the applied

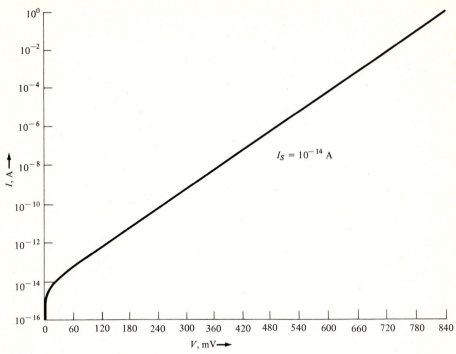

FIGURE 1.5
Silicon-diode VI characteristic in the forward-bias direction (semilog plot).

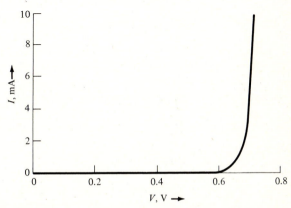

FIGURE 1.6
The diode characteristic from Fig. 1.5 plotted with both scales linear.

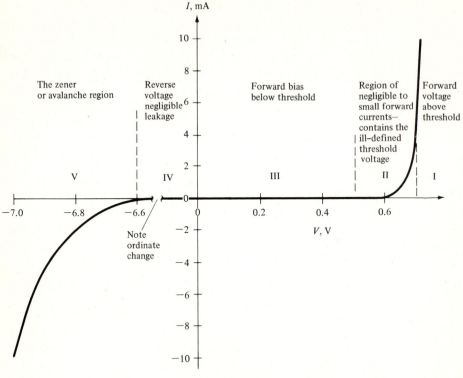

FIGURE 1.7
Definitions of voltage regions for a silicon junction diode.

voltage reaches a certain value called the *threshold voltage*. The threshold voltage depends on the semiconductor material used to make the diode (i.e., whether it is silicon, gallium-arsenide, or germanium) but is nearly independent of other structural details. From a circuit standpoint the existence of a threshold voltage is useful since it simplifies circuit design and analysis.

With this in mind, we divide the voltage axis into the five separate ranges shown in Fig. 1.7 and define five operating regions as follows.

Region I: The high-conduction state When the applied voltage is greater than the threshold voltage (which is 0.6 V in Fig. 1.7), the diode is in the *high-conduction state*. Current is flowing in the easy direction through the diode, its magnitude being limited only by external circuit resistances. For silicon diodes at room temperature, the voltage that must be applied to reach this high-conduction state is 0.6 to 0.8 V.†

---

† In Sec. 1.1.1, when we previously considered $V \gg V_T$, we wanted only to make $e^{V/V_T} \gg 1$. Now we are interested in getting much larger currents so that the device is operating near its optimum current densities. For most ICs this means the value of e^{V/V_T} is about 10^{10} to 10^{13}.

Region II: The transition state   From our examination of the VI characteristic given in Fig. 1.7, we can see that negligible current flows for applied voltages less than about 0.5 V, while 0.7 V of applied voltage is enough to carry the diode into the high-conduction state. Hence, for a silicon diode at room temperature, the voltage range 0.5 to 0.7 V defines a *transition region* in which the forward current changes from negligible to significant values. The threshold voltage for the diode lies somewhere in this transition-voltage range. This definition is, of course, somewhat imprecise since the terms *negligible current* and *significant current* depend on the circuit. As we saw earlier, the diode actually has an exponential VI characteristic, and hence the concept of a threshold voltage is largely a convenient way of describing the appearance of the linear VI plot.

Region III: Low-forward-conduction state   Proceeding in Fig. 1.7 to the voltage range of 0 to 0.5 V, we see that negligible current flows compared to that in the high-conduction state. For silicon diodes the current levels in this region are generally in the nanoampere to tens of microampere range, which we describe by the general term low forward conduction.

Region IV: The reverse-leakage state   As we pass into the range of reverse bias, the current in an idealized *p-n* junction diode quickly saturates at the value I_S. In an IC if the actual current that flows is not completely negligible, we treat it as a *leakage* factor; hence the current is called the *reverse-leakage current*, and we call this region of operation the *reverse-leakage state*. It extends from 0 V to a reverse bias equal to the reverse-breakdown voltage.

Region V: The reverse-breakdown state   As we increase the voltage beyond the onset of reverse breakdown, we enter the *zener* or *avalanche* region† at the extreme left of Fig. 1.7. For our purposes it is unimportant to distinguish between the different mechanisms which lead to breakdown, except to say that in any case substantial power dissipation may occur when large currents are passed through a diode in the reverse direction. To avoid damage to the diode, it is necessary to limit this maximum current to values given by the diode manufacturer.

In the first demonstration at the end of this chapter, D1.1, we study the characteristics of a diode and the effect of temperature on them.

1.1.4 A Qualitative Interpretation of Current Flow in a *p-n* Junction Diode

Before proceeding to develop circuit models for diodes, we provide in this section a brief qualitative discussion of current flow in a *p-n* junction. Our principal objective is to provide some physical understanding of why the junction diode conducts so differently in the two directions. Such a discussion is not essential to our main purpose, which is to lead the reader as directly as possible to the knowledge that he needs to design basic electronic systems with ICs. For this purpose circuit models of transistors and diodes prove to be

---

† If the value of the breakdown voltage is 5 V or less the mechanism is referred to as *zener breakdown*. For diodes whose breakdown voltages are above this, the breakdown is affected by the avalanche process. However, these are still referred to as zener diodes.

completely sufficient. We are, however, mindful of the fact that some students can remember a model better if they are also given a qualitative description of why its characteristics turn out the way they do. In what follows we attempt to provide such a discussion for the junction diode, using concepts that will also be helpful for understanding circuit models for transistors.

The basic characteristics of *p-n* junction devices can be traced to the fact that there are two carriers of electricity in a semiconductor: the *hole*, which is a positive carrier, and the *electron*, which is a negative carrier. In a perfect semiconductor crystal, electrons and holes are present in equal numbers and make roughly equal contributions to the electrical conductivity of the crystal. However, by introducing suitable chemicals (dopants) into the crystal, the density of one of these carriers can be increased at the expense of the other one. We can then obtain material in which either holes or electrons predominate. The former is called *p*-type material (because positive charges dominate the electrical conductivity); the latter, *n*-type (because negative charges dominate).

The simplest device that can be made using both *n*- and *p*-type semiconductor materials is the junction diode. In its ideal form, the junction diode is a piece of single-crystal semiconductor which is *p*-type on one side of a metallurgical boundary (or junction) and *n*-type on the other. Figure 1.3*c* gives a schematic representation of such a structure. Each type of material shown there is electrically neutral, but the *p*-type region contains predominantly holes and the *n* region contains predominantly electrons.

The fabrication of junction diodes and transistors is accomplished by using the process techniques described in Appendix A and the references cited there. For our purposes, these processes will simply be assumed to produce the idealized physical structure shown in Fig. 1.3*c*.

The fact that there are two adjacent regions of different conductivity type implies, among other things, that an electrostatic voltage difference will develop between the *n* and *p* layers in the region where the two materials have a common boundary. The reason for this is approximately as follows. Holes in the *p*-type material are free to move. As a result, some of them that are near the junction can move from the *p* side of the boundary to the *n* side. However, when this occurs, a net positive charge will have been transferred from the *p* to the *n* side. The laws of electrostatics then tell us that the *n* side will develop a potential that is *positive* with respect to the *p* side. Similarly, electrons are free to move from the *n* to the *p* side, and when they do, the *p* side will become *negative* with respect to the *n* side. The motion of both electrons and holes across the junction therefore produces an electrostatic voltage difference across the boundary.

The voltage difference thus developed is called the *junction electrostatic voltage drop*. Its magnitude is about 0.7 V for typical conditions in a silicon diode at room temperature, and its polarity is such as to make the *n* region positive with respect to the *p* region.

The junction electrostatic voltage drop is important because it defines (in a semi-quantitative way) the amount of energy that must be given to a hole to cause it to move from the *p* to the *n* region. That is, the junction electrostatic voltage drop is itself *developed* by holes that move from the *p* to the *n* region (and electrons moving oppositely), but once developed it inhibits further motion of holes and electrons. Holes that remain in the *p* layer must then be given an energy of 0.7 V to be able to move from the *p* to the *n* region, and similarly for the electrons.

The source of this energy is the battery that is applied externally to promote current flow. According to the oversimplified model just presented, an external battery of 0.7 V, connected to make the *p* layer positive with respect to the *n* layer, will completely cancel the effect of the internal junction electrostatic voltage drop. A large current can then flow. In

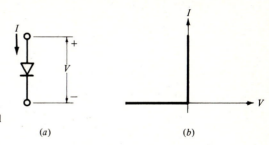

FIGURE 1.8
The ideal diode. (*a*) Symbol for an ideal
diode; (*b*) *VI* plot for an ideal diode.   (*a*)   (*b*)

effect the junction electrostatic voltage drop is the *threshold voltage* which must be overcome to produce current flow. The polarity of the external voltage for easy current flow is
positive on the *p* layer, so the *n* layer is what we have previously called the *cathode* and the *p*
layer is the *anode.*

It also follows from this description that an applied voltage of the *opposite* polarity (*n*
positive with respect to *p*) will simply add to the junction electrostatic voltage drop and
therefore further inhibit the flow of carriers across the junction. This suggests that in the
oversimplified ideal case *no* current will flow as a result of reverse bias.

Of course, this model is very much oversimplified since *some* carriers can obtain
enough thermal energy from their normal surroundings to surmount any barrier, however
high. Current can therefore flow for forward-bias voltages that are less than 0.7 V, and a
small reverse current can also be permitted.

To summarize, a junction electrostatic voltage drop on the order of 0.7 V develops at
the interface between the *p* and *n* layers. This voltage inhibits the flow of carriers across the
junction unless an external voltage is applied to compensate for the effect of the junction
electrostatic voltage. In the presence of such a voltage, holes can flow freely into the *n*
region and electrons can flow freely into the *p* region.

1.1.5   The Ideal-Diode Circuit Model

Having discussed the basic form of the *VI* characteristics of junction diodes, we
now turn to the problem of constructing diode circuit models, which we can then
use to compute currents and voltages in circuits that contain junction diodes.

The simplest model for a junction diode is the *ideal diode,†* represented by
the diode symbol with a clear *arrowhead* in Fig. 1.8*a*. The *VI* characteristic assigned to this element is shown in Fig. 1.8*b*. It is basically a simplification of the
theoretical-diode *VI* plot as it appears on *linear VI* scales, in which the threshold
voltage is zero and the reverse-breakdown voltage is infinite.

Because the ideal *VI* characteristic is made up of two straight-line pieces,
this model is referred to as a *piecewise linear approximation.* For current flowing in
the forward direction, the model is a short circuit; i.e., it has zero voltage drop
across it. In the reverse direction, it is an open circuit; i.e., zero current flows for
any reverse bias.

† By way of clarification the *ideal* diode is a circuit element. When we discussed the *idealized p-n
junction* diode, we were implying that some ideal physical constants were involved in the
physics leading to Eq. (1.1).

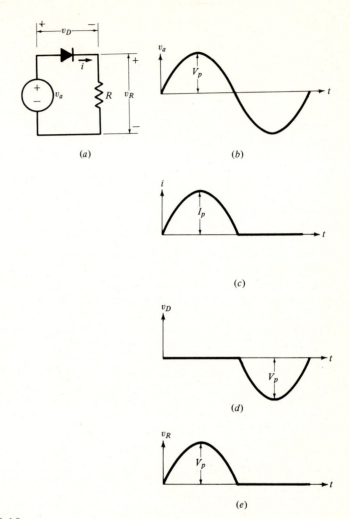

FIGURE 1.9
Half-wave rectifier circuit and waveforms. (*a*) Circuit and reference polarities;
(*b*) applied-voltage waveform; (*c*) current waveform; (*d*) diode-voltage waveform;
(*e*) resistor-voltage waveform.

The ideal-diode model is a sufficiently accurate approximation to be used in many problems. As an example let us consider the following. We are given the simple unfiltered half-wave rectifier circuit in Fig. 1.9*a*, which consists of a real diode D connected to an ac (60 Hz) source of 115 V rms with a 750-Ω resistor as a load. The waveform of the applied voltage v_a is† shown in Fig. 1.9*b*. We wish to

---

† For the instantaneous value of variables we use lowercase letters, while capital letters are used
  for the dc value of the same variable; i.e., v_a is really $v_a(t)$. If $v_a(t)$ is a constant value, we refer to
  its value as V_a.

find the peak value of the current waveform i; this is I_p of Fig. 1.9c. For this purpose, we substitute the ideal-diode model for the actual diode and then reason as follows. Since the diode voltage drop is zero in the forward direction, we must have

$$I_p = \frac{(1.414)(115)}{0.75} = 217 \text{ mA}$$

where the factor 1.414 converts the 115-V-rms voltage to its peak value of 163 V.

The waveforms for the voltage across the diode (V_D) and the resistor (V_R) are shown in Fig. 1.9d and e, respectively. When the applied voltage is negative, the current is zero since the diode is "blocking." Hence all the applied voltage is across the diode. When the applied voltage is positive, zero voltage drop exists across the ideal diode and all the applied voltage shows up across the resistor.

Of course this is only an approximation, since the forward voltage drop across a real diode when it is conducting 217 mA is not zero. In fact, if the diode has the VI characteristic shown in Fig. 1.5, we see that 0.8 V must be applied to the diode for it to conduct 217 mA. Knowing this, should we go back and recompute I_p or not? To answer this question, we observe that the calculated current flow would be reduced only by 0.5 percent (0.8 V/163 V) by incorporating the 0.8-V diode drop in the calculation. Hence, if we require our result only to a 1 percent precision, or if the resistor tolerance is 1 percent (or poorer), or if the accuracy to which V_p can be determined is 1 percent (or poorer), then we would not bother to recompute the result. In effect, the ideal-diode model is suitable for circuit calculations when the actual-diode voltage drop is small compared to the voltages that drive the circuit.

1.1.6 Diode Model Consisting of an Ideal Diode in Series with a Voltage Source

In most ICs the voltage across a forward-biased diode carrying a current in the order of microamperes to milliamperes is *not* so small as to be negligible compared to the other voltage drops in the circuit. In these cases the ideal-diode model is unsuitable and must be modified to permit realistic computations. As an example, let us reconsider the circuit of Fig. 1.9a, with a peak voltage V_p of 1.5 V instead of the 163 V used before. If we assume again that the diode drop was 0 V in the forward direction, we would calculate a peak current of

$$I_p = \frac{1.5}{0.75} = 2 \text{ mA}$$

From Fig. 1.6 we see that for 2 mA to flow through the diode we require a forward voltage close to 0.7 V. Now a drop of 0.7 V across the diode would be 47 percent

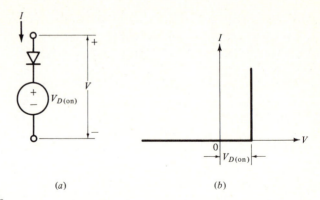

(a)   (b)

FIGURE 1.10

A more elaborate model for a silicon diode. (*a*) Circuit elements for the model consisting of an ideal diode in series with a dc voltage source; (*b*) the *VI* characteristic for the model in part (*a*).

(0.7 V/1.5 V) of the voltage applied to the circuit, and cannot be ignored if we want an error less than 47 percent in our computed result.

To take care of this we modify the ideal-diode model by adding a voltage source $V_{D(on)}$ in series with the ideal diode, as shown in Fig. 1.10*a*. The *VI* characteristic of our new model is shown in Fig. 1.10*b* and now contains the elements that are necessary to represent the concept of a threshold voltage. That is, for applied voltages less than the threshold voltage, $V_{D(on)}$, zero current flows. Equivalently, for any value of forward current in the diode, the forward voltage drop across the diode will be equal to the threshold voltage. The model is still piecewise linear.

Returning to our numerical example, with $V_p = 1.5$ V, we find on assuming a $V_{D(on)}$ of 0.7 V that

$$I_p = \frac{1.5 - 0.7}{0.75} = 1.07 \text{ mA}$$

Figure 1.11 shows the waveforms of the voltage v and the current i. Note that the current flow begins in our modeled circuit when v_a is $+0.7$ V.

Of course, situations arise where the diode voltage must be known even more precisely than provided by the model of Fig. 1.10*a*. As an example, consider once again the circuit of Fig. 1.9*a* but with V_p equal to 0.7 V. If we used the model of Fig. 1.10*a* with a $V_{D(on)}$ of 0.7 V, we would always have zero current through the diode. This approximate answer may be adequate in some applications, but if we want more precision, we will have to either change our approach or change our model. If we knew the range of *currents* which were to flow in the diode, we could revise our model by simply changing the value assigned to $V_{D(on)}$ in Fig. 1.10. We could be even more exact by using the exponential form for the *VI* relation given in Eq. (1.3), but the algebra then becomes messy and a digital computer could best be used to perform the calculation.

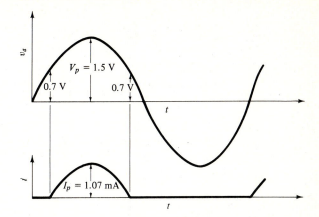

FIGURE 1.11
Applied-voltage and diode-current waveforms for the circuit in Fig. 1.9*a* for
$V_p = 1.5$ V and $V_{D(\text{on})} = 0.7$ V.

1.2 ANALYZING DIODE CIRCUITS BY GRAPHICAL TECHNIQUES

So far our analysis of diode circuits has been completely algebraic. In this section we will introduce a complementary graphical technique for determining the diode current in a simple circuit and compare this with results obtained using circuit models for the diode.

For simplicity we consider a basic dc circuit consisting of a diode, resistor, and a battery, as shown in Fig. 1.12*a*. This circuit is a dc version of the instantaneous-value circuit given in Fig. 1.9*b* with V_p of 1.5 V. Utilizing electronic circuit conventions, the circuit is redrawn in Fig. 1.12*b*. To solve for the dc current I, we use Kirchhoff's voltage law to obtain

$$V_{CC} = V_D + IR \qquad (1.7)$$

Solving for the current we have

$$I = \frac{V_{CC} - V_D}{R} = \frac{V_{CC}}{R} - \frac{V_D}{R} \qquad (1.8a)$$

Equation (1.8*a*) gives one relation between the two unknown circuit quantities I (the diode current) and V_D (the diode voltage). In addition, we know that the diode current and voltage must satisfy the diode equation

$$I = I_s(e^{V_D/V_T} - 1) \qquad (1.8b)$$

so Eqs. (1.8*a*) and (1.8*b*) give a solvable system of two equations in the two unknowns. Unfortunately, Eq. (1.8*b*) is a *nonlinear* equation (that is, I is not proportional to V_D), so we cannot combine Eqs. (1.8*a*) and (1.8*b*) to obtain a simple closed expression for I.

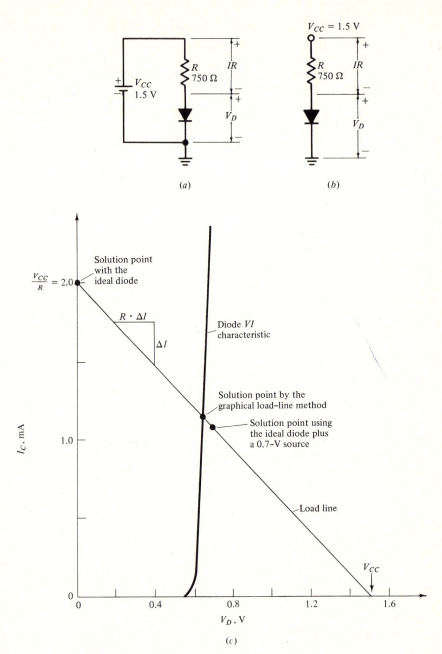

FIGURE 1.12
Graphical load-line solution to the diode plus series-resistor problem. (*a*) Schematic showing power source as a battery; (*b*) IC convention showing positive power supply at top of schematic; (*c*) the graphical solution in the VI plane.

Of course, some definite current *does* flow, and we want to find out what this is. We can find the current by

1 Using a computer or trial-and-error numerical methods to find the values of I and V_D that simultaneously satisfy Eqs. (1.8a) and (1.8b) (see Prob. P1.21)

2 Using one of the diode models described in the previous section *in place of* Eq. (1.8b)

3 Solving Eqs. (1.8a) and (1.8b) *graphically*

We use this latter technique in the following section.

1.2.1 The Load-Line Method

To obtain a graphical solution to Eqs. (1.8a) and (1.8b), we start by obtaining the data that describe the VI characteristic of the diode [experimental data or one point and Eq. (1.8b)]. Let us assume this is given to us in the graphical form shown in Fig. 1.12c. Now we can also plot Eq. (1.8a) on the same set of coordinates (V_D and I), this giving us the downward-sloping straight line in Fig. 1.12c. This line simply provides a graphical representation of the effects of the supply voltage V_{CC} and the load resistor R in the solution of the equations. This line is commonly referred to as a *load line*.

The diode characteristic curve as given by the data shown in Fig. 1.12c must, of course, intersect the load line at only one point. This point of intersection then gives the values of I and V_D that simultaneously solve Eqs. (1.8a) and (1.8b).

The construction is carried out in Fig. 1.12c, where we see that the simultaneous solution is obtained for $V_D = 0.66$ V and $I = 1.12$ mA. This current and voltage satisfy both Kirchhoff's voltage law and the diode law, so we have a solution for the current. The load-line technique will also be used in analyzing transistor circuits later.

1.2.2 Analytical Solutions with Linear Models Compared to Graphical Solutions

The problem solved in Sec. 1.2.1 can also be solved to various levels of approximation by algebraic means, using appropriate piecewise models that were discussed in Secs. 1.1.5 and 1.1.6. There we discussed two models: the ideal-diode model of Fig. 1.8 and the ideal diode in series with a voltage source as in Fig. 1.10.

To use the ideal-diode model for the present problem, we simply set $V_D = 0$ in Eq. (1.8a) and obtain

$$I = \frac{V_{CC}}{R} = \frac{1.5}{0.75} = 2.0 \text{ mA}$$

Using the graphical solution of Sec. 1.2.1 as a standard of comparison, we find that the ideal-diode solution gives a 79 percent error (2 mA instead of 1.12 mA). This problem can also be solved quite simply by considering how the *graphical* solution would change if the VI characteristic for an ideal diode were used in Fig. 1.12c. (Try it!)

The ideal-diode–plus–voltage-source model can also be used in the algebraic analysis, in which case we substitute 0.7 V for V_D whenever I has a *positive* value. The solution of Eq. (1.8a) then becomes

$$I = \frac{V_{CC} - V_D}{R} = \frac{1.5 - 0.7}{0.75} = 1.07 \text{ mA}$$

Again, a graphical method can be employed by using a graphical representation of the VI characteristic for the ideal diode in series with a voltage source. Compared to the graphical solution of Sec. 1.2.1, we now have only a 4 percent error. The reason the second method is better is that silicon diodes in ICs at room temperature have forward voltages closer to 0.7 V than 0 V for the current levels involved.

Table 1.1 summarizes these results. For dc calculations in ICs, the ideal-diode–plus–0.7-V source will be an adequate model for the diode.

Table 1.1   COMPARISON OF SOLUTIONS TO FIG. 1.12a

Diode model	Ideal diode	Ideal-diode–plus–0.7-V source	Graphical method
Value of current calculated	2 mA	1.07 mA	1.12 mA

The reader might now be thinking, "Very well, but how does this effort relate to our basic problem of determining the transfer characteristic of a transistor inverter shown?" The answer is that the general process that we follow to compute the transfer characteristic of the inverter is exactly the same as the process used to compute the current in the example just completed. To be specific, we need *circuit models* that can represent the transistor with reasonable accuracy. In addition, we must write appropriate circuit equations for the inverter that show how the applied voltages and circuit resistances affect the circuit currents. We must then find a graphical or algebraic solution of these equations that satisfies both the transistor model and Mr. Kirchhoff.

We will develop the necessary transistor models in Sec. 1.4. As we will see there, bipolar transistors themselves contain p-n junctions, and therefore the models for bipolar transistors contain ideal diodes that we have already presented in this section.

EXERCISE

E1.3 Solve for the current I, and the voltage V_D in Fig. 1.12, using $V_{CC} = 1.2$ V and $R = 2.0$ kΩ. Use the diode plot of Fig. 1.6. Do this (a) graphically, and (b) by using diode models with $V_{D(on)}$ of (1) 0 V, (2) 0.7 V, and (3) 0.6 V.

1.3 SUMMARY OF DIODE DISCUSSION

The main concepts presented so far in this chapter are:

For a junction diode biased in the forward direction, the relationships of Eqs. (1.3a) and (1.5b) are useful for diodes and transistors in ICs:

$$I = I_s e^{V/V_T} \qquad (1.3a)$$

$$V = (60 \text{ mV}) \log \frac{I}{I_S} \qquad (1.5b)$$

Thus by Eq. (1.5b) we see that the voltage across a diode increases (decreases) by 60 mV for a tenfold increase (decrease) in the current (I) through it.

In our considerations of ICs, if a reverse voltage is applied across a diode, negligible current flows until the breakdown voltage is reached.

In ICs a diode will have negligible current flowing in the forward direction if the voltage across the diode is less than the threshold voltage, which is about 0.7 V.

An ideal diode in series with a voltage source of 0.7 V serves as an adequate model for the real diode used in ICs. Applying this model and elementary circuit laws allows one to easily account for the effect of a diode in an IC.

PART II

1.4 BIPOLAR-TRANSISTOR MODELS AND CHARACTERISTICS

In the following treatment of bipolar transistors, we will present a general model that we can then simplify to obtain working models for various situations. We present only enough of the basic physics of operation to show why the various elements in the circuit model are required. The references at the end of the chapter present a more complete picture for readers who wish to pursue this topic. Our goal here is to be able to analyze circuits, such as the inverter circuit of Fig. 1.13, which is a simplified version of the circuit in Fig. 1.1.

1.4.1 Reference Directions and Circuit Schematic Symbols for Transistors

Before proceeding to develop a circuit model for the transistor, it is useful first to define the reference directions for currents and voltages that we will use to characterize the transistor, together with the symbol that we use to represent the transistor in circuit schematics.

The symbol used in circuit schematics for an *n-p-n* transistor is shown in Fig. 1.14a, where the arrow on the emitter lead points in the direction that emitter current normally flows. As shown in Fig. 1.14b for an *n-p-n* transistor, currents are

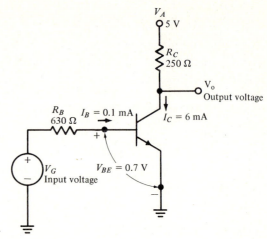

FIGURE 1.13
The simplified inverter circuit of Fig. 1.1.

referenced to be positive when they flow in at the collector and the base and out at the emitter, and the voltages are referenced positive when the base-emitter junction is forward-biased and the base-collector junction is reverse-biased.

We have chosen these reference polarities so that for most operating conditions the variables will have positive values. This will help us check any numerical work because, using this convention, a negative value of a variable will act as a flag for either an error in computation or unusual operating conditions. Moreover, if we consistently use this notation and orientation, we will find it helpful in understanding the dc operation of a circuit containing many transistors.

The reference directions are easily remembered if it is noted that dc currents flow "downhill" from the top of the schematic, where the most positive dc voltages are, to the bottom of the schematic, where the most negative dc voltages are found.

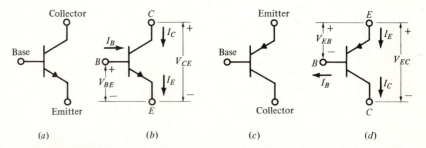

(a)   (b)   (c)   (d)

FIGURE 1.14
Symbols and reference directions for *n-p-n* and *p-n-p* transistors. (a) Symbol for an *n-p-n* transistor; (b) reference directions for currents and voltages for an *n-p-n* transistor; (c) symbol for a *p-n-p* transistor; (d) reference directions for currents and voltages for a *p-n-p* transistor.

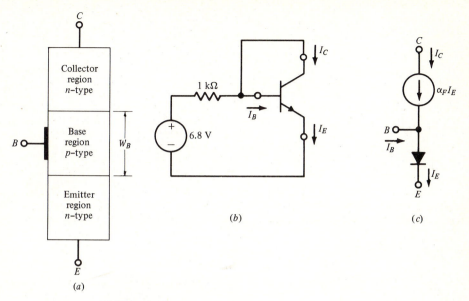

FIGURE 1.15

Transistor structure and operation in the normal direction. (*a*) The cross section of an *n-p-n* transistor showing the two *n* regions separated by a *p* region; (*b*) a circuit to forward-bias the base-emitter junction with V_{BC} set to zero; (*c*) model for a transistor biased as in part (*b*).

For *p-n-p* transistors which are also used in some ICs, the conventional symbol is shown in Fig. 1.14*c*, where we have placed the emitter with the arrow pointing downward at the top. The reason for this is to keep consistent the idea that all currents flow downhill. Hence the reference directions for a *p-n-p* are shown in Fig. 1.14*d*. If we reference the two voltage variables as shown, the emitter-base voltage and the emitter-collector voltage, both are positive for normal operation.

EXERCISE

E1.4 Using Kirchhoff's voltage law, determine V_G and V_A in Fig. 1.13, given that $I_B = 0.1$ mA, $I_C = 6$ mA, $V_{CE} = 3.5$ V, and $V_{BE} = 0.7$ V. Note that in this problem the terminal conditions of the transistor are given.

In the next several sections we will see how to obtain the relationships between the terminal conditions of a transistor.

1.4.2 Operating Principle for Junction Transistors

The basic principle of operation for a junction transistor can be explained by referring to Fig. 1.15*a*. Here we show an *n-p-n* transistor, which consists of a central *p*-type layer called the *base*, sandwiched between *n*-type layers that are called the *emitter* and *collector*, respec-

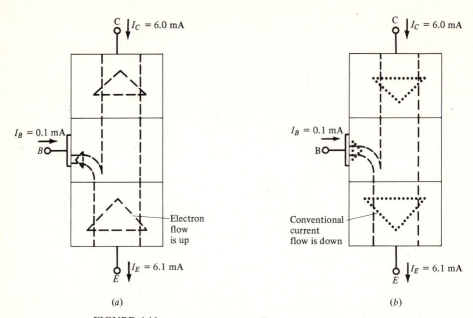

FIGURE 1.16
Electron flow and conventional current flow in an *n-p-n* transistor. (*a*) Internal electron flow is up (dashed arrowheads); (*b*) conventional current flow is down (dotted arrowheads).

tively. From the construction shown in Fig. 1.15*a*, it will be clear that the emitter and collector regions form *p-n* junctions on opposite sides of the base.

Both junctions can be biased by applying external voltages to the leads that are attached to the three layers. For our purposes it is convenient to first consider the biasing arrangement shown in Fig. 1.15*b*, where the collector and base regions are wired together externally and a forward bias is applied across the base-emitter junction. The result of the forward bias will be to cause electrons to flow from the emitter into the base. Holes also flow from the base into the emitter, but in a properly designed structure this flow can be made small compared to the flow of electrons. Now some of these electrons that came from the emitter region will *recombine* with holes in the base, much as H^+ and $(OH)^-$ ions will recombine to form H_2O molecules when they are mixed together in a solution. However, if the base is thin enough, many of the electrons that flow into the base from the emitter *diffuse* to the collector region and hence do not recombine with holes in the base.

Therefore, the electron flow† that crosses the emitter-base junction is divided into two components in the base. If the base is very narrow, a large fraction of the electrons flow into the collector layer and are returned through the collector lead. The remaining small fraction are lost by recombination, these electrons returning via the base lead as suggested in the cross-sectional view in Fig. 1.16*a*. In Fig. 1.16*b* we show the internal flow of mobile charges as conventional current. Recall that *conventional* current flow goes in the opposite direction of *electron* flow.

---

† The convention that current flow is opposite to electron flow must be reckoned with in applying these concepts.

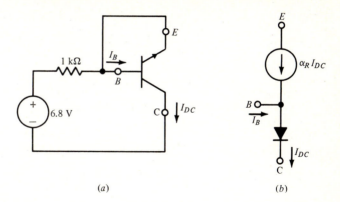

(a)   (b)

FIGURE 1.17
The transistor in the inverted connection. (a) A circuit to forward-bias the base-collector junction; (b) model for the transistor in part (a).

In Fig. 1.15c we show a model that can represent these current flows. A *junction diode* is drawn to represent the base-emitter diode, and a *current source* is drawn to represent the current collected by the collector. In general, the collected current turns out to be a fixed fraction of the emitter current, the fraction being defined by the symbol α_F:

$$\alpha_F = \frac{I_C}{I_E}\bigg|_{V_{CB}=0} \qquad (1.9a)$$

Applying Kirchhoff's current law to the model shown in Fig. 1.15c, we also see that the base current is

$$I_B = (1 - \alpha_F)I_E \qquad (1.9b)$$

for this bias condition.

The fraction of emitter current that reaches the collector can be made to approach unity by making the *base width* (W_B in Fig. 1.15a) very small. Thus one of the most critical dimensions of the transistor is its base width, which must be sufficiently small so that an efficient transfer of electrons from emitter to collector can occur. This transfer mechanism is commonly called *transistor action*.

Transistor action can also be obtained in the opposite direction (from the collector to the emitter) if the biases are arranged properly. To treat a case parallel to that just given, we show in Fig. 1.17a a transistor with the base-collector junction forward-biased and the emitter region tied to the base. The model for the configuration clearly requires a junction diode to represent the base-collector junction and a current source to represent the transfer of electrons from collector to emitter. The model is shown in Fig. 1.17b and is identical in form to that shown in Fig. 1.15c for the opposite connection of biases.

In a truly symmetrical transistor the current transfer will be the same in the collector-to-emitter transfer as it is for the emitter-to-collector process, so α for the two cases will be equal. However, transistors are usually not symmetrical, so in most cases it is necessary to define a new constant of proportionality in the current generator in Fig. 1.17b. This constant is given the symbol α_R, where

$$\alpha_R = \frac{-I_E}{-I_C}\bigg|_{V_{BE}=0,\ V_{BC}=\text{a forward bias value}}$$

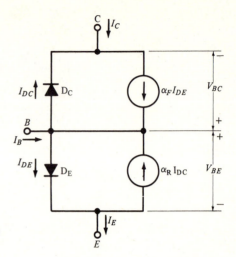

FIGURE 1.18
Ebers-Moll model for a bipolar *n-p-n*
transistor.

We will see shortly that a complete model for the transistor can be developed by
simply adding the two partial models we have just discussed. Two diodes will be used to
represent the two junctions, and two controlled current sources to represent the transfer of
currents from emitter to collector and vice versa.

1.4.3   The Ebers-Moll Model of a Bipolar Transistor

The basis for the models which we use in studying transistor circuits is referred to
as the *Ebers-Moll model* and is shown in Fig. 1.18. It contains two diodes (D_E and
D_C) and two controlled current sources whose "contract" is to deliver currents
that are proportional to the currents flowing in the diagonally located diodes. As
we saw in the previous section, these current sources represent the effects of
transistor action. Without them the model would simply reduce to two back-to-
back diodes.

The diodes in the model are idealized *p-n* junction diodes, and therefore the
diode currents are given by Eq. (1.1):

$$I_{DE} = I_{ES}(e^{V_{BE}/V_T} - 1) \qquad (1.10a)$$

$$I_{DC} = I_{CS}(e^{V_{BC}/V_T} - 1) \qquad (1.10b)$$

The values I_{ES} and I_{CS} are saturation-current values that are constant at a given
temperature. The reference directions V_{BE} and V_{BC} are chosen so that when they
are positive the diodes are forward-biased.

The currents flowing at the emitter and collector terminals can be written
directly from the model in Fig. 1.18 as follows:

$$I_E = I_{DE} - \alpha_R I_{DC} \qquad (1.11a)$$

$$I_C = -I_{DC} + \alpha_F I_{DE} \qquad (1.11b)$$

where I_{DE} and I_{DC} are the diode currents defined in Eqs. (1.10). By combining Eqs. (1.10) and (1.11), the terminal currents can also be expressed in terms of the applied voltages V_{BE} and V_{BC} as

$$I_E = I_{ES}(e^{V_{BE}/V_T} - 1) - \alpha_R I_{CS}(e^{V_{BC}/V_T} - 1) \qquad (1.11c)$$

$$I_C = \alpha_F I_{ES}(e^{V_{BE}/V_T} - 1) - I_{CS}(e^{V_{BC}/V_T} - 1) \qquad (1.11d)$$

The base current can be determined from Kirchhoff's current law and is found to be

$$I_B = I_E - I_C \qquad (1.12)$$

Hence, using Eqs. (1.11a) and (1.11b),

$$I_B = I_{DE}(1 - \alpha_F) + I_{DC}(1 - \alpha_R) \qquad (1.13a)$$

or using Eqs. (1.11c) and (1.11d),

$$I_B = (1 - \alpha_F)I_{ES}(e^{V_{BE}/V_T} - 1) + (1 - \alpha_R)I_{CS}(e^{V_{BC}/V_T} - 1) \qquad (1.13b)$$

Hence given the four transistor constants I_{ES}, I_{CS}, α_F, and α_R and also given the voltages applied across the base-emitter and base-collector diodes, V_{BE} and V_{BC}, we can determine the currents I_E, I_C, and I_B.

Equations (1.11c) and (1.11d) describe what is called the Ebers-Moll model of the transistor. This mathematical model provides very useful general relationships that apply to a transistor under any bias condition. However, as was true for diodes, we find that some key approximations can greatly simplify the model while still keeping it accurate enough for many applications. Moreover in circuit analysis we will find that we will often want to treat one of the terminal currents (I_B) as an independent variable rather than having all the independent variables be voltages. Our modified models will permit us to do this.

1.4.4 Normal-mode Active-region Models

We define the *normal-mode active region* for the following conditions:

1   V_{BE} is equal to or larger than the threshold voltage of diode D_E.
2   V_{BC} is less than the threshold voltage of diode D_C.

In this latter case I_{DC} can be treated as zero. Hence in Fig. 1.18 the diode D_C and the current source $\alpha_R I_{DC}$ can be dropped out of the model to give us the model shown in Fig. 1.19a. Under these conditions the expression for I_C becomes

$$I_C = \alpha_F I_E = \alpha_F \frac{I_B}{1 - \alpha_F} \qquad (1.14)$$

We can now define a new constant β_F, where

$$\beta_F = \frac{\alpha_F}{1 - \alpha_F} = \frac{I_C}{I_B} \qquad (1.15)$$

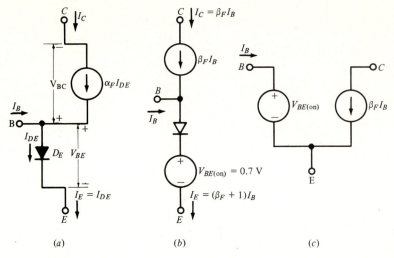

$$(a) \qquad\qquad (b) \qquad\qquad (c)$$

FIGURE 1.19
Normal-mode active-region model. (a) A simplified version of the circuit of
Fig. 1.18; (b) replacing D_E by an ideal diode and a voltage source; (c) rearranging
the circuit elements of part (b) produces this.

Hence

$$I_C = \beta_F I_B \qquad (1.16)$$

The term β_F is called the *dc current gain* of a transistor and is a parameter
most commonly given on discrete-transistor data sheets. On transistor data sheets
this parameter is called h_{FE}. Nevertheless, common practice is to refer to this as
beta.

In digital ICs, β_F centers around 60, while for the linear ICs it is around 200
(except for "super beta" transistors, where it can be 10,000). If necessary, the value
of α_F can be determined from β_F by transposing Eq. (1.15) to obtain

$$\alpha_F = \frac{\beta_F}{\beta_F + 1} \qquad (1.17)$$

For dc circuit analysis we can follow the principles developed in Sec. 1.1.6,
where we replaced diode D_E by an ideal diode and a 0.7-V voltage source. This is
done in Fig. 1.19b. Furthermore, if we are only using the model when the terminal
conditions are such that $I_B > 0$, we eliminate the ideal diode. Finally, since the
location of the bottom terminal of the current generator has no effect on the
terminal voltages, we can rearrange it to obtain the model for the normal-mode
active region given in Fig. 1.19c.

1.4.5 Inverted-mode Active-region Model

In certain applications of transistors we will find that the collector is forward-
biased and the emitter is reverse-biased. This condition is referred to as *inverted*

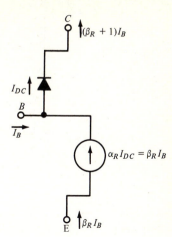

FIGURE 1.20
Inverted-mode active-region model.

operation. The operating conditions are then

1 V_{BC} is equal to or larger than $V_{BC(\text{on})}$.
2 V_{BE} is less than $V_{BE(\text{on})}$.

Hence I_{DE} is treated as zero, and we have the model given in Fig. 1.20.

In a completely analogous fashion as in the normal-mode active-region model, we express $-I_E$ as

$$-I_E = \frac{\alpha_R}{1 - \alpha_R} I_B = \beta_R I_B \qquad (1.18)$$

For digital ICs, β_R is usually less than 1 and can be as low as 0.01. We shall find in Chap. 3 that a low value of β_R is desirable for the input transistors in a popular family of digital ICs.

1.4.6 Cutoff-region Model

If both diodes are biased below their threshold values, that is, $V_{BE} < V_{BE(\text{on})}$ and $V_{BC} < V_{BC(\text{on})}$, we are in the cutoff condition, where only leakage currents flow. If these are negligible for our circuit considerations (a leakage current in one circuit is a flood-level flow in another!), the model is three leads connected to an open circuit, as shown in Fig. 1.21.

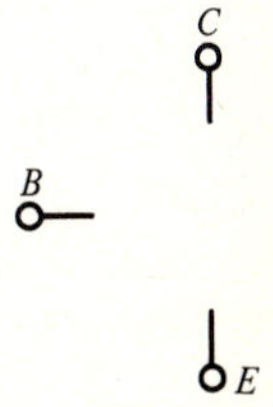

FIGURE 1.21
Cutoff-region model.

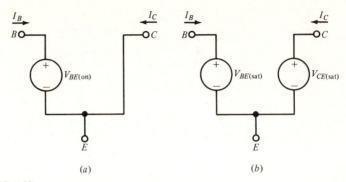

(a)   (b)

FIGURE 1.22
Saturated-mode model. (a) An approximate model for a transistor in saturation;
(b) a more exact model for a transistor in saturation.

1.4.7   Saturation-region Model

If the voltage across both diodes equals or exceeds the threshold voltages (that is, $V_{BE} \geq V_{BE(on)}$ and $V_{BC} \geq V_{BC(on)}$) we are in a saturation condition. Using Kirchhoff's voltage law, we see that the voltage from collector to emitter in saturation is

$$V_{CE(sat)} = V_{BE(on)} - V_{BC(on)} \qquad (1.19a)$$

If we assume that

$$V_{BC(on)} \approx V_{BE(on)}$$

then

$$V_{CE(sat)} \approx 0 \text{ V} \qquad (1.19b)$$

Equation (1.19b) suggests the saturation-region circuit model shown in Fig. 1.22a. The expression for I_B is, from Eq. (1.13a),

$$I_B = I_E - I_C = I_{DC}(1 - \alpha_R) + I_{DE}(1 - \alpha_F) \qquad (1.20)$$

Since I_{DC} and I_{DE} are both positive, and also since α_R and α_F are both less than unity, we have $I_B > 0$. Thus the only condition that we have on any of the terminal currents is that I_B is not zero or negative. Therefore in saturation I_C and I_E can take on any values (including zero and negative) that are consistent with the diode law and Kirchhoff's current law.

For most applications that we deal with, the collector current will be positive (flowing into the collector terminal). We refer to this as the *normal* saturation mode. Furthermore, the approximation $V_{CE(sat)} \approx 0$ will not always be adequate, and we will find that a low-valued voltage source (labeled $V_{CE(sat)}$ in Fig. 1.22b) should be inserted in the collector-emitter branch to give a better model. For consistency in notation, the voltage source in the base-emitter branch is called $V_{BE(sat)}$.

In Table 1.2 we summarize the results of our models and give the values of the three voltages V_{BE}, V_{BC}, and V_{CE} and the collector and emitter currents for an *n-p-n* transistor in the three different regions of operation. Note that when not in saturation the external circuit determines the value of I_B, and then the value of I_C is $\beta_F I_B$.

Table 1.2 VOLTAGE AND CURRENT VALUES AND EXPRESSIONS FOR THE THREE REGIONS OF OPERATION FOR AN *n-p-n* **IC TRANSISTOR**

Region	V_{BE} (V)	V_{BC} (V)	V_{CE} (V)	I_B	I_C	I_E
Active	≈ 0.7	< 0.5	> 0.2	†	$\beta_F I_B$	$(\beta_F + 1)I_B$
Saturation (normal mode)	≈ 0.7	≈ 0.5	≈ 0.2	†	†	$I_B + I_C$
Cutoff	< 0.5	< 0.5	$\cdots$	0	0	0

† The circuitry external to the transistor must be considered in determining these values.

1.4.8 Measured Transistor Characteristics

The typical electrical characteristics of an *n-p-n* transistor, the 2N4275,† are shown in Figs. 1.23 and 1.24 (see Demonstration D1.2). Although this transistor is a *discrete* device, many of its characteristics are similar to those which would be found in an IC.‡ In what follows, we present these characteristics in the order in which we have just discussed the separate operating regions.

Normal active region In Fig. 1.23*a* we show the *collector family characteristic,* that is, the collector current vs. collector-emitter voltage with the base current held at a different constant value for each curve. In the region where the collector current is positive and the collector-emitter voltage is in the range of $1 \text{ V} < V_{CE} < 10 \text{ V}$, the collector current of the transistor is almost constant for a given value of the base current. This is the normal active region, where the value of the collector current is primarily determined, i.e., controlled, by the base current.

To compare these characteristics with those predicted by our model for the normal active region, we return to Fig. 1.19 and Eq. (1.16). This equation reads

$$I_C = \beta_F I_B$$

which shows that the collector current should be constant for constant base

† The "name" 2N4275 is only a registration number issued by an industry association and only tells us by the 2N portion that the device is a three-terminal transistor (either a bipolar or field effect!). The 4275 means that there were 4,274 transistor types registered before this one.
‡ Since many different transistors are used in even one circuit, IC manufacturers do not give even typical transistor characteristics. However, some very simple ICs can be used that provide insight into actual internal transistor parameters (see Demonstration D8.1).

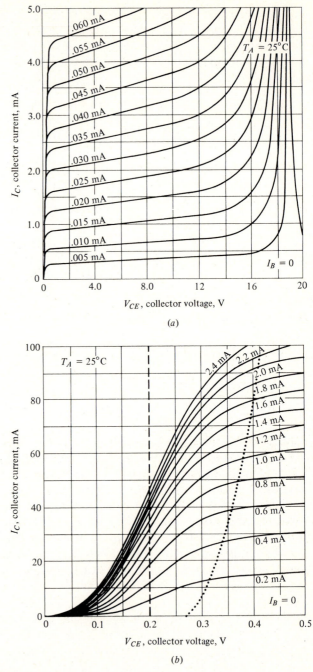

(a)

(b)

FIGURE 1.23
Collector characteristics of a transistor (a) at high voltages and low currents, (b) at low voltages and high currents.

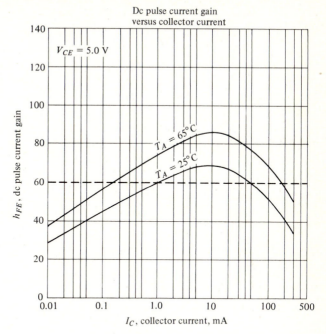

FIGURE 1.24
Current gain as a function of collector current.

current (as long as the transistor is operated in the normal active region). We can obtain a value for β_F directly from the data in Fig. 1.23a by simply taking the ratio of I_C to I_B for one of the curves shown in Fig. 1.23a.

For the purposes of this text, we will assume that β_F is constant. However, in reality β_F is generally dependent on the value of the collector current I_C. The form of the curve for our present example (2N4275) is shown in Fig. 1.24, where at an ambient temperature of 25°C β_F has a peak value of 70 at a collector current of 10 mA. The curve also shows that $\beta_F \geq 60$ for collector currents in the range of 1 to 50 mA, so a reasonable choice for *constant* β_F to approximate the transistor is $\beta_F = 60$ at $T_A = 25°C$. The effects of temperature are also described in the plot of Fig. 1.24, where it is seen that β_F (65°C) $> \beta_F$ (25°C) for the above range of values of I_C. In all these curves, β_F is measured at a sufficiently large value of V_{CE} (5 V) to ensure that the transistor is indeed operating in the normal active mode.

The value of the base-emitter voltage in the active region, $V_{BE(on)}$, is found (see Demonstration D1.2, part *a*, item 4) to be close to the value of 0.7 V for collector currents in the range of 1 to 50 mA.

To summarize, then, the normal-active-mode model consists of a diode connected between base and emitter leads and a current generator connected between collector and base leads. The current generator is described by the parameter β_F, which can be obtained directly from transistor data sheets for discrete devices. For our purposes, β_F will be assumed to be constant, though in fact it varies with collector current.

The saturation region Turning now to an examination of the saturation region, we show in Fig. 1.23b the collector family with current and voltage scales changed to display the device behavior at low V_{CE}. We notice in Fig. 1.23b that at low collector-to-emitter voltages the collector current is strongly dependent upon the collector-to-emitter voltage. This is the saturation region of the transistor. The dotted line in the figure indicates what we have chosen (somewhat arbitrarily) as the boundary line of this region. For collector-to-emitter voltages to the right of this boundary line, we are again in the region where the collector current is nearly independent of the collector-to-emitter voltage, and therefore in the normal active region.

Another way of looking at saturation is to measure a transistor's characteristics in the circuit shown in Fig. 1.25a. Here the collector current I_C is *required by the circuit* to be 10 times the base current I_B. Under these conditions, Fig. 1.25b shows the plot of V_{CE} versus I_C for a typical 2N4275 transistor obtained by varying I_B from 0.01 to 50 mA.

If we assume that we would be always using a transistor in saturation at or near the conditions where the collector current is 10 times the base current, we could refer to the solid lines in Fig. 1.25b to find the actual collector-to-emitter saturation voltage. As indicated by the dashed straight line in Fig. 1.25b, we can also approximate this curve over a wide range of collector current (0.1 through 50 mA) with a straight horizontal line. The horizontal line implies that $V_{CE(sat)}$ is a constant 0.2 V. We have also shown this approximation as a dashed line in Fig. 1.23b. While this approximation is admittedly coarse, it nevertheless provides an adequate characterization for the circuits we will study. The critical fact is to determine that the transistor is actually operating in the normal-mode saturation region. We shall discuss this point more fully later.

Turning now to the input characteristic in the saturated region, we show in Fig. 1.25c a plot of the base-emitter saturation voltage under the condition that the collector current is always 10 times the base current. Again, we approximate V_{BE} by a constant value of 0.7 V over the range of collector currents from 0.1 to 50 mA. The chosen value of 0.7 V is shown as the dashed line in Fig. 1.25c. The fact that this voltage is the same as the $V_{D(on)}$ of a silicon diode is not surprising since the base-emitter junction of a silicon transistor is a silicon diode.

SUMMARY OF CHARACTERISTICS OF A TYPICAL n-p-n TRANSISTOR

Table 1.3 gives the range of $V_{BE(sat)}$, $V_{CE(sat)}$, and β_F for typical operating conditions for the 2N4275. For transistors in ICs, the range of variation of a single parameter over a given chip is expected to be much less than is indicated in Table 1.3.

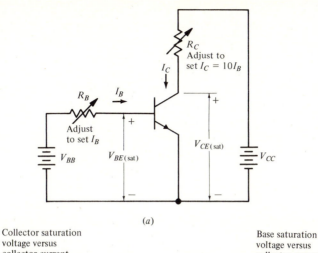

(a)

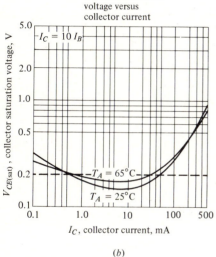

(b)

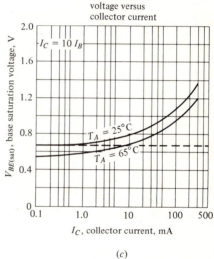

(c)

FIGURE 1.25
Saturation-region characteristics. (a) Circuit for measuring saturation voltages; (b) collector saturation voltage vs. collector current; (c) base saturation voltage vs. collector current.

Table 1.3 ELECTRICAL SPECIFICATIONS
FOR A 2N4275 AT 25°C

	Min	Typical	Max	I_C	I_B
$V_{BE(sat)}$	0.72	0.80	0.85	10 mA	1 mA
$V_{CE(sat)}$	⋯	0.14	0.20	10 mA	1 mA
$V_{CE(sat)}$	⋯	0.28	0.50	100 mA	10 mA
β_F	35	70	120	10 mA	
β_F	18	45	⋯	100 mA	
β_F	⋯	50	⋯	1 mA	

However, the variation of a single parameter among different *packaged* ICs could be much larger than that found in discrete transistors. Generally, this is unimportant because the circuit properties that can be measured at the IC *terminals* (which are related to transistor and resistor values) are guaranteed by the manufacturer.

Table 1.4 gives some "rules of thumb" indicating the effects of temperature T (in °C) on β_F and $V_{BE(\text{sat})}$ for n-p-n IC transistors. In words, these rules say that β_F increases by about 0.7 percent/°C and $V_{BE(\text{sat})}$ decreases by approximately 2 mV/°C. $V_{BE(\text{on})}$ also follows this rule. Both "rules" should be applied near the current where the β_F peaks.

1.4.9   Fitting a Model to Measured Data

The approximations that we made to obtain the dashed lines in Figs. 1.24 and 1.25 can, of course, be transformed into parameter values for our circuit models. First we should draw the VI characteristics that reflect the approximations we have agreed to use. The collector family is drawn in dotted lines in Fig. 1.26a. Two factors are reflected in it: (1) the vertical line at 0.2 V gives us a constant $V_{CE(\text{sat})}$ of 0.2 V, and (2) in the active region the collector circuit is always 60 times the base current, so that β_F is 60.

The $V_{BE} I_B$ approximation is shown in Fig. 1.26b. Here I_B is zero for V_{BE} less than 0.7 V. After the base-emitter voltage exceeds this threshold, infinite current would flow as the straight-up line indicates. This does not cause any difficulties, as all practical circuits have some limiting resistance external of the transistor that would act to set a finite value on I_B.

To summarize, for the normal-mode active-region model of Fig. 1.19c and the saturated-mode model of Fig. 1.22b, we would use $\beta_F = 60$, $V_{BE(\text{on})} = V_{BE(\text{sat})} = 0.7$ V, and $V_{CE(\text{sat})} = 0.2$ V.

EXERCISES

E1.5   Determine the collector saturation voltage from Fig. 1.23b for $I_C/I_B = 40$ at $I_C = 20$ mA and $I_C = 40$ mA.

E1.6   Given that a silicon n-p-n transistor is to be modeled as in Fig. 1.18 with the

Table 1.4   TEMPERATURE   COEFFICIENTS OF n-p-n-TRANSISTOR PARAMETERS

$$\beta_F(T) = \beta_F \text{ (at 25°C)}[1 + 0.007(T - 25)]$$
$$V_{BE(\text{sat})}(T) = V_{BE(\text{sat})} \text{ (at 25°C)} - (2 \text{ mV})(T - 25)$$
where T = temperature, °C

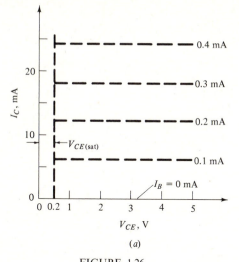

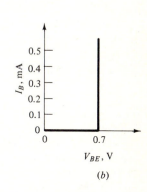

FIGURE 1.26

Characteristics for a simple transistor model. (*a*) Collector characteristics if $V_{CE(sat)} = 0.2$ V and $\beta_F = 60$; (*b*) base characteristic if $V_{BE(on)} = V_{BE(sat)} = 0.7$ V.

following parameters:

$$\alpha_F = 0.98$$

$$\alpha_R = 0.0125$$

   (*a*) If $V_{BC} = -1$ V and $I_B = 0.1$ mA, in what region is the transistor and what are I_C and I_E? What would a good guess be for the value of V_{BE} for these conditions? Based on this, what is the value of V_{CE}?

   (*b*) If now $V_{BE} = -1$ V and $I_B = 0.79$ mA, in what region is the transistor and what are I_C and I_E? What would your guess be for the value of V_{BC}?

1.5   TRANSISTOR-INVERTER ANALYSIS

We began this chapter with a brief discussion of the transfer characteristic of the transistor inverter shown in Fig. 1.1, promising at that point to compute the transfer characteristic after suitable circuit models for the transistor had been developed. The models described in the preceding section permit us to make the required calculation and hence to compare theory with experiment.

   In what follows, we compute the transfer characteristic, showing in the process that, as the input voltage V_{in} is increased from 0 to $+3$ V, the transistor moves from the *cutoff* region to the *normal active* region and then to the *saturated* region. The calculations therefore require us to use three separate models and to find the two *transition points* at which circuit operation moves the transistor from one region to the next. The transition points turn out to be especially important

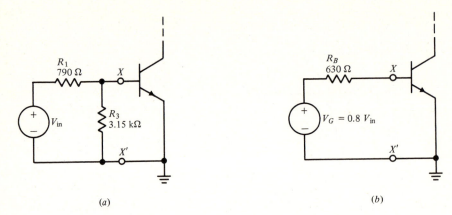

(a)   (b)

FIGURE 1.27
Input-circuit simplifications. (a) The input circuit of Fig. 1.1; (b) the Thévenin
equivalent to part (a).

because, as we will see, the entire transfer characteristic can be drawn once the
transition points are specified.

After using our simple circuit models to compute the transfer characteristic,
we will compare it with the experimental characteristic given in Fig. 1.2; we will
then perform the calculation graphically to see how a graphical analysis compares
with the analytical one.

The method that we use to make the computation is as follows. First, we
examine the circuits connected to the base and collector of the transistor to see if
they can be simplified (using either the Thévenin or the Norton theorem). Then,
with simplified source and load networks, we determine how both the operating
state of the transistor and the resultant output voltage vary as the input voltage
changes from 0 to $+3$ V.

1.5.1   Thévenin Equivalent to Simplify Input and Output Networks

We begin by applying the Thévenin theorem twice: once to the source network
and again to the load network.

The source network is the portion of the circuit to the left of the base of the
transistor in Fig. 1.1, which is redrawn in Fig. 1.27a. The Thévenin equivalent is
shown in Fig. 1.27b. It consists of a voltage source V_G, which is simply the open-
circuit voltage obtained from the source network, and a series resistance R_B,
which is simply the dc resistance seen looking back into the source network with
V_{in} set to zero.

Following this prescription, we see from the simple voltage-divider action of
R_1 and R_3 that the open-circuit voltage is

$$V_G = \frac{R_3}{R_1 + R_3} V_{in} = \frac{3.15}{0.79 + 3.15} V_{in} = 0.8 V_{in}$$

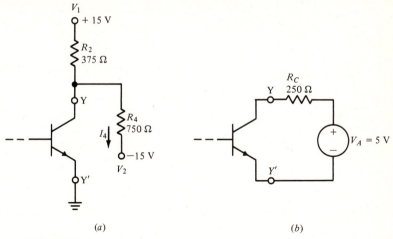

(a)   (b)

FIGURE 1.28
Output-circuit simplification. (a) The output circuit of Fig. 1.1; (b) the Thévenin equivalent of part (a).

Hence, as V_{in} varies from 0 to 3 V, V_G varies from 0 to 2.4 V. Similarly, replacing V_{in} by a short circuit, the Thévenin resistance R_B is seen to be simply the parallel combination of R_1 and R_3:

$$R_B = R_1 \| R_3 = \frac{R_1 R_3}{R_1 + R_3} = \frac{(0.79)(3.15)}{0.79 + 3.15} = 0.63 \text{ k}\Omega$$

Let us consider now the Thévenin representation of the load network. This network consists of all the elements seen to the right and above the collector of the transistor in Fig. 1.1. The network is redrawn in Fig. 1.28a, and the Thévenin equivalent to this is shown in Fig. 1.28b.

The determination of V_A, the open-circuit voltage of the load network, is more complicated than that of V_G, but it is still straightforward. We proceed by finding the current I_4 that flows when an open circuit exists to the left of $Y - Y'$ (that is, $I_C = 0$). From this we determine the voltage that exists at $Y - Y'$. This is the required open-circuit voltage V_A.

From Kirchhoff's voltage law and Ohm's law, we have

$$V_A = V_2 + R_4 I_4$$

$$= V_2 + \frac{R_4}{R_2 + R_4} (V_1 - V_2)$$

$$= -15 + \frac{0.75}{0.375 + 0.75} [15 - (-15)]$$

$$= 5 \text{ V}$$

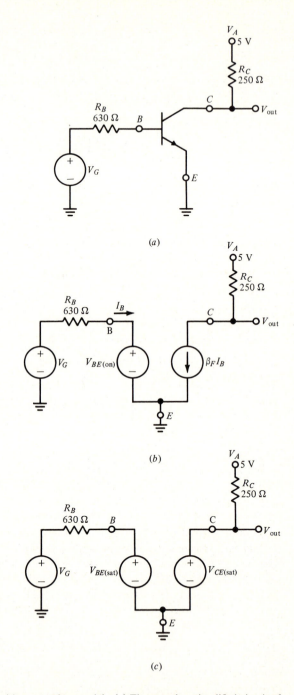

(a)

(b)

(c)

FIGURE 1.29
The inverter circuit with appropriate models. (a) The complete simplified circuit of
Fig. 1.1; (b) the circuit of part (a) with the transistor replaced by its active-region
model; (c) the circuit of part (a) with the transistor replaced by its saturation-
region model.

The Thévenin equivalent resistance for the load network R_C is

$$R_C = R_2 \| R_4 = (0.375) \| (0.75) = 0.25 \text{ k}\Omega$$

Now with the circuit of Fig. 1.1 simplified to that in Fig. 1.29a, we can use our transistor models to obtain the transfer characteristic of the inverter.

1.5.2 Analytical Solution of the Inverter Problem Using Basic Transistor Models

For the analytical determination of the transfer characteristic, we use the basic transistor models developed in Sec. 1.4. These are shown embedded in the circuit of Fig. 1.29. When necessary we will use the values $\beta_F = 60$, $V_{BE(\text{sat})} = V_{BE(\text{on})} = 0.7$ V, and $V_{CE(\text{sat})} = 0.2$ V.

To begin the calculation, we note that with $V_G < 0.7$ V the base-emitter diode is cut off. Hence, for $V_G = 0$, $I_B = 0$ in the circuit of Fig. 1.29a, and the transistor is *cut off* $(I_C = 0)$. Furthermore, the transistor remains cut off until $V_G = 0.7$ V. The collector current is zero and the collector output voltage is 5 V over this entire range. In Fig. 1.30 we show this behavior in the plots of I_B, I_C, and V_{out} as a function of V_G and V_{in}.

When V_G is increased beyond 0.7 V, however, base current begins to flow and the transistor moves from the cutoff region to the *normal active* region. Hence the coordinates $(V_G = 0.7$ V, $V_{\text{out}} = 5$ V) mark the first *transition point* for the transfer function of this circuit. These points are also referred to as breakpoints since they represent breaks in a straight line as we shall soon see. This is denoted in Fig. 1.30 by the label EOC which stands for *edge of cutoff*.

To compute the transfer characteristic for $V_G > 0.7$ V, we note first that the base current is (see Fig. 1.26b)

$$I_B = \frac{V_G - V_{BE(\text{on})}}{R_B} \qquad (1.21)$$

This base current will then produce the controlled collector current $\beta_F I_B$ as long as $V_{CE} > V_{CE(\text{sat})}$. Assuming this to be true for the moment, we see that a base current I_B causes the output voltage to be reduced from V_A according to Kirchhoff's voltage law as

$$V_{\text{out}} = V_A - \beta_F I_B R_C \qquad (1.22a)$$

Now by including Eq. (1.21) we can also rewrite Eq. (1.22a) in the form

$$V_{\text{out}} = V_A - \beta_F (V_G - V_{BE(\text{on})}) \frac{R_C}{R_B} \qquad (1.22b)$$

which shows that V_{out} decreases *linearly* with increasing V_G as long as the transistor is in the normal active mode.

As I_B (or V_G) is increased, however, a second *transition point* will be reached when $V_{\text{out}} = V_{CE(\text{sat})}$. This point will occur at a critical value of collector current, where we must change from a normal active model to a *saturation-region* model.

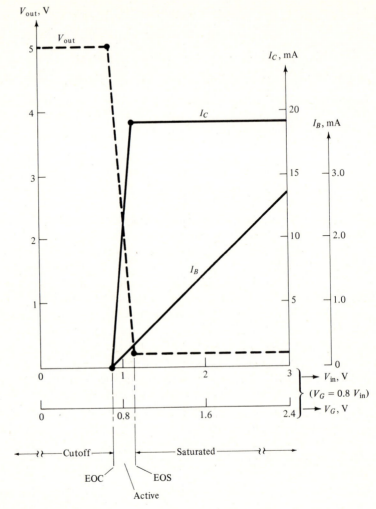

FIGURE 1.30
Plots of I_B, I_C, and V_{out} calculated from models as a function of V_G (and V_{in}) for the circuit of Fig. 1.29a.

We can calculate the values of I_C and I_B for this condition as follows (the subscript EOS means *edge of saturation*):

$$I_{C(EOS)} = \frac{V_A - V_{CE(sat)}}{R_C} \qquad (1.23)$$

$$= \frac{5 - 0.2}{0.25} = 19.2 \text{ mA}$$

and

$$I_{B(EOS)} = \frac{I_{C(EOS)}}{\beta_F} \tag{1.24}$$

$$= \frac{19.2}{60} = 0.32 \text{ mA}$$

Now, using this value of $I_{B(EOS)}$ in the circuit of Fig. 1.29b, we find

$$V_{G(EOS)} = V_{BE(on)} + I_{B(EOS)}R_B \tag{1.25}$$

$$= 0.7 + (0.32)(0.63) = 0.9 \text{ V}$$

Hence the transistor is cut off when $V_G \leq 0.7$ V and saturated when $V_G \geq 0.9$ V. It operates in the normal active mode for $0.7 < V_G < 0.9$ V.

In Fig. 1.30 we show the complete plot of calculated values of V_{out}, I_C, and I_B for V_G varying from 0 to 2.4 V. The transition points EOC and EOS are also shown. Note carefully that as a result of our model each plot consists of a series of connected straight-line segments. Note that, because the computed transfer characteristic consists of a series of connected straight-line segments, it can be drawn *completely* when the transition points are specified. Also note that even though I_B *increases* as V_G increases, I_C *saturates* when the transistor saturates.

Since $V_{in} = 1.25V_G$ we can show the dependent variables as a function of V_{in} by just changing the scale on the horizontal axis as shown on the upper set of horizontal coordinates in Fig. 1.30.

In review then, the inverter transfer characteristic consists of three straight-line segments, representing the three operating regions: (1) the cutoff region, where the transistor is cut off and the output voltage has its highest value; (2) the normal active region, where the slope of V_{out} versus V_{in} is constant and negative (the magnitude of this slope is called the *voltage gain* and will be discussed in greater detail in Chap. 7); and (3) the saturation region, where the output is at its lowest value and is equal to the saturation voltage of the transistor Q_1.

1.5.3 Comparison of Theoretical and Experimental Transfer Characteristics

The transfer characteristic computed with our simple models is compared to the actual experimental transfer characteristic for the inverter in Fig. 1.31 (see Demonstration D1.3). The most noticeable differences between theory and experiment are that the experimental curve changes smoothly from the cutoff region to the normal active region and from the normal active region to the saturation region. The abrupt changes apparent in the computed transfer characteristic arise from the fact that we have represented each diode in the transistor model by an ideal-diode–plus–0.7-V voltage source. Had we used the VI relation of Eq. (1.3) for each diode instead, we would have obtained a theoretical curve that is even closer to the experimental one. However, this is hardly necessary since the quality of agreement exhibited in Fig. 1.31 is satisfactory for our purposes.

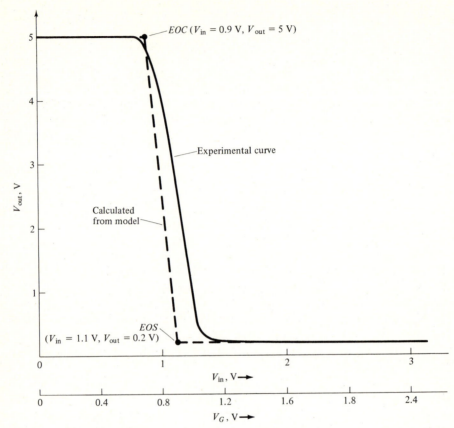

FIGURE 1.31

V_{out} versus V_G and V_{in} determined from experimental data compared to that calculated from models in Fig. 1.29.

1.5.4   Graphical Analysis of the Transistor Inverter

The load-line technique used in the analysis of diodes in Sec. 1.2.1 can be applied to our inverter as well. For this purpose we simply treat the transistor as a device with a nonlinear VI characteristic and construct a load line as shown in Fig. 1.32. If we know I_B, we can then determine I_C and V_{CE} (which is V_{out}) from the load line.

To determine how I_B depends on V_G, we require a plot of I_B versus V_{BE}, with either V_{CE} or I_C as a variable parameter.

Curves of I_B versus V_{BE} that cover all conditions of I_C or V_{CE} are not generally available, so that the load-line technique for determining I_B is not feasible. However, we can use *as an approximation* the in-saturation data given in the curves of Fig. 1.25b, which we have plotted in Fig. 1.33 with I_B as the ordinate and

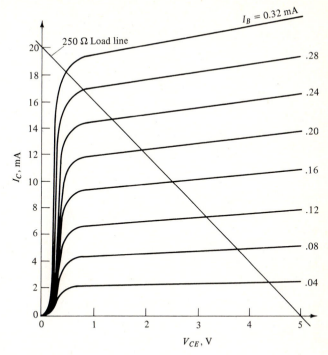

FIGURE 1.32
The collector characteristic with the load line for a graphical solution of V_{out}.

V_{BE} as the abscissa. Here we simply have a diodelike characteristic, so we can find I_B using the method of Sec. 1.2.1. For V_G (volts) at 2.2, 2, 1.5, 1.0, and 0.8, we have shown the base-circuit load lines and their intersections on the I_B versus V_{BE} curve. For each such value of base current, we can go to the collector characteristic and from the collector-circuit load line determine the other variables. Table 1.5 shows the values of the various variables in the circuit for these points in saturation plus several points in the active region. For these active-region points, the assumption is made that the base-emitter voltage is constant at 0.65 V. The resulting graphically determined transfer function is shown as a solid curve in Fig. 1.34. For comparison we have shown the results of our preceding model analysis as a dashed line.

We see that the two approaches give reasonably good agreement. The more complex graphical method would only be used if the precise shape of transfer characteristic was needed. However, the computational method based on circuit models can also be improved greatly by computer-oriented techniques which in effect change the model after each calculation. Because computers are available, graphical techniques have lost much of their applicability, although they are still useful in situations where models are not well developed.

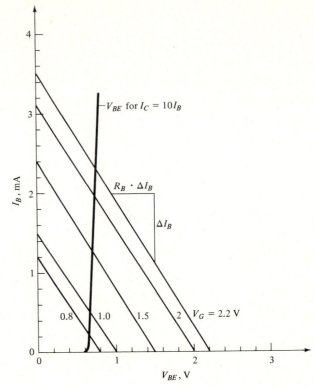

FIGURE 1.33
The graphical construction for determining I_B as a function of V_G.

Table 1.5

V_{in} (V)	V_G (V)	V_{BE} (V)	I_B (mA)	I_C (mA)	V_{out} (V)
2.75	2.2	0.8	2.3	19	0.14
2.5	2.0	0.75	2.0	19	0.14
1.875	1.5	0.72	1.2	19	0.18
1.25	1.0	0.7	0.5	19	0.22
1.00	0.8	0.65	0.2	12.4	1.9
0.86	0.69†	0.65	0.06	3.8	4.0
0.84	0.67†	0.65	0.035	2.1	4.4
0.81	0.65†	0.65	0.000	0.0	5.0

† Calculated by using $V_G = 0.65 + R_B I_B$.

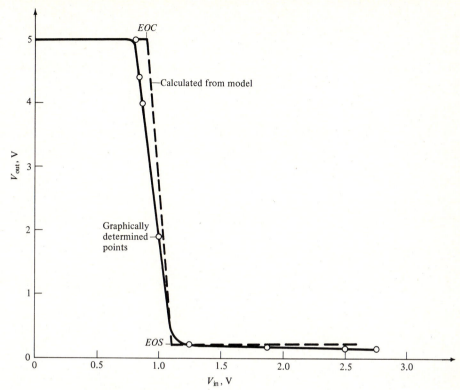

FIGURE 1.34

Comparison of graphical solution of V_{out} versus V_{in} to that obtained from the models in Fig. 1.29.

EXERCISES

E1.7   Use the collector family VI curve of Fig. 1.23 and graphically determine the base current required to drive an inverter to a 0.2-V collector saturation voltage with $V_1 = 5$ V, $R_2 = 200 \ \Omega$, and $R_4 \rightarrow \infty$, in Fig. 1.1.

Repeat the above, using the analytical technique with the simple models used in the text to calculate $I_{B(EOS)}$ when the output voltage is 0.2 V. Use $\beta_F = 60$.

E1.8   Use the Norton equivalent for the output part of the circuit in Fig. 1.28b to determine V_{out} when $I_C = 10$ mA.

E1.9   Use the analytical methods and values for models of Sec. 1.5.2 to determine the coordinates of the critical points of the transfer function for the circuit in Fig. 1.29a, where now $R_B = 10$ kΩ and $R_C = 1$ kΩ.

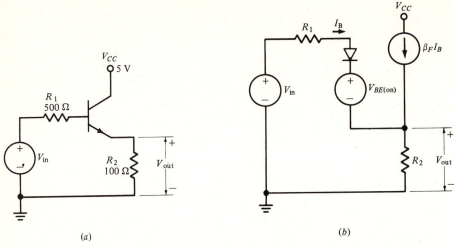

FIGURE 1.35
Emitter-follower circuit and model. (a) Schematic for an emitter follower;
(b) circuit model for analyzing the emitter follower in the active region.

1.6 EMITTER-FOLLOWER ANALYSIS

We have now analyzed the transistor inverter in sufficient detail to fully understand its operation in digital ICs. However, since bipolar transistors are also quite
commonly used in the *emitter-follower* configuration in ICs, we will apply our
bipolar models to the analysis of the emitter follower in this section.

The main features of the inverter circuit of Fig. 1.1 are that the input signal is
applied to the base, the emitter is grounded, and the load is placed in series with
the collector. If we place the load R_2 between the emitter and ground, we have the
emitter-follower circuit of Fig. 1.35a. We will find that emitter followers are widely
used both in digital and linear ICs.

As we shall see in more detail in Chap. 7, the voltage gain of an emitter
follower is always less than unity (though usually very close to unity). Nevertheless, this circuit is useful since current gain is obtained with it.

In the following we will derive the expression for the output voltage as a
function of the input voltage. As in the analysis of the inverter, we assume that
initially V_{in} is zero. We also assume V_{CC} sufficiently large so that the collector
diode will always be reverse-biased. Thus the transistor is not saturated.

Now looking at the circuit and transistor model in Fig. 1.35b, we see that if
I_B is zero, I_C and I_E are zero. Hence the voltage applied across the ideal diode is

$$V_{in} - V_{BE(on)}$$

As long as V_{in} is less than $V_{BE(on)}$, the ideal diode is cut off and our initial assumption that I_B is zero holds. As no emitter current flows, the output voltage is zero.
To see what happens when $V_{in} \geq V_{BE(on)}$, we can use Kirchhoff's voltage law for

the *loop* containing the V_{in} and ideal diode. Assuming that a positive I_B flows, we have

$$V_{in} = I_B R_1 + V_{BE(on)} + I_E R_2 \qquad (1.26)$$

With the assumption that V_{CC} is large enough so the transistor cannot saturate, we have $I_C = \beta_F I_B$, so

$$I_E = I_C + I_B = \beta_F I_B + I_B = (\beta_F + 1)I_B \qquad (1.27a)$$

On substituting Eq. (1.27a) in Eq. (1.26), we have

$$V_{in} = I_B[R_1 + (\beta_F + 1)R_2] + V_{BE(on)} \qquad (1.27b)$$

or

$$I_B = \frac{V_{in} - V_{BE(on)}}{(\beta_F + 1)R_2 + R_1} \qquad (1.28)$$

Thus for $V_{in} > V_{BE(on)}$, a positive base current given by Eq. (1.28) flows and the transistor is in the active region. The output voltage is now given by

$$V_{out} = I_E R_2 \qquad (1.29)$$

and on substituting Eq. (1.27a) and (1.28) in (1.29), we obtain

$$V_{out} = \frac{R_2(\beta_F + 1)}{R_2(\beta_F + 1) + R_1}(V_{in} - V_{BE(on)}) \qquad (1.30)$$

If $R_2(\beta_F + 1) \gg R_1$, which is a reasonable approximation for the numerical case in Fig. 1.35a [since $(61)(100) \gg 500$], then the first factor in Eq. (1.30) can be assumed to be unity. Under this assumption,

$$V_{out} \approx V_{in} - V_{BE(on)} \qquad (1.31)$$

Equation (1.31) shows that the output voltage is a nearly direct replica of the input voltage except for the term $V_{BE(on)}$. The effect of this is to shift the output level down from V_{in} by $V_{BE(on)}$. This *level-shifting* feature of emitter followers is advantageously employed in one form of logic circuits discussed in Chap. 3.

When the input voltage V_{in} is very large compared to $V_{BE(on)}$, we can make a further approximation in Eq. (1.31) to obtain $V_{out} \approx V_{in}$.

Thus the output voltage at the emitter *follows* the input voltage; hence, the name emitter follower. The "following" holds as long as the transistor is not saturated. In Prob. P1.30 we determine what the input voltage is that brings the transistor to the edge of saturation.

EXERCISE

E1.10 For the modified emitter-follower circuit of Fig. E1.10, find the value of V_{in} that gives a V_{out} of 1.22 V. Assume β_F is 60 and that $V_{BE(on)} = 0.7$ V and $V_{D(on)} = 0.7$ V.

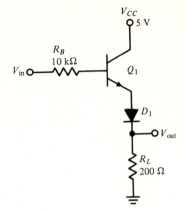

FIGURE E1.10
An emitter-follower circuit with an added diode.

1.7 MOS FIELD-EFFECT TRANSISTORS

We now turn our attention to a second type of transistor that finds wide application in ICs, the MOS field-effect transistor (for brevity this transistor is often called a MOSFET). The name MOS comes from the distinguishing feature of its structure, which consists of layers of metal, oxide, and semiconductor, hence MOS.

The basic fabrication of a MOSFET is described in Appendix A (see Fig. A.9). Its importance in ICs arises from three interesting properties. First, because of its very small size, it is possible to pack three to five MOSFETs into the area required by the smallest bipolar transistor. A second advantage that MOSFETs have is that, by making appropriate circuit connections, it is possible to use the MOSFET as a high-value "resistor," which also occupies a very small area. A third advantage is that the resistance measured between one electrode (the gate) and any of the other two electrodes is very high. The very small capacitance that this electrode has to the rest of the structure can then be used as the "storage" element in *memory circuits*. The information is stored as charge in this capacitance (see second half of Demonstration D1.5). Hence the number of circuit functions that can be performed on a given chip of silicon is much higher if MOSFETs are used instead of bipolar transistors; and therefore, circuit designs using MOSFETs are widely used for large-scale integration (LSI).

The above remarks hold for MOS circuits made with either *p*- or *n*-channel devices (these terms are defined in Appendix A). However, it is also possible to incorporate both *p*- and *n*-channel devices on the same chip. Such complementary MOS (CMOS) digital circuits do not possess the same high-density advantage over the bipolar-transistor circuits that single-polarity MOS circuits do. They do, however, have the useful property of drawing practically zero dc power when operating at low frequency. Both MOS and CMOS ICs will be discussed further in Chap. 3. In this section we will investigate the characteristics of the MOSFETs and their application to an inverter circuit.

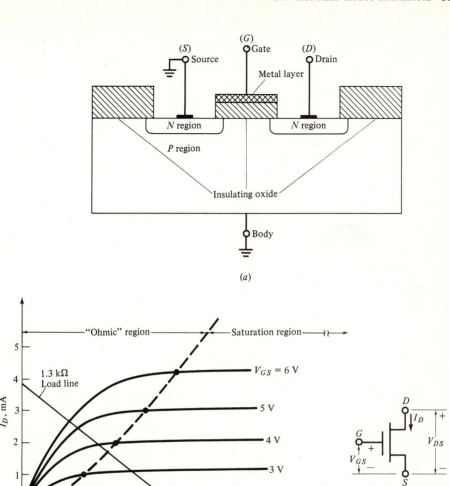

FIGURE 1.36
MOSFET transistor cross section, characteristics, and schematic conventions.
(*a*) Cross-sectional view of an *n*-channel MOSFET; (*b*) drain family characteristic;
(*c*) circuit schematic symbol and reference polarities for the MOSFET in part (*a*).

1.7.1 MOSFET Characteristics

We begin by referring to the cross section of an *n*-channel MOSFET shown in
Fig. 1.36*a* and the $I_D\,V_{DS}$ characteristics for the same device shown in Fig. 1.36*b*.
This is a four-terminal device having terminals labeled *source*, *gate*, *drain*, and
body. We will study its characteristics with the body and source tied together. The

conventional schematic symbol for the resulting three-terminal device is shown in Fig. 1.36c.

With the source and body of the transistor grounded, the application of a positive voltage to the drain (D) ensures that the p-n body-drain junction is reverse-biased. Since the source (S) is connected to the body, the source-body junction is at zero bias. The result is that with zero voltage at the gate (G) only reverse (leakage) currents flow between the drain and the source. Thus we have a cutoff condition for the transistor.

Now, if we put a sufficiently positive voltage on the gate relative to the source, some very interesting things take place. A positive voltage applied between the gate and the body will produce an electric field which tends to repel holes from and attract electrons to the surface of the silicon directly underneath the gate electrode. As the gate-source voltage (V_{GS}) is increased, this effect becomes sufficiently pronounced so that the surface of the silicon under the gate changes from p- to n-type. Thus, when a sufficiently large voltage is applied between the gate and the body, a *conducting channel* of electrons is formed between the two formerly separated n regions, allowing drain current (I_D) to flow between the drain and the source. The gate-to-source voltage (V_{GS}) that must be applied to form the channel is called the threshold voltage (V_{th}). From Fig. 1.36b we see that V_{th} for this device is $+1$ V.

From the drain family characteristic (Fig. 1.36b), we see that when V_{DS} is low the channel behaves like a normal ohmic resistance (that is, I_D is proportional to V_{DS}). The *value* of the resistance is controlled by the gate-source voltage V_{GS}. This is due to the fact that as V_{GS} becomes more positive than V_{th}, the number of electrons that accumulate on the surface, and hence the surface conduction, increases.

We also observe from Fig. 1.36b that, for any given value of V_{GS}, as V_{DS} becomes more positive, I_D eventually levels off or saturates.† This behavior can be understood by referring to Fig. 1.37. With V_{GS} equal to 4 V and V_{DS} at 0.1 V, as in Fig. 1.37a, there is 4 V between the gate and source and almost 4 V between the gate and drain, resulting in a nearly constant voltage difference between the gate and every point of the silicon surface underneath the gate. Hence there is a uniform accumulation of electrons along the surface.

With V_{DS} increased to 2 V, the density of electrons in the conduction channel will now not be constant along the x direction, having fewer electrons at the drain end since the gate-to-drain voltage will be significantly reduced there (to 2 V in our example). This situation is shown in Fig. 1.37b.

If we increase V_{DS} to 3 V, V_{GD} is now equal to the threshold voltage $(V_{th} = 1$ V$)$ and the channel is now "pinched off" at the drain. That is, electron accumulation right at the drain edge of the channel has been reduced to zero, and the channel will terminate just where the drain n region meets the surface of the silicon (see Fig. 1.37c). Both theory and experiment show that the drain current

---

† In FETs we speak of drain-*current* saturation with increasing drain voltage, whereas in *bipolar transistors* we had collector-to-emitter *voltage* saturation with increasing base current.

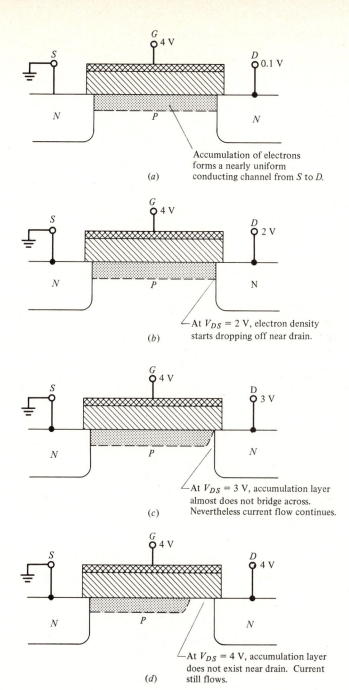

FIGURE 1.37

A representation of electron densities in the channel region of an n-channel MOSFET with $V_{GS} = 4$ V in all cases and $V_{th} = 1$ V. (a) For $V_D = 0.1$ V the gate-to-channel voltage is nearly the same along the channel, and so the density of electrons is nearly uniform along the channel; (b) for $V_D = 2$ V the electron density starts dropping at the drain end of the channel; (c) for $V_D = 3$ V the *accumulation* layer of electrons just comes up to the n region of the drain; (d) for $V_D = 4$ V the accumulation layer does not go all the way to the n region of the drain. Nevertheless current still flows *at a saturated value.*

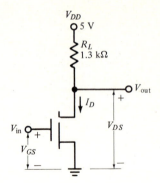

FIGURE 1.38
A MOSFET inverter with a resistive load.

saturates at this point. For the device in question, with $V_{GS} = 4$ V, the current saturates at 2 mA (see Fig. 1.36b).

For higher values of V_{DS} the transistor operates with essentially constant drain current. The channel now ends before the drain region, as shown in Fig. 1.37d, but current still flows because of the injection of electrons into the region between the end of the channel and the drain; however, the current is now independent of V_{DS}.

The dashed line in Fig. 1.36b, where $V_{DS} = V_{GS} - V_{th}$, separates the *ohmic region* from the *saturation region*. With $V_{DS} < (V_{GS} - V_{th})$ we are in the ohmic region, where the drain current is a strong function of the drain-source voltage. With $V_{DS} > (V_{GS} - V_{th})$ we are in the saturation region, where the drain current is a constant value independent of the drain-source voltage. These characteristics of Fig. 1.36b are obtained in Demonstration D1.5.

1.7.2   MOSFET Inverter

In Fig. 1.38 we show a MOSFET inverter with a 1.3-kΩ load resistor and a 5-V supply. The appropriate load line is drawn on the characteristic curves of Fig. 1.36b. From this we can easily determine the voltage transfer characteristic for the inverter of Fig. 1.38, which is drawn as curve A in Fig. 1.41.

As an inverter this circuit has useful properties, although it can be improved significantly by using a specially connected MOSFET in place of the fixed resistor as a load. The circuit connection is shown in Fig. 1.39a, where the gate is directly connected to the drain. The VI plot of this two-terminal device is shown as a dashed line on the characteristic curves of Fig. 1.39b and is obtained by simply drawing a curve through the points defined by the condition $V_{GS} = V_{DS}$. If we wanted to approximate the curve by a straight line, we would have an effective resistance of approximately 1.3 kΩ, as shown by the dotted line. We could then use as a model for the MOSFET resistor the circuit model shown in Fig. 1.39c. It contains an ideal diode, a 1.3-kΩ resistor, and a voltage source equal to the threshold voltage ($V_{th} = 1$ V).

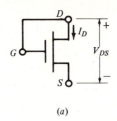

(a)

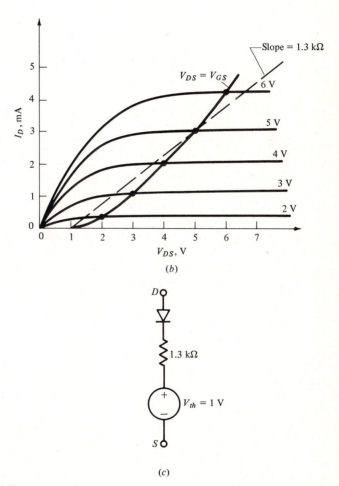

(b)

(c)

FIGURE 1.39

A MOSFET used as a resistor. (a) Connection for a MOSFET as a resistor; (b) the resulting VI characteristic shown in the drain family; (c) circuit model for the dotted characteristic shown in part (b).

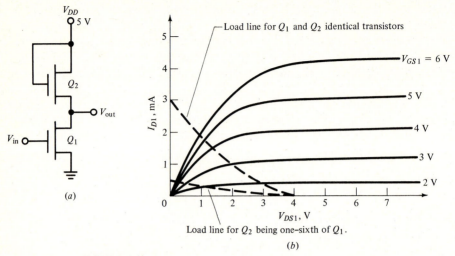

FIGURE 1.40
The MOSFET inverter using a MOSFET load resistor. (*a*) The circuit of an all-MOSFET inverter; (*b*) load lines of MOSFETs as loads.

In Fig. 1.40*a* we show the inverter circuit with Q_2 acting as a load for the inverter Q_1. Assuming Q_1 and Q_2 are identical, the load line is the upper dashed curve of Fig. 1.40*b*. This load line is obtained from Fig. 1.39*b* by noting the voltage across the "load resistor" at any given drain current. Then, by subtracting this voltage from V_{DD}, we obtain the voltage across the inverting transistor. For example, at $I_D = 1$ mA, $V_{DS(2)} = 3$ V; therefore $V_{DS(1)} = 2$ V. The complete transfer function is shown as curve $B1$ in Fig. 1.41. If we used the circuit model in Fig. 1.39*c* to represent Q_2 in Fig. 1.40*a*, we would compute the transfer function

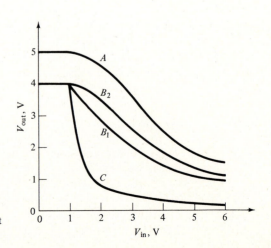

FIGURE 1.41
Transfer functions for three different MOSFET inverters.

shown as curve B2 in Fig. 1.41. In any case, Q_1 is cut off when V_{in} is less than the threshold voltage, which is 1 V.

If we use as Q_2 a MOSFET that has a channel length† six times the channel length of Q_1, then the channel resistance will be six times greater, and for any given V_{DS} the drain current will be reduced by a factor of 6. We then have the lower load line shown in Fig. 1.40b. The transfer characteristic for this MOSFET inverter circuit is shown in Fig. 1.41 as curve C. If the "resistance" of Q_2 is further increased, the transition of the output voltage from the HIGH- to LOW-voltage state in the transfer characteristic will become very abrupt. We shall see in Chap. 3 that such an abrupt transition is very desirable.

EXERCISE

E1.11 Determine V_{out} for the modified circuit of Fig. 1.38 (use $V_{DD} = 6$ V and $R_L = 2$ kΩ) by using the MOSFET characteristics of Fig. 1.36b when

(a) $V_{in} = 5$ V

(b) $V_{in} = 1.5$ V

1.8 SUMMARY OF TRANSISTOR DISCUSSION

The main concepts presented in the second part of this chapter are:

\#   The three regions that a bipolar transistor normally operates in were described. These are called the cutoff mode, the active mode, and saturated mode.

\#   Corresponding to each mode, a simple circuit model was found whose VI characteristic(s) was adequate as long as we used it in the appropriate region.

\#   These three models were employed in analyzing the transfer characteristic of an inverter circuit that contains one bipolar transistor.

\#   The characteristics of a MOS field-effect transistor (MOSFET) were described. An inverter circuit made using a MOSFET as a load resistor was studied.

The material in this chapter is useful for the dc analysis and design of digital and linear circuits. In order to lay the groundwork for digital circuit specifications, we will in the next chapter introduce the fundamentals of logic design that we will use throughout the rest of our discussion on digital ICs.

DEMONSTRATIONS

D1.1   Silicon-diode VI characteristic   Using a low-power silicon diode,‡ make the following measurements:

(a) At room temperature measure V_D at 0.5, 5, and 50 mA.

---

† The linear dimension between the source and drain in Fig. 1.36a is the channel length.
‡ 1N754A or equivalent.

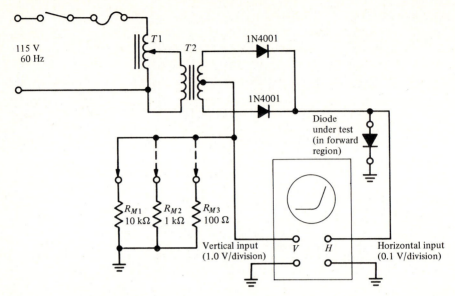

FIGURE D1.1
VI curve tracer. $T1$ = variable transformer (Ohmite Model VTO2N or equivalent). $T2$ = 110 V: 16-V center-tap transformer (Signal Model 16-1 or equivalent). The resistors are for current monitoring. For 1 mA/div vertical scale, R_{M2} = 1 kΩ.

(b) Repeat part (a) at a temperature of 100°C. This can be done by immersing the diode in boiling water and waiting for the reading to stabilize. An alternate method is to apply heat to the cathode end of the diode with a low-power soldering iron and gradually increase the power to the iron by means of a variable transformer. When a drop of water on the diode first sputters, take your reading.

(c) Repeat part (a), but now cool the diode by embedding it in an ice bath. Wait several minutes so the temperature of the chip stabilizes before taking your reading.

(d) Use the data from parts (a) to (c) to plot V_D (on an expanded scale) versus temperature and determine the average slope of $\Delta V_D/\Delta T$ at the three different currents.

(e) Use the data at 5 mA to determine I_S in Eq. (1.3) at three temperatures. What is your estimate for the temperature coefficient of I_S?

(f) Use a commercial curve tracer (or one made as in Fig. D1.1) to display the forward characteristic to see that the value of the threshold is a value also set by circuit consideration.

(g) Use the same curve tracer as in part (f) to determine the incremental slope $(\Delta V_D/\Delta I_D)$ of the diode VI characteristic around 50 mA. This is the diode *body* resistance.

(h) Interchange diode leads going to the curve tracer and adjust the voltage so that the reverse characteristic going into breakdown is seen. If a model for the

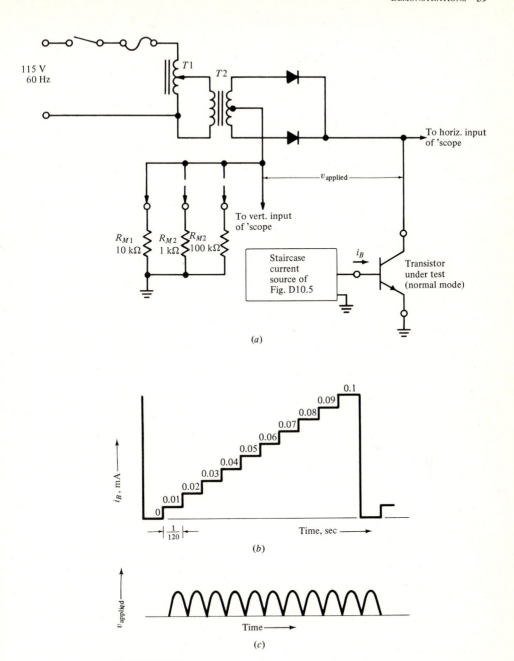

FIGURE D1.2
Schematic for a transistor curve tracer. (*a*) Circuit diagram; (*b*) base-current wave-form for a transistor curve tracer; (*c*) collector-applied voltage for a transistor curve tracer.

zener region is to be made using a voltage source and a resistor, what values would you use for your diode?

D1.2 Bipolar-transistor VI characteristics With the use of a commercial curve tracer (or the circuit of Fig. D1.2, which is Fig. D1.1 with the staircase current generator of Fig. D10.5 [Chap. 10] into the base of the transistor under test), display the VI characteristic of a low-power n-p-n switching transistor:†

(a) Determine and then plot the following parameters as a function of collector current. Use I_C values of 0.5, 1, 2, 5, 10, 20, and 50 mA.

    (1) β_F at $V_{CE} = 1$ V
    (2) $V_{CE(\text{sat})}$, $I_B = 0.1I_C$
    (3) $V_{BE(\text{sat})}$ at $I_B = 0.1I_C$
    (4) $V_{BE(\text{on})}$ at $V_{CE} = 1$ V

(b) Reverse the emitter and collector leads of the transistor to determine and plot for $I_E = 0.1$, 1, and 10 mA and β_R at $V_{EC} = 1$ V.

(c) In the saturation region, the transistor can be modeled by a voltage source and a resistor. Estimate a value of this voltage source which is the intercept on the voltage axis and the "saturation" resistance r_{sat}, where r_{sat}, the approximate "body" resistance, is $\Delta V_{CE}/\Delta I_C$ in saturation.

D1.3 Bipolar inverter With an XY oscilloscope and a sine-wave signal generator set at a 5-V peak-to-peak output:

(a) Show the transfer function ($v_{\text{out 1}}$ versus v_{in}) of the inverter of Fig. D1.3a. Connect the X input to V_{in} and the Y input to $V_{\text{out 1}}$. The deflection sensitivity of both X and Y channels should be set a 5 V full scale.

(b) With the aid of the low-power soldering iron and ice, as in Demonstration D1.1, determine the transfer characteristic at $\approx 100°$C and $\approx 0°$C. Note the shift in the transition region. What is the major cause of this?

(c) Connect a similar inverter as a load as shown in Fig. D1.3b and determine the modified transfer characteristic, $V_{\text{out 1}}$ versus V_{in}.

D1.4 Diode-connected transistor as a thermometer A transistor‡ connected as a diode, as in Fig. D1.4, can be shown to have a linear $V_{BE(\text{on})}$ versus temperature variation. Using a DVM that resolves 1-mV changes, measure V_{BE} at room temperature (determined with a thermometer), ice water (0°C), and boiling water (100°C). If a transistor§ in a metal can is used, one can easily determine another point (181°C) which occurs when normal electronic solder applied to the can starts to melt under the heat of a soldering iron. Determine the slope of the V_{BE} versus T curve and compare to the result given in Prob. P1.8.

       † 2N4275 or 2N5134.
       ‡ 2N4275 or 2N2369A.
       § 2N2369A.

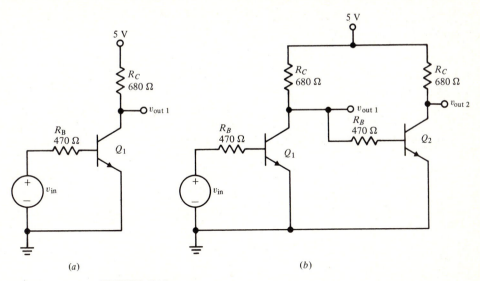

(a)   (b)

FIGURE D1.3
RTL inverter circuits. (a) A circuit for determining the transfer function of an
unloaded inverter; (b) a circuit for investigating a transistor inverter that is loaded
with an identical inverter.

D1.5 MOSFET characteristics Use an n-channel enhancement-mode
MOSFET transistor† in the circuit of Fig. D1.5, which is the same as the circuit of
Fig. D1.2 except that the gate is driven from a voltage staircase (see Fig. D10.5).

To obtain this with a circuit like that in Fig. D10.5, place a 2-kΩ resistor to
ground at the base-step output so that the 0.5 mA per step-current source is
converted to a 1 V per step-voltage staircase.

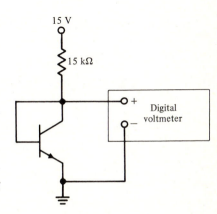

FIGURE D1.4
Determining the temperature variation of
$V_{BE(on)}$.

† Use a gate-protected M116 (manufactured by Siliconix) or nonprotected 2N4351.

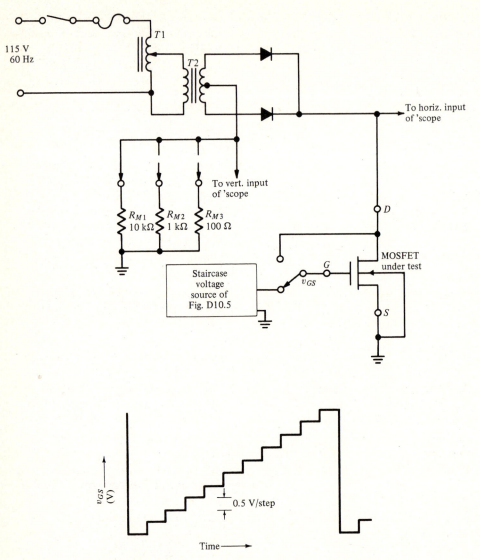

FIGURE D1.5
MOSFET connections to a transistor curve tracer.

Determine for the unit under test:
(a) V_{th}, the threshold voltage
(b) $r_{DS(on)}$ for $V_{GS} = 10$ V at the operating point $V_{DS} = 0$ V; $I_D = 0$; where
$r_{DS(on)} = \Delta V_{DS}/\Delta I_D$.

Apply a 4-V dc source to the gate and determine the drain-to-source VI characteristic. Trace it with a grease pencil on the scope. Now apply an 11-V dc source on the gate and then remove it so the gate is floating. Determine the time required for the charge on the gate electrode to leak off and the trace to pass through the 4-V curve. For a gate capacitance of approximately 4 pF, this measurement determines a gate leakage resistance r_{leakage}, in teraohms.†

$$r_{\text{leakage}} = \frac{\text{time in seconds to go to 4-V characteristic}}{4 \text{ pF}}$$

In MOSFET ICs less leakage occurs than in discrete MOSFETs since the effect of leakage at the terminals is eliminated and the areas of the devices are smaller. Hence this mechanism makes *dynamic* MOSFET circuits possible, where information is stored on capacitors having fractional picofarad capacitances.

Connect the gate to the drain, and from the $V_{DS} I_D$ characteristic determine the MOSFET resistor characteristic. What values of resistance and offset value would be reasonable to use for your device?

REFERENCES

A general introductory text covering a large number of topics in electrical and electronic engineering is that of SMITH, R. J.: "Circuits, Devices and Systems," 2d ed., chaps. 1, 2, 9, 10, and 12, Wiley, New York, 1971.

Another introductory text amplifying on the material in this chapter is OLDHAM, W. G., and S. E. SCHWARTZ: "An Introduction to Electronics," chaps. 1–3, 5–8, and 13, Holt, New York, 1972.

For an introductory text that provides a foundation of semiconductor physics as well as circuit models and analysis, see GIBBONS, J. F.: "Semiconductor Electronics," chaps. 1–7, 9, and 11, McGraw-Hill, New York, 1966.

PROBLEMS

P1.1 Manufacturers' data sheets for *p-n* diodes do not give a value of I_S but rather the diode forward voltage for a given current. Assuming a diode obeys the relationship of Eq. (1.3), find:

(a) The value of I_S if you are given that a diode requires 720 mV when conducting 1 mA in the forward direction of 27°C.

(b) What forward voltage will be across the same diode if it's conducting 0.1 μA at 27°C.

*P1.2 Given a diode that obeys Eq. (1.3) with I_S at 27°C (300 K) of 10^{-14} A and where I_S increases 17 percent for a 1°C increase in temperature, find:

(a) V_D at 1 mA at 27°C.

(b) I_S at 28°C.

(c) V_D at 1 mA at 28°C. (*Hint:* Express in terms of V_D at 27°C plus a small correction.)

(d) The temperature coefficient of V_D (that is, $\Delta V_D/\Delta T$ at 1 mA and 27°C).

† 1 teraohm = 10^{12} Ω.

P1.3 For the circuit in Fig. P1.3 find the diode current with V_{in} of 4 V, using as a diode model:

(a) An ideal diode.

(b) An ideal diode in series with a $V_{D(on)}$ of 0.7 V.

(c) Do part (b) with V_{in} varying from 0 to 4 V. Plot I versus V_{in} showing break-points and slopes.

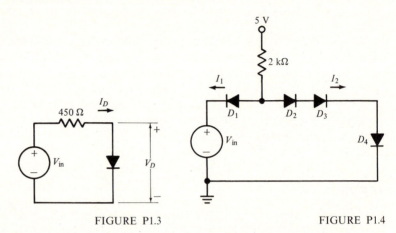

FIGURE P1.3   FIGURE P1.4

P1.4 (a) For the diode gate circuit of Fig. P1.4, determine and then plot values of I_1 and I_2 versus V_{in} as V_{in} varies from 0 to 5 V. For all diodes use the model with $V_{D(on)}$ of 0.7 V. Show all breakpoints and slopes.

(b) When V_{in} exactly equals 1.4 V, determine values for I_1 and I_2.

P1.5 A microammeter that is 100 μA full scale has a dc resistance of 600 Ω. In the circuit of Fig. P1.5 it is used with a resistor R_1 of 100 kΩ to make a voltmeter of 10 V full-scale reading.

If one accidentally applied a 1000-V potential across terminals A and A', what must be the specifications (that is, I_D at V_D) on diodes D_1 and D_2 so that the current through the microammeter is equal or less than 10 times its rated full-scale value?

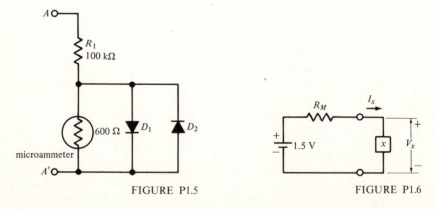

FIGURE P1.5   FIGURE P1.6

*P1.6 The ohmmeter portion of a vacuum-tube voltmeter (VTVM) consists of the circuit shown in Fig. P1.6.

(a) Show that the expression for the ohms reading Ω_X, in terms of the voltage across the unknown and the resistance R_M, is

$$\Omega_X = \frac{V_X}{I_X} = \frac{R_M}{(1.5/V_X) - 1}$$

(b) Use the result of part (a) with $R_M = 10\ k\Omega$ to find the ohms of a forward-biased diode as the unknown "resistor." Assume $V_{D(on)} = 0.7$ V.

(c) Repeat part (b) but with $R_M = 1\ k\Omega$.

P1.7   This problem should be solved using Norton and Thévenin equivalent circuits.

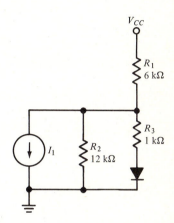

FIGURE P1.7

Given the circuit of Fig. P1.7, find the diode current I_D, with $V_{D(on)} = 0.7$ V at:

(a) $V_{CC} = 12$ V, $I_1 = 1.25$ mA.

(b) $V_{CC} = -6$ V, $I_1 = -1.25$ mA.

*P1.8   Given the following relationship for the saturation current of a junction diode,

$$I_S = ce^{-(E_{go}q/kT)}$$

where c = constant depending on dimensions and fabrication
            procedures used in making diode

$E_{go} = 1.2$ V for silicon

k, T, q = defined in Eq. (1.2)

(a) Prove that for a diode biased with a constant dc current,

$$\frac{dV}{dT} = \frac{V - E_{go}}{T}$$

As shown in Demonstration D1.4 this can be used as a basis for an electronic thermometer.

(b) Use this result to recalculate the answer to Prob. P1.2d and compare results obtained there.

P1.9   This problem is to determine the transistor operating point by making use of either the analytical or the load-line technique. The transistor in Fig. P1.9a can be in one of three states: active, saturated, or cutoff.

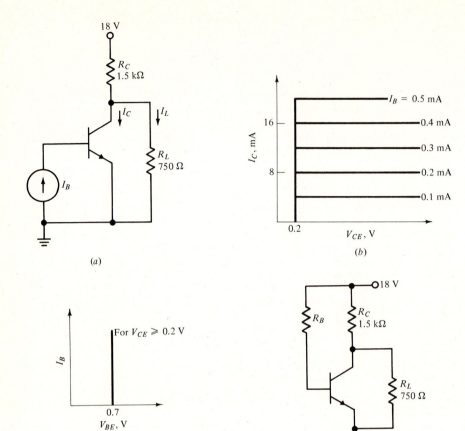

FIGURE P1.9

Use the transistor characteristics given in Fig. P1.9b and c to find:

(a) β_F, $V_{CE(\text{sat})}$, $V_{BE(\text{on})}$, and $V_{CE(\text{sat})}$.

Next find I_C, V_{RL}, and I_L in Fig. P1.9a (*Hint:* Make use of the Thévenin theorem.):

(b) For $I_B = 0.1$ mA.

(c) For $I_B = 0.5$ mA.

(d) For $I_B = 0$ mA.

(e) Find the three different values of R_B so that the transistor in Fig. P1.9d operates at the same currents corresponding to parts (b) to (d).

(f) Find the value of R_B that puts the transistor just at the edge of saturation (EOS).

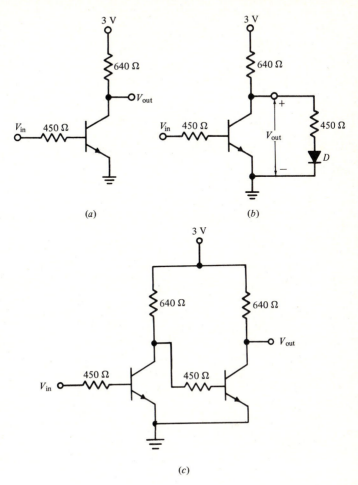

FIGURE P1.10

P1.10   This problem is useful for resistor-transistor-logic (RTL) inverters and gate
analysis. (*Hint:* Make use of Prob. P1.3.)

$$V_{BE(\text{sat})} = V_{BE(\text{on})} = 0.7 \text{ V}$$

$$V_{CE(\text{sat})} = 0.2 \text{ V}$$

$$\beta_F = 20$$

Find for Fig. P1.10a:

(a) $V_{\text{in(EOC)}}$ and $V_{\text{out(EOC)}}$.
(b) $V_{\text{in(EOS)}}$ and $V_{\text{out(EOS)}}$.
(c)   Plot V_{out} versus V_{in} (with V_{in} from 0 to 3 V).
(d)   Repeat parts (a) to (c) by using Fig. P1.10b where now $V_{D(\text{on})} = 0.7$ V.
(e)   Make use of part (d) to plot V_{out} versus V_{in} for the circuit in Fig. P1.10c for V_{in}
from 0 to 3 V.

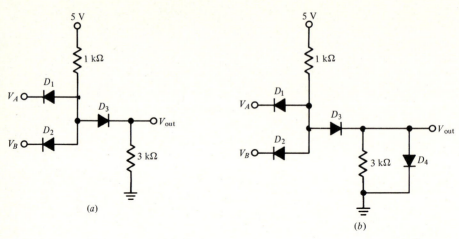

FIGURE P1.11

P1.11   This problem is related to diode gates and related circuits. Use $V_{D(on)} = 0.7$ V. For
the circuit in Fig. P1.11a, plot V_{out} versus V_B.
(a) $V_A = 5$ V, with V_B varying from 0 to 5 V.
(b) $V_A = 0$ V, with V_B varying from 0 to 5 V.
(c)   Repeat parts (a) and (b) by using Fig. P1.11b.

P1.12   For the diode gate and transistor-inverter circuit in Fig. P1.12, plot I_{in}, I_B, I_C, and
V_{out} versus V_{in} for V_{in} varying from 0 to 5 V. Make use of the results of Prob. P1.4
to determine I_B. Use $V_{D(on)} = 0.7$ V, $V_{CE(sat)} = 0.2$ V, $V_{BE(sat)} = V_{BE(on)} = 0.7$ V, and
$\beta_F = 50$.

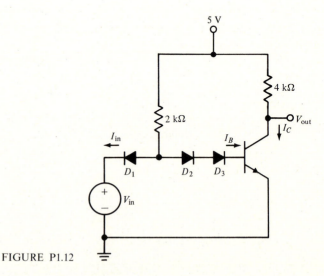

FIGURE P1.12

P1.13   The emitter-follower circuit given in Fig. 1.35 is modified so that $R_1 = 20$ kΩ,
$R_2 = 1$ kΩ, and $V_{CC} = 5$ V. Determine the transfer characteristic for $V_{in} = 0$ V to

$V_{in} = V_{in(EOS)}$. Use $\beta_F = 50$, $V_{BE(on)} = 0.7$ V, and $V_{CE(sat)} = 0.2$ V. Plot I_B and V_{out} versus V_{in} and clearly label the transition points with their values.

P1.14   For the inverter circuit in Fig. P1.14:

(a) Use the transistor characteristics given in Fig. P1.9b and c to determine I_C and I_{RL}.

(b) To what value should R_B be changed to get Q_1 at the EOS.

(c) Place a diode pointing down in series with R_L. The diode $V_{D(on)}$ is 0.7 V. Repeat part (a).

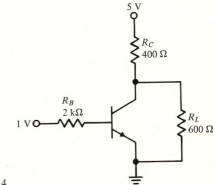

FIGURE P1.14

P1.15   Given the circuit in Fig. P1.15, where all transistors (Q_1, Q_2, and Q_3) have the following parameters: $\beta_F = 20$, $V_{BE(on)} = V_{BE(sat)} = 0.7$ V, and $V_{CE(sat)} = 0.2$ V.

(a) With $R_1 = 500$ kΩ and switch S open, determine the base and collector currents in Q_1, Q_2, and Q_3 and the output voltage V_{out}. Show the steps in your work, then summarize your results in tabular form along with the results in parts (b) and (c).

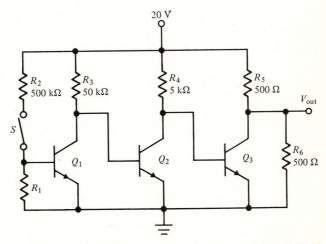

FIGURE P1.15

(b) Repeat part (a) but with switch S closed.

(c) Repeat part (a) but with switch S closed and $R_1 = 25$ kΩ.

P1.16   Given the three-transistor circuit of Fig. P1.16 with $\beta_F = 20$, $V_{BE(on)} = V_{BE(sat)} = 0.7$ V, and $V_{CE(sat)} = 0.2$ V:

(a) If $R_E = 0$ and Q_1 is cut off, in what region will Q_3 be?

(b) If $R_E = 0$ and Q_1 is saturated, what will the emitter current in Q_3 be?

(c) Assume Q_1 is in the correct state so that when $R_E = 0$, Q_3 is saturated. We then increase R_E until Q_3 is just at the EOS. What value of R_E will we need? (*Hint:* Treat the voltage at the emitter of Q_3 as a variable.)

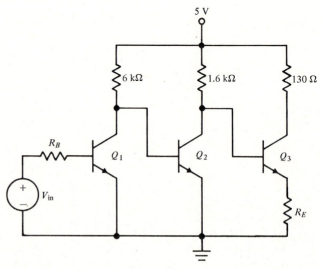

FIGURE P1.16

P1.17   Given a light-emitting-diode (LED) circuit of Fig. P1.17 where $I_D = 20$ mA $\pm$ 20 percent. Using a 5 percent tolerance resistor with a power supply that is regulated ± 10 percent, find the nominal (i.e., design center) values of R and V_{CC} that do this with minimum power drain. You are given that $V_{D(on)}$ of the LED is (for this problem) greater than 2.8 V and less than 3.2 V.

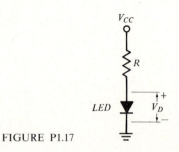

FIGURE P1.17

P1.18   The particular interconnection of two transistors Q_1 and Q_2 in Fig. P1.18 is called the *Darlington* connection.

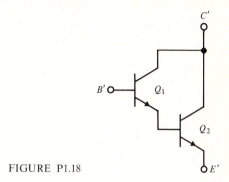

FIGURE P1.18

(a) Prove that when both transistors are biased in the active region they effectively can be modeled as a single transistor whose current gain is

$$\beta_F = \beta_{F2}(\beta_{F1} + 1) + \beta_{F1}$$

where β_{F1} and β_{F2} are the current gains of Q_1 and Q_2, respectively.

(b) For β_{F1} and β_{F2} very much larger than 1, what is a good approximation for β_F?

(c) What is the value of $V_{BE(on)}$ of the composite transistor under the conditions of part (a)?

P1.19 This problem relates to the use of resistors for the *biasing* of discrete transistors in the normal active region. Given the following transistor parameters and desired circuit conditions: $\beta_F = 50$, $V_{BE(on)} = 0.7$ V, $V_{RE} = 0.93$ V, $I_C = 100$ mA, $V_{CE} = 7.27$ V. It is desired that $R_{B1} \parallel R_{B2} = 140\ \Omega$. Find R_C, R_E, R_{B1}, and R_{B2}.

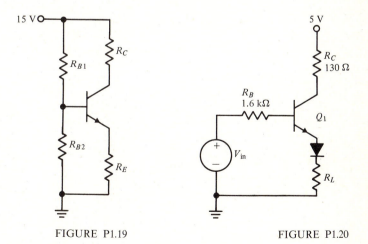

FIGURE P1.19   FIGURE P1.20

P1.20 In the emitter-follower circuit of Fig. P1.20; $\beta_F = 40$, $V_{BE(on)} = V_{D(on)} = 0.7$ V, $V_{CE(sat)} = 0.2$ V:

(a) For $V_{in} = 4.6$ V find I_C so that Q_1 is at EOS. [*Hint*: At EOS, $V_{CB} = -0.5$ V (that is, $V_{BC} = 0.5$ V).]

(b) Find the value of R_L that goes with the conditions of part (a).

*_P1.21_ Solve for I in Eq. (1.8a) by the following _iterative_ method: Assume that the diode current I is 2 mA initially, which corresponds to assuming that V_D is 0 V. Find the value of V_D that corresponds to 2 mA by means of Eq. (1.5b), using $I_S = 10^{-14}$ A. Now solve for a new value of I in Eq. (1.8a) by using this value of V_D. Repeat this process until the amount that I changes between steps is equal to or less than 1 percent of your last computed value.

*_P1.22_ (a) The transistor connection of Fig. P1.22a is commonly used in ICs as the equivalent to a diode, as shown in Fig. P1.22b. For a transistor with $\beta_F = 50$ and $V_{BE(\text{on})} = 0.7$ V, what are the values of I_C, I_B, I_E, V_{BE}, V_{BC}, V_{CE}, and V_D when I_D is 5.1 mA?

(b) If a circuit is wanted that approximates the behavior of several diodes in series (including a "fractional" diode), the circuit in Fig. P1.22c is used. Show that this circuit can be modeled by the circuit in Fig. P1.22d where the "diodes'" "on voltage" is

$$V_{D(\text{on})} = V_{BE(\text{on})}\left(1 + \frac{R_2}{R_1}\frac{\beta_F + 2}{\beta_F + 1}\right)$$

and the diode series resistance is

$$R_D = \frac{R_2}{\beta_F + 1}$$

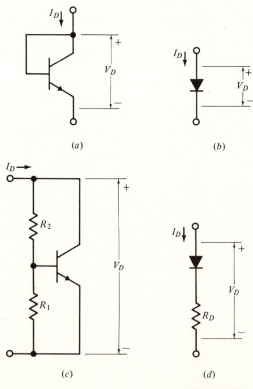

(a)   (b)

(c)   (d)

FIGURE P1.22

(*Hint:* First solve for I_B as a function of V_D by making use of Thévenin's theorem.)

(c) Use the results of part (*b*) to solve for the values in the circuit model in Fig. P1.22*d* that are obtained in the circuit of Fig. P1.22*a*.

P1.23 (a) Determine graphically and then plot the voltage transfer function for the MOSFET inverter circuit described in Exercise E1.11.

(b) Repeat part (*a*) but now use a similar transistor as the load to replace the 2-kΩ load resistor.

2

BASIC LOGIC DESIGN

When digital ICs were introduced in the early 1960s, people began to talk about a "computer on a chip." This is now no exaggeration. Circuits now exist in a single chip which are able to add, subtract, multiply, and divide as well as perform other mathematical and logic functions. Therefore, any investigation of digital ICs would not be complete without some study of logic design as well as circuit design.

We noted in the Introduction that digital systems are based on circuits which process signals that can have one of two levels or states. We have seen in Chap. 1 that the transistor inverter to be used in a digital circuit has two states. It is either OFF (in the cutoff region) or ON (in the saturated region). We can associate these states or regions with the binary states of a 1 (logic one) or a 0 (logic zero). For those unfamiliar with the binary number system, we have included in Appendix C a description of how to relate the more familiar decimal system to the binary system and vice versa.

Logic design with binary quantities has some peculiarities, but it also presents some useful opportunities. For instance, the answer to a question can only be YES or NO, it can never be MAYBE! This has led to the development of a particular form of algebra, where the value of a variable can only have one of two

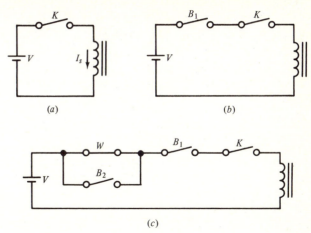

FIGURE 2.1
The car-starter problem. (*a*) The simple case with just an ignition key; (*b*) a simple modification including the driver's seat belt; (*c*) the more complicated case of including the passenger and his seat belt.

values (either 1 or 0). As with normal algebra there are basic rules for the reduction of complex logic expressions to simplified or minimized forms.

These minimization aids are also useful when considering a circuit design which will equate with the logic expression. For example, consider the simple car starter diagramed in Fig. 2.1*a*. When the ignition key (K) is turned to the start position, a current I_s flows through a solenoid which activates the starter motor. Now, in order to reduce traffic fatalities, seat belts are installed in a fleet of cars. Moreover it is required that the driver has his seat belt on and tight before the car can start. A simple modification to the car starter is shown in Fig. 2.1*b*, where B_1 is not closed until a minimum tension is applied to the seat belt. In words we say that both B_1 *and* K have to be closed for the starter circuit to operate. Here we use the word *and* in the *logic* sense, wherein the starter operating is conditional upon this statement.

Now, consider the further complication that all passengers in the car must have their seat belts on and tightened before the car can start. However, it is also required that without a passenger in a seat the state of that seat belt is to have no effect on the starting of the car. For simplicity we only consider the case of a two-seat car, with provision for a driver and one passenger. A solution to the problem is shown in Fig. 2.1*c*, where B_2 is the passenger seat belt and W is a tension switch that is normally closed but which opens when the passenger seat is occupied. Now the condition is that either W *or* B_2, *and* both B_1 *and* K, have to be closed to activate the starter motor. This modification to our original circuit is not as obvious as that in Fig. 2.1*b*. The techniques for obtaining this solution will be described in this chapter.

Logic expressions, like circuit equations, are best understood when they are arranged on paper as a logic diagram. Also, like circuit components, there are definite symbols for the basic logic expressions. Following the section on Boolean algebra and minimization techniques, the symbols for the basic logic operations are then described. Finally, it is necessary to implement the logic design as a circuit design. In this chapter, this is simply done with diodes and resistors.

2.1 BOOLEAN ALGEBRA

A special algebra applicable to the binary system was invented by George Boole (1815–1864). It is not intended in this text to give a full description of Boolean algebra. However, this form of algebra can be very useful to a logic and/or circuit designer, and some basic familiarity with it is essential for using digital ICs. We also find many simple applications of Boolean algebra around the home. In this section, we examine the rules and relationships of this algebra.

In Boolean algebra only two states of the variable are permitted. The two states are represented by a 1 (logic one) and a 0 (logic zero), although TRUE and FALSE, ON and OFF, or HIGH and LOW, are also names given to the two states. By definition those values are exclusive, that is, using the 0 and 1 representations,

$$\text{If } A \neq 0 \qquad \text{then } A = 1$$

and

$$\text{If } A \neq 1 \qquad \text{then } A = 0$$

There are three basic operations to be performed with this algebra.†

1 The OR, represented by a plus sign between variables, such as $A + B$, which we read as *A or B*.

2 The AND, represented by a dot sign between variables, such as $A \cdot B$, which we read as *A and B*.

3 The NOT, represented by a bar over the variable, such as $\bar{A}$, which we read as *not A*.

The relationship of Boolean algebra to digital circuits can readily be seen by making use of switches. A closed switch will be used to represent a 1 and an open switch a 0. This is shown in Fig. 2.2.

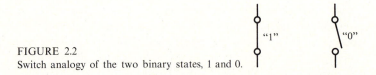

FIGURE 2.2
Switch analogy of the two binary states, 1 and 0.

† These three operations are sometimes referred to as addition, multiplication, and negation.

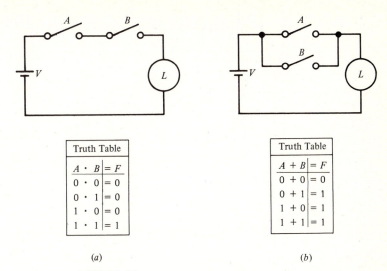

FIGURE 2.3
Basic logic functions. (*a*) The AND function; (*b*) the OR function.

In Fig. 2.3, the switches A and B are used to control the current to the lamp L. In the following development we will agree that a statement is TRUE (or represented by 1) if the lamp is lit. With the lamp not lit, the statement is FALSE (or represented by 0).

A common household switch arrangement with a lamp is shown in Fig. 2.3*a*. It can be seen that it is necessary that both switch A *and* switch B be closed for the lamp to be lit. This is the AND statement and is written† as

$$A \cdot B = F \qquad (2.1)$$

Equation (2.1) is read

F is equivalent to A and B

The AND function is similar to a series connection of the switches.

Another switch circuit is shown in Fig. 2.3*b*. Here the lamp is lit by *either* switch A *or* switch B being closed. That is, the switching function can be performed by either A *or* B. However, the lamp is also lit by *both* switches being closed, so in addition to the either-or condition, we also have a *both* condition. This is called the INCLUSIVE-OR statement and is described by the equation

$$A + B = F \qquad (2.2)$$

Equation (2.2) is read

F is equivalent to A or B

We see that the OR function is equivalent to a parallel connection of the switches.

† As in regular algebra the dot is often dropped, giving $F = AB$.

The third basic operation is the NOT statement. This is also known as *inverting* or *complementing*. It is represented by the equation

$$\bar{A} = F \qquad (2.3)$$

and is read

F is equivalent to not A

(the terms *A not* or *A bar* are also used). Using the fact that only two states of the variable are allowed, we can see that

$$\text{If } A = 1 \qquad \text{then } \bar{A} = 0$$

or

$$\text{If } A = 0 \qquad \text{then } \bar{A} = 1$$

Further common examples of the use of switches to illustrate the basic properties of Boolean algebra are given in Demonstration D2.1.

2.1.1 Truth Tables

Logic statements can be displayed graphically with a *truth table*, which is simply a systematic listing of the values for the dependent variable in terms of all the possible values of the independent variable.

Truth tables are shown in Fig. 2.3 for the simple cases illustrated there. It should be noted that since we are working with a binary system there are 2^N number of combinations, where N is the number of independent variables being considered.

In the cases shown in Fig. 2.3, where there are two independent variables (A and B), the truth table lists the four possible combinations. The truth of each combination can be checked with the aid of the switch-and-lamp diagram. In the truth table of Fig. 2.3b, the first statement shows that with A and B represented by a 0 (i.e., an open switch) the function F is FALSE (or a 0). In the second statement, switch B has been closed (a 1), with the result that F is now TRUE (represented by a 1). The state of the two variables is interchanged in the third statement, with the similar result that F is a 1. For the final statement, both A and B are TRUE as is the dependent variable F. The conclusion is that with *either A or B* or *both* TRUE (or a 1), the statement is TRUE.

A similar order of listing for the independent variables appears in Fig. 2.3a. However, this is an AND function, and the first three statements are shown to be FALSE (or a 0). Only for the fourth statement, a 1 at both A *and B*, is the statement TRUE.

2.1.2 Basic Theorems

With the three basic operations (AND, OR, and NOT) it is possible to deduce a set of basic theorems. These are shown in Fig. 2.4, along with the switch analogy. In each case, the OR function is represented by a parallel connection of the switches, with the series connection representing the AND function.

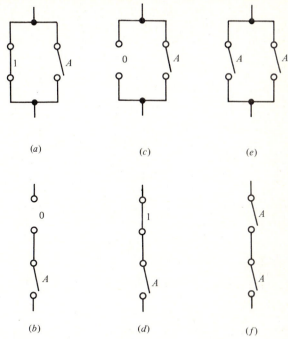

FIGURE 2.4
Switch analogy of some basic Boolean theorems. (*a*) Theorem 1A: $1 + A = 1$. (*b*) Theorem 1B: $0 \cdot A = 0$. (*c*) Theorem 2A: $0 + A = A$. (*d*) Theorem 2B: $1 \cdot A = A$. (*e*) Theorem 3A: $A + A = A$. (*f*) Theorem 3B: $A \cdot A = A$.

We will later find these theorems useful in simplifying a complex logic expression for a clearer understanding of the logic to be performed. There is also an economic benefit in the saving in components when implementing logic with ICs.

Theorem 1A $1 + A = 1$ With TRUE representing the 1 state and FALSE the 0 state, we see from Fig. 2.4*a* that the function is TRUE irrespective of the setting of the switch *A*.

Theorem 1B $0 \cdot A = 0$ The function shown in Fig. 2.4*b* is always FALSE notwithstanding the position of switch *A*.

Notice in the application of both these theorems that the switch *A* is not needed. In Theorem 1A the statement is always TRUE, and in Theorem 1B the statement is always FALSE. Both theorems show a minimized expression and a redundant part.

Theorem 2A $0 + A = A$ As shown in Fig. 2.4c, the function is only dependent upon the setting of the switch A.

Theorem 2B $1 \cdot A = A$

This is a companion theorem of 2A, with the same conclusion that the function is only dependent upon the setting of the switch A. This is illustrated in Fig. 2.4d.

Theorem 3A $A + A = A$

Theorem 3B $A \cdot A = A$

Simplification of a Boolean algebraic term can be performed as shown by the theorems demonstrated in Fig. 2.4e and f. Both functions are dependent only on A, and each of the expressions may be reduced to a simple A. Again, these minimized expressions indicate an economy in components.

Theorem 4A $A + \bar{A} = 1$

Theorem 4B $A \cdot \bar{A} = 0$

These two theorems involve the inversion of the variable. By visualizing the switch analogy, we can see that the OR function is TRUE irrespective of the setting of switch A. Either one of the two switches must be closed at all times. The AND function is FALSE, as one of the two switches is open at all times.

Theorem 5 $\bar{\bar{A}} = A$ Double negation or inversion on a variable produces the original variable.

The commutative, associative, and distributive laws of ordinary algebra also hold with Boolean algebra and are included in the following theorems:

Theorem 6A $A + B = B + A$ (commutative)

Theorem 6B $AB = BA$ (commutative)

Theorem 7A $A + (B + C) = (A + B) + C$ (associative)

Theorem 7B $A(BC) = (AB)C$ (associative)

Theorem 8A $A(B + C) = AB + AC$ (distributive)

Theorem 8B $(A + B)(A + C) = A + BC$ (distributive)

The commutative and associative theorems are straightforward and follow the rules of ordinary algebra. Normal algebra can also be used to confirm the

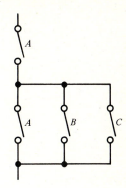

FIGURE 2.5
Switch analogy showing $A(A + B + C) = A$.

relationship shown in Theorem 8A. However, notice the laws of ordinary algebra are not maintained in Theorem 8B:

$$(A + B)(A + C) = AA + AC + BA + BC$$
$$= A(A + C + B) + BC$$
$$= A + BC$$

The last step can readily be seen with the aid of switches, as in Fig. 2.5. Here we see that where a common term, such as A, is included in a complex AND function, as $A(A + B + C)$, the whole function may be reduced to the common term A. Also notice the single switch A can do the same job that four switches were doing. This is another example of why we look for ways to minimize the number of terms in our expression and also to minimize the number of variables in a term. In Chap. 4 we will discuss such methods in detail.

Proof of theorems by use of a truth table   The distributive theorems can also be demonstrated with the use of a truth table. With three independent variables there must be 2^3 or eight combinations, and these are conveniently listed in the systematic manner shown in Fig. 2.6. That is, they are listed as a binary number in

C	B	A	$A + B$	$A + C$	$(A + B)(A + C)$	BC	$A + BC$
0	0	0	0	0	0	0	0
0	0	1	1	1	1	0	1
0	1	0	1	0	0	0	0
0	1	1	1	1	1	0	1
1	0	0	0	1	0	0	0
1	0	1	1	1	1	0	1
1	1	0	1	1	1	1	1
1	1	1	1	1	1	1	1

FIGURE 2.6
Proof of Theorem 8B: $(A + B)(A + C) = A + BC$.

increasing order, with A being the least significant bit. With the value of both terms of Theorem 8B in agreement for every possible combination of values which the variables may have, the theorem is confirmed. In a similar manner to that of Fig. 2.6, the truth table may also be used to demonstrate the proof of Theorem 8A.

DeMorgan's theorems Two final theorems which are extremely useful in the reduction of complex Boolean expressions to simple and workable proportions are the DeMorgan theorems.

Theorem 9A $(\overline{A + B}) = \bar{A}\bar{B}$

Theorem 9B $(\overline{AB}) = \bar{A} + \bar{B}$

Here the expression is complemented by changing any OR function to an AND ($+$ to $\cdot$) and any AND function to an OR ($\cdot$ to $+$); then each variable in the function is inverted.

The proof of these two theorems is shown, through the use of the truth table, in Fig. 2.7a and b.

Summary We conclude this section with a table of frequently used theorems:

Theorem 1A	$1 + A = 1$	Theorem 1B	$0A = 0$
Theorem 2A	$0 + A = A$	Theorem 2B	$1A = A$
Theorem 3A	$A + A = A$	Theorem 3B	$AA = A$
Theorem 4A	$A + \bar{A} = 1$	Theorem 4B	$A\bar{A} = 0$
Theorem 5	$\bar{\bar{A}} = A$		
Theorem 6A	$A + B = B + A$	Theorem 6B	$AB = BA$
Theorem 7A	$A + (B + C) = (A + B) + C$	Theorem 7B	$A(BC) = (AB)C$
Theorem 8A	$A(B + C) = AB + AC$	Theorem 8B	$(A + B)(A + C) = A + BC$
Theorem 9A	$(\overline{A + B}) = \bar{A}\bar{B}$	Theorem 9B	$(\overline{AB}) = \bar{A} + \bar{B}$

EXAMPLES

As mentioned earlier, for a clear understanding of logic expressions and to save on the number of components, we look for ways to minimize the number of terms in an expression and also to minimize the number of variables in a term. The following example will serve to illustrate this point.

EXAMPLE 2.1 Simplify $\bar{A}\bar{B} + A\bar{B} + \bar{A}B$.

SOLUTION 1

$$\bar{A}\bar{B} + A\bar{B} + \bar{A}B = \bar{B}(\bar{A} + A) + \bar{A}B \quad \text{(Theorem 8A)}$$
$$= \bar{B}(1) + \bar{A}B \quad \text{(Theorem 4A)}$$
$$= \bar{B} + \bar{A}B \quad \text{(Theorem 2B)} \quad ////$$

B	A	$A + B$	$\overline{A + B}$	$\bar{A}$	$\bar{B}$	$\bar{A} \cdot \bar{B}$
0	0	0	1	1	1	1
0	1	1	0	0	1	0
1	0	1	0	1	0	0
1	1	1	0	0	0	0

(*a*)

B	A	AB	$\overline{AB}$	$\bar{A}$	$\bar{B}$	$\bar{A} + \bar{B}$
0	0	0	1	1	1	1
0	1	0	1	0	1	1
1	0	0	1	1	0	1
1	1	1	0	0	0	0

(*b*)

FIGURE 2.7
Proof of DeMorgan's theorems. (*a*) Theorem 9A: $(\overline{A + B}) = \bar{A}\bar{B}$;
(*b*) Theorem 9B: $\overline{AB} = \bar{A} + \bar{B}$.

Notice that we may add $\bar{A}\bar{B}$ to the initial expression without changing its value. We then have

SOLUTION 2

$$\begin{aligned} \bar{A}\bar{B} + A\bar{B} + \bar{A}B &= \bar{A}\bar{B} + A\bar{B} + \bar{A}B + \bar{A}\bar{B} && \text{(Theorem 3A)} \\ &= \bar{B}(\bar{A} + A) + \bar{A}(B + \bar{B}) \\ &= \bar{B}(1) + \bar{A}(1) \\ &= \bar{B} + \bar{A} \end{aligned}$$

////

The initial expression contains three terms and four variables. The final result contains just two terms and two variables.

As an example of the use of DeMorgan theorems in simplifying a logic expression, consider the following example:

EXAMPLE 2.2 Simplify $\overline{\overline{(\bar{A} + B)} + \overline{(A + \bar{B})} + (\overline{\bar{A}B})(\overline{A\bar{B}})}$.

SOLUTION

$$\overline{\overline{(\bar{A} + B)} + \overline{(A + \bar{B})} + (\overline{\bar{A}B})(\overline{A\bar{B}})}$$

$$= (\overline{\overline{\bar{A} + B}})(\overline{\overline{A + \bar{B}}}) + (\overline{\overline{\bar{A}B}}) + (\overline{\overline{A\bar{B}}}) \qquad \text{(Theorems 9A \& 9B)}$$

$$= (\bar{A} + B)(A + \bar{B}) + \bar{A}B + A\bar{B} \qquad \text{(Theorem 5)}$$

$$= \bar{A}A + \bar{A}\bar{B} + BA + B\bar{B} + \bar{A}B + A\bar{B} \qquad \text{(Theorem 8A)}$$

$$= \bar{A}\bar{B} + AB + \bar{A}B + A\bar{B} \qquad \text{(Theorem 4B)}$$

$$= \bar{A}(\bar{B} + B) + A(B + \bar{B}) \qquad \text{(Theorem 8A)}$$

$$= \bar{A}(1) + A(1) \qquad \text{(Theorem 4A)}$$

$$= \bar{A} + A \qquad \text{(Theorem 2B)}$$

$$= 1 \qquad\qquad\qquad ////$$

The result shows that an involved logic expression can sometimes be reduced to very simple requirements. Finally, we will make use of a truth table to solve the problem given at the beginning of this chapter involving the automobile seat belts and starting the car.

EXAMPLE 2.3   In a two-seat automobile it is required that both driver and passenger have their seat belts on and tightened before the car can start. However, it is also required that without a passenger in the car the state of that seat belt is to have no effect on the starting of the car. Show a switch diagram suitable for a car starter, including the ignition key.

SOLUTION   Let K represent the ignition switch which when on has the state 1 and when off has the state 0. Let B_1 and B_2 represent, respectively, the driver and passenger seat belt which when on and tightened has the state 1 but otherwise is 0. Let W represent the tension switch which is normally closed (that is, $W = 1$) but which opens (that is, $W = 0$) when the passenger seat is occupied. The car will start only if $F = 1$. We now proceed to complete the truth table as shown below.

W	B_2	B_1	K	F	Comments
0	0	0	0	0	
0	0	0	1	0	
0	0	1	0	0	
0	0	1	1	0	
0	1	0	0	0	
0	1	0	1	0	
0	1	1	0	0	
0	1	1	1	1	Driver and passenger properly buckled
1	0	0	0	0	
1	0	0	1	0	
1	0	1	0	0	
1	0	1	1	1	Driver only, buckled. No passenger
1	1	0	0	0	
1	1	0	1	0	
1	1	1	0	0	
1	1	1	1	1	No passenger, but tension on seat belt

Hence

$$F = WB_2B_1K + W\bar{B}_2B_1K + \bar{W}B_2B_1K$$

When simplifying by algebra,

$$F = W(B_1 K) + B_2(B_1 K) = (W + B_2)(B_1 K)$$

We can now draw the switching diagram, which is shown in Fig. 2.1c.   ////

EXERCISES

E2.1 Simplify the carry output of a binary adder which has the following form:

$$K = CB\bar{A} + C\bar{B}A + \bar{C}BA + CBA$$

E2.2 Simplify the following switching function:

$$W + \overline{W}XYZ + \overline{W}X\bar{Y}\bar{Z} + \overline{W}XY\bar{Z} + \overline{W}X\bar{Y}Z$$

E2.3 An automatically controlled electric subway car can only leave a station if all the doors are closed and the next section of track is clear. However, should there be trouble with closing the doors, the car can be moved under manual control with open doors; but a clear track is required. Use a truth table to determine all the combinations for moving the car and express this as a simplified Boolean function.

2.2 KARNAUGH MAPS

The algebraic reduction of a switching function is not always easy and generally requires a certain element of intuition or luck. Many techniques have been developed to aid in this reduction. One of the most helpful is the Karnaugh map. This is a matrix array of all the possible combinations of the independent variables. The word *literal* is sometimes used in place of variable.

The coordinates of a two-variable map are shown in Fig. 2.8. This figure has four squares, each of which corresponds to one particular combination of the two variables. In Fig. 2.8 the columns contain the two states of A ($\bar{A}$ and A, or 0 and 1), while the rows contain the two states of B. Two ways of indicating this are shown in Fig. 2.8.

A simple example will help prove the usefulness of the Karnaugh map.

EXAMPLE 2.4   Simplify $F = A + \bar{A}B$.

SOLUTION   For illustrative purposes the expression is rewritten

$$F = A + \bar{A}B$$

$$= f_1 + f_2$$

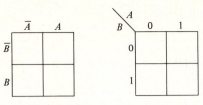

FIGURE 2.8
The coordinates of a two-variable Karnaugh map.

In Fig. 2.9a we show the truth table $f_1 = A$. There are two TRUE statements, and these are located on the map at $\bar{B}A$ and BA. The truth table for $f_2 = \bar{A}B$ is shown in Fig. 2.9b, and the one TRUE statement is located at $B\bar{A}$. In Fig. 2.9c the truth table is shown for the complete statement $F = A + \bar{A}B$, and the TRUE statements are indicated on the map. A grouping of two squares in either the columns or the rows will indicate a redundant variable. That is, in the column $A = 1$, we have

$$\bar{B}A + BA = (\bar{B} + B)A \qquad \text{(Theorem 8A)}$$

$$= A \qquad \text{(Theorem 4A)}$$

Similarly, in the row $B = 1$, we have

$$B\bar{A} + BA = B(\bar{A} + A)$$

$$= B$$

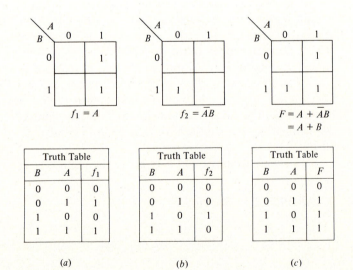

(a)   (b)   (c)

FIGURE 2.9
Solution of Example 2.4. (a) $f_1 = A$; (b) $f_2 = \bar{A}B$; (c) $F = A + \bar{A}B = A + B$.

B \ A	00	01	11	10
0				
1				

FIGURE 2.10
The coordinates of a three-variable Karnaugh map.

We now see that the original statement

$$F = A + \bar{A}B$$

can be reduced or simplified to

$$F = A + B \qquad ////$$

The coordinates of a three-variable Karnaugh map are shown in Fig. 2.10. A map of three variables contains eight squares ($2^3 = 8$). For ease of reduction the listing of the variable is important. There must be only a one-variable change between any two adjacent squares. The listing has a special cyclic symmetry. That is, the columns are listed as 00-01-11-10. The leftmost digit refers to the leftmost variable.

Notice also that the cyclic order of listing is continuous in both the horizontal and vertical directions. The form of the map may be pictured as two cylinders, as shown in Fig. 2.11.

EXAMPLE 2.5   Simplify $F = A\bar{B}C + A\bar{B}\bar{C} + \bar{A}BC + ABC$.

SOLUTION   Again we rewrite the expression as

$$F = A\bar{B}C + A\bar{B}\bar{C} + \bar{A}BC + ABC$$
$$= f_1 + f_2 + f_3 + f_4$$

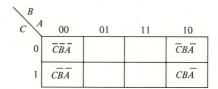

B \ A	00	01	11	10
0	$\bar{C}\bar{B}\bar{A}$			$\bar{C}B\bar{A}$
1	$C\bar{B}\bar{A}$			$CB\bar{A}$

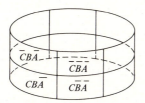
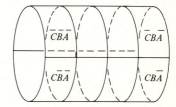

FIGURE 2.11
Cyclic properties of the three-variable Karnaugh map.

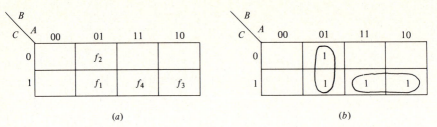

(a)   (b)

FIGURE 2.12
Two representations of $F = A\bar{B}C + A\bar{B}\bar{C} + \bar{A}BC + ABC$ on a Karnaugh map.

We start by constructing a three-variable Karnaugh map and identifying within the map each of the terms of the expression, as shown in Fig. 2.12. A grouping of terms is then made to eliminate the redundant variable. The vertical grouping of two adjacent squares eliminates the variable C, and the horizontal grouping of two adjacent squares eliminates the variable A. We therefore have a simplified form:

$$F = A\bar{B} + BC \qquad ////$$

Generally, the indicated squares of Fig. 2.12a are marked with a 1, indicating them to be true. It is implied that the open squares are marked with a 0, or false. However, the marking of the 0s is generally not necessary. This is shown in Fig. 2.12b.

Of course, for this problem an algebraic reduction could also have been possible.

$$F = A\bar{B}C + A\bar{B}\bar{C} + \bar{A}BC + ABC$$
$$= A\bar{B}(C + \bar{C}) + BC(\bar{A} + A)$$
$$= A\bar{B} + BC$$

For completeness we also use a truth table to show the equivalence of the original expression (F_1) and its simplified form (F_2). Both F_1 and F_2 are in agreement for every possible combination of values which the variable may have. Therefore, the expressions are equivalent.

C	B	A	F_1	$A\bar{B}$	BC	F_2
0	0	0	0	0	0	0
0	0	1	1	1	0	1
0	1	0	0	0	0	0
0	1	1	0	0	0	0
1	0	0	0	0	0	0
1	0	1	1	1	0	1
1	1	0	1	0	1	1
1	1	1	1	0	1	1

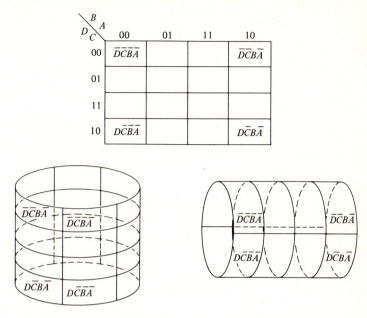

FIGURE 2.13
Cyclic properties of the four-variable Karnaugh map.

As has been shown with the Karnaugh map of Fig. 2.12, the grouping of the 1s leads to the elimination of a redundant variable. The grouping of two adjacent cells eliminates one of the two 3-variable terms and leaves a term with two variables, i.e.,

$$ABC + \bar{A}BC = BC$$

A four-variable map, containing 16 squares ($2^4 = 16$), is shown in Fig. 2.13. Again notice the listing of the variables. This is important to readily identify the redundant variable. Notice also that the cyclic order of listing may be continued by repeating the pattern to the right or left, up or down.

The recognition of basic patterns is readily gained by practice, and some representative patterns are shown in Fig. 2.14. It should be noted that all four variables are required to uniquely define one square in a four-variable map, as illustrated in Fig. 2.14*a* and *b*.

We have seen that the grouping of two adjacent squares eliminates one variable, and this is shown again in Fig. 2.14*c* to *f*. In Fig. 2.14*e*, a study of the map will show that within the grouping $\bar{A}$ and CD are required terms, B and $\bar{B}$ are redundant, or

$$\bar{A}\bar{B}CD + \bar{A}BCD = \bar{A}CD$$

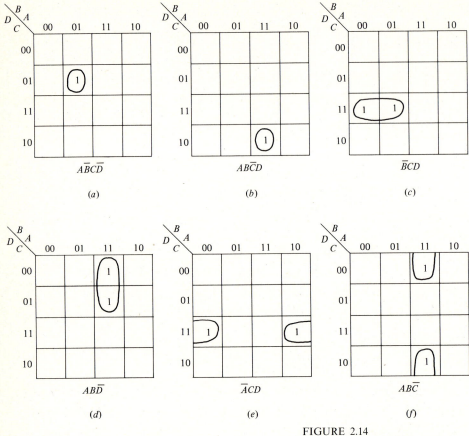

FIGURE 2.14
Some representative Karnaugh maps.

(*Continued on facing page*)

It logically follows that the grouping of four adjacent squares should eliminate two variables, as illustrated in Fig. 2.14g to j. Referring to the map of
Fig. 2.14g, in the grouping $\bar{B}$ and D are the common or required terms, with both
states of A ($\bar{A}$ and A) and C (C and $\bar{C}$) not required, or

$$\bar{A}\bar{B}CD + A\bar{B}CD + \bar{A}\bar{B}\bar{C}D + A\bar{B}\bar{C}D = \bar{B}D$$

A grouping of eight adjacent squares is shown in Fig. 2.14k and l. With a
grouping of eight squares, three of the variables are eliminated. For the map
shown in Fig. 2.14k, only $\bar{B}$ is the required term; the variables A, C, and D, are
redundant.

To illustrate reduction with a four-variable map, consider the following
example:

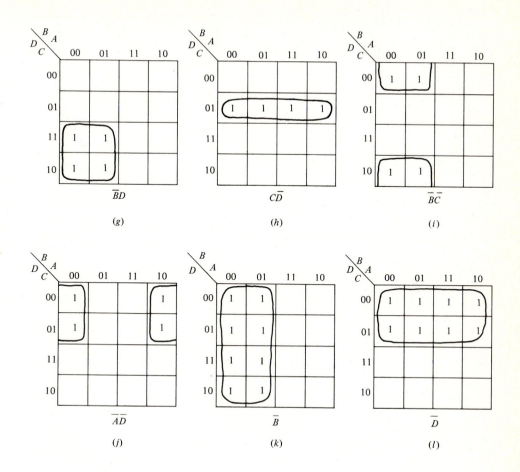

$\overline{B}D$

(g)

$C\overline{D}$

(h)

$\overline{B}\overline{C}$

(i)

$\overline{A}\overline{D}$

(j)

$\overline{B}$

(k)

$\overline{D}$

(l)

EXAMPLE 2.6 Simplify $F = ABD + \bar{A}C\bar{D} + \bar{A}\bar{C}\bar{D} + A\bar{B}D + A\bar{B}C\bar{D}$.

SOLUTION   This is shown in the Karnaugh map of Fig. 2.15a. The grouping of adjacent 1s is shown in Fig. 2.15b and c, as well as the simplified solution, illustrating a point that often there is no unique solution to the problem. The designer must choose the best solution based on some other criteria. Sometimes this is obtained by grouping the 0s in a NOT statement, as shown in Fig. 2.15d. The implication here is that the simplified function is obtained using DeMorgan's theorem from $\text{NOT}(AB\bar{D} + A\bar{C}\bar{D} + \bar{A}D)$:

$$F = \text{NOT}(AB\bar{D} + A\bar{C}\bar{D} + \bar{A}D)$$

$$= \overline{AB\bar{D} + A\bar{C}\bar{D} + \bar{A}D}$$

$$= (\overline{AB\bar{D}})(\overline{A\bar{C}\bar{D}})(\overline{\bar{A}D})$$

$$= (\bar{A} + \bar{B} + D)(\bar{A} + C + D)(A + \bar{D}) \qquad ////$$

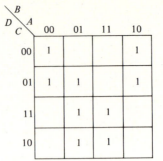

$$F = ABD + \bar{A}C\bar{D} + \bar{A}C\bar{D} + A\bar{B}D + A\bar{B}C\bar{D}$$

(a)

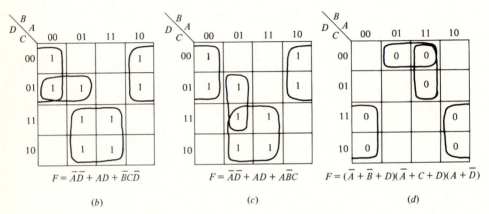

$$F = \bar{A}\bar{D} + AD + \bar{B}C\bar{D}$$

(b)

$$F = \bar{A}\bar{D} + AD + A\bar{B}C$$

(c)

$$F = (\bar{A} + \bar{B} + D)(\bar{A} + C + D)(A + \bar{D})$$

(d)

FIGURE 2.15
Solutions to $F = ABD + \bar{A}C\bar{D} + \bar{A}C\bar{D} + A\bar{B}D + A\bar{B}C\bar{D}$ by use of Karnaugh maps.

The Karnaugh map can be used for the simplification of five- and even six-variable functions. Generally, however, the use of the Karnaugh map is limited to no more than four variables. When the function has more than four variables, there are other minimization techniques that allow for a more straightforward solution. Algorithms for the computer solution of multivariable functions also exist.

Later, in Chap. 4, it will be shown that it is convenient to arrange a logic equation in one of two standard forms. These have been demonstrated in the last example. By grouping the 1s, the equation has been arranged as a sum of products.

$$F = \bar{A}\bar{D} + AD + \bar{B}C\bar{D}$$

or

$$F = \bar{A}\bar{D} + AD + A\bar{B}C$$

In the first equation, F is the sum of the products $\bar{A}\bar{D}$, AD, and $\bar{B}C\bar{D}$; in the second, F is the sum of the products $\bar{A}\bar{D}$, AD, and $A\bar{B}C$. The *sum of products* is called the OR*ing of several* AND *terms.*

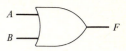

Truth Table		
A	B	F
0	0	0
0	1	1
1	0	1
1	1	1

FIGURE 2.16
Logic symbol of OR gate with truth table.

By grouping the 0s, the resulting equation is in the form of a product of sums.

$$F = (\bar{A} + \bar{B} + D)(\bar{A} + C + D)(A + \bar{D})$$

Here, F is the product of three summations: $(\bar{A} + \bar{B} + D)$, $(\bar{A} + C + D)$ and $(A + \bar{D})$. The *product of sums* is called the AND*ing of several* OR *terms.*

EXERCISES

E2.4   Use the Karnaugh map to simplify the expression of Exercise E2.1.

E2.5   Use the Karnaugh map to simplify the expression of Exercise E2.2.

E2.6   Use the Karnaugh map to express the following function in a minimum sum-of-products form:

$$F = (\bar{A}\bar{B} + B\bar{D})\bar{C} + BD(\overline{\bar{A}\bar{C}}) + \bar{D}(\bar{A} + \bar{B})$$

E2.7   Use the Karnaugh map to express the following function in a minimum product-of-sums form:

$$F = A\bar{B}CD + D(\bar{B}\bar{C}D) + (A + C)B\bar{D} + \bar{A}(\bar{B} + C)$$

2.3   LOGIC SYMBOLS

In this section we shall describe how the three basic logic functions are represented in a logic diagram. We have seen that the INCLUSIVE-OR statement is described by the equation $A + B = F$ and demonstrated in the truth table of Fig. 2.16. As seen from the truth table, the statement is true only if A or B or both are at a 1.

The equation can also be described with the *logic symbol* for an OR gate, also shown in Fig. 2.16. The requirement is that one or both of the inputs (A or B) be at a 1 for the output (F) to be at a 1.

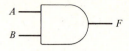

Truth Table		
A	B	F
0	0	0
0	1	0
1	0	0
1	1	1

FIGURE 2.17
Logic symbol of AND gate with truth table.

Another basic logic function is the AND statement $AB = F$, demonstrated in the truth table of Fig. 2.17. The statement is true only if both A and B are at a 1.

The *logic symbol* for an AND gate is also shown in Fig. 2.17. The requirement here is that both the inputs (A and B) be at a 1 for the output (F) to be at a 1.

Finally, the complementing or inverter statement shows that F is equivalent to NOT A.

$$\bar{A} = F$$

The truth table and *logic symbol* are shown in Fig. 2.18, the small circle at the output indicating inversion.

From these three basic logic symbols, two other commonly used symbols may be derived.

The inverter can be combined with the OR gate of Fig. 2.16 to form the NOT(A or B) statement, or NOR (A, B). The symbol and the truth table for the NOR gate is shown in Fig. 2.19. The truth table is similar to that of Fig. 2.16, but now F is inverted. It will be shown in Chap. 4 that with the NOR function by itself one is able to synthesize any logic function desired. We refer to systems that use NOR gates exclusively as NOR *logic systems*.

Similarly the inverter may be combined with the AND gate to form the NOT(A and B) or NAND (A, B) statement. The symbol and truth table for the NAND gate is shown in Fig. 2.20. Again, the only change from the truth table of

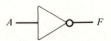

Truth Table	
A	F
0	1
1	0

FIGURE 2.18
Logic symbol of INVERTER with truth table.

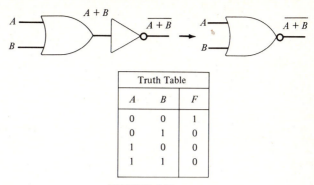

FIGURE 2.19
Logic symbol of NOR gate with truth table.

Fig. 2.17 is in the inversion of *F*. The NAND function is also able to satisfy all systems needs by itself, and systems built exclusively with NAND blocks are referred to as NAND *logic systems*.

2.4 DIODE LOGIC

Earlier in the chapter we used switches to illustrate Boolean algebraic functions. However, the use of switches to perform digital logic is not just an illustrative tool. In its more sophisticated form it is known as *relay logic*, which is developed to a very high degree in the automatic telephone switching system. However, electromechanical switches are too slow for many applications and require a good deal of maintenance, which can be expensive.

The introduction of low-cost semiconductor components led to the development of economical high-speed electronic digital systems. An understanding of the basic logic blocks used in digital systems can be obtained by studying OR and AND functions implemented with just diodes and resistors.

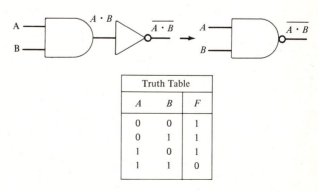

FIGURE 2.20
Logic symbol of NAND gate with truth table.

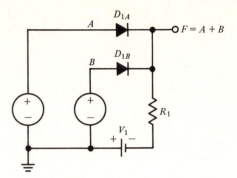

FIGURE 2.21
Circuit diagram of diode OR gate.

2.4.1 Diode OR Gate

The simple diode logic gate comprises just two diodes and a resistor. The gate circuit shown in Fig. 2.21 has two inputs and one output. The voltage levels at the input are chosen such that:

A voltage of 0.5 V represents a logic 0 (0) and is indicated as the LOW level (L);

A voltage of 1.5 V represents a logic 1 (1) and is indicated as the HIGH level (H).

With a LOW at both A and B inputs, both diodes D_{1A} and D_{1B} are forward-biased, and current flow is from the input, through the diodes, and then through the resistor R_1 to the negative supply voltage. The voltage at F is lower than that at A and B by the voltage drop across the diode. Using silicon diodes, this is about 0.7 V, so that the voltage at F is then about -0.2 V.

Now with a HIGH at, say, A, and a LOW at B, the current through D_{1A} is increased. The voltage at F is increased to 0.8 V, that is, about 0.7 V less than that at A. Diode D_{1B} is now reverse-biased by about 0.3 V. Of course, with a HIGH at B and a LOW at A, the voltage at F is still 0.8 V, and diode D_{1A} is reverse-biased by 0.3 V.

With a HIGH at both A and B, both diodes are conducting, but F is still about 0.8 V. We conclude that the output F responds to a HIGH or 1 at either A or B, or both. This, then, is an INCLUSIVE-OR gate.

Notice, however, that the voltage levels for the 0 and the 1 at the input are 0.5 and 1.5 V, respectively, but at the output the voltage levels are -0.2 and 0.8 V. Because the voltage levels do not reproduce themselves, they are said to be incompatible. In fact, the voltage levels have shifted down by 0.7 V in going through the gate. In a series string of gates, losing 0.7 V going through each gate could be troublesome in even an elementary logic system. It is thus very desirable to have the voltage levels at the output be the same as those at the input.

A simple solution to the problem is shown in Fig. 2.22. The level-shifting battery acts to raise the voltage levels by the same amount they were lowered in going through the diodes. A battery, however, could be bulky and bothersome. A

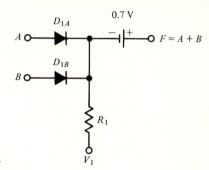

FIGURE 2.22
Circuit diagram of modified diode OR gate.

better and simpler solution is shown in Fig. 2.23. As long as current flows through
diode D_2 from the positive supply voltage through R_2 and D_2, then through R_1 to
the negative supply voltage, the voltage drop across D_2 will counteract the voltage
drop across D_1. Now, with a LOW or 0.5 V at the inputs A and B, there will be a
LOW or 0.5 V at the output F. With a HIGH or 1.5 V at either A or B, or both,
there will be a HIGH or 1.5 V at F. That is, the voltage levels are now compatible.

2.4.2 Diode AND Gate

With the same components of three diodes and two resistors we can also construct
an AND gate; this is shown in Fig. 2.24. Again, diode D_2 acts as a level-shifting
diode and is conducting current at all times. With a LOW at both A and B inputs,
current is flowing in all diodes, and the voltage at F is also at a LOW or 0.5 V. The
common anodes of the diodes would be at 1.2 V.

Now, with a HIGH at A and a LOW at B, current continues to flow through
D_{1B}, and the common anodes remain at 1.2 V. Diode D_2 is still conducting, and
the voltage at F remains 0.5 V. Only D_{1A} is reversed by 0.3 V. However, with a

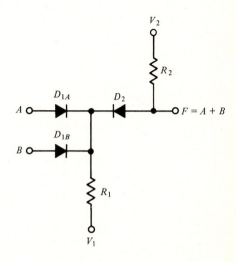

FIGURE 2.23
Circuit diagram of compatible diode OR gate.

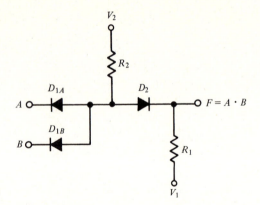

FIGURE 2.24
Circuit diagram of diode AND gate.

HIGH at both A and B inputs, the voltage at the common anodes is increased to 2.2 V, the current in D_2 is increased a little, and F is now at 1.5 V. This, then, is an AND gate, in that F responds only to a HIGH or 1 at both A and B.

The function of the diode AND circuit is illustrated with LEDs in Demonstration D2.2. In Demonstration D2.3 a pulse generator and an oscilloscope are used to demonstrate further the properties of diode logic gates.

EXERCISES

E2.8 For the circuit shown in Fig. 2.23,

$$V_1 = -5 \text{ V} \qquad R_1 = 1 \text{ k}\Omega$$
$$V_2 = +5 \text{ V} \qquad R_2 = 1 \text{ k}\Omega$$
$$V_{D(on)} = 0.7 \text{ V}$$

Determine the voltage at F and the current in the diode D_2 for the following input voltages:

  (a) $A = +1.0 \text{ V}; B = +1.0 \text{ V}.$
  (b) $A = +4.0 \text{ V}; B = +1.0 \text{ V}.$

E2.9 For the circuit shown in Fig. 2.24,

$$V_1 = -2 \text{ V} \qquad R_1 = 5 \text{ k}\Omega$$
$$V_2 = +4 \text{ V} \qquad R_2 = 2 \text{ k}\Omega$$
$$V_{D(on)} = 0.7 \text{ V}$$

Determine the voltage at F and the current in the diode D_2 for the following input voltages:

  (a) $A = 0.2 \text{ V}; B = 0.2 \text{ V}.$
  (b) $A = +4 \text{ V}; B = +4 \text{ V}.$

We have been representing a 1 as 1.5 V and a 0 as 0.5 V, where the 1 is more positive than the 0. This is known as *positive logic*.

Negative logic is just the reverse. The 1 has a level negative compared to the 0. That is, a 1 may be 0.5 V, and a 0 is 1.5 V.

Using this as a representation of negative logic, let us reexamine Fig. 2.23. Now with a 0 at both A and B inputs, the common cathodes are at 0.8 V, and F is at 1.5 V, or a 0. With a 1 at A and a 0 at B, the common cathodes are still at 0.8 V, and F is still at 1.5 V. We must have a 1 at both A *and* B to obtain a 1 at F. That is, the OR circuit in positive logic becomes an AND circuit in negative logic. It is left as an exercise for the reader to show that for negative logic the circuit as shown in Fig. 2.24 is an OR circuit.

We may now make the general statement:

	Positive logic		*Negative logic*
	AND	⟷	OR
	OR	⟷	AND

The general convention with digital ICs is that positive logic is intended unless otherwise noted. This convention will be followed in this text.

The ability to change the function of a gate from AND to OR, or vice versa, by a change of logic-level representation can help reduce the total number of gate circuits required to implement a system. This idea will be developed more fully in Chap. 4.

2.5 SUMMARY

\#   Values of the variables in the *binary system* are limited to just two states, logic one (1) and logic zero (0). These can also be represented as TRUE and FALSE, ON and OFF, or HIGH and LOW.

\#   The basic operations in *Boolean* or *switching algebra* are limited to

OR	as	*A or B*	(that is, $A + B$)
AND	as	*A and B*	(that is, AB)
NOT	as	*not A*	(that is, $\bar{A}$, also known as inverting or complementing)

\#   The *truth table* is useful in determining the state of a dependent variable as a function of all the states of the independent variables. The truth table is a systematic listing of all states of the independent variables to determine the state of the dependent variable in a logic problem.

\#   *Boolean theorems* are useful in the reduction of complex logic expressions to simpler and more meaningful terms.

\#   Generally, the reduction of complex logic statements with four or less variables is simplified by the use of the *Karnaugh map*.

\#   Definite *graphic symbols* are used to describe the logic operations of OR, NOR, AND, NAND, and NOT.

\#   The operations of OR and AND can be done with simple *diode logic*, using just diodes and resistors. The addition of a *level-shifting diode* makes the voltage levels at the input and output of the logic gate compatible.

\#   The function of a logic gate is dependent upon whether the *positive* or *negative logic* representation is used. Positive logic is most generally used.

DEMONSTRATIONS

D2.1   Logic relationships demonstrated with switches and lights (LEDs)   In this chapter we have been mostly concerned with number systems and Boolean algebra as a background for logic circuit design. With the aid of a power supply, resistors, switches, and a LED† (or lamp‡), the truth of the Boolean theorems can be demonstrated by using circuits shown in Fig. D2.1. Double-pole as well as single-pole switches will be required.

(*a*) An example of AND (Fig. D2.1*a*): similar to two switches controlling a lamp; one at the wall (*A*) controlling an outlet into which a lamp is plugged; the other (*B*) is in the lamp.

(*b*) An example of OR (Fig. D2.1*b*): similar to the switch control of the overhead light in an automobile. *A* and *B* are switches in the left and right door jamb that close when the associated door is not shut. *C* is a switch that can be manually operated inside the automobile.

(*c*) An example of NOT (Fig. D2.1*c*): when switch *A* is open ($A = 0$), the LED lights up ($F = 1$); when *A* is closed, no current flows through the LED ($F = 0$).

(*d*) An example of the EXCLUSIVE-OR (Fig. D2.1*d*): this combines concepts of parts (*a*) and (*b*). If $A = 1$ and $B = 0$, the top branch supplies current to LED; if $A = 0$ and $B = 1$, the bottom branch is closed.

(*e*) Further example of EXCLUSIVE-OR (Fig. D2.1*e*): by using two single-pole–double-throw switches, we can implement the same function as in part (*d*). $A = 1$ and $B = 1$ when both switches are in "up" position. This is similar to the situation that exists in wiring in a house where a stairway light is to be controlled by one switch at the top of the stairwell and another at the bottom.

(*f*) An example of the digital comparator (Fig. D2.1*f*): by eliminating the twist in the connections of part (*e*), the output is the complement of the output in part (*e*).

(*g*) An example of NOR (Fig. D2.1*g*): this is an extension of part (*c*). Only if both *A* and *B* are open ($A = 0, B = 0$) is $F = 1$.

(*h*) An example of generating the function $\bar{A} \cdot \bar{B}$ (Fig. D2.1*h*): this gives the same function as part (*g*). For $A = 0$ (switch $\bar{A}$ is closed) and $B = 0$ (switch $\bar{B}$ is closed), $F = 1$.

---

† Litronix Red-Lit 50 or 209 or equivalent.
‡ Sylvania 5ESB or equivalent.

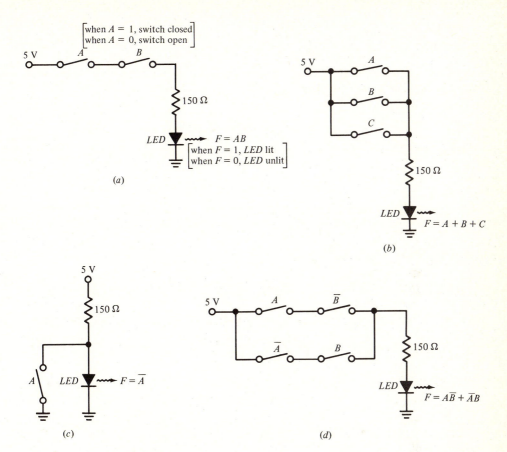

FIGURE D2.1

(*a*) Series switches form the AND functions. (*b*) Parallel switches form the OR functions. (*c*) Switch in parallel with LED forms the NOT function. (*d*) EXCLUSIVE-OR formed from single-pole–single-throw switches. (*e*) EXCLUSIVE-OR formed (as per house wiring) with two single-pole–double-throw switches. (*f*) Digital comparator formed with single-pole–double-throw switches. (*g*) NOR circuit: only if both *A* and *B* are open (=0) is $F = 1$. (*h*) Same Boolean function as part (*g*) using series switches. (*i*) NAND circuit: only if both *A* and *B* are closed (=1) is $F = 0$. (*j*) A HALF ADDER circuit.

(*Continued on next page*)

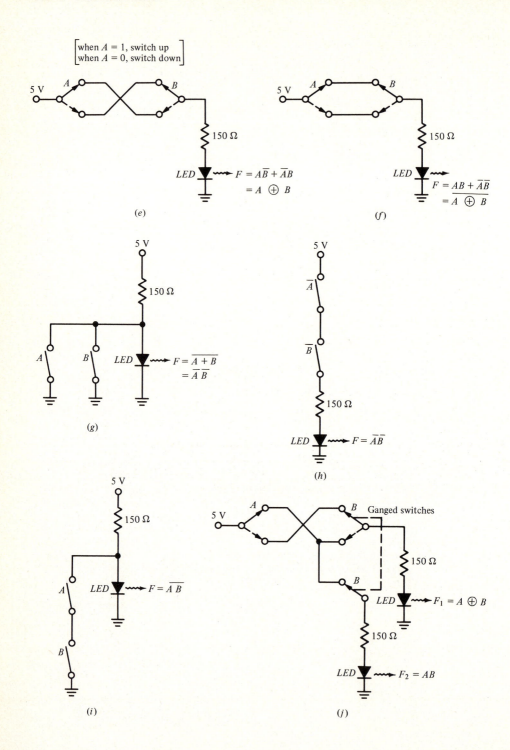

$$\left[\begin{array}{l}\text{when } A = 1,\ \text{switch up}\\ \text{when } A = 0,\ \text{switch down}\end{array}\right]$$

$F = A\bar{B} + \bar{A}B$
$\quad = A \oplus B$

(e)

$F = \overline{AB + \bar{A}\bar{B}}$
$\quad = \overline{A \oplus B}$

(f)

$F = \overline{A + B}$
$\quad = \bar{A}\,\bar{B}$

(g)

$F = \bar{A}\bar{B}$

(h)

$F = \overline{A}\,B$

(i)

Ganged switches

$F_1 = A \oplus B$

$F_2 = AB$

(j)

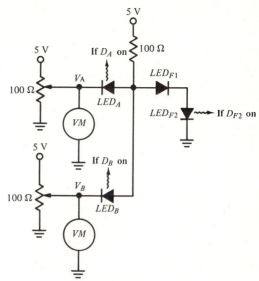

FIGURE D2.2
Visual diode AND gate.

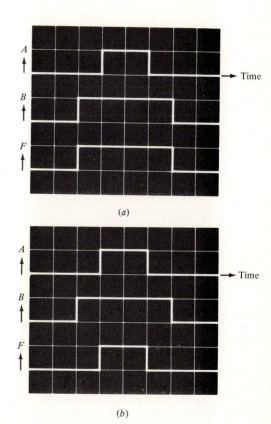

FIGURE D2.3
Oscilloscope displays for diode gates.
(*a*) OR gate waveforms; (*b*) AND gate
waveforms.

(*i*) An example of NAND (Fig. D2.1*i*): this is also an extension of part (*c*). Only if both *A* and *B* are closed ($A = 1$, $B = 1$) is $F = 0$.

(*j*) An example of HALF ADDER (Fig. D2.1*j*): this circuit is an extension of parts (*e*) and (*d*) so that two outputs F_1 and F_2 are obtained; F_1 is the EXCLUSIVE-OR, while F_2 is the AND obtained by *ganging* another single-pole switch in the circuit.

D2.2   A visual diode AND gate   The circuit in Fig. D2.2 shows a diode AND gate where a visual display is generated. All diodes are LEDs. With V_A (or V_B) set LOW, LED_A (or LED_B) is lit, while LED_1 and LED_2 are unlit and $F = 0$. Only with both V_A and V_B HIGH ($A = 1$ and $B = 1$) is D_{F2} lit ($F = 1$).

D2.3   Waveforms in a diode gate   The simple circuit shown in Fig. 2.23 can be used with a pulse generator and an oscilloscope to demonstrate the properties of the diode OR gate made with silicon diodes. The trace on the CRT display is shown in Fig. D2.3*a*. With the diodes and the battery polarity reversed, the circuit becomes an AND gate. The resulting display is shown in Fig. D2.3*b*. Circuit values used were $V_1 = -5$ V, $V_2 = 5$ V, $R_1 = 1$ $k\Omega$, and $R_2 = 1$ $k\Omega$.

REFERENCES

The fundamentals of logic design including the elements of Boolean algebra and the use of Karnaugh maps are found in MALEY, G. A.: "Manual of Logic Circuits," chaps. 1 and 2, Prentice-Hall, Englewood Cliffs, N.J., 1970; and WICKES, W. E.: "Logic Design with Integrated Circuits," chaps. 1–3, Wiley, New York, 1968.

For a more advanced treatment of these subjects, see PEATMAN, J. B.: "The Design of Digital Systems," chaps. 2 and 3, McGraw-Hill, New York, 1972; and HILL, F. J., and G. R. PETERSON: "Introduction to Switching Theory and Logical Design," chaps. 2, 4, and 6, Wiley, New York, 1968.

A handy little booklet covering simple logic design including many examples is "Computer Lab Workbook," Digital Equipment Corporation, Maynard, Mass., 1969.

PROBLEMS

P2.1   Convert from the decimal to the binary number system:
(*a*) 14; (*b*) 18; (*c*) 25; (*d*) 46; (*e*) 83; (*f*) 124; (*g*) 270.

P2.2   Convert from the binary to the decimal number system:
(*a*) 10111; (*b*) 01000; (*c*) 11010; (*d*) 101010; (*e*) 110011; (*f*) 101110; (*g*) 11011011.

P2.3   Convert the numbers listed in Prob. P2.1 from the decimal to the octal number system.

P2.4   Convert from the octal to the decimal number system:
(*a*) 13; (*b*) 16; (*c*) 25; (*d*) 56; (*e*) 77; (*f*) 120; (*g*) 243.

P2.5   Convert the numbers listed in Prob. P2.2 from the binary to the octal number system.

P2.6 Convert the numbers listed in Prob. P2.4 from the octal to the binary number system.

P2.7 Convert from the decimal number system to the 8421 and 2421 BCD codes (see Table C.2 in Appendix C):

(*a*) 642; (*b*) 579; (*c*) 438; (*d*) 216; (*e*) 813; (*f*) 475.

P2.8 Convert from the designated BCD codes to the decimal number system:

(*a*) 0110 0011 1000 0010   (8421)
(*b*) 1101 1011 0010 1100   (2421)
(*c*) 0111 1001 0101 0110   (8421)
(*d*) 1110 0100 1100 0011   (2421)

P2.9 The circuit shown in Fig. P2.9 consists of four 2-position switches, a lamp, and a power source.

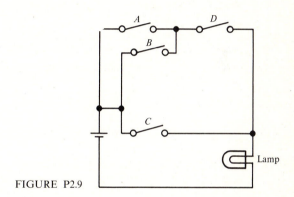

FIGURE P2.9

(*a*) Make a truth table of all the possible combinations of switch positions, assign a 1 for the condition of a switch being closed and for the condition of the lamp being lit.

(*b*) From the results in part (*a*) write a Boolean expression to indicate the lamp being lit.

P2.10 Given the statements (*a*) through (*d*) below, make a truth table and write an expression for the function *F* which represents *NO* talking in class. Assign a 1 for a student present and a 1 for no talking in class.

(*a*) *Allen* never talks in class.

(*b*) *Bill* talks only when Chuck is present.

(*c*) *Chuck* talks all the time.

(*d*) *David* talks only when Allen is present.

P2.11 Obtain a truth table from the Boolean expression

$$AB(C + D) + AC\bar{D}$$

P2.12 Make a truth table from the Boolean expression

$$F = X[\bar{W}Y + Z(Y + \bar{X})]$$

and draw a switch-and-lamp diagram for the function *F* that is true when the lamp is on.

P2.13   Given the circuit in Fig. P2.13, obtain a truth table and write a Boolean expression for the condition that the lamp is on. Let the true state represent a closed switch and the lamp on.

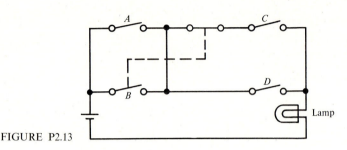

FIGURE P2.13

P2.14   Prove by the use of the truth table the expression

$$A\bar{B} + AC = (A + D)(\bar{B} + C)(A + \bar{D})$$

P2.15   Prove the equalities by applying the Boolean theorems.
   (*a*) $AB\bar{D} + A\bar{B}\bar{D} + AB\bar{C} = A\bar{D} + AB\bar{C}$
   (*b*) $ACD + A\bar{C}D + \bar{A}D + BC + B\bar{C} = B + D$
   (*c*) $WY\bar{Z} + \bar{W}X\bar{Z} + \bar{X}\bar{Y}\bar{Z} = \bar{Z}(W + X + \bar{Y})(\bar{W} + \bar{X} + Y)$
   (*d*) $(Y + \bar{Z})(W + X)(\bar{Y} + Z)(Y + Z) = YZ(W + X)$

P2.16   Simplify by applying the Boolean theorems:
   (*a*) $\bar{A}BCD + CD + A\bar{B}CD$
   (*b*) $A(\bar{B}C + \bar{A}D) + A(B\bar{C} + A\bar{D})$
   (*c*) $\bar{A}\bar{B}E + \bar{C}E(B\bar{E} + A\bar{C}\bar{E}) + A\bar{E} + A\bar{E}C$
   (*d*) $\bar{V}Z(V + \bar{Z}) + VWY + YZ(W + Y) + VW\bar{Y}$

P2.17   Write a simplified complement expression by applying the Boolean theorems.
   (*a*) $B(A\bar{D} + C)(C + D)(A + \bar{B})$
   (*b*) $A[B\bar{C} + C(B + \bar{D})]$
   (*c*) $\bar{R}S\bar{T} + R\bar{S}T + RST$
   (*d*) $W\bar{Z}R + T(WR + \bar{Z}) + \bar{T}Z$

P2.18   A modern university has agreed to set up a blue-ribbon committee to make key decisions. The committee is composed of a dean, a department chairman, a professor, and a student. A two-thirds majority rules, but just to make things fair the votes are partitioned as follows: four votes each for the dean and department chairman, three votes for the professor, and one vote for the student. Each member of the committee has a switch he closes to "yes" for all of his share of the voting and opens to vote "no." There are no split votes. Design a minimum-element switching circuit which will light a lamp if and *only if* a measure passes.

P2.19   Use Karnaugh maps to prove the equalities in Prob. P2.15.

P2.20   Use Karnaugh maps to simplify the expressions given in Prob. P2.17.

P2.21   Use Karnaugh maps to simplify the complement of the expression in Prob. P2.16.

P2.22   Minimize the following functions by using the Karnaugh map.
   (*a*) $\bar{A}B\bar{C} + AB\bar{C} + \bar{A}\bar{B}C + ABC + A\bar{B}C$
   (*b*) $ABC + \bar{A}B\bar{C} + \bar{A}\bar{B}C + \bar{A}\bar{B}\bar{C} + \bar{A}BC + AB\bar{C}$
   (*c*) $A\bar{B} + A\bar{C} + \bar{A}B + \bar{A}C + B\bar{C} + \bar{A}\bar{B}C$
   (*d*) $C(A\bar{B} + AB\bar{C} + A\bar{B}C) + A\bar{C}$

P2.23   Given the truth table shown below, determine the minimum form of the function *F*.

A	B	C	D	F
0	0	0	0	0
0	0	0	1	1
0	0	1	0	1
0	0	1	1	0
0	1	0	0	0
0	1	0	1	1
0	1	1	0	0
0	1	1	1	1
1	0	0	0	0
1	0	0	1	1
1	0	1	0	0
1	0	1	1	1
1	1	0	0	0
1	1	0	1	1
1	1	1	0	0
1	1	1	1	1

P2.24   Use a truth table and map to write a minimal expression for fighting, knowing that
   (*a*) Al will start a fight with anyone, but only if someone else is watching.
   (*b*) Bill will only start a fight with Charlie, but only then if no one else is around.
   (*c*) Charlie will start a fight with anyone, but not if Donald is around.
   (*d*) Donald never starts a fight.

P2.25   A car will not start when the ignition switch is turned on if either
   (*a*) The doors are closed and the seat belts are unbuckled.
   (*b*) The seat belts are buckled and the parking brake is on.
   (*c*) The parking brake is off and the doors are not closed.
   Use a truth table to write a minimal expression allowing the car to start when the switch is turned on.

P2.26   Implement with a minimal number of NAND gates, assuming input lines are uncomplemented:
   (*a*) $F = A\bar{B} + \bar{A}B$
   (*b*) $F = \bar{B}C(A + \bar{D}) + BD(A + \bar{C})$
   (*c*) $F = \bar{A} + B(\bar{C} + DE)$
   (*d*) $F = ABD + B\bar{D} + AC\bar{D}$

P2.27   Repeat Prob. P2.26 by using NOR gates. Complemented inputs are now available.

P2.28   Analyze the circuit in Fig. P2.28 and write a minimal expression for *F* in the sum-of-products form.

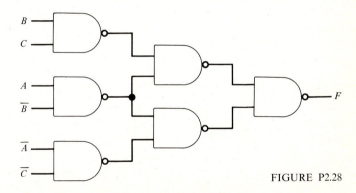

FIGURE P2.28

P2.29 Using NOR gates only, implement the minimized expression obtained in Prob. P2.28.

P2.30 Analyze the circuit in Fig. P2.30 and write the minimal expression.

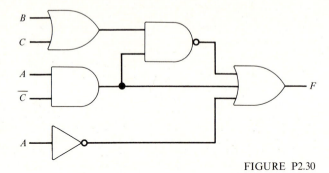

FIGURE P2.30

P2.31 Using only a minimum number of NOR gates, implement the function

$$F = \bar{B}\bar{C} + \bar{A}C + \bar{B}D$$

P2.32 Implement a minimum logic network of NOR gates that receives the 8421 BCD representation of a number on four input lines A, B, C, and D and provides a logic 1 output for an input representation that is ≥ 5.

P2.33 Given the circuit in Fig. P2.33a and the diode characteristics in Fig. P2.33b:
(a) Plot V_F versus V_A for V_A from 0 to 5 V, with $V_1 = 5$ V and $V_B = 0$ V.
(b) Repeat part (a) except let V_B take on values of 1, 2, 3, 4, and 5 V as V_A varies from 0 to 5 V.
(c) Assuming positive logic, what function does this gate represent?

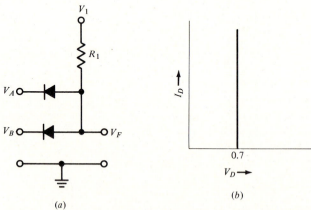

FIGURE P2.33

P2.34 Repeat Prob. P2.33 with the diodes in the opposite direction (reverse the diodes) and use values of $V_1 = -5$ V, $-5 \leq V_A \leq 0$, with V_B taking on values of -5, -4, -3, -2, -1 and 0 V.

P2.35 Given the function $F = AC + BD + BC + AD$, implement the function using the minimum number of two input diode logic OR and AND gates.

P2.36 Repeat Prob. P2.35 by using negative logic.

3
LOGIC GATE CIRCUITS

The design of the electronics of a digital system begins with the work of a logic designer, who first evolves both the logic equations which the system must solve and a tentative system design with logic blocks. The conversion of the logic design to the actual *hardware* of the system is the job of the circuit designer, who must realize the required logic with the necessary electronic circuits. Since compromises must usually be made and alternatives evaluated, the interaction of these two designers is essential in the development of an efficient working system. The circuit designer's task is performed by the creative application of the fundamentals of electronic circuits presented in Chap. 1 and the principles of logic described in Chap. 2.

The foundation of any digital system is the logic gate. Hence it is appropriate that we discuss in this chapter the various electronic circuits which can be used as basic logic gates.

A variety of gate circuits exist since a logic circuit has to satisfy more requirements than just the logic equation. For example, in a space satellite application, the power requirements are very stringent, and a low-power consumption for the logic block is a necessity. A low-power consumption is also a prime requirement of an electronic digital wristwatch. A battery life of 1 year is considered minimum. In a large scientific computer, on the other hand, high-speed operation is required, so the time taken for a signal to *propagate* through the logic block must be short, and this criterion is more important than power consumption. In digital control of electrical machinery, the "noisy" environment will have to be considered, and a larger voltage difference between the 1 level and the 0 level will be required as compared to a "quieter" environment. In all systems, cost is important, and this will enter into the design. Other factors to be considered might be compatibility with other logic blocks and the flexibility of the block to perform other logic operations.

A powerful stimulus to the proliferation of digital systems is the economy and abundance of many types of integrated logic circuits. A large selection of inexpensive circuits in turn arose from the feasibility of packing a large number of logic functions on a small area of silicon, which helped to create a new generation of computing machinery. Furthermore, with a decreasing number of pin connections found in complex ICs, the reliability of a digital system has continually improved.

Many of the circuit designs used in digital ICs represent variations of older designs in which discrete components (transistors, diodes, and resistors) were used. However, many ICs are using "new" components. An example is the input transistor used in transistor-transistor logic (TTL), which has two or more separate emitters but only one base and one collector. These circuits became practical only with the advent of ICs.

The successful design of a digital system requires a thorough knowledge of the input, output, and transfer characteristics of the basic logic circuit. In this chapter, various circuits are developed and described which perform a simple logic function. The complete design of digital ICs is beyond the scope of this text, though the material presented here will serve as an introduction to many aspects of a more sophisticated design.

We have already seen in Chap. 2 how to perform the basic logic operations of OR and AND with diodes and resistors. However, the operation of negation (logic NOT) is required to complete the set of basic logic operations. This is not possible with just diodes and resistors. An inversion of signal polarity is required for the NOT operation, which is provided by the transistor inverter in IC logic circuits.

Logic circuits which require transistors to saturate for their basic operation are described as *saturating logic circuits*. Not surprisingly, *nonsaturating logic circuits* describe those in which the transistors are not required to saturate for the basic operation of the circuit. We have therefore chosen to separate this chapter into two major divisions.

In Part I we describe those bipolar-transistor digital ICs which depend upon the transistors saturating for their basic operation. These include resistor-transistor logic (RTL), diode-transistor logic (DTL), and transistor-transistor logic (TTL).

In Part II we turn our attention to the nonsaturating logic circuits. We describe the bipolar-transistor emitter-coupled logic (ECL) circuits and the metal-oxide-semiconductor (MOS) digital ICs, including complementary MOS (CMOS) circuits.

We conclude the chapter with a short section on medium-scale integration (MSI) and large-scale integration (LSI).

PART I

3.1 SATURATING LOGIC CIRCUITS

The basic saturating-transistor logic circuit is built around a simple transistor-inverter stage. The normal or static condition of the transistor in such a circuit is either OFF (in the cutoff region) or ON (the transistor is saturated). Of course, the transistor must pass through the active region to make this transition. We shall see later that it is desirable that this transition region be as narrow as possible.

3.1.1 Transistor Inverter

A typical transistor-inverter stage is shown in Fig. 3.1a. Given the characteristics of the transistor, as noted in the figure, we can compile the following table. It is assumed that β_F for the transistor is constant throughout the active region. The analysis follows that of Sec. 1.5.2.

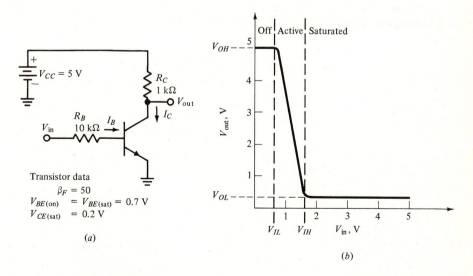

(a)

(b)

FIGURE 3.1

(a) Circuit diagram of transistor inverter; (b) voltage transfer characteristic of unloaded transistor inverter.

V_{in} (V)	I_B (mA)	I_C (mA)	V_{out} (V)
0	0	0	5.0
0.5	0	0	5.0
1.0	0.03	1.5	3.5
1.5	0.08	4.0	1.0
2.0	0.13	4.8	0.2
3.0	0.23	4.8	0.2
4.0	0.33	4.8	0.2
5.0	0.43	4.8	0.2

The voltage transfer characteristic of the inverter, shown in Fig. 3.1*b*, is plotted from the values given in the table.

From Fig. 3.1*b*, notice that the transistor is OFF when V_{in} is less than 0.7 V. As V_{in} increases, the transistor enters the active region, where

$$I_C = \beta_F I_B \qquad (3.1)$$

However, with $V_{in} \geq 1.66$ V, the transistor saturates since the collector current is limited to

$$I_{C(EOS)} = \frac{V_{CC} - V_{CE(sat)}}{R_C} \qquad (3.2)$$

The *breakpoints* of the voltage transfer characteristic indicate where the transistor operating region changes from *cutoff* to *active* and from *active* to *saturated*. At the input, the threshold voltage for the LOW level is called V_{IL}; for the HIGH level, it is called V_{IH}. For the output voltage, the threshold voltages are represented by V_{OL} and V_{OH}, respectively. These voltage levels are shown graphically in Fig. 3.2.

For the input, any voltage less than 0.7 V is recognized as a LOW-input level, and any voltage greater than 1.66 V indicates a HIGH-input level. Input

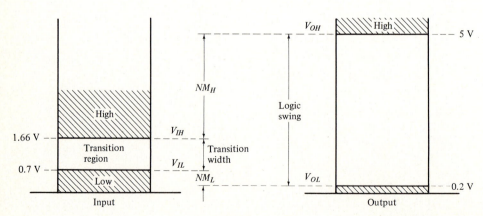

FIGURE 3.2
Logic-level diagram.

voltage levels between 0.7 to 1.66 V are to be avoided because they lead to output voltage levels that are ambiguous.

The difference between the two output voltages $(V_{OH} - V_{OL})$ determines the *logic swing* of the circuit. That is,

$$\text{Logic swing} = V_{OH} - V_{OL} \qquad (3.3)$$

$$= 5.0 - 0.2 = 4.8 \text{ V}$$

3.1.2 Noise Margins

Notice in Fig. 3.2 that the output levels are compatible with the input levels in that V_{OL} and V_{OH} are located safely within the acceptable input LOW and input HIGH regions, respectively. The input LOW threshold voltage (V_{IL}) is 0.7 V, but the output LOW-level voltage (V_{OL}) is 0.2 V. Hence there is a *safety margin* of 0.5 V $(V_{IL} - V_{OL})$. That is, a noise generator with an amplitude to 0.5 V could be inserted in series at the input, as shown in Fig. 3.3a, without having any effect on the output. This difference $(V_{IL} - V_{OL})$ determines the LOW-*level noise margin* (NM_L) for the circuit. That is,

$$NM_L = V_{IL} - V_{OL} \qquad (3.4)$$

$$= 0.7 - 0.2 = 0.5 \text{ V}$$

Similarly, the difference $(V_{OH} - V_{IH})$ is the HIGH-*level noise margin* (NM_H), which we can represent as a generator, as shown in Fig. 3.3b. Thus

$$NM_H = V_{OH} - V_{IH} \qquad (3.5)$$

$$= 5.0 - 1.66 = 3.34 \text{ V}$$

We note from Fig. 3.2 that an increased noise-margin capability is obtained as either V_{OH} and V_{OL} move away from each other or as V_{IH} and V_{IL} move toward each other. Thus, with a larger logic swing or a narrower *transition width*, the noise margins will improve. The transition width is defined by the equation

$$\text{Transition width} = V_{IH} - V_{IL} \qquad (3.6)$$

$$= 1.66 - 0.7 = 0.96 \text{ V}$$

for the inverter of Fig. 3.1a.

3.1.3 Fanout

One transistor-inverter stage is generally not operated in isolation by itself; rather connection will be made from the output of one inverter to the input of one or more similar circuits. This is shown in Fig. 3.4, where Q_0 is the driving gate and Q_1 through Q_n are the driven or load gates. There is an upper limit to n, the number of load gates that can be connected to such an inverter output, and this is known as N, the *fanout*† of the inverter.

† The term *fanout* is also commonly used for n, the number of load gates—not necessarily the limiting number.

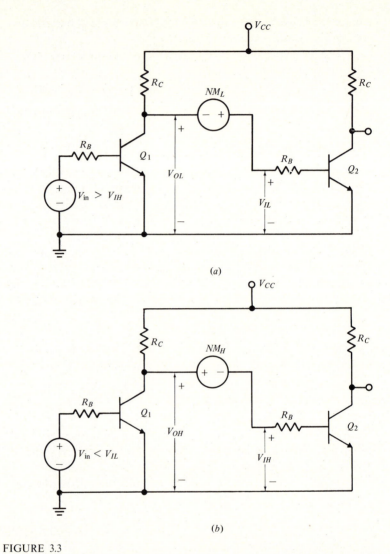

FIGURE 3.3
Representation of the noise margin as a voltage source in a transistor-inverter circuit. (a) The LOW noise margin (NM_L). While Q_1 is ON, Q_2 is now OFF, but at the borderline of OFF. (b) The HIGH noise margin (NM_H). While Q_1 is OFF, Q_2 is now ON, but at the borderline of ON.

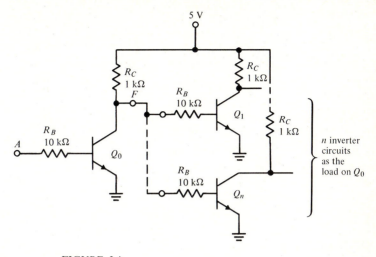

FIGURE 3.4
Transistor-inverter circuit with n other inverter inputs as a load.

To compute the fanout N for the circuit shown in Fig. 3.4, we first observe that with a HIGH level at point A, transistor Q_0 is saturated and the output at point F will be at a LOW level. With a $V_{CE(sat)}$ of 0.2 V for Q_0, all the load gates will be in the OFF state. Now consider a LOW level at A. Transistor Q_0 will be OFF with a resulting HIGH level at F. In this case it is required that the voltage at F be sufficient to ensure all the load gates Q_1 through Q_n are saturated. Each load gate added to F will require a certain base current to cause saturation of the load transistor. In fact, we have already seen that for the transistor to be in or at the edge of saturation, the collector current must be

$$I_{C(EOS)} = \frac{V_{CC} - V_{CE(sat)}}{R_C} \qquad (3.7)$$

Assuming the β_F of the transistor to be a constant, then the base current required for the transistor to be at the edge of saturation is

$$I_B = I_{B(EOS)}$$

where,

$$I_{B(EOS)} = \frac{I_{C(EOS)}}{\beta_F} = \frac{V_{CC} - V_{CE(sat)}}{\beta_F R_C} \qquad (3.8)$$

Also, with just one load gate, the base current for the load transistor in Fig. 3.4 is

$$I_B = \frac{V_{CC} - V_{BE(sat)}}{R_C + R_B}$$

Now each load gate input consists of the same-valued resistor (that is, $R_B = 10$ kΩ) in series with the base of similar transistors. We can consider the V_{BE}

of each transistor as a fixed voltage source of the same value (that is, 0.7 V). For a fanout of N, there are N similar circuits connected in parallel.

For N transistors as loads, the load may be reduced to one voltage source $V_{BE(sat)}$ and a series resistor R_B/N. Thus the total of all the base currents is

$$I_{BT} = \frac{V_{CC} - V_{BE(sat)}}{R_C + (R_B/N)} \qquad (3.9)$$

For a maximum fanout of N, we require

$$I_{BT} \geq NI_{B(EOS)}$$

Hence, utilizing Eqs. (3.8) and (3.9),

$$\frac{V_{CC} - V_{BE(sat)}}{R_C + (R_B/N)} \geq N \frac{V_{CC} - V_{CE(sat)}}{\beta_F R_C} \qquad (3.10a)$$

that is,

$$N \leq \beta_F \frac{V_{CC} - V_{BE(sat)}}{V_{CC} - V_{CE(sat)}} - \frac{R_B}{R_C} \qquad (3.10b)$$

We can use the data of Fig. 3.4 to calculate the fanout of the inverter stage. The result is

$$N \leq 50 \frac{5 - 0.7}{5 - 0.2} - \frac{10}{1}$$

$$= 34.9 \to 34$$

In order to ensure saturation, we have to round down 34.9 to 34.

3.1.4   Effect of Noise Margin on Fanout

It should be stressed that what we have just calculated is an *upper* limit for the number of load gates. A lower fanout is obtained when a noise margin for the load gates is included in our calculation. To show this, let us consider the HIGH-level noise margin (NM_H).

In Fig. 3.4, all the base current for the load gates flows through the collector resistor R_C from the 5-V supply to the inputs of the load gates. An increase in the number of load gates, and consequent increase in total base current, therefore lowers the voltage at F. A limit in the fanout is set when the voltage at F is insufficient to cause the transistor of the load gate to saturate. Recall from Fig. 3.1b that this voltage is V_{IH} and is equal to 1.66 V.

Let us now include a noise margin in our calculations in which we arbitrarily set NM_H to be 0.5 V. The minimum HIGH-level voltage at F is then, from Eq. (3.5),

$$V_{OH} = V_{IH} + NM_H$$

$$= 1.66 + 0.5 = 2.16 \text{ V}$$

That is, with V_F at 2.16 V we have built in a 0.5-V safety margin at the input of the load gates. Using this voltage at V_F we can calculate the current (I_{BT}) in resistor R_C. We can also calculate the base current (I_B) for a load transistor with 2.16 V at its input. The ratio of these two currents will yield the fanout of the inverter circuit for a NM_H of 0.5 V. Specifically,

$$I_{BT} = \frac{V_{CC} - V_{OH}}{R_C} \qquad (3.11a)$$

$$= \frac{5 - 2.16}{1} = 2.84 \text{ mA}$$

$$I_B = \frac{V_{OH} - V_{BE(\text{sat})}}{R_B} \qquad (3.11b)$$

$$= \frac{2.16 - 0.7}{10} = 0.15 \text{ mA}$$

$$N = \frac{I_{BT}}{I_B} \qquad (3.11c)$$

$$= \frac{2.84}{0.15} = 19.4 \rightarrow 19$$

Hence, incorporating a reasonable noise margin, we have reduced the fanout of the inverter from 34 to 19. An increased NM_H would, of course, reduce the fanout even further.

3.1.5   The Base Overdrive Factor

The transistor inverter is a very simple circuit, but still there are many variables involved. Chief among these is the β_F of the transistor. We noted in Chap. 1 that this parameter can vary over a considerable range, especially for ICs. In the design of the biasing of a transistor inverter, to ensure the transistor saturating, a minimum value of β_F would be used. Then for the typical case, the base current will be more than sufficient to saturate the transistor. This introduces the idea of a *base overdrive factor*, denoted by k.

The base current required to bring a transistor to the edge of saturation has been defined as $I_{B(\text{EOS})}$. Then,

$$I_{B(\text{EOS})} = \frac{I_{C(\text{EOS})}}{\beta_F} \qquad (3.12a)$$

Now from Eq. (3.11*b*),

$$I_B = \frac{V_{OH} - V_{BE(\text{sat})}}{R_B}$$

then

$$k = \frac{I_B}{I_{B(\text{EOS})}} \qquad (3.12b)$$

We see that k is the ratio of the actual base current and the base current required to just saturate the transistor.

EXAMPLE 3.1   Use the circuit shown in Fig. 3.4 to determine the base overdrive factor k for a load transistor when 10 load gates are connected to F ($n = 10$).

$$V_{BE(\text{sat})} = 0.7 \text{ V} \qquad \beta_F = 50$$

$$V_{CE(\text{sat})} = 0.2 \text{ V}$$

SOLUTION   We will first use Eq. (3.7) to solve for the collector saturation current of the load transistor:

$$I_{C(\text{EOS})} = \frac{5 - 0.2}{1} = 4.8 \text{ mA}$$

Now the base current of the transistor at the edge of saturation is given by Eq. (3.12a):

$$I_{B(\text{EOS})} = \frac{4.8}{50} = 0.096 \text{ mA}$$

With a fanout of 10, the total base current (I_{BT}) at the output F is obtained from Eq. (3.9):

$$I_{BT} = \frac{5 - 0.7}{1 + \dfrac{10}{10}} = 2.15 \text{ mA}$$

The base current to each load transistor is therefore

$$I_B = \frac{2.15}{10} = 0.215 \text{ mA}$$

We can now use Eq. (3.12b) to determine the base overdrive factor:

$$k = \frac{0.215}{0.096} = 2.24 \qquad \qquad ////$$

We should note that when k is 1, the transistor is operating at the edge of saturation. Typically, the value for k will range from 2 to 10.

3.1.6 Worst-case Design

It should be realized that variations in transistor characteristics, power-supply voltages, resistor values, logic levels, and so on, will play a significant role in actual circuit performance. In worst-case analysis, the circuit is analyzed by using the particular combination of parameter values which will result in an extreme value of the desired variable.

It is not our intention in this book to do a thorough worst-case analysis of each circuit, even though this is an important aspect of design. However, we will want to see what effect the variation of the circuit parameters will have on the fanout of our inverter circuit. For this purpose, we can rewrite the basic equation for the fanout Eq. (3.10b), where we have incorporated the overdrive factor k.

$$N \le \frac{\beta_F}{k}\left(\frac{V_{CC} - V_{BE(sat)}}{V_{CC} - V_{CE(sat)}}\right) - \frac{R_B}{R_C}$$

To make it a worst-case analysis, we choose values for the variables that will minimize the right-hand side of the equation, i.e.,

$$N \le \frac{\underline{\beta_F}}{k}\left(\frac{\underline{V_{CC}} - \overline{V_{BE(sat)}}}{\overline{V_{CC}} - \underline{V_{CE(sat)}}}\right) - \frac{\overline{R_B}}{\underline{R_C}}$$

where the bar above the parameter indicates the maximum value and the bar below indicates the minimum value. As an illustration, with a nominal V_{CC} of 5 V and a 10 percent variation, we have $\overline{V}_{CC} = 5.5$ V and $\underline{V}_{CC} = 4.5$ V.

Generally, in an IC all resistor tolerances vary together; they are said to track. If one resistor on the chip is 10 percent higher than the design-center value, then all resistors in the circuit will be 10 percent higher than their design-center value. Therefore, resistor tolerance will not enter into the above relation.

We do have a problem with V_{CC}, however, since it cannot be at its minimum and maximum value at one and the same time. To treat this case, we simply calculate two values for N: one when V_{CC} is at a minimum and the other when V_{CC} is at a maximum. The worst-case fanout will then be the smaller N.

To continue the numerical example, typical minimum and maximum values of the necessary parameters are given below.

$$4.5 < V_{CC} < 5.5 \qquad 0 < V_{CE(sat)} < 0.5$$
$$20 < \beta_F < 100 \qquad 0.5 < V_{BE(sat)} < 1.0$$
$$k = 1$$

Assuming the V_{CC} supply is at a minimum value, we obtain

$$N \le 20\,\frac{4.5 - 1}{4.5 - 0} - \frac{10}{1}$$

or

$$N \le 5.5$$

On the other hand, if the V_{CC} supply is at a maximum value, we find

$$N \le 20\,\frac{5.5 - 1}{5.5 - 0} - \frac{10}{1}$$

or

$$N \le 6.4$$

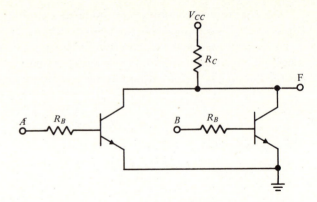

FIGURE 3.5
Circuit diagram of an RTL gate.

For the worst-case analysis we thus select $V_{CC} = 4.5$ V, and we then obtain a fanout of 5.

It is not unusual when using a worst-case philosophy to find that there is no acceptable solution to the problem, e.g., one may obtain zero or negative answers for the fanout. In these cases it is necessary to either change the ratio of the resistor values in the circuit or set other limits for the transistor parameters and supply voltages.

3.1.7   Logic Circuit

We are now in a position to use inverter circuits to build a logic gate. The circuit to be studied is shown in Fig. 3.5, where two inverters are connected to a common-collector resistor to form a simple logic gate. A set of truth tables for the logic gate is shown in Table 3.1.

Table 3.1   TRUTH TABLE FOR LOGIC GATE

Voltage level HIGH = H LOW = L			Positive logic H = 1 L = 0			Negative logic L = 1 H = 0		
A	B	F	A	B	F	A	B	F
L	L	H	0	0	1	1	1	0
L	H	L	0	1	0	1	0	1
H	L	L	1	0	0	0	1	1
H	H	L	1	1	0	0	0	1
	(a)			(b)			(c)	

Table 3.1*a* shows the four states of the independent variables *A* and *B*, with the resulting condition of the dependent variable *F*. For example, with both inputs LOW, both transistors are OFF and the output is HIGH. With a HIGH at one or both inputs, one or both transistors are saturated, and in each case the output is LOW.

In Table 3.1*b*, *positive logic* is used to substitute a 1 and a 0 for the *H* and *L*, respectively, of Table 3.1*a*. We see that the output *F* is a 1 only when both inputs *A* and *B* are a 0, that is,

$$F = \bar{A} \cdot \bar{B}$$

or invoking DeMorgan's theorem

$$F = \overline{A + B}$$

Hence, *F* is identical to NOT(*A* OR *B*). Therefore, for *positive logic*, this is a NOR gate, that is, NOR(*A*, *B*).

Negative logic is used in Table 3.1*c* to describe Table 3.1*a*. Here we find

$$F = A\bar{B} + \bar{A}B + \bar{A}\bar{B}$$
$$= \bar{A}(B + \bar{B}) + \bar{B}(A + \bar{A})$$
$$= \bar{A} + \bar{B}$$
$$= \overline{A \cdot B}$$

That is, *F* is equivalent to NOT(*A* AND *B*) or NAND(*A*, *B*). Hence, for *negative logic*, the circuit is a NAND gate.

EXERCISES

E3.1 The transistors used in the inverter circuit shown in Fig. 3.4 have the following characteristics:

$$V_{BE(\text{sat})} = 0.7 \text{ V} \qquad \beta_F = 50$$
$$V_{CE(\text{sat})} = 0.2 \text{ V}$$

With a fanout of 5 determine:
(*a*) The LOW and HIGH voltage levels at *F*.
(*b*) The LOW and HIGH noise margins at *F*.
(*c*) The base overdrive factor for the load transistors.

E3.2 The circuit to be used is similar to that shown in Fig. 3.4, but with the following changes:

$$R_B = 1 \text{ k}\Omega \qquad V_{CC} = 3 \text{ V}$$
$$R_C = 2 \text{ k}\Omega$$

The transistors have the following characteristics:

$$0.6 < V_{BE(sat)} < 0.9 \qquad 2.7 < V_{CC} < 3.3$$
$$0.1 < V_{CE(sat)} < 0.4 \qquad 10 < \beta_F < 100$$

Use the worst-case design considerations to determine:

(a) The maximum collector current for the transistors in saturation.

(b) The minimum base current to ensure the transistors are saturated.

(c) The maximum number of loads, with all load transistors at the edge of saturation (that is, $k = 1$).

3.2   RESISTOR-TRANSISTOR LOGIC

The logic gate described in the previous section and shown in Fig. 3.5, falls under the generic term *resistor-transistor logic*, or RTL for short. Such a circuit was among the first commercially available digital ICs, introduced by Fairchild Semiconductor in 1961. A portion of a typical data sheet for an RTL gate is shown in Fig. 3.6. The gate is packaged in many different arrangements. For example, in Fig. 3.6, we show a quadruple 2-input gate. That is, the package contains four independent gate circuits, each gate has two inputs and one output, but all gates share the common V_{CC} and ground pins. Another circuit that goes with this family is a dual 4-input gate consisting of two gate circuits, each having four inputs and one output. Another common package is the triple 3-input gate. For all these circuits, the V_{CC} is 3.6 V, the base resistor R_1 is 450 Ω, and the collector resistance R_2 is 640 Ω. The maximum guaranteed fanout for the gate is 5.

A typical voltage transfer characteristic for this gate is shown in Fig. 3.7. We have already seen that V_{OL} is determined by the collector saturation voltage of the transistor. From the figure we note that $V_{OL} \simeq 0.15$ V. The input threshold (V_{IL}) is set by the base-emitter turn-on voltage of the transistor; this is about 0.6 V for this gate. The HIGH-output level has been shown to be a function of the number of load gates, and this is shown on the transfer characteristic.

From Fig. 3.7 we are able to determine the logic swing at the output of the driving gate. With a load of five other gates,

$$\text{Logic swing} = V_{OH} - V_{OL}$$
$$= 1.15 - 0.15 = 1.0 \text{ V}$$

We may also determine the typical noise margins for this circuit from the same figure. For a load of five other gates,

$$NM_L = V_{IL} - V_{OL}$$
$$= 0.60 - 0.15 = 0.45 \text{ V}$$
$$NM_H = V_{OH} - V_{IH}$$
$$= 1.15 - 0.70 = 0.45 \text{ V}$$

QUAD 2-INPUT GATES

MC724P · MC824P
MC724AP · MC824AP

PLASTIC MRTL MC700P/800P series **MOTOROLA**

JUNE 1970

This monolithic device consists of four 2-input positive logic NOR gates. Each may be used independently, or cross-coupled to form bistable elements.

The MC724AP and MC824AP are compatible with MTTL and MDTL devices. Extra electrical tests are performed to insure that the devices will drive two MDTL loads and at least one MTTL load using any MTTL family.

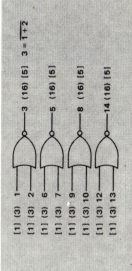

Number in Parenthesis Indicates mW MRTL Loading Factor.
Number in Brackets Indicates MRTL Loading Factor.

t_{pd} = 12 ns
P_D = 100 mW (input high)
　　　30 mW (inputs low)

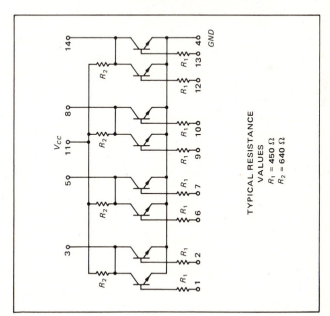

TYPICAL RESISTANCE VALUES
R_1 = 450 Ω
R_2 = 640 Ω

FIGURE 3.6
Data sheet for RTL gate (*Motorola Inc.*).

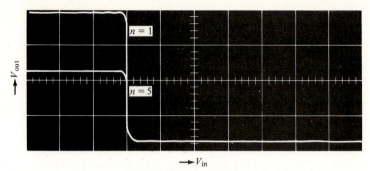

V_{out}

$n = 1$

$n = 5$

$\longrightarrow V_{in}$

Vertical scale: 0.5 V/division
Horizontal scale: 0.2 V/division

FIGURE 3.7
Voltage transfer characteristic of MC824P RTL gate.

A method of experimentally obtaining the voltage transfer characteristic of a logic gate is given in Demonstration D3.2. The particular description is not for an RTL gate, but a simple change of the V_{CC} to 3.6 V is all that is needed to obtain some practical information on the design of this type of logic gate.

3.2.1   Power Dissipation of RTL Gates

An important specification contained in the manufacturer's data sheet is the *power dissipation* of the gate. Since a relatively simple system can contain over a thousand gates, the power requirements become very important. For a single power-supply system this power must be obtained from the V_{CC} supply. Also, the small dimensions of the device make for a high-component packing density. If the power density becomes too high, expensive cooling facilities may be required to keep the chip temperatures within a safe operating region.

For our example in Fig. 3.6, the power dissipation in the whole package with all inputs at a HIGH level is 100 mW. Thus, PD_H is 25 mW per gate. With the inputs at a LOW level, the power dissipation of the four gates is 7.5 mW per gate.

Sometimes the average power dissipation (PD_{av}) of the gate is given. This number is obtained by assuming the input to the gate spends as much time at the LOW level as it does at the HIGH level, that is, there is a 50 percent *duty cycle*, and hence for our example,

$$PD_{av} = \frac{25 + 7.5}{2} = 16 \text{ mW}$$

3.2.2   Propagation Delay in RTL Gates

Another important parameter of a logic gate is the *propagation delay time*. Due to circuit capacitances and the finite switching speed of the transistors used, there is a delay from the time a signal is applied to the input of a logic gate until the desired

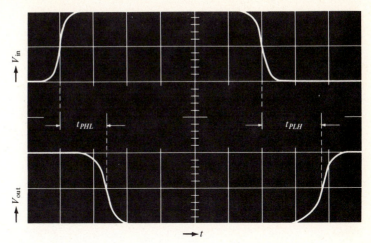

FIGURE 3.8
Waveforms showing effect of propagation delay time. Horizontal scale: 10 ns/cm.

change appears at the output of the gate. This is shown in the oscilloscope display in Fig. 3.8. The propagation delay time (t_{pd}) is the average of the turn-on delay time $(t_{PHL}$, sometimes written $t_{pd(0)}$, the output voltage is changing from HIGH to a LOW) and the turn-off delay time $(t_{PLH}$ or $t_{pd(1)}$, the output is changing from LOW to HIGH).

$$t_{pd} = \frac{t_{PHL} + t_{PLH}}{2}$$

From the data sheet for the RTL gate (Fig. 3.6) the typical propagation delay time† is 12 ns. The measured propagation delay time from Fig. 3.8 is 16 ns.

It is sometimes possible to decrease the propagation delay time of a logic gate by increasing the operating currents of the transistors in the circuit. However, this does increase the power consumption of the gate. The product of propagation delay time and power dissipation $(t_{pd} \times PD)$ therefore serves as a useful figure of merit on the performance of a gate. Normally a minimum value of the product is desired. For this gate, the figure of merit is given as (12 ns) (16 mW) = 192 pJ. Power dissipation and propagation delay are the subject of investigation in Demonstrations D3.3 and D3.4, respectively.

3.2.3 Low-power RTL Gates

For many applications, high speed is not a basic requirement. Especially with small portable equipment, the basic requirement is low-power dissipation. Hence, a low-power version of this RTL gate is available for these applications. With a

---

† Delay times in digital ICs are usually given in nanoseconds, abbreviated as ns. 1 nanosecond = 10^{-9} second.

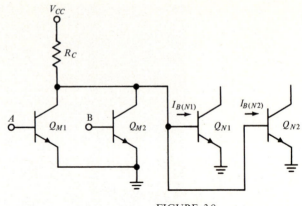

FIGURE 3.9
Circuit diagram of DCTL gates.

V_{CC} of 3.6 V, the transistor operating current is reduced by increasing the base and collector resistors to 1.5 and 3.6 kΩ, respectively. This results in a typical t_{pd} of 27 ns, but the power dissipation with the inputs HIGH is only 5 mW per gate.

3.2.4   Direct-coupled Transistor Logic

One method of reducing the propagation delay time is to eliminate the base resistor (R_B) and make a direct connection from the collector of a driving transistor to the base of the load transistor. This is known as *direct-coupled transistor logic* (DCTL) and is shown in Fig. 3.9. The simplicity of the circuit is a real advantage. However, there is also a severe disadvantage; unless the input characteristics of the load transistors are well matched under all conditions, one of them will tend to take most of the current available from the driving transistor. This is known as "current hogging," a problem that is illustrated in Fig. 3.10. Notice that since the base-emitter junctions for the two load transistors Q_{N1} and Q_{N2} are in parallel, the same V_{BE} will appear across both junctions. If these two transistors

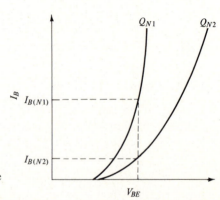

FIGURE 3.10
The input characteristic of a DCTL gate
showing the "current-hogging" problem.

have matched input VI characteristics, then the same base current will flow in each transistor. However, a more likely prospect is that the two transistors will not be exactly matched. The condition is then as shown in Fig. 3.10. Here, for the same V_{BE}, the base current of Q_{N1} is greater than that of Q_{N2}. Transistor Q_{N1} is then said to "hog" the current. Indeed, $I_{B(N2)}$ may be insufficient to saturate Q_{N2}.

This defect can be diminished by placing a resistor in series with the base, so that the base current is less dependent upon the V_{BE} characteristic of the transistor. We then again have the typical RTL circuit, as shown in Fig. 3.5.

EXERCISES

E3.3 Use the voltage transfer characteristic shown in Fig. 3.7, with a fanout of 1, to determine:
- (a) The logic swing.
- (b) The noise margins.
- (c) The transition width.

E3.4 For the low-power RTL gate, $R_B = 1.5$ kΩ, $R_C = 3.6$ kΩ, and $V_{CC} = 3.6$ V. The transistors used in the gate have the following characteristics:

$$V_{BE(sat)} = 0.65 \text{ V} \qquad \beta_F = 30$$

$$V_{CE(sat)} = 0.15 \text{ V}$$

With a fanout of 4, determine:
- (a) The logic swing.
- (b) The noise margins.
- (c) The base overdrive factor for the load transistors.

3.3 DIODE-TRANSISTOR LOGIC

A disadvantage of RTL gates is the small logic swing. We have seen in the examples that the logic swing for RTL circuits is only about 1 V typical. In the presence of extraneous noise, this logic swing gives too small a noise margin for many applications. *Diode-transistor logic* (DTL) was developed to satisfy a need for a larger swing than that obtainable with RTL. A simple form of the DTL gate is shown in Fig. 3.11. It should be noted that this is basically a diode AND circuit followed by an inverter. The input is similar to Fig. 2.24. The diodes D_{1A}, D_{1B}, and D_2 with resistors R_1 and R_2 comprise a *level-shifting*† diode AND gate. The output stage consists of an inverter circuit, transistor Q_1, and resistor R_3.

---

† D_2 is always conducting irrespective of the voltages at A and B; thus it looks like a 0.7-V battery with its positive terminal tied to the point where the anodes of diodes A and B are located. Hence, the voltage at the cathode of D_2 and the base of Q_1 is approximately equal to the lowest of the two input voltages (V_A or V_B). Consequently, the diode D_2 *shifts* the *signal level* down by the amount gained in the diode gate.

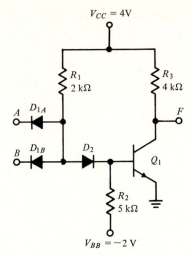

FIGURE 3.11
Circuit diagram of a basic DTL gate.

With either or both inputs A and B at a LOW level, say, 0 V, the level at the base of Q_1 is also LOW, or 0 V. The transistor is then in the cutoff region, and the output at F is at a HIGH level. With A and B at a HIGH level, the input diodes are reverse-biased and normally transistor Q_1 saturates, so that the output is at a LOW level. This differs from RTL in that now part of the saturation current in Q_1 comes from the external load.

A truth table for the gate is shown in Table 3.2. Using *positive logic*, we see from Table 3.2b that

$$F = A\bar{B} + \bar{A}B + \bar{A}\bar{B}$$

$$= \bar{A} + \bar{B}$$

$$= \overline{A \cdot B}$$

Therefore, for *positive logic*, this is a NAND gate.

Table 3.2 TRUTH TABLE FOR DTL GATE

Voltage level HIGH = H LOW = L			Positive logic H = 1 L = 0		
A	B	F	A	B	F
L	L	H	0	0	1
L	H	H	0	1	1
H	L	H	1	0	1
H	H	L	1	1	0
(a)			(b)		

Similar to the methods described earlier, it can be shown that, using *negative logic*, the gate performs the NOR operation.

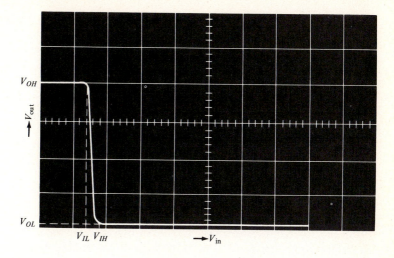

Vertical scale: 1V/division
Horizontal scale: 0.5 V/division

FIGURE 3.12
Voltage transfer characteristic of DTL gate. Vertical scale: 1 V/div; horizontal scale: 0.5 V/div.

Shown in Fig. 3.12 is the voltage transfer characteristic for the circuit of Fig. 3.11. A LOW-output level is still represented by the $V_{CE(sat)}$ of the output transistor. That is, V_{OL} is equal to $V_{CE(sat)}$. At the LOW-level threshold the input voltage must be sufficiently positive to turn off the input diodes and start turning on the output transistor. That is, V_{IL} is given by

$$V_{IL} = V_{BE1(sat)} + V_{D2(on)} - V_{D1(on)} \qquad (3.13)$$

with

$$V_{D2(on)} = V_{D1(on)}$$

$$V_{IL} = V_{BE1(sat)} \qquad (3.14a)$$

If we let $V_{BE(sat)} = 0.7$ V, the noise margin for the LOW level is then

$$NM_L = V_{IL} - V_{OL}$$

$$= 0.7 - 0.2 = 0.5 \text{ V}$$

It may be noted that the LOW-level noise margin may be increased by using two diodes in series in place of the one at D_2. To show this, we simply write out the appropriate equation for V_{IL}:

$$V_{IL} = V_{BE1(sat)} + 2V_{D2(on)} - V_{D1(on)} = V_{BE1(sat)} + V_{D2(on)} \qquad (3.14b)$$

Assuming again that $V_{BE(sat)} = V_{D(on)} = 0.7$ V, the NM_L is now

$$NM_L = 1.4 - 0.2 = 1.2 \text{ V}$$

For the output HIGH level, transistor Q_1 is cut off and V_{OH} is equal to V_{CC}. It should be noted that when the output is loaded with similar gates, V_{OH} is still approximately equal to V_{CC}. The input diodes of the load gates are reverse-biased, and only a negligible "leakage" current can flow through them from the V_{CC} supply. This also means there is no current-hogging problem with DTL. With the output of the driving gate HIGH, all the input diodes connected to that node are reverse-biased and the load gates are isolated from each other. The base current for each transistor of the load gate is determined by its own resistor R_1. The HIGH-level noise margin is again obtained from the transfer characteristic:

$$NM_H = V_{OH} - V_{IH}$$

$$= 4.0 - 0.9 = 3.1 \text{ V}$$

The logic swing is given by the difference of the two output voltage levels. From the voltage transfer characteristic of Fig. 3.12, we see that the logic swing is considerably greater for a DTL gate than for an RTL circuit and is essentially unaffected by the number of load gates.

$$\text{Logic swing} = V_{OH} - V_{OL}$$

$$= 4 - 0.2 = 3.8 \text{ V}$$

The difference between the two input voltage levels indicates the transition width:

$$\text{Transition width} = V_{IH} - V_{IL}$$

$$= 0.9 - 0.7 = 0.2 \text{ V}$$

Another characteristic of the DTL circuit which is different from RTL circuits is the function of the output transistor. With RTL gates, the output is a source of current for the load transistors. However, with the DTL gate, the output transistor is a sink for the dc current in the input diode of the load gates. We therefore speak of the output of RTL circuits being a *current source* and the output of DTL circuits acting as a *current sink*. The general practice is to consider the current *flowing out* of a gate terminal to have a *negative value* and the current *flowing into* a terminal to have a *positive value*. The understanding of the direction of current flow at the gate output is necessary in calculating the fanout of the gate. Therefore this point is worth repeating. With RTL circuits the direction of dc current at the output is *to the load gate*. With DTL circuits the direction of dc current is *to the driving gate*.

3.3.1   Fanout of Basic DTL Gate

A basic DTL circuit with two voltage-offset diodes is shown in Fig. 3.13. To calculate the fanout, we must determine the base current available to the driving transistor Q_{1A}. Then, knowing the β_F of the transistor, we can calculate the collector current. With Q_{1A} saturated, each load gate contributes a current, I_D.

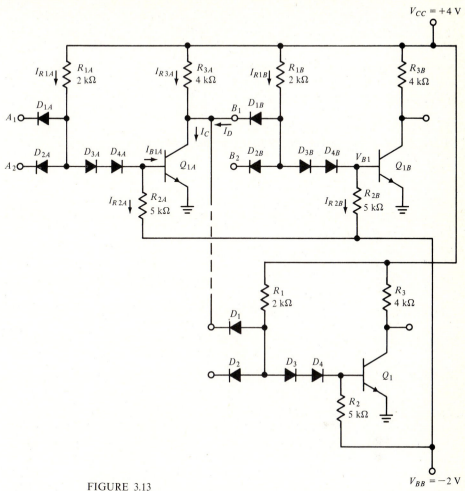

FIGURE 3.13
The basic DTL gate (whose output is at the collector of Q_{1A}) loaded with n other gate inputs.

Notice there is also a contribution to I_C through the collector resistor R_{3A}. The fanout is then the ratio $(I_C - I_{R3A})/I_D$. The details of the fanout calculation are illustrated in the following example.

EXAMPLE 3.2   Use the component values given in Fig. 3.13 to determine the fanout of this basic DTL gate.

$$V_{BE(\text{sat})} = V_{D(\text{on})} = 0.7 \text{ V} \qquad \beta_F = 20$$

$$V_{CE(\text{sat})} = 0.2 \text{ V} \qquad\qquad k = 1$$

SOLUTION   We begin by calculating the base current of transistor Q_{1A} when there is a **HIGH** level at both inputs A_1 and A_2. Assuming both input diodes D_{1A}

and D_{2A} are reversed-biased, we can use Kirchhoff's current law to determine the current I_{R1A}. That is,

$$I_{R1A} = \frac{V_{CC} - 2V_{D(on)} - V_{BE(on)}}{R_{1A}} \tag{3.15}$$

$$= \frac{4 - 1.4 - 0.7}{2} = 0.95 \text{ mA}$$

We again make use of Kirchhoff's current law to determine I_{R2A} since

$$I_{R2A} = \frac{V_{BE(on)} - V_{BB}}{R_{2A}} \tag{3.16}$$

$$= \frac{0.7 + 2}{5} = 0.54 \text{ mA}$$

Therefore

$$I_{B1A} = I_{R1A} - I_{R2A} \tag{3.17}$$

$$= 0.95 - 0.54 = 0.41 \text{ mA}$$

The collector current of Q_1 at the edge of saturation is given by

$$I_C = \beta_F I_B$$

$$= (20)(0.41) = 8.2 \text{ mA}$$

We now turn our attention to the output of the gate, where we note that with Q_{1A} saturated or even at the edge of saturation the input diodes of all the load gates connected to the collector will be forward-biased. The voltage at the base of Q_{1B}, and all other inverting transistors of the load gates, will then be two diode voltage drops below this figure, and consequently we find these transistors are cut off.

$$V_{B1} = V_{CE(sat)} + V_{D(on)} - 2V_{D(on)}$$

$$= V_{CE(sat)} - V_{D(on)} \tag{3.18}$$

$$= 0.2 - 0.7 = -0.5 \text{ V}$$

Knowing the voltage at the anode of D_{1B}, we can simply calculate the current I_{R1B} as

$$I_{R1B} = \frac{V_{CC} - (V_{CE(sat)} + V_{D(on)})}{R_{1B}} \tag{3.19}$$

$$= \frac{4 - (0.2 + 0.7)}{2} = 1.55 \text{ mA}$$

With the load-gate transistor Q_{1B} off, the current I_{R2B} is computed as

$$I_{R2B} = \frac{V_{B1} - V_{BB}}{R_{2B}} \tag{3.20}$$

$$= \frac{-0.5 + 2}{5} = 0.3 \text{ mA}$$

Now the load current I_D is due to the difference between I_{R1B} and I_{R2B}. That is,

$$I_D = I_{D1B} = I_{R1B} - I_{R2B} \tag{3.21}$$

$$= 1.55 - 0.3 = 1.25 \text{ mA}$$

Before determining the fanout we must also calculate the current in the collector load resistor R_{3A} of transistor Q_{1A}:

$$I_{R3A} = \frac{V_{CC} - V_{CE(\text{sat})}}{R_{3A}} \tag{3.22}$$

$$= \frac{4 - 0.2}{4} = 0.95 \text{ mA}$$

Finally, the collector current of Q_{1A} is the sum of load current and the current I_{R3A}. That is,

$$I_C = NI_D + I_{R3A} \tag{3.23}$$

But the collector current has already been determined, for $k = 1$, as 8.2 mA. Therefore, the fanout N is equal to

$$N = \frac{8.2 - 0.95}{1.25} = 5.8 \to 5 \qquad /\!/\!/\!/$$

Note that, since we have not considered any worst-case parameters, this is not the worst-case fanout; rather this might be the fanout if all elements were at their nominal values.

EXERCISE

E3.5 Calculate the fanout of a basic DTL gate similar to that shown in Fig. 3.13, but with the following component values.

$R_1 = 2 \text{ k}\Omega$	$V_{CC} = +4 \text{ V}$	$V_{BE(\text{sat})} = V_{D(\text{on})} = 0.65 \text{ V}$
$R_2 = 20 \text{ k}\Omega$	$V_{BB} = -2 \text{ V}$	$V_{CE(\text{sat})} = 0.15 \text{ V}$
$R_3 = 2 \text{ k}\Omega$	$k = 1$	$\beta_F = 30$

3.3.2 Improved DTL Gate

A variation of the DTL circuit is shown in Fig. 3.14. The offset diode D_3 of Fig. 3.13 has been replaced by a transistor Q_1 acting as an emitter follower (see Sec. 1.6). Note that the voltage difference (level shift) between the base of Q_2 and the base of Q_1 is still $\simeq 1.4$ V.

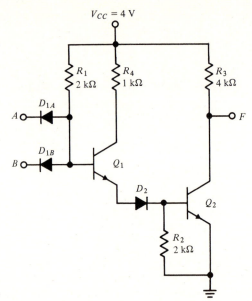

FIGURE 3.14
Circuit diagram of improved DTL gate.

This circuit, compared to that of Fig. 3.13, allows an increase of base current to the output transistor Q_2 for the same resistance values and voltage levels. This implies that the fanout is increased. However, the load at the output of the gate can be anywhere from one gate up to the maximum number of load gates. When lightly loaded the large value of base current to Q_2 can mean a large overdrive factor for this transistor. The large overdrive factor indicates that transistor Q_2 is heavily saturated. This results in a long propagation delay time when Q_2 turns off, and therefore the switching speed of the gate is degraded. A resistor, R_4 in Fig. 3.14, from V_{CC} to the collector of Q_1 serves to limit the current to the base of Q_2 and the subsequent overdrive of Q_2. However, the problem now is that Q_1 also saturates. To see how this is remedied, we must look at the modified DTL circuit described in the next section.

3.3.3 Modified DTL Gate

The most widely used form of an IC DTL gate is another development of Fairchild Semiconductor. The circuit is shown in Fig. 3.15. Notice in this circuit the connection of the resistance R_1 from the collector to the base of transistor Q_1. This form of feedback-biasing is used to prevent Q_1 saturating. That is, as the collector current of Q_1 increases, the collector voltage falls, but the feedback connection causes the base current to be decreased. The collector current now decreases, and we have a closed loop where the collector current, and also the emitter current,

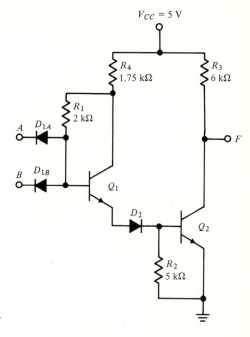

FIGURE 3.15
Circuit diagram of modified DTL gate.

can be controlled to a designed value without Q_1 saturating. By using the transistor parameters of the previous example, we can calculate an operating point for transistor Q_1 and also determine the fanout of this modified gate circuit. Notice for this DTL gate the change in V_{CC} to 5 V and the change in resistor values compared to Fig. 3.14.

Similar to the previous example, with a HIGH level at both A and B inputs, the voltage at the base of Q_1 is equal† to $3V_{BE(on)}$ or 2.1 V. Now the voltage difference between V_{CC} and the base of Q_1 is due to the base current in R_1 and the sum of the base and collector current in R_4. That is,

$$V_{CC} - 3V_{BE(on)} = I_{B1}(R_1 + R_4) + I_{C1}(R_4) \qquad (3.24a)$$

But since Q_1 is not saturated, I_{C1} can be replaced with $\beta_F I_{B1}$. Then

$$V_{CC} - 3V_{BE(on)} = I_{B1}(R_1 + R_4) + \beta_F I_{B1}(R_4) \qquad (3.24b)$$

Using this expression, we can solve for the base current of Q_1 as

$$I_{B1} = \frac{V_{CC} - 3V_{BE(on)}}{(R_1 + R_4) + \beta_F(R_4)} \qquad (3.25)$$

$$= \frac{5 - 2.1}{(3.75) + 20(1.75)} = .075 \text{ mA}$$

† This assumes $V_{BE(sat)} = V_{BE(on)} = V_{D(on)}$.

The collector current of Q_1 simply follows as

$$I_{C1} = \beta_F I_{B1}$$
$$= (20)(0.075) = 1.5 \text{ mA}$$

We can now calculate the voltage at the collector of Q_1 by determining the voltage drop across resistor R_4. However, remember that both the collector current and the base current of the transistor flow in the resistor R_4.

$$V_{C1} = V_{CC} - R_4(I_C + I_B) \qquad (3.26)$$
$$= 5 - 1.75(1.575) = 2.25 \text{ V}$$

The voltage at the base of transistor Q_1 is equal to 2.1 V. The base-collector junction of this transistor is therefore reverse-biased by 0.15 V; that is, the transistor is in the active region, not saturated.

We now continue the calculation to determine the fanout when the gate is driving similar gates. We need to know the base current of transistor Q_2; we can then determine the collector current of Q_2. The emitter current of transistor Q_1 is given by the sum of the base and collector currents; namely,

$$I_{E1} = I_{B1} + I_{C1}$$
$$= 0.075 + 1.5 = 1.58 \text{ mA}$$

The base current of Q_2 is equal to the difference between the emitter current of Q_1 and the current in the resistor R_2.

$$I_{B2} = I_{E1} - \frac{V_{B2}}{R_2} \qquad (3.27)$$

$$= 1.58 - \frac{0.7}{5} = 1.44 \text{ mA}$$

The collector current of Q_2 at the edge of saturation is then

$$I_{C2} = \beta_F I_{B2}$$
$$= (20)(1.44) = 28.8 \text{ mA}$$

For the load gates, we must remember that their inputs are at a LOW and that the base voltage of transistor Q_1 in the load gates is the sum of the saturation voltage of the output transistor of the driving gate and the diode voltage of the input diodes of the driven gate.

$$V_{B1(L)} = V_{CE(\text{sat})} + V_{D(\text{on})} \qquad (3.28)$$
$$= 0.2 + 0.7 = 0.9 \text{ V}$$

This voltage is insufficient to turn on transistor Q_1 through diode D_2. With Q_1 in the load gates off, the load current (I_D) is from V_{CC} through R_4, R_1, and the input diodes D_1 to the collector of the output transistor in the driving gate.

$$I_D = \frac{V_{CC} - (V_{CE(sat)} + V_{D(on)})}{R_1 + R_4} \qquad (3.29)$$

$$= \frac{5 - 0.9}{3.75} = 1.09 \text{ mA}$$

As in the previous example, we must also include the current in the collector resistor (R_3) of the driving gate by using Eq. (3.22).

$$I_{R3} = \frac{V_{CC} - V_{CE(sat)}}{R_3}$$

$$= \frac{5 - 0.2}{6} = 0.8 \text{ mA}$$

We may now calculate the fanout for this circuit in a manner similar to the previous example. As in Eq. (3.23),

$$I_{C2} = NI_D + I_{R3}$$

or

$$N = \frac{I_{C2} - I_{R3}}{I_D}$$

$$= \frac{28.8 - 0.8}{1.09} = 25.6 \rightarrow 25$$

The modified DTL circuit of Fig. 3.15 therefore has a superior fanout to the basic DTL circuit, but unlike the circuit of Fig. 3.14, only the output transistor saturates.

EXERCISE

E3.6 The circuit for the high-threshold-logic DTL gate is similar to that shown in Fig. 3.15 except the diode D_2 is replaced with a zener diode having a breakdown voltage of 6 V. The higher ON voltage of D_2 also means the V_{CC} has to be increased to 15 V. To avoid an excessive power dissipation, the values of the resistor are also increased.

$R_1 = 12 \text{ k}\Omega$	$V_{CC} = 15 \text{ V}$	$V_{CE(sat)} = 0.2 \text{ V}$
$R_2 = 5 \text{ k}\Omega$	$\beta_F = 20$	$V_{BE(sat)} = 0.7 \text{ V}$
$R_3 = 15 \text{ k}\Omega$		$V_{D1(on)} = 0.7 \text{ V}$
$R_4 = 3 \text{ k}\Omega$		$V_{D2(on)} = 6.0 \text{ V}$

**EXPANDABLE DUAL
4-INPUT "NAND" GATE**

MC930 · MC830
MC961 · MC861

MDTL MC930/830 series **MOTOROLA**

ISSUE A

Add Suffix F for ceramic flat package (Case 607).
Suffix L for ceramic dual in-line package (Case 632).
Suffix P for plastic dual in-line package (Case 646).
MC830 series only.

**EXPANDABLE DUAL
3-2 INPUT "NAND" GATE**

MC930 · MC830
MC961 · MC861

Add Suffix G for TO-100 metal package (Case 603-02).

This gate element, in the 14-pin flat and dual in-line packages, consists of two expandable 4-input NAND gate circuits. Since the metal can (G suffix) has only 10 pins, that circuit consists of one 3-input and one 2-input expandable gate. The elements may be cross-coupled to form a bistable multivibrator, or the outputs may be connected in parallel to perform the logic "OR" function.

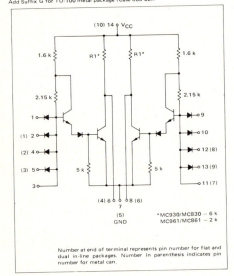

Number at end of terminal represents pin number for flat and dual in-line packages. Number in parenthesis indicates pin number for metal can.

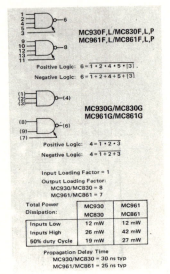

MC930F,L/MC830F,L,P
MC961F,L/MC861F,L,P

Positive Logic: $6 = \overline{1 \cdot 2 \cdot 4 \cdot 5 \cdot [3]}$

Negative Logic: $6 = \overline{1 + 2 + 4 + 5 + [3]}$

MC930G/MC830G
MC961G/MC861G

Positive Logic: $4 = \overline{1 \cdot 2 \cdot 3}$

Negative Logic: $4 = \overline{1 + 2 + 3}$

Input Loading Factor = 1

Output Loading Factor:
MC930/MC830 = 8
MC961/MC861 = 7

Total Power Dissipation:	MC930 MC830	MC961 MC861
Inputs Low	12 mW	12 mW
Inputs High	26 mW	42 mW
50% duty Cycle	19 mW	27 mW

Propagation Delay Time
MC930/MC830 = 30 ns typ
MC961/MC861 = 25 ns typ

ELECTRICAL CHARACTERISTICS

Characteristic	Symbol	Pin Under Test	MC930, MC961 TEST LIMITS −55°C Min Max	+25°C Min Max	+125°C Min Max	Unit	MC830, MC861 TEST LIMITS 0°C Min Max	+25°C Min Max	+75°C Min Max	Unit
Output Voltage	V_{OL}	6	- 0.40	- 0.40	- 0.45	Vdc	- 0.45	- 0.45	- 0.50	Vdc
	V_{OH}	6	2.50 -	2.60 -	2.50 -		2.60 -	2.60 -	2.50 -	
Short-Circuit Current MC930/MC830	I_{SC}	6	- −1.34	- −1.34	- −1.30	mAdc	- −1.30	- −1.30	- −1.25	mAdc
MC961/MC861		6	- −4.00	- −4.00	- −3.90	mAdc	- −3.90	- −3.90	- −3.75	mAdc
Reverse Current	I_R	1 2 4 5	- 2.0	- 2.0	- 5.0	μAdc	- 5.0	- 5.0	- 10	Adc
Output Leakage Current	I_{CEX}	6	- -	- 50	- -	μAdc	- -	- 100	- -	Adc
Forward Current	I_F	1 2 4 5	- −1.60	- −1.60	- −1.50	mAdc	- −1.40	- −1.40	- −1.33	mAdc
Power Drain Current (Total Device) MC930/MC830	I_{PDH}	14	- -	- 6.5	- -	mAdc	- -	- 8.0	- -	mAdc
MC961/MC861	I_{PDH}	14	- -	- 10.7	- -		- -	- 13.1	- -	
All Types	I_{max}	14	- -	- 5.5	- -		- -	- 8.0	- -	
Switching Times MC930/MC830	t_{pd+}	1,6	- -	25 80	- -	ns	- -	25 60	- -	ns
	t_{pd-}	1,6	- -	10 30	- -		- -	10 30	- -	
MC961/MC861	t_{pd+}	1,6	- -	15 60	- -		- -	15 60	- -	
	t_{pd-}	1,6	- -	10 30	- -		- -	10 30	- -	

Pins not listed are left open.

©MOTOROLA INC., 1972

FIGURE 3.16
Data sheet of DTL gate (*Motorola Inc.*).

For this gate determine:

    (*a*) The logic swing.
    (*b*) The breakpoint voltages.
    (*c*) The noise margins.
    (*d*) The fanout for $k = 1$.

3.3.4   930 Series DTL Gate

A typical example of the modified DTL gate is contained on the data sheet shown in Fig. 3.16. The MC930 is† a dual expandable 4-input gate. The expander input (at pin 3) allows for the number of inputs to the gate to be extended by connecting additional diodes to this node. For example, this gate can perform a 7-input NAND function by connecting each anode of three additional diodes to the expander input. It should be noted that the characteristics of the expanding diodes should be similar to the input diodes of the basic gate.

For most digital ICs we will find that the positive power supply (V_{CC}) is applied at the highest-numbered pin (pin 14 in this case) while ground is tied to the pin diagonally opposite this terminal (which is pin 7 in the case of a 14-pin package).

The electrical characteristics of all gates in the 930 Series are similar; it will therefore be useful and interesting to review the data for this gate. The 830 Series is similar to the 930 Series, but the 830 family operates over a more restricted temperature range of 0 to 75°C. The normal operating V_{CC} for both series of DTL gates is 5.0 V $\pm$ 10 percent.

Notice from the output voltage characteristic for the 930 Series that as the ambient temperature is varied from -55 to $+125°C$ there is little change in either the output LOW voltage (V_{OL}) or the output HIGH voltage (V_{OH}). Both these measurements are made with V_{CC} at the lower limit to obtain the worst-case characteristic. By specifying the maximum V_{OL} and the minimum V_{OH}, a minimum value for the logic swing of the gate is indicated. That is, from Eq. (3.3),

$$\text{Logic swing} = V_{OH} - V_{OL}$$

$$= 2.6 - 0.4 = 2.2 \text{ V}$$

Using the values for V_{IL} and V_{IH} given elsewhere on the data sheet, a value for the noise margin of the gate may also be obtained. At 25°C, V_{IL} is equal to 1.1 V and V_{IH} is equal to 2.0 V. Then using Eqs. (3.4) and (3.5),

$$NM_L = V_{IL} - V_{OL}$$

$$= 1.1 - 0.4 = 0.7 \text{ V}$$

$$NM_H = V_{OH} - V_{IH}$$

$$= 2.6 - 2.0 = 0.6 \text{ V}$$

---

† The same IC is often supplied by several manufacturers, with the only changes in some of the code numbers that refer to the circuit. As an example, consider the case of the dual 4-input DTL gate of the "930" family. This is referred to as the 9930 by Fairchild Semiconductor, MC930 by Motorola, and 15930 by Texas Instruments.

Table 3.3   COMPARISON OF THE ELECTRICAL CHARACTERISTICS OF THE 930 DTL GATE AT AN AMBIENT TEMPERATURE OF 25°C

		Calculated nominal values	Typical measured values	Worst-case values
Output voltage	V_{OL}	0.2 V	0.2 V	0.4 V max
	V_{OH}	5.0 V	4.8 V	2.6 V min
Input voltage	V_{IL}	1.4 V	1.2 V	1.1 V max
	V_{IH}	1.4 V	1.5 V	2.0 V min
Short-circuit current	I_{SC}	−0.83 mA	−1.0 mA	−1.34 mA max
Reverse current	I_R	⋯	1.0 nA	2.0 μA max
Output leakage current	I_{CEX}	⋯	1.0 μA	50 μA max
Forward current	I_F	− 1.09 mA	−1.2 mA	−1.6 mA max
Power-drain current (per gate)	I_{PDH}	2.38 mA	2.0 mA	3.25 mA max
	I_{PDL}	1.09 mA	1.2 mA	

The transition width completes the data associated with the voltage transfer characteristic of the gate. From Eq. (3.6),

$$\text{Transition width} = V_{IH} - V_{IL}$$

$$= 2.0 - 1.1 = 0.9 \text{ V}$$

It should again be emphasized that these are worst-case characteristics. In Table 3.3 a comparison is made between the worst-case values obtained from the data sheet and some experimentally measured values for a typical 930 DTL gate. Also included in the table are the calculated values (using nominal element values) for the same parameters. These were derived in Sec. 3.3.3.

The short-circuit current (I_{SC}) This current indicates the maximum current from an output HIGH level should the output be accidentally shorted to ground. The test is made with the V_{CC} at the upper limit to obtain the worst-case characteristic. Incidentally, this test also indicates a minimum value for the collector load resistance, R_3, of the output transistor. The full V_{CC} voltage appears across this resistor, and the maximum I_{SC} at 25°C is given as − 1.34 mA.† Therefore

$$\bar{I}_{SC} = \frac{\bar{V}_{CC}}{\underline{R}_3}$$

Hence,

$$\underline{R}_3 = \frac{5.5}{1.34} = 4.1 \text{ k}\Omega$$

Now the nominal value of R_3 is 6 kΩ. We may therefore calculate the maximum variation allowed for this resistor. Moreover, with all resistors of a

† The generally accepted standard is that current flowing out of a terminal is a negative value.

monolithic IC having the same characteristics, the variation for resistor R_3 also indicates the variation for all resistors of this circuit.

$$R_3 - (\Delta)(R_3) = \underline{R}_3$$

where Δ is equal to the fractional change in resistance,

$$\Delta = \frac{6 - 4.1}{6} = 0.32$$

A possible variation of ± 32 percent is therefore indicated for the diffused resistors of this circuit.

Reverse current   The *reverse current* (I_R) is a measure of the leakage current of the input diodes of the gate. The *output leakage current* (I_{CEX}) is a measure of the leakage current at the collector of the output transistor of the gate. Both these leakage currents are sufficiently small to be insignificant in most circuit design problems.

Forward current   The *forward current* (I_F) is similar in magnitude to the load current (I_D) used in this text, but the current flowing out of a terminal is defined as being a negative value. Moreover, I_F as specified in Fig. 3.16 is a worst-case parameter, measured with V_{CC} at the upper limit and the input to the gate at ground.

Power-drain current   The power-drain current, the total current from the V_{CC} supply, is measured for two cases: one called a HIGH condition and the other a LOW condition, referring to the input to the gate. In one case all the inputs are left open, a HIGH level, and in the other, at least one of the inputs is grounded, a LOW level. With all inputs to the gate at a HIGH level, the input diodes are off and both transistors Q_1 and Q_2 are conducting. The total current is the HIGH-level power-drain current (I_{PDH}). With one input diode forward-biased by a LOW-level input, both transistors Q_1 and Q_2 are cut off, and essentially the only current from the V_{CC} supply is the load current (I_D). This is indicated as the LOW-level power-drain current (I_{PDL}). However, in the data sheet of Fig. 3.16, I_{max} is specified. This is measured under the same conditions as I_{PDL} but with V_{CC} at its absolute maximum value of 8 V.

For a single gate, with the normal operating V_{CC} of 5 V, the calculated power dissipation with the inputs LOW is (5 V)(1.09 mA) = 5.45 mW; with the inputs HIGH the power dissipation is (5)(2.38 mA) = 11.9 mW. The average power dissipation is obtained by assuming a 50 percent duty cycle for the gate.

$$PD_{av} = \frac{5.45 + 11.9}{2} = 8.7 \text{ mW}$$

This value may be compared with the total power dissipation (for two gates) of 19 mW given on the data sheet.†

---

† The component design-center values, in particular R_1 and R_4, for the Motorola circuit are slightly different from those shown in Fig. 3.15.

From the data sheet we also learn that the typical propagation delay time is 30 ns. We can use the values listed on the data sheet to determine the speed-power product for the 930 DTL gate. This figure of merit is

$$t_{pd} PD_{av} = (30 \text{ ns})(9.5 \text{ mW}) = 285 \text{ pJ}$$

Compared to the RTL gate, the power dissipation is about the same, but the RTL gate is about a factor of 2 faster than the DTL gate. However, the fanout, logic swing, and noise margin are much greater for the more widely used DTL circuit.

3.4   TRANSISTOR-TRANSISTOR LOGIC

In our discussion of the circuit model of a bipolar transistor in Sec. 1.4.3, we saw how the characteristics of the base-emitter junction of a transistor are similar to those of a *p-n* junction diode. This fact is used to advantage in *transistor-transistor logic* (TTL) circuits. The input diodes of a DTL gate are replaced by a multi-emitter transistor in the TTL gate; this is shown in Fig. 3.17. For the TTL gate, the collector-base junction of Q_1 provides an offset voltage, replacing an offset diode. Because the emitter occupies the least area of a transistor, the addition of more emitters poses no problem. In fact, the replacing of many diodes by a multi-emitter

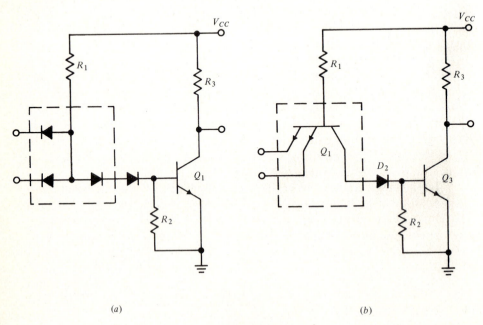

(a)   (b)

FIGURE 3.17
Circuit diagram of the basic (a) DTL and (b) TTL gate.

transistor saves silicon area. Moreover, the switching speed is improved because the use of a transistor at the input speeds up the process of taking transistor Q_3 out of saturation.

The operation of a TTL gate is very similar to the DTL gate. The logic function of each is the same; that is, with positive logic they are both basic NAND gates. We will first consider the operation of the circuit shown in Fig. 3.17b, with a LOW level $(V_{CE(sat)})$ at the gate input. In this case, there is current flowing from V_{CC} through R_1 and the forward-biased base-emitter junction of transistor Q_1, then out of the input terminal of the gate. The transistor Q_1 is operating in the normal manner. However, the only collector current that can flow is the leakage current of the offset diode D_2, and this is very small. This small collector current is much less than the product $\beta_F I_B$. The result is that transistor Q_1 saturates. Since Q_1 is saturated, the voltage at the collector of Q_1 is then ~ 0.2 V above the voltage at its emitter. Hence, there is insufficient voltage to forward-bias the diode D_2 and the base-emitter junction of transistor Q_3. Therefore, Q_3 is off. The output of the gate is then at V_{CC}, a HIGH logic level.

Now consider all inputs at a HIGH level (V_{CC}). In this case the base-emitter junctions of Q_1 are all reverse-biased. Current flows from V_{CC}, through R_1, then through the forward-biased base-collector junction of transistor Q_1, then through D_2 into the base-emitter junction of Q_3. This current is sufficient to cause Q_3 to saturate, resulting in a LOW logic level at the output of the gate.

Note that since, by Kirchhoff's voltage law, the sum of all the junction voltages around a transistor must equal zero; that is,

$$V_{CE} = V_{BE} + V_{CB} \qquad (3.30a)$$

$$= V_{BE} - V_{BC}$$

then,

$$V_{BC} = V_{BE} - V_{CE} \qquad (3.30b)$$

With the nominal saturation voltages for Q_1 of 0.7 V and 0.2 V for $V_{BE(sat)}$ and $V_{CE(sat)}$, respectively, then $V_{BC(sat)}$ is equal to 0.5 V.

3.4.1 The Input Transistor

The operation of the input transistor Q_1 in Fig. 3.17b is interesting. As we have seen, with a LOW level at the gate input, Q_1 operates in the forward mode. The flow of external current of the transistor is into the collector and out of the emitter. However, with all the gate inputs at a HIGH level, transistor Q_1 operates in an inverse manner. Current flow in the transistor is into the emitters and out of the collector. That is, in the inverse mode the base-collector junction becomes the emitting junction and the base-emitter junction is the collecting junction. Notice the *only* condition for inverse operation is when *all* the inputs are HIGH. Even with just one emitter LOW, Q_1 operates in the forward mode.

For an understanding of the inverse operation it is useful to refer to the schematic of a multi-emitter transistor shown in Fig. 3.18, where the indicated flow

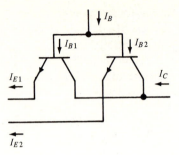

FIGURE 3.18
Current flow in multi-emitter transistor.

of current is for the normal or forward mode. In the inverse active region with the emitter acting as a collector, then,

$$-I_{E1} = \beta_R I_{B1} \qquad (3.31a)$$

where β_R is the inverse beta (current gain) of the transistor, but

$$I_{B1} = \frac{I_B}{M}$$

where M is equal to the number of emitters in the transistor; therefore,

$$-I_{E1} = \frac{\beta_R I_B}{M} \qquad (3.31b)$$

Also for the collector current in the inverse mode, since the collector is acting as an emitter,

$$-I_C = (\beta_R + 1)I_{B1} + (\beta_R + 1)I_{B2} \qquad (3.32)$$
$$= (\beta_R + 1)(I_{B1} + I_{B2}) = (\beta_R + 1)I_B$$

With the simple TTL circuit shown in Fig. 3.17b, this inverse operation can be a problem when one gate is driving another. If transistor Q_1 should have a high inverse beta (β_R), a large current could flow into the input terminal; coming from a collector load resistance of high value, this could reduce the voltage of a HIGH logic level and decrease the HIGH-level noise margin. There are two solutions to the problem. In the fabrication of the IC, the emitter and collector regions of the transistors are preferentially doped so the β_R is generally much less than 1. The other solution is in the design of the circuit. The output of the gate is taken from an emitter-follower circuit which effectively reduces the output load resistance seen by the input of the driven gate. More will be said of this design technique later.

An additional advantage of the input transistor Q_1 has been briefly mentioned earlier. With the input to the gate at a HIGH level, transistor Q_3 is saturated. When the input moves from a HIGH to a LOW level, the current in the input transistor reverses. The transistor Q_1 is then in the forward active mode, which speeds up the turn off of Q_3. It is not until Q_3 is completely off that Q_1 saturates. This is one reason for the superior switching speed of TTL compared to DTL.

Further familiarity with the basic TTL gate circuit can be obtained from Demonstration D3.1.

EXERCISE

E3.7   The following characteristics apply to the TTL gate shown in Fig. 3.17:

$$R_1 = 4 \text{ k}\Omega \qquad V_{CC} = 5 \text{ V} \qquad V_{CE(sat)} = 0.2 \text{ V}$$

$$R_2 = 5 \text{ k}\Omega \qquad \beta_F = 20 \qquad V_{BE(sat)} = V_{D(on)} = 0.7 \text{ V}$$

$$R_3 = 4 \text{ k}\Omega \qquad \beta_R = 0.5$$

With the inputs to the gate joined together and connected to a similar gate, what is the direction and value of the current at the input terminals:
(a) With a HIGH level at the input to the test gate.
(b) With a LOW level at the input to the test gate.

3.4.2   The Output Stage

A popular form of a TTL circuit is shown in Fig. 3.19 as a 2-input NAND gate. Of particular interest is the output stage, commonly referred to as a *totem-pole* output. We have just seen the need for a TTL gate to have a low output resistance

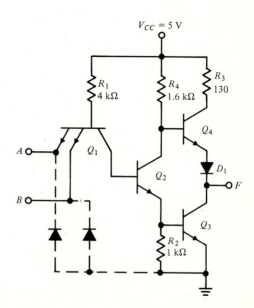

FIGURE 3.19
Circuit diagram of a TTL gate.

with a HIGH-output level. The totem-pole circuit has a low output resistance for both a HIGH and LOW level. All the output circuits we have seen with digital gates have had a low output resistance for a LOW level. The resistance at the collector of a saturated output transistor of many digital ICs is approximately 10 Ω. However, for the HIGH level, the output resistance has been the collector load resistance. For the modified DTL gate this has been 6 or 2 kΩ; for RTL, either 3.6 kΩ or 650 Ω. With the totem-pole output stage, the output HIGH is obtained from an emitter follower (Q_4), and the typical output resistance is less than 100 Ω.

A further advantage of this output stage is the rapid charge and discharge of any capacitance at the gate output, through the low output resistance obtained with the totem-pole circuit. Where the LOW-output level is determined by $V_{CE(sat)}$ of a saturated transistor, the transition time for the output to change from a HIGH to a LOW is generally very fast. The discharge of any load capacitance is through the saturation resistance of the output transistor. However, only with a totem-pole output stage is the transition from a LOW to a HIGH comparably as fast. The load capacitance is then charged through the output resistance of the emitter-follower circuit.†

For the circuit shown in Fig. 3.19 the transistor Q_2, besides serving as a level shifter similar to the diode D_2 in Fig. 3.17b, also acts as a *phase splitter*. That is, the voltage phase at the collector is 180° out of phase with that at the emitter, since the collector has the inverted signal. When there is a LOW level at the emitter of Q_2 to turn off Q_3, there is a HIGH level at the collector of Q_2 to turn on Q_4, and vice versa. The diode D_1 is required to avoid an indeterminate output level, as can be seen from the following explanation. With a LOW level at the output, both transistors Q_2 and Q_3 are saturated. Their collector voltages, referred to ground, are approximately 0.9 and 0.2 V, respectively. In the absence of D_1, these voltages appear at the base and emitter, respectively, of transistor Q_4. The voltage from base to emitter is sufficient to turn Q_4 on. Current will then be diverted from the collector of Q_2 to the base of Q_4. Consequently, transistor Q_2 will come out of saturation and both Q_3 and Q_4 will be conducting. Whether the output level is LOW or HIGH is indeterminate. However, including diode D_1 provides the necessary voltage offset to ensure that in the static state Q_4 is off when Q_2 and Q_3 are on.

The function of the resistor R_3 in this design can be best seen by considering what would happen if R_3 was zero; i.e., the collector of Q_4 is connected directly to V_{CC}. With both gate inputs HIGH, the output is at a LOW, and Q_3 is saturated. With one of the inputs going LOW, transistor Q_2 turns off and its collector rises and turns on Q_4. Now if Q_3 is slow in coming out of saturation, there is the distinct possibility of both Q_3 and Q_4 being on, and we briefly have almost a short circuit from V_{CC} to ground. The inclusion of R_4 serves to limit this current, but

---

† The rapid transition of the voltage waveform at the output of the totem-pole circuit can cause excessive "ringing" on the signal lines connecting one TTL gate to another. Most TTL ICs have clamping diodes at their input terminals (shown dotted in Fig. 3.19) to minimize this ringing.

notice that the current can approach about 30 mA as the output switches from a LOW to a HIGH.† That is,

$$I = \frac{V_{CC} - (V_{CE3(sat)} + V_{D1(on)} + V_{CE4(sat)})}{R_3} \tag{3.33}$$

$$= \frac{5 - (0.2 + 0.7 + 0.2)}{0.13} = 30 \text{ mA}$$

This large change of current can cause large voltage transients ("spikes") on the V_{CC} supply lines.

A disadvantage of the totem-pole output is that *collector logic*‡ is not allowed. With a collector load resistance at the output, as with RTL and DTL, a useful technique (that sometimes can save on the number of logic gates used) is to connect one or more collector outputs together to form a logic function. With the totem-pole output, and a low output resistance for both the HIGH and LOW, this is not allowed. One would have a low-resistance source voltage (at the HIGH) feeding into a large current sink (a LOW), with an unpredictable resulting output voltage level.

3.4.3 Fanout of TTL Gate

When considering the fanout of a DTL gate driving similar gates, we saw that the output of the driving gate acts as a current sink for the load gates. Our concern at that time was how much current the output transistor of the driving gate could sink. We have a similar situation in considering the fanout of a TTL gate to other TTL gates when the output of the driving gate is at a LOW level. However, because of the bilateral action of the input transistor in a TTL gate, we must also consider what effect the fanout has on the output of the driving TTL gate at a HIGH level.

With the output of a driving gate, similar to that shown in Fig. 3.19, at a LOW level, we need to calculate the base current of Q_3 in the driving gate. For this logic state, transistors Q_2 and Q_3 are both saturated and the base-collector junction of Q_1 is forward-biased. Referred to ground, the voltage at the base of Q_1 is therefore

$$V_{B1} = 2V_{BE(sat)} + V_{BC(sat)} \tag{3.34}$$

with $V_{BE(sat)} = 0.7$ V and $V_{BC(sat)} = 0.5$ V,

$$V_{B1} = 1.9 \text{ V}$$

---

† R_4 also limits the short-circuit dc current that would flow if the output was grounded accidentally while any input is LOW.

‡ Collector logic is one form of *wired logic*. For *n-p-n* transistors and positive logic, it is referred to as *wired-AND* (see Sec. 4.5.2).

The voltage across R_1 is therefore known, and the base current of Q_1 can readily be determined; namely,

$$I_{B1} = \frac{V_{CC} - V_{B1}}{R_1} \tag{3.35}$$

$$= \frac{5 - 1.9}{4} = 0.78 \text{ mA}$$

For these operating conditions, transistor Q_2 is saturated and its collector voltage is $V_{CE(\text{sat})}$ above the emitter voltage which, in turn, is equal to $V_{BE(\text{sat})}$ of Q_3. The collector saturation current of Q_2 can then be calculated as

$$I_{C2} = \frac{V_{CC} - (V_{CE(\text{sat})} + V_{BE(\text{sat})})}{R_4} \tag{3.36}$$

$$= \frac{5 - (0.2 + 0.7)}{1.6} = 2.56 \text{ mA}$$

If we assume $\beta_R = 0.2$ for transistor Q_1, a β_F of only 2.8 is required of transistor Q_2 for it to saturate. That is, for Q_2,

$$\beta_F = \frac{I_{C2}}{I_{B2}} = \frac{I_{C2}}{(\beta_R + 1)I_{B1}}$$

$$= \frac{2.56}{(1.2)(0.78)} = 2.8$$

Again referring to Fig. 3.19, the base current of Q_3 is equal to the emitter current of Q_2 less the current in R_2. The emitter current of Q_2 is the sum of the base and collector currents of Q_2, and the current through R_2 must be sufficient to develop $V_{BE(\text{sat})}$ across the base-emitter junction of Q_3:

$$I_{B3} = I_{B2} + I_{C2} - \frac{V_{BE(\text{sat})}}{R_2} \tag{3.37}$$

$$= 0.93 + 2.56 - \frac{0.7}{1} = 2.79 \text{ mA}$$

Assuming $\beta_F = 20$, we can now calculate the collector current of the output transistor Q_3 in the driving gate:

$$I_{C3} = \beta_F I_{B3}$$

$$= (20)(2.79) = 55.8 \text{ mA}$$

To determine the load current† (I_E) we will assume there is no collector current in the load-gate transistor Q_1 when it is in the forward mode. The load current is then simply the current through resistor R_1 to the forward-biased base-emitter

---

† The analogous current was I_D in DTL, but now it is an emitter current, so we refer to it as I_E.

junction of Q_1. With one gate driving a similar gate, the input voltage to the load gate is $V_{CE(\text{sat})}$, and the load current is

$$I_E = \frac{V_{CC} - (V_{CE(\text{sat})} + V_{BE(\text{sat})})}{R_1} \tag{3.38}$$

$$= \frac{5 - (0.2 + 0.7)}{4} = 1.02 \text{ mA}$$

Notice with the totem-pole output circuit and Q_3 conducting, transistor Q_4 is cut off, and there is no contribution to the collector current of Q_3 from transistor Q_4. With the output LOW the fanout (N_L) is then the ratio of the collector saturation current of the output transistor Q_3 and the load current of the driven gate:

$$N_L = \frac{I_{C3}}{I_E} \tag{3.39}$$

$$= \frac{55.8}{1.02} = 54$$

We must now see how the fanout of a gate is affected when the input transistor of a load gate is operating in the inverse mode. We saw in Sec. 3.4.1 how the current flowing into the input terminal of the load gate reduced the voltage of the output HIGH level. Now we will arbitrarily set a minimum output HIGH voltage level of 2.4 V. With driving-gate transistor Q_4 conducting, the voltage at the base of Q_4 is two diode voltage drops above the voltage at the output; that is, $2(0.7 \text{ V}) + 2.4 \text{ V} = 3.8 \text{ V}$. For this operating condition, transistor Q_2 is cut off and the voltage drop across resistor R_4 is due to the base current of Q_4. The emitter current of Q_4 is equivalent to $(\beta_F + 1)(I_{B4})$; then, assuming $\beta_F = 20$,

$$I_{E4} = (\beta_F + 1)\frac{V_{CC} - V_{B4}}{R_4} \tag{3.40}$$

$$= 21\frac{5 - 3.8}{1.6} = 16.8 \text{ mA}$$

Before giving this as the source current from the gate we should check the collector saturation current of Q_4. With Q_4 saturated, the voltage at its collector is $V_{CE(\text{sat})}$ above that at the emitter. The collector saturation current is

$$I_{C4} = \frac{V_{CC} - (V_{CE(\text{sat})} + V_{D(\text{on})} + V_{OH})}{R_3} \tag{3.41}$$

$$= \frac{5 - (0.2 + 0.7 + 2.4)}{0.13} = 13.1 \text{ mA}$$

The base current of Q_4 is

$$I_{B4} = \frac{V_{CC} - V_{B4}}{R_4}$$

$$= \frac{5 - 3.8}{1.6} = 0.75 \text{ mA}$$

and then

$$I_{E4} = I_{B4} + I_{C4} \tag{3.42}$$

$$= 0.75 + 13.1 = 13.8 \text{ mA}$$

Therefore, for these parameter values, the maximum source current from the output of the gate, with a minimum V_{OH} of 2.4 V, is not 16.8 mA, but rather it is 13.8 mA.

The input current to a load gate, with the input at a HIGH level, is $\beta_R(I_{B1}/M)$. We will assume a typical β_R of 0.2. The base current of the input transistor Q_1 acting in the inverse mode has already been determined. From Eq. (3.35), I_{B1} is 0.78 mA. With the output at a HIGH level the fanout (N_H) is the ratio of the source current of Q_4 in the driving gate and the input current of Q_1 in the load gate.

$$N_H = \frac{MI_{E4}}{\beta_R I_{B1}} \tag{3.43}$$

$$= \frac{(2)(13.8)}{(0.2)(0.78)} = 177$$

For the TTL gate circuit shown in Fig. 3.19, and with the typical values used in these calculations, the fanout is 54, determined by the LOW-output condition. The β_R of the input transistor is generally sufficiently low that the conclusion $N_H > N_L$ is common in most IC TTL gate circuits.

3.4.4 Series 54/74 TTL Gate

The most widely used form of TTL circuit is shown on the data sheet in Fig. 3.20. In this section we will review the electrical characteristics given on the data sheet and compare them with typical measured values as well as calculated nominal values. For purposes of comparison these values are listed in Table 3.4.

The recommended operating V_{CC} for the Series 54 TTL gate† is 5.0 V ± 10 percent. This is similar to the 930 Series DTL gates. Also similar is the operating ambient temperature range of −55 to +125°C. The maximum fanout for a gate driving a similar gate is given as 10.

---

† Series 74 TTL gate provides the same function but over a smaller temperature range of 0 to 70°C.

CIRCUIT TYPES SN5400, SN7400
QUADRUPLE 2-INPUT POSITIVE NAND GATES

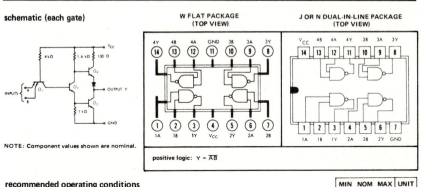

schematic (each gate)

NOTE: Component values shown are nominal.

W FLAT PACKAGE (TOP VIEW)

J OR N DUAL-IN-LINE PACKAGE (TOP VIEW)

positive logic: $Y = \overline{AB}$

recommended operating conditions

		MIN	NOM	MAX	UNIT
Supply Voltage V_{CC}:	SN5400 Circuits	4.5	5	5.5	V
	SN7400 Circuits	4.75	5	5.25	V
Normalized Fan-Out From Each Output, N			10	
Operating Free-Air Temperature Range, T_A:	SN5400 Circuits	−55	25	125	°C
	SN7400 Circuits	0	25	70	°C

electrical characteristics over recommended operating free-air temperature (unless otherwise noted)

	PARAMETER	TEST FIGURE	TEST CONDITIONS[†]		MIN	TYP[‡]	MAX	UNIT
$V_{in(1)}$	Logical 1 input voltage required at both input terminals to ensure logical 0 level at output	1			2			V
$V_{in(0)}$	Logical 0 input voltage required at either input terminal to ensure logical 1 level at output	2					0.8	V
$V_{out(1)}$	Logical 1 output voltage	2	V_{CC} = MIN, V_{in} = 0.8 V, I_{load} = −400 μA		2.4	3.3		V
$V_{out(0)}$	Logical 0 output voltage	1	V_{CC} = MIN, V_{in} = 2 V, I_{sink} = 16 mA			0.22	0.4	V
$I_{in(0)}$	Logical 0 level input current (each input)	3	V_{CC} = MAX, V_{in} = 0.4 V				−1.6	mA
$I_{in(1)}$	Logical 1 level input current (each input)	4	V_{CC} = MAX, V_{in} = 2.4 V				40	μA
			V_{CC} = MAX, V_{in} = 5.5 V				1	mA
I_{OS}	Short-circuit output current[§]	5	V_{CC} = MAX	SN5400	−20		−55	mA
				SN7400	−18		−55	
$I_{CC(0)}$	Logical 0 level supply current	6	V_{CC} = MAX, V_{in} = 5 V			12	22	mA
$I_{CC(1)}$	Logical 1 level supply current	6	V_{CC} = MAX, V_{in} = 0			4	8	mA

switching characteristics, V_{CC} = 5 V, T_A = 25°C, N = 10

	PARAMETER	TEST FIGURE	TEST CONDITIONS		MIN	TYP	MAX	UNIT
t_{pd0}	Propagation delay time to logical 0 level	65	C_L = 15 pF,	R_L = 400 Ω		7	15	ns
t_{pd1}	Propagation delay time to logical 1 level	65	C_L = 15 pF,	R_L = 400 Ω		11	22	ns

[†] For conditions shown as MIN or MAX, use the appropriate value specified under recommended operating conditions for the applicable device type.

[‡] All typical values are at V_{CC} = 5 V, T_A = 25°C.

[§] Not more than one output should be shorted at a time.

FIGURE 3.20
Data sheet of TTL gate (*Texas Instruments Inc.*).

Input voltage   The gate-input-voltage threshold level listed on the data sheet are the minimum and maximum characteristics, which is emphasized by the measurement being made at the lower limit of V_{CC}. For the gate to make the transition from a HIGH- to a LOW-output level, it is necessary that the base-emitter junction of both transistors Q_2 and Q_3 and the base-collector junction of Q_1 be forward-biased. In this case, with $V_{BE(sat)}$ equal to 0.7 V and $V_{BC(sat)}$ equal to 0.5 V, the voltage at the base of Q_1 must be 1.9 V. The gate-input threshold level will then be one diode voltage drop less than 1.9 V, that is, 1.2 V. As we are assuming a fixed value for the forward-biased junction voltages, V_{IH} and V_{IL} will both be equal to 1.2 V. For any input voltage less than this, transistor Q_1 is operating in the forward mode with a resulting HIGH level at the gate output. With any input voltage greater than 1.2 V, transistor Q_1 operates in the inverse mode, and the result is a LOW level at the output of the gate. Notice the typical value of 1.2 V lies nicely between the minimum V_{IH} and the maximum V_{IL} given† on the data sheet.

Output voltage   The maximum V_{OL} and the minimum V_{OH} for the 5400 TTL gate are similar to the values given for the 930 DTL gate. A typical value for V_{OL} is the $V_{CE(sat)}$ of the output transistor Q_3, that is, 0.2 V. The value for V_{OH} given on the data sheet is with a source current of 0.4 mA. That is, with transistor Q_3 off, there is current of 0.4 mA coming from the emitter of Q_4 through the diode D_1. For this condition the base current of Q_4 is determined with $\beta_F = 20$ as

$$I_{B4} = \frac{I_{E4}}{\beta_F + 1}$$

$$= \frac{0.4}{21} \simeq 0.02 \text{ mA}$$

Table 3.4   COMPARISON OF THE ELECTRICAL CHARACTERISTICS OF THE SN 5400 TTL GATE AT AN AMBIENT TEMPERATURE OF 25°C

		Calculated nominal values	Typical measured values	Worst-case values
Input voltage	V_{IL}	1.2 V	1.2 V	0.8 V max
	V_{IH}	1.2 V	1.4 V	2.0 V min
Output voltage	V_{OL}	0.2 V	0.2 V	0.4 V max (0.2 V typical)
	V_{OH}	3.1 V	3.1 V	2.4 V min (3.3 V typical)
Low-level input current	I_{IL}	−1.0 mA	−1.1 mA	−1.6 mA max
Short-circuit output current	I_{OS}	−34 mA	−34 mA	−55 mA max
V_{CC} supply current (per gate)	I_{CCL}	3.3 mA	3.6 mA	5.5 mA max (3 mA typical)
	I_{CCH}	1.0 mA	1.1 mA	2.0 mA max (1 mA typical)

† The conversion between our notation and the data sheet is: $V_{IH} \equiv V_{in(1)}$, $V_{IL} \equiv V_{in(0)}$, $V_{OH} \equiv V_{out(1)}$, $V_{OL} \equiv V_{out(0)}$.

With an output HIGH level, transistor Q_2 is also off, so the voltage drop across R_4 is solely due to I_{B4}. The voltage at the base of Q_4 can then be calculated; namely,

$$V_{B4} = V_{CC} - I_{B4}R_4 \tag{3.44}$$

$$= 4.5 - (0.02)(1.6) = 4.47 \text{ V}$$

The output voltage is two diode voltage drops below the base voltage of Q_4.

$$V_{OH} = V_{B4} - 2V_{BE} \tag{3.45}$$

$$= 4.47 - 1.4 = 3.07 \text{ V}$$

The calculated output voltages agree with the typical values on the data sheet.

Low-level input current The logic 0 level input current ($I_{IL} \equiv I_{\text{in}(0)}$) given on the data sheet is again a worst-case characteristic. The nominal load current I_E, which is similar to I_{IL}, has already been determined by Eq. (3.38).

Short-circuit output current The short-circuit output current (I_{OS}) is measured from an output HIGH level with the gate output shorted to ground. This current is an output source current, and the calculation is therefore similar to Eq. (3.40), except the output voltage level is now 0 V.

$$I_{E4} = (\beta_F + 1)\frac{V_{CC} - (V_{OH} + 2V_{BE(\text{on})})}{R_4}$$

$$= 21\frac{5 - 1.4}{1.6} = 47 \text{ mA}$$

Before giving this as the value for I_{OS}, we should again check the collector saturation current of Q_4, this time with V_{OH} as 0 V. Using Eq. (3.41),

$$I_{C4} = \frac{V_{CC} - (V_{CE(\text{sat})} + V_{D(\text{on})} + V_{OH})}{R_3}$$

$$= \frac{5 - (0.2 + 0.7)}{0.13} = 31.6 \text{ mA}$$

Now the base current of Q_4 is

$$I_{B4} = \frac{V_{CC} - V_{B4}}{R_4}$$

$$= \frac{5 - 1.4}{1.6} = 2.2 \text{ mA}$$

and then

$$I_{E4} = I_{B4} + I_{C4}$$

$$= 31.6 + 2.2 = 33.8 \text{ mA}$$

Hence, with the output shorted to ground, for nominal circuit values the output transistor saturates and I_{OS} is equal to -34 mA. Recall that the current out of a terminal has a negative value.

V_{CC} **supply current**   To find $I_{CCL}(\equiv I_{CC(0)})$, the V_{CC} supply current when the output of the gate is at a LOW level, we must calculate the collector current of Q_2, and the base current of Q_1 when Q_1 is operating in the inverse mode. Transistor Q_3 will also be conducting for this condition, but because Q_4 is cut off, there will be no current to the collector of Q_3 from the V_{CC} of this particular gate. The base current of Q_1, in the inverse mode, was calculated in Eq. (3.35); namely,

$$I_{B1} = \frac{V_{CC} - V_{B1}}{R_1}$$

$$= \frac{5 - 1.9}{4} = 0.78 \text{ mA}$$

The collector saturation current of Q_2 was determined in Eq. (3.36); that is,

$$I_{C2} = \frac{V_{CC} - (V_{CE(\text{sat})} + V_{BE(\text{sat})})}{R_4}$$

$$= \frac{5 - (0.2 + 0.7)}{1.6} = 2.56 \text{ mA}$$

The I_{CCL} for a single gate is the sum of these two currents:

$$I_{CCL} = I_{B1} + I_{C2}$$

$$= 0.78 + 2.56 = 3.34 \text{ mA}$$

The $I_{CCH}(\equiv I_{CC(1)})$ is measured with a HIGH level at the output of the gate. In this condition, both transistors Q_2 and Q_3 are cut off. Transistor Q_4 will be conducting, but with no connection at the output of the gate the only current in Q_4 will be the collector leakage current of Q_3, which is very small. The only current from V_{CC} is the base current of Q_1 operating in the forward mode. For the typical case this is the load current I_E. A value for I_E is given by Eq. (3.38):

$$I_E = \frac{V_{CC} - (V_{CE(\text{sat})} + V_{BE(\text{sat})})}{R_1}$$

$$= \frac{5 - (0.2 + 0.7)}{4} = 1.02 \text{ mA}$$

This, then, is I_{CCH} for a single gate:

$$I_{CCH} = I_E$$

$$= 1.02 \text{ mA}$$

Power dissipation   With the nominal value of V_{CC} at 5 V, we may calculate the average power dissipation for a single gate. We again assume a 50 percent duty cycle for the gate.

$$PD_{av} = V_{CC} \frac{I_{CCL} + I_{CCH}}{2}$$

$$= 5\frac{3 + 1}{2} = 10 \text{ mW}$$

The typical propagation delay times given on the data sheet will be used to determine the speed-power product for the 5400 TTL gate.

$$t_{pd}PD_{av} = (9 \text{ ns})(10 \text{ mW}) = 90 \text{ pJ}$$

For the popular DTL and TTL circuits, the power dissipation per logic gate is about the same, but the TTL gate is about a factor of 4 faster. However, the logic levels of the two circuits are compatible as is the single power supply, which is usually 5 V. The units may therefore be intermixed in a logic circuit or system.

54H and 54L TTL gate   There are several variations of interest in the Series 54 TTL gates. The circuit of these various gates is essentially similar to that shown in Fig. 3.19, but the values of the resistors are changed. The 54H is a high-speed series of logic gates with a typical propagation delay time of 6 ns. The resistor values for these gates are about one-half those shown in Fig. 3.19. The result is that the power dissipation of the Series 54H is about twice that for the Series 54. The power dissipation of the 54H is typically 22 mW per gate. The Series 54L is a low-power version of the TTL gate, with a typical power dissipation of only 1 mW per gate. The resistor values are about 10 times those shown in Fig. 3.19. The propagation delay time of the 54L gate is typically 33 ns.

With this we conclude the part on saturating logic gate circuits. A short review of the characteristics of the circuits so far described will be found in Sec. 3.8, in Table 3.5, where a comparison is made between these saturating digital ICs and other types of IC logic gates.

EXERCISE

E3.8   With a circuit similar to that shown in Fig. 3.19, the nominal element values of a Series 54L gate are shown below:

$$R_1 = 40 \text{ k}\Omega \qquad V_{CC} = 5 \text{ V} \qquad V_{CE(sat)} = 0.2 \text{ V}$$
$$R_2 = 12 \text{ k}\Omega \qquad \beta_F = 20$$
$$R_3 = 500 \text{ }\Omega \qquad \beta_R = 0.5 \qquad V_{BE(sat)} = V_{D(on)}$$
$$R_4 = 20 \text{ k}\Omega \qquad \qquad = 0.7 \text{ V}$$

(a) Calculate the fanout of the gate for a LOW- and HIGH-output level (assume $V_{OH} = 2$ V).

(b) Calculate the V_{CC} supply currents I_{CCL} and I_{CCH} and determine the average power dissipation of a single gate.

PART II

3.5   NONSATURATING LOGIC CIRCUITS

With saturating circuits, one of the logic levels is nicely defined by the transistor collector-emitter saturation voltage but at the cost of extra time delay due primarily to the saturation delay time of the transistor inverter. An obvious solution is to avoid saturation. Three possible solutions are shown in Fig. 3.21.

The transistor inverter of Fig. 3.21a is an all-silicon circuit. Current in diodes D_{2A} and D_{2B} sets the *clamp* voltage to about 1.4 V. With the transistor Q_1 cut off, the collector voltage is at V_{CC}, the diode D_1 is reverse-biased, and the clamp voltage is essentially disconnected from the transistor collector. When the transistor turns on, the voltage at the collector falls until D_1 becomes forward-biased. Then the collector voltage is one diode voltage drop less than the clamp voltage. That is, V_{CE} of the transistor is about 0.7 V. The collector is clamped at this voltage, and the transistor is held out of saturation, providing the diodes D_{2A} and D_{2B} are always forward-biased. Notice with D_1 forward-biased, the collector current of Q_1 subtracts from the diode current of D_2. It is therefore necessary to maintain a large standby current in diode D_2. Therefore, this scheme does require more current than the simple saturating inverter.

The circuit of Fig. 3.21b makes use of the low voltage drop of a germanium diode (0.3 V, compared to 0.7 V for a silicon diode). With the transistor Q_1 cut off, the collector voltage is at V_{CC} and D_1 is reverse-biased. When the transistor turns on, the collector voltage falls until D_1 becomes forward-biased and the collector is

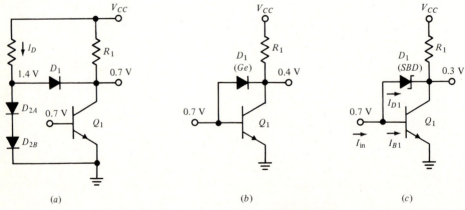

(a)   (b)   (c)

FIGURE 3.21
Nonsaturating transistor-inverter circuits.

clamped at about 0.3 V less than the base voltage. The base-collector junction of Q_1 is forward-biased by about 0.3 V, but with a silicon junction only a very small current flows and the transistor is effectively still out of saturation. In this circuit, the extra base current which would cause the transistor to saturate is diverted through the diode D_1 and then through the collector of Q_1. The drawback to this scheme is the technical problem of integrating a germanium diode into a silicon technology. This has been done, but not on a mass-fabrication scale.

The circuit of Fig. 3.21c is exactly the same as Fig. 3.21b, but the germanium diode has been replaced by a Schottky barrier diode (SBD). By the proper choice of metal deposited on the silicon, a metal-silicon junction diode can be formed with characteristics very similar to a germanium diode. The process is relatively simple, and digital ICs using this device are available.

The Series 54S TTL Gate circuit is essentially the same as the Series 54H already described, but SBDs are connected between the collector and base of each transistor to prevent the transistor from saturating. The power dissipation of a 54S gate is the same as a similar 54H gate, namely, 22 mW per gate, but the typical propagation delay time is reduced from 6 to 3 ns. A low-power version of the SBD clamped TTL gate is also available, the 54LS. The typical propagation delay time of this gate is 10 ns (approximately the same as the standard Series 54), but the power dissipation is only 2 mW per gate, compared with 10 mW per gate for the Series 54.

EXAMPLE 3.3   Determine the current in a forward-biased SBD connected across the base-collector junction of the output transistor of a simple TTL gate, as shown in Fig. 3.21c. Assume

$$I_{in} = 1.44 \text{ mA} \qquad R_1 = 6 \text{ k}\Omega$$

$$V_{CC} = 5 \text{ V} \qquad \beta_F = 20$$

SOLUTION   For the new circuit, let I_{D1} be the current in the SBD and I_{B1} the new base current for the output transistor Q_1. The input current to the new output stage is 1.44 mA. That is,

$$I_{in} = I_{B1} + I_{D1} = 1.44 \text{ mA} \qquad (3.46)$$

With the SBD forward-biased, the current through the collector load resistor R_1 can be calculated:

$$I_{R1} = \frac{V_{CC} - V_{CE}}{R_1} \qquad (3.47)$$

$$= \frac{5 - 0.3}{6} = 0.78 \text{ mA}$$

Now with no other load at the collector we can determine the collector current. By Kirchhoff's current law,

$$I_{C1} = I_{R1} + I_{D1}$$

The transistor is not saturated so the base current is given by

$$I_{B1} = \frac{I_{C1}}{\beta_F} = \frac{I_{R1} + I_{D1}}{\beta_F} \qquad (3.48)$$

Combining Eqs. (3.46) and (3.48), we can obtain a value for I_{D1}; that is,

$$\frac{I_{R1} + I_{D1}}{\beta_F} + I_{D1} = I_{\text{in}}$$

$$\frac{0.78 + I_{D1}}{20} + I_{D1} = 1.44 \text{ mA}$$

hence

$$I_{D1} = 1.33 \text{ mA} \qquad ////$$

Notice the SBD has no effect on the fanout of the gate. The output transistor can still sink the maximum fanout current:

$$I_{C1} = \beta_F I_{B1}$$

$$= (20)(1.44) = 28.8 \text{ mA}$$

The current through the SBD is now zero. That is, as the loading on the output varies from no load to maximum fanout, the current through the SBD changes from 1.33 mA to zero.

3.6 EMITTER-COUPLED LOGIC

Another solution to the saturation-delay-time problem avoids saturation with the circuit shown in Fig. 3.22. This is the so-called current switch and is the basic gate circuit of *emitter-coupled logic* (ECL), also referred to as *current-mode logic* (CML). The basic circuit consists of two identical transistors with their emitters connected together to a common current source I_E. Notice, at the input of this circuit, the LOW level is -1 V and the HIGH level is $+1$ V. The base of transistor Q_2 is connected directly to ground. The *reference voltage* V_R is therefore 0 V. Now, V_{in} is either less than V_R (that is, $V_{\text{in}} = -1$ V), or V_{in} is greater than V_R (that is, $V_{\text{in}} = +1$ V). With V_{in} equal to -1 V, the base-emitter junction of Q_2 is forward-biased and the common emitter is at -0.7 V with respect to ground. The base-emitter junction of Q_1 is therefore reverse-biased by 0.3 V. That is, Q_1 is off and its collector voltage is at V_{CC}. All the current I_E is from Q_2, and assuming β_F is very large, the collector voltage of Q_2 is approximately $V_{CC} - I_E R_{C2}$.

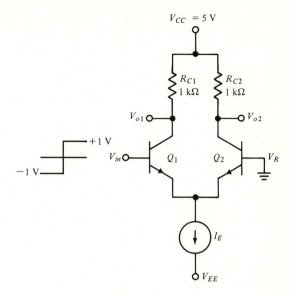

FIGURE 3.22
Circuit diagram of the current switch.

With V_{in} at a **HIGH** level, the base-emitter junction of Q_1 is forward-biased, the common emitter voltage is $+0.3$ V, and now the base-emitter junction of Q_2 is reverse-biased by about 0.3 V. All the emitter current is then from Q_1, and its collector voltage is approximately $V_{CC} - I_E R_{C1}$. With Q_2 off, its collector voltage is at V_{CC}.

With the value of R_{C1} equal to that of R_{C2} the voltage swing at the two collectors is equal, but the outputs are complementary. The output at the collector of Q_2 is commonly referred to as the *in-phase* output, because the voltage moves in the same direction as the input voltage. The output at the collector of Q_1, which moves in the opposite direction to that of the input, is known as the *out-of-phase* output.

Notice at the midpoint of the transition, V_{in} is equal to V_R. The current I_E then divides equally in Q_1 and Q_2. Now recall from Chap. 1 the current through a diode changes a factor of 10 for each change of 60 mV across the junction. With the voltage at the base of Q_1 just 60 mV less than that at the base of Q_2, the current in Q_2 is 10 times that in Q_1, and vice versa. As V_{in} becomes greater than V_R, all the current in Q_2 is switched into Q_1. Hence, the term *current switch*. The transition width $(V_{IH} - V_{IL})$ is about 120 mV, is essentially independent of the transistor parameters, and is centered about the reference voltage V_R.

By a judicious choice of resistance values, this logic block can be made nonsaturating. For the example shown in Fig. 3.22, with an emitter current of 2 mA, the voltage levels at the transistor collector are 5 V when the transistor is cut off and 3 V when it is conducting. At these voltage levels the transistors are out of saturation, but they can cause a problem because we are unable to directly connect the output of one gate to the input of another. However, we have already seen how diodes or the base-emitter junction of a transistor can be used as voltage-level shifters.

Because the transistors do not saturate in ECL circuits, it is possible to fabricate transistors with high current gain ($\beta_F = 100 - 300$). The steady-state base current being so much smaller than the collector current leads to a general approximation, for these circuits, that the collector current is equal to the emitter current. In the practical case, the current source (I_E) is simply a comparatively large-valued resistor (R_E) connected from the junction of the two emitters to V_{EE}. With the difference between V_R and V_{EE} much greater than $V_{BE(on)}$, the emitter current is then approximately given as

$$I_C \approx I_E \approx \frac{V_R - V_{EE}}{R_E} \qquad (3.49)$$

At the output of the current switch, the voltage of the two logic levels is

$$V_{OH} = V_{CC} \qquad (3.50a)$$

$$V_{OL} = V_{CC} - I_C R_C = V_{CC} - \frac{R_C}{R_E}(V_R - V_{EE}) \qquad (3.50b)$$

The voltage levels at the output of an ECL gate are then uniquely defined, essentially independent of the transistor parameters. To a good approximation the LOW level is determined by the ratio of the collector and emitter resistances. We have seen that with an IC the ratio of resistances within the circuit can be controlled better than 2 percent, though the absolute value of the resistances may have a variation of ± 30 percent from the nominal value.

3.6.1   Analysis of ECL Gate

An example of an ECL gate circuit is shown in Fig. 3.23. Transistors Q_1 and Q_2 form the basic current switch. An additional input is provided by paralleling Q_1 with another transistor Q_3. An output is obtained from the common collectors of Q_1 and Q_3 through the voltage-level shifting of the base-emitter junction of the emitter follower Q_4. In a similar fashion, an output is obtained from the collector of Q_2 at the emitter of Q_5.

With a HIGH level at either A or B, transistor Q_1 or Q_3 is conducting and Q_2 is cut off. Consequently, there is a HIGH level at the emitter of Q_5, and the logic expression is

$$\text{OR output} = A + B$$

There is a LOW level at the emitter of Q_4, or

$$\text{NOR output} = \overline{A + B}$$

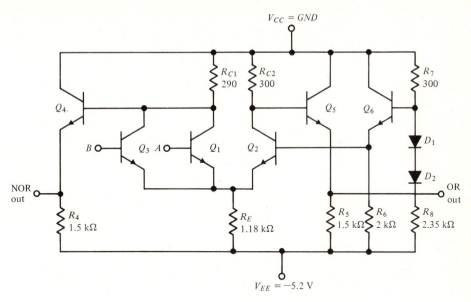

FIGURE 3.23
Circuit diagram of an ECL gate.

With a LOW level at both A and B, both Q_1 and Q_3 are cut off and Q_2 is conducting. Now there is a LOW level at the OR output and a HIGH level at the NOR output. Logically, the gate is described as a 2-input OR-NOR gate.

Notice for this circuit V_{CC} is chosen to be ground and V_{EE} is at -5.2 V. This is because the output voltage levels are closely related to V_{CC} and show less variation in voltage with V_{CC} connected to ground rather than to a voltage supply.

The reference voltage for the current switch is taken from an emitter follower (Q_6). The diodes of the biasing circuit for this transistor help to stabilize the current in R_E against variations in temperature. Any change with temperature of the V_{BE} of Q_6 and Q_2 is compensated by a similar change across the two diodes.

Now we wish to calculate the voltage levels corresponding to the logic LOW- and HIGH-output levels. However, first we must determine the reference voltage, the voltage at the base of Q_2. We will ignore the effects of base current in these calculations and assume 0.75 V for both the forward-biased base-emitter voltage ($V_{BE(on)}$) and the voltage drop across the forward-biased diode ($V_{D(on)}$). With V_{CC} at ground and V_{EE} at -5.2 V, we can determine the voltage at the base of Q_6.

$$V_{B6} = V_{CC} - \frac{R_7}{R_7 + R_8}[V_{CC} - (V_{EE} + 2V_{D(on)})] \qquad (3.51)$$

$$= 0 - \frac{0.3}{0.3 + 2.35}[0 - (-5.2 + 1.5)]$$

$$= -0.42 \text{ V}$$

The voltage at the base of Q_2 is

$$V_{B2} = V_{B6} - V_{BE6(on)} \tag{3.52}$$

$$= -0.42 - 0.75 = -1.17 \text{ V}$$

The reference voltage for the current switch is therefore -1.17 V.

With a LOW level at both A and B, transistor Q_2 is conducting and the common-emitter voltage of the current switch is -1.92 V. Now all the current in R_E is coming from the emitter of Q_2, so the collector current of Q_2 is approximately

$$I_{C2} = I_{E2} = \frac{V_{E2} - V_{EE}}{R_E} \tag{3.53}$$

$$= \frac{-1.92 + 5.2}{1.18} = 2.78 \text{ mA}$$

We can now determine the voltage at the collector of Q_2 and then the LOW-level output voltage at the OR output, the emitter of Q_5.

$$V_{E5} = V_{CC} - I_{C2}R_{C2} - V_{BE5(on)} \tag{3.54}$$

$$= 0 - (2.78)(0.3) - 0.75 = -1.59 \text{ V}$$

For this logic condition, the voltage at the collector of Q_1 and Q_3 is 0 V, and the HIGH-output voltage level at the NOR output is -0.75 V.

Now directly connecting the output of one gate to the input of another, with a HIGH level (-0.75 V) at the base of Q_3, transistor Q_2 is cut off and Q_3 is now conducting all the current in the emitter resistor R_E. The common-emitter voltage is then -1.5 V. The collector current of Q_3 is therefore

$$I_{C3} = I_{E3} = \frac{V_{E3} - V_{EE}}{R_E} \tag{3.55}$$

$$= \frac{-1.5 + 5.2}{1.18} = 3.14 \text{ mA}$$

The voltage at the collector of Q_3 is

$$V_{C3} = V_{CC} - I_{C3}R_{C1} \tag{3.56}$$

$$= 0 - (3.14)(0.29) = -0.91 \text{ V}$$

The LOW-output voltage level at the NOR output is therefore -1.66 V. With transistor Q_2 cut off, the HIGH-output voltage at the OR output is -0.75 V.

We therefore have for this circuit:

At the OR output: $V_{OH} = -0.75 \text{ V}$ $V_{OL} = -1.59 \text{ V}$

At the NOR output: $V_{OH} = -0.75 \text{ V}$ $V_{OL} = -1.66 \text{ V}$

The logic swing is almost symmetrical about the reference voltage (-1.17 V). Thus, we have almost equal noise margins for the HIGH- and LOW-output voltage levels.

Notice that in this circuit the voltage-level shifting is by the V_{BE} of an emitter follower (Q_4 and Q_5). These emitter followers act as low-output-resistance voltage sources. The input to the gate is to the base of high-beta transistors (Q_1 and Q_3). The base current is very small, and the input resistance is very high. We therefore have the possibility of a very high fanout. The fan-in can be increased by paralleling more transistors at the gate input.

3.6.2   MECL II Series ECL Gate

The preceding analysis was made on the basic gate circuit of the MECL II Series that was introduced by Motorola Semiconductor. The data sheet shown in Fig. 3.24 is for the MC1210, a typical gate circuit of the MECL II Series. The MC1210 is a quad 2-input NOR gate. Because of a limitation in the number of pins to the package, the OR outputs are not brought out. Notice that the one reference voltage is supplied to all four gates. When wire-ORing,† by connecting the output of two gates together, the resistors in the emitter circuit of the output emitter follower appear in parallel. One of the *output pulldown* resistors may therefore be omitted by using a 1211 or 1212 gate without affecting the electrical characteristics of the gate, but with a consequent saving in power-drain current.

We will improve our familiarity with the gate circuit if we calculate values for some of the parameters and compare them with the electrical characteristics given on the data sheet. We will also be able to make some interesting conjectures regarding the design of the gate circuit.

Power-supply drain current   In general with other logic families the 1200 Series has one, two, or three gates to a package. Some gates have both OR and NOR outputs; some like the 1210 have only the NOR outputs. Therefore in calculating the power-drain current it is advantageous to consider separately the three portions of the gate circuit: the current switch, the output emitter followers, and the reference voltage supply.

To determine the current in the current switch, we consider the two cases of a LOW and a HIGH at the gate input. With LOW levels at the inputs to the gate, transistor Q_2 (Fig. 3.23) is conducting, and we have already determined this emitter current in Eq. (3.53); namely,

$$I_{E2} = 2.78 \text{ mA}$$

With a HIGH level at the B input, transistor Q_2 is cut off and Q_3 is conducting; the current through the current switch is then obtained from Eq. (3.55):

$$I_{E3} = 3.14 \text{ mA}$$

---

† For ECL gates with positive logic, tying emitters together of emitter followers that provide the OR function gives the wired-OR; that is, whenever an individual output would go HIGH, the composite goes HIGH. This is analogous to the wired-AND (collector logic) discussed in Sec. 4.5.2.

QUAD 2-INPUT GATE   MECL II MC1000/1200 series   *MOTOROLA*

MC1010 thru MC1012
MC1210 thru MC1212

Provide the NOR output function. These devices contain an internal bias reference insuring that the threshold point is always in the center of the transition region over the temperature range.

Emitter follower output configurations differ for these three circuits as shown in the circuit schematic.

MC1010/MC1210 CIRCUIT SCHEMATIC

MC1011/MC1211
Omit pulldown resistors
on pins 3 and 6

MC1012/MC1212
Omit all output
pulldown resistors

Resistor values are nominal.

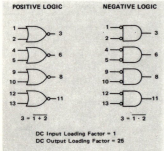

POSITIVE LOGIC   NEGATIVE LOGIC

$3 = \overline{1 + 2}$ $3 = \overline{1 \cdot 2}$

DC Input Loading Factor = 1
DC Output Loading Factor = 25

Power Dissipation: MC1010/MC1210 — 115 mW typical
MC1011/MC1211 — 95 mW typical
MC1012/MC1212 — 65 mW typical

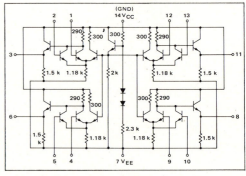

Characteristic	Symbol	Pin Under Test	MC1210-1212 Test Limits							MC1010-1012 Test Limits						
			−55°C		+25°C		+125°C			0°C		+25°C		+75°C		
			Min	Max	Min	Max	Min	Max	Unit	Min	Max	Min	Max	Min	Max	Unit
Power Supply Drain Current	I_E	7														
MC1210/MC1010			-	-	-	32	-	-	mAdc	-	-	-	32	-	-	mAdc
MC1211/MC1011			-	-	-	26	-	-		-	-	-	26	-	-	
MC1212/MC1012			-	-	-	18	-	-		-	-	-	18	-	-	
Input Current	I_{in}	1	-	-	-	100	-	-	µAdc	-	-	-	100	-	-	µAdc
		2	-	-	-	100	-	-	µAdc	-	-	-	100	-	-	µAdc
Input Leakage Current	I_R	Inputs*	-	-	-	0.2	-	1.0	µAdc	-	-	-	0.2	-	1.0	µAdc
"NOR" Logical "1" Output Voltage‡	V_{OH}‡	3	-0.990	-0.825	-0.850	-0.700	-0.700	-0.530	Vdc	-0.895	-0.740	-0.850	-0.700	-0.775	-0.615	Vdc
		3	-0.990	-0.825	-0.850	-0.700	-0.700	-0.530	Vdc	-0.895	-0.740	-0.850	-0.700	-0.775	-0.615	Vdc
"NOR" Logical "0" Output Voltage	V_{OL}	3	-1.890	-1.580	-1.800	-1.500	-1.720	-1.380	Vdc	-1.830	-1.525	-1.800	-1.500	-1.760	-1.435	Vdc
		3	-1.890	-1.580	-1.800	-1.500	-1.720	-1.380	Vdc	-1.830	-1.525	-1.800	-1.500	-1.760	-1.435	Vdc
Switching Times			Typ	Max	Typ	Max	Typ	Max		Typ	Max	Typ	Max	Typ	Max	
Propagation Delay (Fan-Out = 3)	t_{1+3-}	3	4.0	7.5	4.5	7.5	6.0	9.0	ns	4.0	7.5	4.5	7.5	5.5	8.5	ns
	t_{1-3+}		5.0	7.0	5.0	7.0	6.0	9.0		5.0	7.0	5.0	7.0	5.5	8.0	
(Fan-Out = 15)	t_{1+3-}		18	-	18	-	22	-		18	-	18	-	20	-	
	t_{1-3+}		6.0	-	6.0	-	9.0	-		6.0	-	6.0	-	7.0	-	
Rise Time (Fan-Out = 3)	t_{3+}		4.0	7.5	4.0	7.0	5.0	8.0		4.0	7.0	4.0	7.0	4.5	7.5	
Fall Time (Fan-Out = 3)	t_{3-}		6.0	8.5	6.0	8.0	7.0	10		6.0	8.0	6.0	8.0	6.5	9.0	

FIGURE 3.24
Data sheet of ECL gate (*Motorola Inc.*).

The average current-switch supply current is therefore one-half the sum of I_{E2} and I_{E3}, or

$$\frac{2.78 + 3.14}{2} = 2.96 \text{ mA}$$

Now for the emitter followers. When the NOR output is at a HIGH level the current in the emitter follower Q_4 is simply found as

$$I_{E4} = \frac{V_{OH} - V_{EE}}{R_4} \tag{3.57a}$$

$$= \frac{-0.75 + 5.2}{1.5} = 2.96 \text{ mA}$$

With a LOW level at the NOR output, the emitter current of Q_4 is

$$I_{E4} = \frac{-1.66 + 5.2}{1.5} = 2.36 \text{ mA} \tag{3.57b}$$

The average emitter-follower current is therefore

$$\frac{2.96 + 2.36}{2} = 2.66 \text{ mA}$$

To determine the reference-voltage supply current we must add the emitter current of Q_6 to the current in the diodes D_1 and D_2. The voltage at the emitter of Q_6 is the reference voltage (-1.17 V). We may solve for the emitter current of Q_6 as

$$I_{E6} = \frac{V_R - V_{EE}}{R_6} \tag{3.58a}$$

$$= \frac{-1.17 + 5.2}{2} = 2.02 \text{ mA}$$

For the diode current

$$I_{D1} = I_{D2} = \frac{V_{CC} - (V_{EE} + 2V_{D(on)})}{R_7 + R_8} \tag{3.58b}$$

$$= \frac{0 + 5.2 - 1.5}{0.3 + 2.35} = 1.4 \text{ mA}$$

We therefore have the total reference-voltage supply current:

$$I_{VR} = I_{E6} + I_{D1} \tag{3.58c}$$

$$= 2.02 + 1.4 = 3.42 \text{ mA}$$

Now the typical total drain current for an MC1210 package consisting of four current switches, four emitter followers, and one voltage reference, is equal to

$$4(2.96 + 2.66) + 3.42 = 25.9 \text{ mA}$$

This compares with the 32 mA *maximum* at 25°C given on the data sheet of Fig. 3.24.

The average power dissipation for an MC1210, using these nominal values, is

$$(5.2)(25.9) = 134 \text{ mW}$$

This is to be compared with the typical value of 115 mW given on the data sheet.

Input current   The input current to the gate is the base current of either Q_1 or Q_3 for the circuit shown in Fig. 3.23. We have already determined that with a **HIGH** level at the base of Q_3 the emitter current of Q_3 is 3.14 mA [Eq. (3.55)]. Now, from the data sheet, the maximum input current is 100 μA at 25°C. We may therefore use these figures to determine a likely minimum beta (β_F) for the transistors used in these MECL II gate circuits.

$$I_B = \frac{I_E}{\beta_F + 1}$$

or

$$\beta_F = \frac{3.14}{0.1} - 1 \simeq 30$$

As has already been indicated, the typical β_F of these transistors is more likely to be about 150. However, we can use a value for β_F of 30 and a base current for the input transistor of 100 μA to later determine a value for the fanout of the gate.

Noise margin   We have seen that the midpoint for transition of the gate input voltage is equal to the reference voltage of the current switch. Further, we have seen that with a difference of 60 mV between the input and the reference voltage the current in the two transistors of the current switch is in the ratio 10 : 1. Now if we allow a transition width of 240 mV, the current in the off-going transistor will be only 0.01 percent of that in the on-going transistor. The input voltage levels are given below for a reference voltage of -1.17 V.

$$V_{IH} = V_R + 0.12 \text{ V} = -1.05 \text{ V}$$
$$V_{IL} = V_R - 0.12 \text{ V} = -1.29 \text{ V}$$

The manufacturer's minimum and maximum values are

$$V_{IH(min)} = -1.02 \text{ V}$$
$$V_{IL(max)} = -1.32 \text{ V}$$

Now the typical noise margin is the magnitude of the difference between the input and output voltage levels for the two logic states.

$$NM_H = V_{IH} - V_{OH}$$

$$= 1.02 - 0.75 = 0.27 \text{ V}$$

$$NM_L = V_{OL} - V_{IL}$$

$$= 1.66 - 1.32 = 0.34 \text{ V}$$

Fanout   The input current of the load gates is supplied by the emitter follower at the output of the driving gate. The limitation to the fanout is set by the high noise margin going to zero. That is, with maximum fanout, the *output* HIGH level is reduced to the minimum *input* HIGH level. For this logic condition, both transistors Q_1 and Q_3 (Fig. 3.23) of the driving gate are off. The increased voltage drop across the resistance, R_{C1}, in the driving gate is therefore due to the base current necessitated by the extra load current from the output emitter follower. We will calculate the fanout of a typical MC1210, using the nominal values shown in Fig. 3.23.

With a minimum V_{IH} of -1.02 V, the voltage at the base of Q_4 is -0.27 V. The maximum base current of Q_4 is, therefore,

$$I_{B4} = \frac{V_{CC} - V_{B4}}{R_{C1}} \tag{3.59}$$

$$= \frac{0 + 0.27}{0.29} = 0.93 \text{ mA}$$

The maximum output current (I_{output}) from transistor Q_4 is equal to the emitter current of Q_4 minus the current through the resistor R_4. Assuming β_F equal to 30,

$$I_{\text{output}} = (\beta_F + 1)I_{B4} - \frac{V_{IH} - V_{EE}}{R_4} \tag{3.60}$$

$$= (31)(0.93) - \frac{-1.02 + 5.2}{1.5} = 26.1 \text{ mA}$$

Now taking the worst-case condition for the gate input current as 100 μA, the fanout of the gate is given by

$$N = \frac{I_{\text{output}}}{I_{\text{input}}} \tag{3.61}$$

$$= \frac{26.1}{0.1} = 261$$

The dc output loading factor given in the data sheet for the MC1210 is 25. This figure for the fanout takes into consideration a complete worst-case analysis, including the temperature variations of -55 to $+125°$C. However as with all extremely fast logic gates, the maximum fanout is not due to the dc loading factor, but rather it is limited by the total capacitive load that a gate can drive in a given time. The total capacity at the output of the driving gate is the sum of the load-gate capacitances. If the maximum dc fanout were used, the resulting large load capacity would increase the propagation delay time of the gate, which would then be so large that the speed advantage of the ECL would be lost.

Speed-power product With a fanout of 3, the typical propagation delay time of the MC1210 is given on the data sheet as 5 ns. This increases to 12 ns for a fanout of 15.

The typical power dissipation of the complete package, given on the data sheet, is 115 mW. For our purposes, we will assume an average power dissipation of $\frac{115}{4}$ mW for each of the gates in the package. For a fanout of 3, the speed-power product of this MECL II gate is therefore

$$t_{pd}PD_{av} = (5 \text{ ns})(29 \text{ mW}) = 145 \text{ pJ}$$

Faster versions of this ECL gate are also available, including MECL III, MECL 10,000, and FSC 9500 Series. The circuit for the MECL III gate is similar to that shown in Fig. 3.24, but the values of the resistances are changed and faster switching transistors are used. The typical propagation delay time of an MECL III gate is 1 ns, but the average power dissipation is 55 mW per gate.

In conclusion, the ECL circuit has many advantages: complementary outputs are available, it is very fast, it has a large fanout and a large fan-in, and it is relatively independent of transistor parameters. A major disadvantage is that it does dissipate much more power than a saturating-type logic circuit. The small logic swing can be a disadvantage in an electrically noisy environment.

EXERCISE

E3.9 Determine the HIGH-level noise margin for the MC1210 NOR output, with a fanout of 25. Assume $\beta_F = 30$, $V_{BE(on)} = 0.75$ V, and the maximum gate input current is 100 μA. Use $V_{IH(min)} = -1.02$ V.

3.7 MOSFET CIRCUITS

MOS transistors used in digital circuits have many advantages. Some of these have already been given in Chap. 1. The predominant advantage is the small surface area of the silicon occupied by this device. An additional advantage is the high input resistance at the gate terminal, which is typically 10^{12} Ω. This means that in the static or dc state almost no current is drawn by the gate input circuit. In a digital circuit application, where one MOS device is driven by another, the dc fanout is almost unlimited. However, the gate of the MOSFET is capacitively coupled to the source and body regions. The fanout limitation of a MOS digital circuit is therefore determined by the propagation delay time associated with charging the load-gate capacitance. This effect causes one major disadvantage of MOSFET circuits: they are slower than bipolar-transistor circuits.

The characteristics of MOSFETs used in digital circuits are similar to those described in Sec. 1.7. There an *enhancement-mode* MOSFET is described. That is, at zero bias ($V_{GS} = 0$ V) the transistor is cut off with only leakage current flowing at the drain connection. By increasing the voltage on the gate, the channel between the source and drain is enhanced, and the transistor conducts when $V_{GS} \geq V_{th}$. Another type of MOSFET is the *depletion-mode* device. With these devices the channel is formed at zero bias and the transistor is conducting. Now decreasing the voltage applied to the gate depletes the channel and the transistor cuts off. Enhancement-mode or depletion-mode devices may be manufactured with either p or n channels.

3.7.1 MOSFET Inverter

Very small and simple logic circuits are possible when a MOS device is used as a load resistor for a MOSFET inverter. As we saw in Sec. 1.7.2 the load transistor, with the gate and drain connected together, is operated in the saturated region, and the inverting transistor is switched from cutoff to the ohmic region. An all-MOS inverter circuit using n-channel devices is shown in Fig. 3.25a. The voltage transfer characteristic, shown in Fig. 3.25b, is that of curve C in Fig. 1.41. Notice this is very similar to the voltage transfer characteristic of the bipolar transistor inverter of Fig. 3.1b.

With V_{in} less than V_{th} (1 V for this example), the inverting transistor (Q_1) is cut off. Then, only leakage current flows through the load transistor (Q_2). At these

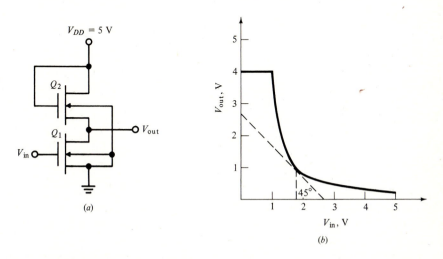

(a)

(b)

FIGURE 3.25
(a) Circuit diagram of IC MOSFET inverter; (b) voltage transfer characteristic of MOSFET inverter.

low currents the voltage appearing across Q_2 is just the threshold voltage of the device. The output HIGH level is then given as

$$V_{OH} = V_{DD} - V_{th(2)} \qquad (3.62)$$

For this example, where $V_{DD} = 5$ V and $V_{th} = 1$ V, $V_{OH} = 4$ V.

The LOW level is a result of the resistive divider action of the load transistor, operated in the saturated region, and the inverter in the ohmic region. A large ratio between these two resistance values results in V_{OL} being very close to 0 V. That is, from the example shown in Fig. 3.25b,

$$V_{OL} = 0.4 \text{ V}$$

The gate has a large logic swing,

$$V_{OH} - V_{OL} = 3.6 \text{ V}$$

The input LOW threshold voltage for the gate is equal to the threshold voltage for the inverting transistor; that is,

$$V_{IL} = V_{th(1)} = 1 \text{ V} \qquad (3.63)$$

The input HIGH threshold voltage is not as well defined in MOSFET circuits as in the bipolar-transistor digital circuits we have described. However, notice in Fig. 3.25b that a line with a slope of 45° is tangential to the transfer characteristic at a point where the gain of the circuit is equal to 1. To the right of this point, the gain becomes less than 1; while to the left, the gain rapidly becomes greater than 1. We may therefore define this point as the input HIGH threshold voltage. In the example,

$$V_{IH} = 1.8 \text{ V}$$

The typical noise margins are then

$$NM_H = V_{OH} - V_{IH} = 4 - 1.8 = 2.2 \text{ V}$$

$$NM_L = V_{IL} - V_{OL} = 1 - 0.4 = 0.6 \text{ V}$$

Notice these steady-state values are approximately the same as obtained with the TTL gate circuit described in Sec. 3.4.4.

3.7.2   MOSFET Gates

Shown in Fig. 3.26a is a 2-input NOR gate using just MOS transistors. A 2-input NAND gate is shown in Fig. 3.26b. Notice the simplicity of these circuits. For the NOR gate, except for the input terminals, the inverting transistors are connected in parallel and share a common load transistor. In the NAND gate circuit the inverting transistors are connected in series, again with a single load transistor. For the parallel connection a HIGH level at either A or B, or both, will cause the output of the gate to go to a LOW level. With the transistors connected in

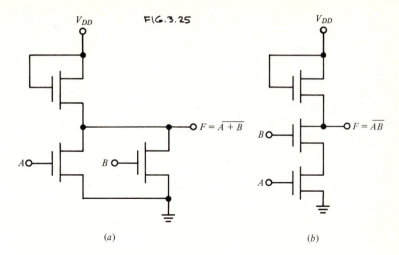

FIG.3.25

$F = \overline{A + B}$

$F = \overline{AB}$

(a)

(b)

FIGURE 3.26
(a) Circuit diagram of MOS 2-input NOR gate; (b) circuit diagram of MOS 2-input NAND gate.

series, a HIGH level is required at both A and B for the gate output to change to a LOW level.

The voltage transfer characteristic of these simple NOR and NAND gates is typically similar to that shown in Fig. 3.25b.

The dc fanout is limited only by the maximum time-delay requirements of the gate. The propagation-time-delay problem for these circuits is similar to that for RTL circuits. Capacitive loads, in this case the load gates, are discharged to a LOW level with relative high speed through the "on resistance" of the inverter transistor. However, the charging to a HIGH level is through the output resistance of the load transistor; this is usually rather high, resulting in a long time delay and limited fanout.

Sometimes in a MOS inverter circuit the gate of the load transistor is connected to a supply voltage (V_{GG}) somewhat greater than the drain supply voltage (V_{DD}). With a larger V_{GS} for the load transistor, more current is made available at the output node for faster switching of the inverter. Also, with the inverting transistor off, the output voltage is more nearly V_{DD}.

3.7.3 Complementary MOS Circuits

The time delay problem of MOS gates can be improved by using *complementary MOS* (CMOS) devices. An n- and a p-channel MOS transistor are fabricated next to each other in the same silicon chip. These CMOS circuits combine relatively fast switching speeds with an extremely low dc power consumption.

An inverter circuit using CMOS transistors is shown in Fig. 3.27a. For the n-channel device the source is connected to ground, but for the p-channel unit the source is tied to V_{DD}. The two gates are connected as are the two drains. With V_{in}

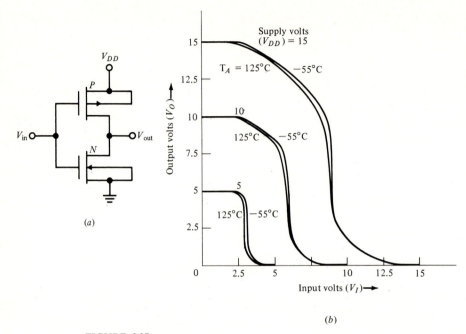

FIGURE 3.27
(a) Circuit diagram of CMOS inverter; (b) voltage transfer characteristic of CMOS inverter.

at 0 V the n-channel transistor is nonconducting ($V_{GS(n)} < V_{th(n)}$), but the full V_{DD} appears from the source to the gate of the p-channel device. With $V_{DD} \gg V_{th(p)}$, the p-channel transistor is on "hard," exhibiting a low resistance, and the output voltage is very close to V_{DD}. With V_{in} at V_{DD} the p-channel transistor is cut off ($V_{GS(p)} = 0$ V), but now with $V_{DD} \gg V_{th(n)}$, the n-channel unit is on "hard," and the output voltage is approximately 0 V. Notice with V_{in} at either a LOW or HIGH level one transistor is off and the other is on. The OFF transistor provides an extremely high resistance path which limits the dc current in the inverter to the nanoampere range. The result is a power dissipation as low as 10 nW. The ON transistor provides a low resistance for charging or discharging any load capacitance, resulting in fast switching and logic levels within a few millivolts of either the supply voltage or ground.

A typical voltage transfer characteristic of a CMOS inverter is shown in Fig. 3.27b. This shows the transfer characteristic for three different values of V_{DD}, namely, 5, 10, and 15 V. Note the almost ideal square characteristic with V_{DD} at 5 V. With $V_{th(n)} = V_{th(p)} \approx \frac{1}{2}V_{DD}$, we have equal noise margins ($NM_L = NM_H$). Additionally we have the very nice feature that the noise margin is approximately 50 percent of V_{DD}. Also shown in Fig. 3.27b is the excellent temperature stability of the transfer characteristic as the temperature of the device is varied from -55 to $+125°C$.

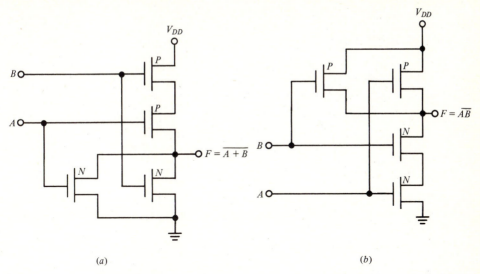

FIGURE 3.28
(a) Circuit diagram of CMOS 2-input NOR gate; (b) circuit diagram of CMOS 2-input NAND gate.

3.7.4 Complementary MOS Gates

A positive-logic NOR gate using CMOS transistors is shown in Fig. 3.28a. With both A and B at a LOW level both p-channel transistors are conducting and the n-channel transistors are cut off; the output is at a HIGH level. With a HIGH level at either A or B, or both, one or both of the p-channel devices turn off and one or both of the n-channel transistors turn on. The result in each case is a LOW voltage at the output. For the NOR gate the n-channel devices are connected in parallel and the p-channel devices are connected in series.

For the NAND gate, shown in Fig. 3.28b, the n-channels are in series and the p-channel transistors are in parallel. With both input levels LOW, both p-channel devices are conducting and the n-channel devices are cut off. The output level is approximately V_{DD} or a HIGH level. Even with only one of the inputs at a LOW level, one of the n-channels is cut off and the output remains at a HIGH level. Both inputs are required to be at a HIGH level for both p-channel transistors to be off and both n-channel transistors to be conducting. The result then is that the output voltage is approximately 0 V, a LOW level.

The voltage transfer characteristic in Fig. 3.27b is from the data sheet of the RCA CD4000A Series of NOR gates. RCA was the initial developer of CMOS logic circuits. The static electrical characteristics of these gates† are shown in Fig. 3.29a. Notice at 25°C, V_{OL} is always less than 10 mV and V_{OH} is guaranteed to

---

† These are all positive-logic NOR gates. The CD4000A, 4001A, 4002A, and 4025A are, respectively, a dual 3-input plus inverter, a quad 2-input, a dual 4-input, and a triple 3-input NOR gate. The suffix D or K refers to the type of package, either dual in-line or flat pack.

(a)   STATIC ELECTRICAL CHARACTERISTICS (All inputs .... $V_{SS} \leq V_I \leq V_{DD}$)
(Recommended DC Supply Voltage ($V_{DD} - V_{SS}$) .... 3 to 15 V)

CHARACTERISTIC	SYMBOL	Vo Volts	VDD Volts	CD4000AD... −55°C Min	Typ	Max	25°C Min	Typ	Max	125°C Min	Typ	Max	CD4000AE... −40°C Min	Typ	Max	25°C Min	Typ	Max	85°C Min	Typ	Max	UNITS	Fig. No.
Quiescent Device Current	I_L		5	−	−	0.05	−	0.001	0.05	−	−	3	−	−	0.5	−	0.005	0.5	−	−	15	µA	
			10	−	−	0.1	−	0.001	0.1	−	−	6	−	−	5	−	0.005	5	−	−	30		
Quiescent Device Dissipation/Package	P_D		5	−	−	0.25	−	0.005	0.25	−	−	15	−	−	2.5	−	0.025	2.5	−	−	75	µW	
			10	−	−	1	−	0.01	1	−	−	60	−	−	50	−	0.05	50	−		300		
Output Voltage Low-Level	V_{OL}	$V_I=V_{DD}$ $I_O=0A$	5	−	−	0.01	−	0	0.01	−	−	0.05	−	−	0.01	−	0	0.01	−	−	0.05	V	1.5 1.6
			10	−	−	0.01	−	0	0.01	−	−	0.05	−		0.01	−	0	0.01	−		0.05		
High-Level	V_{OH}	$V_I=V_{SS}$ $I_O=0A$	5	4.99	−	−	4.99	5	−	4.95	−	−	4.99	−	−	4.99	5	−	4.95	−	−	V	1.7
			10	9.99	−	−	9.99	10	−	9.95	−	−	9.99	−	−	9.99	10	−	9.95	−	−		
Noise Immunity (Any Input) For Definition, See Appendix	V_{NL}	3.6 / $I_O=0$	5	1.5	−	−	1.5	2.25	−	1.4	−	−	1.5			1.5	2.25		1.4			V	−
		7.2	10	3	−	−	3	4.5	−	2.9	−	−	3			3	4.5		2.9				
	V_{NH}	0.95	5	1.4	−	−	1.5	2.25	−	1.5	−	−	1.4			1.5	2.25		1.5			V	
		2.9	10	2.9	−	−	3	4.5	−	3	−	−	2.9			3	4.5		3				
Output Drive Current N-Channel	I_{DN}	$V_I=V_{DD}$ 0.4*	5	0.5	−	−	0.40	1	−	0.28	−	−	0.35			0.3	1	−	0.24			mA	1.8 1.10
		0.5	10	1.1	−	−	0.9	2.5	−	0.65	−	−	0.72			0.6	2.5	−	0.48				
P-Channel	I_{DP}	$V_I=V_{SS}$ 2.5#	5	−0.62	−	−	−0.5	−2	−	−0.35	−	−	−0.35			−0.3	−2		−0.24			mA	1.9 1.11
		9.5	10	−0.62	−	−	−0.5	−1	−	−0.35	−	−	−0.3			−0.25	−1		−0.2				
Input Current	I_I			−	−	−	−	10	−	−	−	−					10					pA	

DYNAMIC ELECTRICAL CHARACTERISTICS at $T_A = 25°C$, $C_L = 15pF$, and input rise and fall times = 20 ns

(b)   Typical Temperature Coefficient for all values of $V_{DD} = 0.3\%/°C$.

CHARACTERISTIC	SYMBOL	TEST CONDITIONS VDD (Volts)	CD4000AD, AF, AK CD4001AD, AF, AK CD4002AD, AF, AK CD4025AD, AF, AK Min	Typ	Max	CD4000AE, CD4001AE CD4002AE, CD4025AE Min	Typ	Max	UNITS	CHARACTERISTIC CURVES & TEST CIRCUITS Fig. No.
Propagation Delay Time: High-to-Low Level	t_{PHL}	5	−	35	50	−	35	80	ns	1.13
		10	−	25	40	−	25	55		
Low-to-High Level	t_{PLH}	5	−	35	95	−	35	120	ns	1.13
		10	−	25	45	−	25	65		
Transition Time: High-to-Low Level	t_{THL}	5	−	65	125	−	65	200	ns	1.14
		10	−	35	70	−	35	115		
Low-to-High Level	t_{TLH}	5	−	65	175	−	65	300	ns	1.14
		10	−	35	75	−	35	125		
Input Capacitance	C_I	Any Input	−	5	−	−	5	−	pF	−

FIGURE 3.29
(a) Static characteristics of CMOS gate; (b) dynamic characteristics of CMOS gate
(RCA Corporation).

be within 1 mV of V_{DD}. The threshold voltage (defined where $I_D = 10 \ \mu A$) for both the *n*- and *p*-channel devices is 1.5 V at 25°C. Typical noise margin, here denoted as noise immunity, is the same for both the LOW and HIGH levels. That is,

$$NM_L(\equiv V_{NL}) = NM_H(\equiv V_{NH})$$

At room temperature the noise margin is typically 45 percent of V_{DD}.

The typical dc input current of 10 pA reflects the high input resistance of MOS devices. The low dc power consumption of CMOS gates is indicated by the typical 10-nW power dissipation per package at $V_{DD} = 10$ V.

The dynamic electrical characteristics of the CD4000A Series are given in Fig. 3.29b. The average propagation delay time for these gates, with $V_{DD} = 10$ V, is typically 25 ns $[t_{pd} = (t_{PHL} + t_{PLH})/2]$.

Because of their small size and simplicity many MOS logic gates can be closely packed in a very small area of silicon to form a single integrated system. This is also possible with bipolar transistors but not with the same packing density. We will further develop this subject after making a comparison of bipolar and MOS digital ICs.

3.8 COMPARISON OF CIRCUITS

Shown in Table 3.5 is a listing of some of the more important characteristics of the popular digital ICs which have been described in this chapter.

The first available digital IC family was RTL; it is easy to fabricate and is relatively fast. However, the small logic swing and low noise margin are distinct disadvantages. While still found in many digital systems, RTL is not presently being used in new designs.

The DTL circuits have been shown to have a large logic swing and high noise margins. They have a reasonable fanout, and the fan-in is large. A limitation to these gates is they are relatively slow.

The TTL circuits are an improvement on DTL in almost every respect. The logic swing is good as are the noise margins. The fanout is high, and the power

Table 3.5 COMPARISON OF TYPICAL ELECTRICAL CHARACTERISTICS FOR WIDELY USED DIGITAL IC GATES

	RTL (900 Series)	DTL (930 Series)	TTL (Series 54)	ECL (10,000 Series)	MOS	CMOS (4000 Series)
V_{OH}/V_{OL} (V)	1.2/0.2	4.8/0.2	3.3/0.2	−0.9/−1.7	11/1.0	10/0
Logic swing (V)	1.0	4.6	3.1	0.8	10	10
NM_H/NM_L (V)	0.5/0.5	3.3/1.0	1.9/1.0	0.3/0.3	4.0/3.0	4.5/4.5
Fanout	5	8	10	16	10	10
Supply voltage (V)	+3.6	+5	+5	−5.2	−12	+10
Power dissipation per gate	16 mW	9.5 mW	10 mW	25 mW	1 mW	10 nW
Prop. delay time (ns)	12	30	9	2	250	25

dissipation is low, and except for ECL, this logic-circuit family has the shortest propagation delay time. At the time of writing (1974) TTL is the most widely used form of digital IC, with about 300 different circuit types available.

Where the highest speed is indicated, then the ECL circuits are to be preferred. In addition the availability of the complementary outputs can mean a decrease in the number of gates required to perform a given function. It should be noted, however, the high speed is obtained at the price of a low logic swing and a high power dissipation.

Since MOS gates are seldom used by themselves, but are generally integrated into a more complex function as a single unit, the data in Table 3.5 is typical for an on-chip MOS gate. Compared to bipolar units the simplicity and low power dissipation of these gates is attractive; however, the long propagation delay time restricts their use to data rates of 1 MHz or less.

The potential for the CMOS devices is very good. The logic swing is practically the full V_{DD}, and the noise margins are good. Like ECL circuits the fanout of CMOS units is more limited by speed considerations than by the dc characteristics. The greatest potential, however, is in their low power consumption as demonstrated by the widespread use of CMOS devices in electronic wristwatches and other battery-operated equipment not requiring very fast switching.

3.9   MSI AND LSI

In describing these digital circuits we have limited all our examples to simple 2-input gates, although we have mentioned that these gates are packaged in many different forms with up to 8 inputs. With some, by using extenders, the number of inputs can be increased until some other criterion, like speed or packing density, becomes the limitation. We shall see in Chap. 5 how simple gates can be connected to form a more advanced function, like a flip-flop or latch circuit. We shall then see in Chap. 6 how these flip-flops may be connected to form even more complex functions, like counting circuits. Notice that we have progressed from a simple circuit to a small system.

Individual gate or flip-flop packages are used where the required logic function is not a common one, or where versatility is required in a system. All the

Table 3.6   SUMMARY OF DEVICE DATA FOR SSI AND MSI UNITS

	SSI	MSI	
Device function	Quad 2-input gate	Decade counter	8-bit shift register
Device number	SN5400	SN5490	SN5491
Chip size (mils)	50 × 60	50 × 115	55 × 110
Chip area (sq. mils)	3000	5750	6050
No. of components	36	102	143
Area/component (sq. mils)	83	56	42
Equivalent no. of gates	4	18	35
Area/gate (sq. mils)	750	320	175
Gates/pin	0.3	1.3	2.5

circuits similar to those which we have so far described in this chapter are examples of *small-scale integration* (SSI). However, if sufficient need can be seen for a particular complex logic function then a complete fabricated IC is indicated. Such examples might be a decade counter or an 8-bit shift register. These circuits, which will be fully described in later chapters, comprise many gates on a single silicon chip, with the intraconnection on the chip to make the given functions. The name *medium-scale integration* (MSI) is used to describe batch-fabricated functional blocks. The average complexity of a bipolar-transistor MSI circuit is about 30 gates. A comparison of a packaged MSI circuit with a simple gate package of the SSI variety is shown in Table 3.6.

Notice the number of components in the MSI circuits is three or four times greater than the SSI package, but the area of the chip is only increased by a factor of 2. This is because a large amount of the silicon surface area is required to make connection to the external pins. A figure of merit for the complexity and efficient use of the silicon area is the gates/pin ratio. For the 8-bit shift register we see an improvement of nearly 10 : 1.

Other examples of bipolar-transistor MSI digital circuits are 4-bit binary full adders, 4-bit up/down counters both binary and decade, and various decoders like 1 out of 16 or 1 out of 10 as well as decoders for seven-segment indicators. All these circuits are combinations of the TTL gates described in Sec. 3.4.4.

The extremely small area used to fabricate a single MOS device in an IC has led to the design of great numbers of repetitive circuits in a silicon chip. *Large-scale integration* (LSI) is used to describe circuits containing more than 100 gates. Examples of these types of circuits are 1024-bit shift registers and 4K (4096-bit) memory elements. In each of these devices the basic circuit element is the MOSFET inverter. Two or more inverters are then intraconnected to make a simple memory or storage element. This storage element is reproduced as many times across the silicon chip as the circuit function requires.

3.10 SUMMARY

\#   Bipolar-transistor logic circuits are characterized as either saturating or nonsaturating. RTL, DTL, and TTL are families of saturating digital ICs. ECL is nonsaturating, as is also the Schottky barrier diode series of TTL (TTL-S).

\#   Basic to saturating logic circuits is the transistor inverter. For ECL, the basic circuit is the current switch.

\#   Important parameters of digital ICs are:

The input HIGH and LOW threshold voltages V_{IH} and V_{IL}
The output HIGH and LOW voltage levels V_{OH} and V_{OL}
The noise margins NM_H and NM_L
The fanout N
The power dissipation PD_{av}
The propagation delay time t_{pd}

\#   The maximum fanout is closely related to the noise margin of the gate as well as the base overdrive factor and worst-case design.

The product $t_{pd} PD_{av}$ is a useful figure of merit, indicating the operating efficiency of the gate.

RTL is termed a current-source logic. DTL and TTL are both current-sink logic.

Compared with ECL, the saturating circuits generally have a greater logic swing, use less power, but are slower. On the other hand, ECL is faster with a smaller logic swing but consumes more power.

MOSFET digital ICs are smaller than their bipolar counterparts and operate with less power but are slower.

CMOS digital ICs dissipate very low dc power and are relatively fast in switching. They do take up more surface area on the silicon chip than the MOSFET circuits.

MSI are functional blocks comprising about 30 logic gates in an IC chip. Most MSI circuits are TTL, but an increasing number are now available in CMOS.

LSI are small-scale systems on a single chip containing 100 or more logic gates. The majority of LSI circuits are made using MOS transistors. Bipolar LSI circuits are usually made in TTL form.

DEMONSTRATIONS

The electrical properties of a logic gate can be characterized by the dc or steady-state conditions and the ac transient conditions. These properties will be developed in the following demonstrations. In the examples of ICs, TTL hex inverters or 2-input NAND packages can be used (Fig. 3.20), but the intention is that any or all of the digital ICs described in this chapter may be utilized.

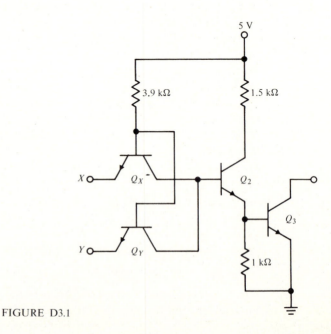

FIGURE D3.1

D3.1 TTL gate made from discrete elements Construct from discrete components† the 2-input TTL NAND gate of Fig. D3.1 with open collector output.

(a) With the X input HIGH (3.5 V) and Y input LOW (0.2 V), measure I_{BX}, I_{BY}, $I_{in(X)}$, and $I_{in(Y)}$. Determine, as in Demonstration D1.2, β_F of all four transistors and β_R of Q_X and Q_Y. Make β measurements at 1 mA. Compare $I_{in(X)}$ to the value you can calculate using the above values and $V_{BE(on)} = V_{BC(on)} = 0.7$ V.

(b) With both inputs HIGH, measure any additional circuit variables and parameters necessary to calculate I_{sink} at the collector of Q_3 when Q_3 is out of saturation ($V_{CE} = 0.4$ V). Compare this to the measured value of I_{sink}.

D3.2 Voltage transfer and VI characteristic To facilitate the demonstration, the horizontal sweep of an oscilloscope is calibrated in terms of volts instead of time. A 1-V peak-to-peak low-frequency sine wave is applied to the external horizontal input, and the variable adjustment on the horizontal control is used to obtain a satisfactory scale of, say, 0.2 V/div, for RTL or ECL or 1 V/div for TTL or DTL.

To display the voltage transfer characteristic a connection is made from point X in Fig. D3.2a to the horizontal input and from point Y to the vertical input of the oscilloscope. The V_{in} is then varied from 0 V to the maximum for the input to the gate. The voltage levels corresponding to V_{OH}, V_{OL}, V_{IL}, and V_{IH} for a fanout of 5 may be noted. The noise margin may then be calculated. By adding additional gate inputs to point Y, the effects of fanout on these parameters may also be demonstrated. By use of a soldering iron and circuit cooler the effects of temperature (primarily on the threshold voltage) can be easily displayed.

If an oscilloscope is not available, the circuit in Fig. D3.2b can be used for determining the same transfer characteristics by removing potentiometer R_2 and connecting the inputs to the load gates at point (Y).

With the aid of the circuit in Fig. D3.2b, the dc currents at the input and outputs can be measured. Using the values of V_{OH}, V_{OL}, V_{IH}, and V_{IL} determined previously, measure the values of the input and output currents that are necessary in order to determine the maximum values of the fanouts, N_H and N_L. These fanout values are what a system could use if (1) all units were identical to the one tested, (2) all units were always at the same temperature as tested, and (3) the power-supply voltage stayed constant at the value used during the test.

VI Characteristics With the aid of a VI plotter (as in Demonstration D1.1) one can show three VI characteristics that give useful information:

(a) $I_{out(sink)}$ versus V_{out} with (all) inputs HIGH. The IC can be cooled and heated to show effects of the temperature change (primarily β_F of Q_3 in the TTL circuit of Fig. 3.19).

(b) $I_{out(source)}$ versus V_{out} with at least one input LOW. This shows the short-circuit protection given by R_3 in Fig. 3.19.

† Using 4-2N4275 or equivalents.

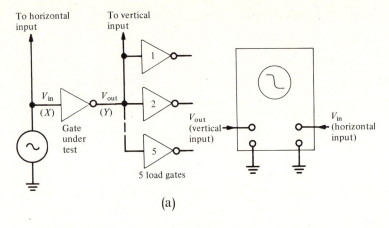

(a)

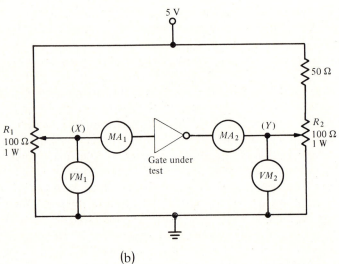

(b)

FIGURE D3.2

(c) I_{in} versus V_{in} shows (1) the effect of β_R in Q_1 of Fig. 3.19, (2) the effect of input clamping diodes (shown dotted in Fig. 3.19), and (3) the current switching of the input "diode" gate.

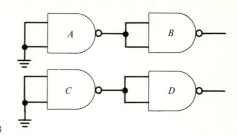

FIGURE D3.3

D3.3 Power dissipation From Fig. D3.3 with V_{in} at 0 V, the output of two gates (A and C) will be at a HIGH level and the output of the other two gates (B and D) will be at a LOW level. The current from the power supply V_{CC} will then be the average power-drain current for the package. The average power dissipation per gate may then be calculated.

D3.4 Propagation delay time A ring oscillator can be made by the three gates (connected as inverters) enclosed within the ring, as shown in Fig. D3.4. The waveform may be observed by connecting point Y to the vertical input of an oscilloscope with the time base triggered by the vertical input signal. The period of the oscillation frequency is equivalent to the total propagation delay time of the three gates. The average propagation delay time per gate is therefore the period of oscillation divided by twice the number of gates within the ring.

When using high-speed gates it may be necessary to cascade more than three gates so as to lengthen the period of oscillation and make it easier to measure. An odd number of gates is always required.

The speed-power product may be calculated for a particular gate circuit and compared with that given in Table 3.5.

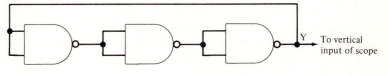

Y   To vertical
input of scope

FIGURE D3.4

REFERENCES

Two recent textbooks with an emphasis on the circuits of digital ICs are MILLMAN, J., and C. C. HALKIAS: "Integrated Electronics," McGraw-Hill, New York, 1972, where bipolar ICs are described in chap. 6 and MOSFET ICs in chap. 10; and OLDHAM, W. G., and S. E. SCHWARZ: "An Introduction to Electronics," Holt, New York, 1972, with description of bipolar ICs in chap. 8 and MOSFET circuits in chap. 13.

The circuit analysis and characteristics of all versions of the Series 54/74 TTL gates are covered in MORRIS, R. L. and J. R. MILLER: "Designing with TTL Integrated Circuits," chaps. 1–3, McGraw-Hill, New York, 1971.

An excellent in-depth survey of digital ICs, from RTL through CMOS circuits, is contained in three issues of the *IEEE Spectrum*. GARRETT, L. S.: Integrated Circuit Digital Logic Families, *IEEE Spectrum*, vol. 7, no. 10, pp. 46–58, October 1970; vol. 7, no. 11, pp. 63–72, November 1970; vol. 7, no. 12, pp. 30–42, December 1970.

At a more advanced level the analysis and design of bipolar digital ICs is ably presented in LYNN, D. K., C. S. MEYER, and D. J. MLTON: "Analysis and Design of Integrated Circuits," chaps. 6–11, McGraw-Hill, New York, 1968. Both the qualitative and quantitative analysis of MOSFET digital circuits appears in CRAWFORD, R. H.: "MOSFET in Circuit Design," McGraw-Hill, New York, 1967.

PROBLEMS

P3.1 (a) Draw a circuit of a single *n-p-n* transistor inverter requiring only one power supply.

(b) Find the value for the collector resistor R_C, given $V_{CC} = 6$ V, $V_{CE(sat)} = 0.2$ V, and $I_{C(sat)} = 10$ mA.

(c) For operating at the edge of saturation, find a value for the base resistor R_B, given $\beta_F = 50$, $V_{BE(sat)} = 0.7$ V, and $V_{IH} = 2.0$ V.

P3.2 (a) Given the circuit shown in Fig. 3.1a, calculate and show the breakpoint voltages $V_{IL}, V_{IH}, V_{OL}, V_{OH}$ on the voltage transfer characteristic (V_{in} versus V_{out}) by using the values of $V_{CC} = 3.0$ V, $V_{BE(sat)} = 0.7$ V, $V_{CE(sat)} = 0.2$ V, $\beta_F = 25$, $R_B = 5$ kΩ, and $R_C = 750$ Ω.

(b) Designate the logic levels graphically, similar to Fig. 3.2.

(c) Calculate and dimension the NM_L, NM_H, logic swing and the transition region on the graph.

P3.3 A single transistor inverter has the following values: $V_{CC} = 5.0$ V, $V_{BE(sat)} = 0.65$ V, $V_{CE(sat)} = 0.1$ V, $\beta_F = 50$, $R_B = 5$ kΩ, $R_C = 600$ Ω.

(a) Compile a table of values for I_B, I_C, and V_{out} for $0 \le V_{in} \le 2.0$ in 0.5-V increments. Include in the table the breakpoint values.

(b) Plot the voltage-transfer-characteristic curve for the values obtained in part (a), and label the cutoff, active, and saturation regions.

(c) Calculate and dimension the NM_L, NM_H, logic swing and the transition region on the curve.

P3.4 (a) Repeat Prob. P3.1a, but now include a base pulldown resistor R_D connected from the transistor base to a negative supply voltage (V_{BB}).

(b) Repeat Prob. P3.1b.

(c) Use the values obtained in part (b) and $R_B = 6.5$ kΩ, $R_D = 20$ kΩ, and $V_{BB} = -3.0$ V to calculate the values of V_{IL} and V_{IH}.

P3.5   (a) Repeat Prob. P3.2a with the addition of a base pulldown resistor $R_D = 10 \text{ k}\Omega$
and $V_{BB} = -1.5$ V.

       (b) Repeat Prob. P3.2b.

       (c) Repeat Prob. P3.2c.

P3.6   Describe the advantages and disadvantages of the base pulldown resistor.

P3.7   (a) Using values for the gate given in Prob. P3.3, repeat Prob. P3.3a, including a
base pulldown resistor with $V_{BB} = -2.0$ V and $R_D = 20 \text{ k}\Omega$.

       (b) Repeat Prob. P3.3b.

       (c) Repeat Prob. P3.3c.

P3.8   Two identical transistor inverters are connected such that one drives the other.

       (a) Determine NM_H, given $V_{CC} = 3.0$ V, $V_{BE(\text{sat})} = 0.7$ V, $V_{CE(\text{sat})} = 0.2$ V,
$\beta_F = 20$, and $R_B = R_C = 750\ \Omega$.

       (b) Calculate the base overdrive factor (k) for the driven gate.

       (c) Identical additional gates are connected to the output of the driving gate such
that V_{OH} is lowered to the level of V_{IH} ($V_{OH} = V_{IH}$). Calculate the ratio of the
current in R_C of the drive gate to the current in one of the base resistors R_B of
the driven gates when the drive gate is off.

       (d) Calculate the base overdrive factor of the driven gates in part (c).

       (e) Calculate the fanout by using Eq. (3.10b) of the text.

       (f) What effect does adding additional driven gates have on the NM_H and the
base overdrive factor? What significance is the ratio of currents found in
part (c)?

*P3.9   A transistor inverter is to drive N identical inverters. The values for the inverter
with a base pulldown resistor are $V_{CC} = 4.0$ V, $V_{BB} = -2.0$ V, $V_{BE(\text{sat})} = 0.68$ V,
$V_{CE(\text{sat})} = 0.17$ V, $R_B = 5 \text{ k}\Omega$, $R_C = 900\ \Omega$, $R_D = 15 \text{ k}\Omega$, and $\beta_F = 25$.

       (a) Derive an equation for N similar to Eq. (3.10b) of the text.

       (b) Solve for the fanout.

       (c) Plot the voltage-transfer-characteristic curve for a load of 0, 1, N identical
inverter gates.

       (d) If all resistors track, does the fanout change for a resistor tolerance of
± 20 percent and if so by how much?

P3.10   It is required that a transistor inverter be designed with the following characteristics: $V_{CC} = 5.0$ V, $V_{BE(\text{sat})} = 0.65$ V, $V_{CE(\text{sat})} = 0.1$ V, $\beta_F = 15$, $N = 6$, and
$PD_{\text{av}} = 20$ mW/gate (max) assuming 50 percent duty cycle.

       (a) Calculate the ratio of R_B to R_C.

       (b) Determine the values of R_B and R_C.

       (c) Plot the voltage-transfer-characteristic curve for an output load of 0, 1, and 6
identical inverter gates.

P3.11   Given the circuit of Fig. 3.4 with the values $V_{CC} = 4.0$ V ± 10 percent,
$0.6 \le V_{BE(\text{sat})} \le 0.8$, $0.1 \le V_{CE(\text{sat})} \le 0.3$, $\beta_F \ge 15$, $R_B = 6$ k$\Omega \pm 20$ percent, and
$R_C = 800\ \Omega \pm 20$ percent. Assuming all resistors track, calculate the worst-case
fanout.

P3.12   A transistor inverter has the following characteristics: $V_{CC} = 6.0$ V ± 10 percent,
$0.65 \le V_{BE(\text{sat})} \le 0.75$, $0.1 \le V_{CE(\text{sat})} \le 0.2$, $20 \le \beta_F \le 100$, $R_B = 12 \text{ k}\Omega \pm 10$ percent,
and $R_C = 950\ \Omega \pm 10$ percent. Assume all resistors track and that $V_{BE(\text{on})}$ has the
same range as $V_{BE(\text{sat})}$.

       (a) Calculate the fanout under worst-case conditions.

       (b) Calculate the worst-case noise margins for an output load of three identical
inverter gates.

*P3.13   An IC chip contains 27 transistor inverters. The full-load average power dissipation of the chip is restricted to 600 mW, based on a 50 percent duty cycle and the operating values of $V_{CC} = 5.5$ V, $V_{BE(on)} = V_{BE(sat)} = 0.7$ V, $V_{CE(sat)} = 0.2$ V, $R_B = 7.5$ kΩ, and $R_C = 1$ kΩ.
   (a) Calculate the fanout.
   (b) If the power supply is regulated to ± 0.1 V, find the value of the worst-case fanout.
   (c) Calculate the minimum β_F under the worst-case conditions of part (b).
   (d) The circuit designer wants to allow for an NM_H equivalent to 25 percent of the no-load NM_H for the worst-case condition in part (b). Use the $\beta_{F(min)}$ of part (c) to calculate the fanout for this case.

P3.14   An RTL quad 2-input gate is to be used by a circuit designer. The gate characteristics are $V_{CC} = 5.0$ V, $V_{BE(sat)} = 0.65$ V, $V_{CE(sat)} = 0.15$ V, $PD = 20$ mW/gate based on 50 percent duty cycle and maximum fanout, $R_B = 5.5$ kΩ, and $R_C = 950$ Ω.
   (a) Determine the maximum fanout for each gate.
   (b) Calculate the maximum power-supply current required per IC package.
   (c) Allow for an NM_H of 20 percent of the no-load NM_H and calculate the adjusted fanout.

P3.15   Given the circuit shown in Fig. 3.5 and the values $V_{CC} = 5.0$ V, $V_{BE(sat)} = 0.7$ V, $V_{CE(sat)} = 0.2$ V, $R_C = 1$ kΩ, and $\beta_F = 50$.
   (a) What logic function does the RTL gate perform, assuming positive logic?
   (b) Determine the maximum allowable value of R_B when driven by a similar gate such that the function can be performed.
   (c) What are the logic levels at V_{out} driving a similar gate using the calculated value of R_B in part (b)?
   (d) What are the breakpoint voltages of V_{out} for fanouts of both 1 and 10, using a recalculated value of R_B so that all transistors are either off or saturated?

*P3.16   (a) Given the $V_{BE} I_B$ characteristic shown in Fig. P3.16a, calculate the base currents for transistors Q_2 and Q_3 in Fig. P3.16b when Q_1 is off and $R_B = 0$. Use $V_{CE(sat)} = 0.1$ V and $\beta_F = 5$.
   (b) Does the circuit in part (a) function as a saturating logic circuit? If not, explain why.
   (c) Determine the minimum value of R_B such that the inverter Q_1 drives the two RTL gates satisfactorily.

*P3.17   The operating characteristics for a dual 4-input RTL IC package are $V_{CC} = 4.5$ V, ± 10 percent, $0.6 \le V_{BE(sat)} \le 0.8$, $0.1 \le V_{CE(sat)} \le 0.3$, $\beta_F \ge 9.1$, $N \le 6$, and $PD_{av} \le 25$ mW/gate with maximum fanout. All resistors track.
   (a) Calculate the worst-case ratio of R_B to R_C.
   (b) Solve for R_B and R_C.

P3.18   Given the circuit in Fig. 3.11 with the following values charged: $V_{CC} = 5.0$ V, $V_{BE(sat)} = V_{D(on)} = 0.65$ V, and $R_2 = 50$ kΩ.
   (a) Identify the type of gate, its logic function, and the function of each diode and transistor in the circuit.
   (b) Calculate the current in resistor R_2 when $V_{BB} = -V_{CC}$ and V_{in} is HIGH.
   (c) Determine the value of resistor R_2 passing the same current calculated in part (b) but with $V_{BB} = 0$.

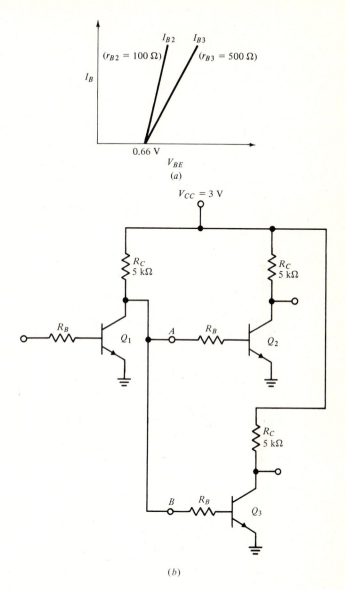

FIGURE P3.16

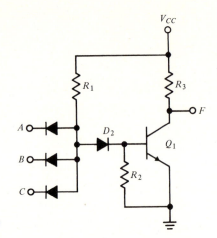

FIGURE P3.19

P3.19 Use the circuit in Fig. P3.19 and the values $V_{CC} = 5.0$ V, $V_{BE(sat)} = V_{D(on)} = 0.65$ V, $V_{CE(sat)} = 0.2$ V, and $R_2 = 10$ kΩ.
(a) Determine the value of the transition width assuming ideal diodes.
(b) Calculate the noise margins for a fanout of 1.
(c) Compare the advantage and disadvantage of this circuit to an RTL gate.
(d) How would you suggest modifying the circuit in order to increase the NM_L?

P3.20 Given the circuit in Fig. P3.19 and the values $V_{CC} = 5.5$ V, $V_{BE(sat)} = V_{D(on)} = 0.7$ V, $V_{CE(sat)} = 0.15$ V, $\beta_F = 20$, $R_2 = 5$ kΩ, and $R_3 = 1$ kΩ.
(a) Calculate the maximum value of R_1 to ensure that Q_1 is saturated when V_{in} is HIGH. (Assume the gate is unloaded.)
(b) Find the fanout of this circuit if $R_1 = 0.5$ R_1 (max) determined in part (a).

P3.21 Modify the circuit in Fig. P3.19 to include two diodes, D_2, in series and the values $V_{CC} = 5.0$ V ± 10 percent, $V_{BE(sat)} = V_{D(on)} = 0.7$ V, $V_{CE(sat)} = 0.2$ V, $\beta_F \geq 10$, $R_1 = R_2 = 4$ kΩ, and $R_3 = 6$ kΩ. Calculate the worst-case fanout. The V_{CC} line is common to all gates.

P3.22 Given the circuit in Fig. 3.14 and the values $V_{CC} = 5.0$ V, $V_{BE(sat)} = V_{D(on)} = 0.7$ V, $V_{CE(sat)} = 0.25$ V, $\beta_F = 25$, $R_1 = 5$ kΩ, $R_2 = R_3 = 3$ kΩ, and $R_4 = 1$ kΩ
(a) Describe the purpose of transistor Q_1.
(b) List the disadvantage of this circuit as a result of the transistor Q_1.
(c) Describe the effect transistor Q_1 has on the noise margins.
(d) Compare the base overdrive factor of transistor Q_2 with and without transistor Q_1 and resistor R_4 in the circuit (resistor R_1 connects directly to diode D_2).

P3.23 (a) Explain the function of diode D_2 in Fig. 3.14.
(b) Use the values in Prob. P3.22 to calculate the fanout of the circuit.
(c) Explain how the fanout is a function of the noise margin.

P3.24 Modify the circuit shown in Fig. 3.15 to a 3-input DTL gate with the following values: $V_{CC} = 4.5$ V, $V_{BE(sat)} = V_{D(on)} = 0.75$ V, $V_{CE(sat)} = 0.1$ V, $\beta_F = 15$, $R_2 = 1$ kΩ, $R_3 = 5$ kΩ, and $N = 10$ to identical gates.
(a) Describe the advantage of the circuit in Fig. 3.15 as compared to Fig. 3.14.
(b) Calculate the noise margins for this gate.
(c) If the current passing through each diode D_{1A} and D_{1B} is 0.7 mA when inputs A and B are LOW and input C is HIGH, calculate the values of resistors R_1 and R_4 to produce an overdrive factor $k = 1$ in Q_2.

P3.25 Use the circuit in Fig. 3.15 and the values $V_{CC} = 5.5$ V, $V_{BE(sat)} = V_{D(on)} = 0.68$ V, $V_{CE(sat)} = 0.25$ V, $\beta_F = 10$, $R_1 = 1.5$ kΩ, $R_2 = 3$ kΩ, $R_3 = 4$ kΩ, and $R_4 = 2$ kΩ. Calculate the PD_{av} of this gate when driving one identical gate, assuming a 50 percent duty cycle.

P3.26 Modify the circuit shown in Fig. 3.15 to replace diode D_2 with a zener diode and assign values of $V_{CC} = 6.0$ V, $V_{BE(sat)} = V_{D1(on)} = 0.7$ V, $V_{D2(on)} = 2.0$ V, $V_{CE(sat)} = 0.2$ V, $\beta_F = 15$, $R_1 = 1$ kΩ, $R_2 = 2$ kΩ, $R_3 = 3$ kΩ, and $R_4 = 2$ kΩ.
(a) Find the maximum collector current in transistor Q_2.
(b) Determine the fanout.
(c) Calculate V_{OH} and V_{OL} for driving one load gate and the maximum number of load gates.

P3.27 (a) Draw a circuit of a TTL gate consisting of a 3-input multi-emitter transistor Q_1, a voltage-offset diode D_1, and an inverter-transistor Q_2 output stage.
(b) Designate the logic operation of this gate, assuming positive logic.
(c) Describe the advantages of the multi-emitter input transistor of the TTL gate compared to the input diodes of the DTL gate.
(d) Describe any disadvantages of using a transistor in place of a diode for an input to a logic gate.

P3.28 Assume the gate in Fig. 3.17b is operated under the conditions $V_{CC} = 5.5$ V, $V_{BE(sat)} = V_{D(on)} = 0.7$ V, $V_{BC(sat)} = 0.5$ V, $V_{CE(sat)} = 0.2$ V, $\beta_F = 20$, $\beta_R = 1.5$, $R_1 = 5$ kΩ, $R_2 = 3$ kΩ, and $R_3 = 1$ kΩ.
(a) Calculate the current in the voltage-offset diode of a driven gate when all the inputs are HIGH, assuming each input to the driven gate is the only output load of each driving gate.
(b) Determine NM_H for the driving gates in part (a).
(c) Calculate the maximum value of β_R for the driven-gate arrangement in part (a).
(d) Find the fanout for a LOW-output condition, assuming the input conditions of part (a).
(e) Calculate the maximum value of β_R for the condition of $NM_H = NM_L$, assuming NM_L remains fixed at maximum.

*P3.29 A 2-input TTL gate shown in Fig. 3.17b has the characteristics $V_{CC} = 4.5$ V, $V_{BE(sat)} = V_{D(on)} = 0.75$ V, $V_{CE(sat)} = 0.1$ V, $\beta_F = 15$, $\beta_R = 0.5$, $R_1 = 5$ kΩ, $R_2 = 4$ kΩ, and $R_3 = 4$ kΩ.
(a) Determine the LOW-level fanout (N_L) under the condition that both inputs are floating; that is, there is no connection made to any input.
(b) Determine the LOW-level fanout under the condition that each input is the only output load of an identical driving gate.
(c) Determine the HIGH-level fanout (N_H) under the condition that both inputs are connected to a driving gate with an output load of maximum fanout at $NM_H = 0.3$ V.

P3.30 A simple TTL inverter shown in Fig. P3.30 has the values $V_{CC} = 3.0$ V, $V_{BE(sat)} = 0.7$ V, $V_{CE(sat)} = 0.1$ V, $\beta_F = 20$, $\beta_R = 1$, $R_1 = 8$ kΩ, and $R_3 = 600$ Ω.
(a) Plot the voltage-transfer-characteristic curve for the gate.
(b) Calculate the LOW-level fanout for the input at V_{CC}.
(c) Calculate the HIGH-level fanout for $NM_H \geq 0.8$ V.
(d) Designate and explain your choice of fanout value for this gate.
(e) From parts (b) and (c), determine how the fanout is affected by adding a second emitter to transistor Q_1 to make a 2-input gate.

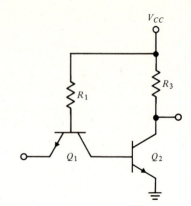

FIGURE P3.30

P3.31 The TTL gate with totem-pole output shown in Fig. 3.19 has the values $V_{CC} = 5.0$ V, $V_{BE(\text{sat})} = V_{D(\text{on})} = 0.75$ V, $V_{CE(\text{sat})} = 0.15$ V, $\beta_F = 15$, $\beta_R = 0.5$, $R_1 = 20$ kΩ, $R_2 = 3$ kΩ, $R_3 = 200$ Ω, and $R_4 = 6$ kΩ.
(*a*) Calculate the LOW-level fanout.
(*b*) Calculate the HIGH-level fanout, assuming NM_H 25 percent of NM_H (max).

P3.32 Consider the gate shown in Fig. 3.19 and the values $V_{CC} = 5.0$ V, $V_{BE(\text{sat})} = V_{D(\text{on})} = 0.7$ V, $V_{CE(\text{sat})} = 0.2$ V, $\beta_F = 10$, $\beta_R = 0.5$, $R_1 = 10$ kΩ, $R_2 = 2$ kΩ, $R_3 = 150$ Ω, and $R_4 = 4$ kΩ.
(*a*) Find the fanout, given $V_{OH} = 3.0$ V.
(*b*) Calculate the PD_{av} value for driving four identical gates, excluding any transient currents.

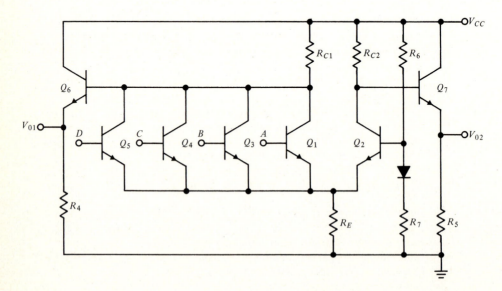

FIGURE P3.33

P3.33   The ECL gate in Fig. P3.33 has the values $V_{CC} = 4.0$ V, $V_{BE(on)} = V_{D(on)} = 0.75$ V, $R_6 = 625\ \Omega$, $R_7 = 1\ k\Omega$, $R_{C1} = 340\ \Omega$, $R_{C2} = 425\ \Omega$, $R_4 = R_5 = 2\ k\Omega$, and $R_E = 850\ \Omega$. Assume $\beta_F \to \infty$.

(a) Determine the voltages at the base of Q_2 for both a HIGH and LOW at one of the gate inputs.

(b) Determine V_{IL}, V_{IH}, V_{OL}, V_{OH}, and logic swing.

(c) Calculate the current in each resistor for both logic states.

(d) Calculate V_{CE} for one of the input transistors when it is conducting with V_{OH} at its base.

P3.34   Given the ECL gate in Fig. P3.34 and the values $V_{CC} = 5.2$ V, $V_{BE(on)} = 0.7$ V, reference voltage $V_R = 4.1$ V, logic swing 0.8 V, symmetrical about the reference voltage, and $I_{E(max)} = 10$ mA for all transistors. Assume that the logic voltage levels V_{in} and V_{out} are compatible.

(a) Calculate the value of resistors R_4 and R_5.

(b) Determine the values of resistors R_{C1}, R_{C2}, and R_E.

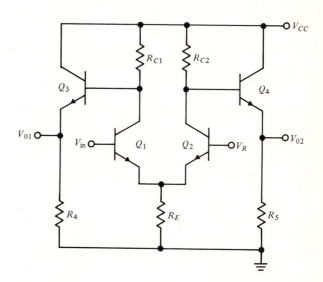

FIGURE P3.34

*P3.35   Consider the ECL gate in Fig. P3.35 and the values $V_{CC} = 3$ V, $V_{EE} = -3$ V, $V_{BE(on)} = 0.7$ V, $R_{C1} = R_{C2} = 500\ \Omega$, and $R_E = 2\ k\Omega$. Assume that the logic voltage levels V_{in} and V_{out} are symmetric about the ground level and that the power dissipation in the unloaded gate is 20 mW when both inputs are LOW. Assume $\beta_F \to \infty$ in parts (a) and (b).

(a) Calculate the values of resistors R_4 and R_5.

(b) Determine the logic swing.

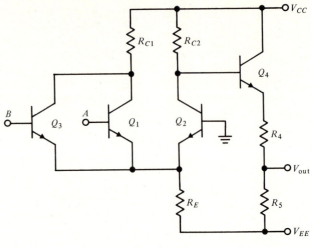

FIGURE P3.35

(c) Assume the gate is used to drive identical gates. Use the values of R_4 and R_5 determined in part (a) and $\beta_F = 40$ to find the fanout for the case where the driven input transistors are driven just hard enough to turn Q_2 off; that is, $I_{C1} = \alpha_F I_E$. (Neglect fan-in effects in the driven gates.)

4

COMBINATIONAL LOGIC DESIGN

In Chap. 2 we studied Boolean algebra which we then used to understand the logic properties of the circuits discussed in Chap. 3. The material in this chapter is a further extension on both Chaps. 2 and 3 where we will learn how to design a network made from standard logic blocks that will generate a desired logic function.

For the purposes of this chapter, the logic design of a system consists of several interrelated tasks. The first of these is to determine the precise logic relationships that will make the system function as intended. Based on this, one then considers various alternative forms of logic circuitry that are economically suitable. The search for the "best" solution from all those that meet the system requirements could require considerable time and expense. Fortunately with ICs, there is usually considerable latitude so that a reasonably small number of alternatives have to be evaluated. Next, one selects the most suitable IC family and then determines the number and kinds of logic gates needed. Last, but not least, one must specify the wiring that ties these logic elements together. We illustrate this by showing the results in the following example.

Let us assume we have been given the problem of designing the logic circuit to control a traffic light that is to be located at the intersection of the two streets shown in Fig. 4.1a. If needed, we can locate automobile detectors D_1 and D_2 in the roadbed as indicated in the figure. These detectors employ *sensors*, amplifiers, and other circuitry, all of which results in generating the logic signals which we call D_1 and D_2. Thus, if a car passes over the detector D_1, a logic level D_1 is positive for a period of time (let us say 20 s) afterward. As long as the car stays over the detector, D_1 stays positive. D_2 operates similarly. The fire-station dispatcher generates a dc logic signal D_3 by means of a switch such that when the fire engines require priority, the traffic light will be red to both directions of approaching traffic. Otherwise, in normal usage, the traffic light for one direction is red, while for the other it is green. Thus G_1 is 1 when the light controlling the one-way street is green, and G_2 is 1 when the light controlling the other street is green.

Figure 4.1b shows a truth table for G_1 and G_2 as a function of D_1, D_2, and D_3. The first three rows in the truth table are the only ones specified so far. For the last two rows, both X and Y are "don't care" conditions that we can select so as to optimize the system. We find (in Prob. P4.27) that to minimize the number of SSI packages we should make the following decisions:

First, we choose $X = 0$ and $Y = 1$, and, concurrently, we decide to use NAND gates that permit the wired-AND. Doing this gives us the logic relationships shown in Fig. 4.1c, which can be realized by the logic circuitry shown in Fig. 4.1d. This permits us to get the logic circuitry in the smallest number of packages (one 14-pin dual in-line package). However, a more important result of this selection of don't cares is that it allows us to implement the system with only one detector D_1. The saving made by having to install only one detector greatly exceeds the small saving made in reducing the number of ICs. Thus the proper system choice is the usual overriding factor.

Since no extremes in performance are required here, a possible choice of circuitry would be† the 930 type DTL. The interconnection wiring of the gates (all inverters for this example) is shown in Fig. 4.1e.

Thus in Fig. 4.1 we have indicated the inputs to the system design and some of the intermediate results to the final answer. We find that by the proper selection of system implementation, significant savings can be made.

In the preceding example we mentioned timing (the 20-s delay) but did not go into it in detail.‡ We omitted any reference as to obtaining this since in this chapter we will restrict ourselves to considering the class of networks whose output function is only dependent on the combination of the existing values of the input variables. We refer to these circuits whose output value at time t is solely determined by the inputs to the network at time t as *combinational* circuits. Circuits whose output value is dependent upon the value at the inputs at a previous

---

† An open-collector TTL inverter (type 7405) with three external pullup resistors could also be used (see Fig. P4.37).

‡ For simplicity we also omitted any consideration of controlling a yellow light as the intermediate light between green and red.

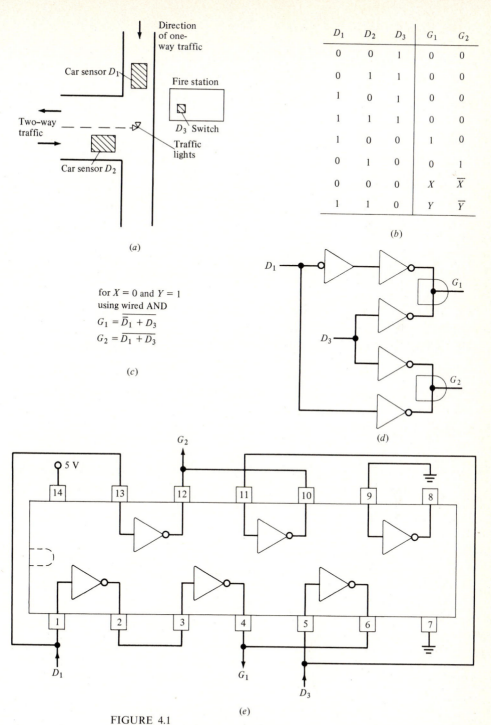

FIGURE 4.1
The traffic-controller problem and solution. (*a*) Plan of the intersection. (*b*) Truth table for G_1 and G_2. (*c*) Expressions for the control of the traffic lights. (*d*) Logic diagram for the expressions in part (*c*). (*e*) Wiring for the circuit that produces the logic diagram in part (*d*). The inverters used permit collector logic.

193

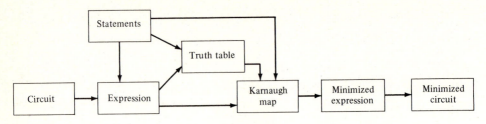

FIGURE 4.2
The possible steps involved in logic analysis and design.

time are referred to as *sequential* circuits. We will discuss sequential circuits in Chaps. 5 and 6.

In this chapter we will cover those aspects of logic design that relate to using SSI and MSI. Having these techniques well in hand is a necessary condition if one is to do LSI design. We find that in any kind of logic design an understanding of the circuit operation from the volts and milliampere view will aid us in getting to an efficient design.

To obtain an overview of what we will cover in this chapter, we have constructed a flowchart in Fig. 4.2. In Chap. 2 we covered some of these techniques; i.e., we saw how to go from a word statement of a problem (the top entry in Fig. 4.2) and generate a minimized expression by utilizing Karnaugh maps.

In this chapter we will introduce some techniques for doing this task while weighing some of the criteria used in determining a suitable minimum solution in circuit form (the rightmost block in Fig. 4.2). Before we do these syntheses, we begin by studying logic circuit *analysis*, where we take an existing logic circuit (the leftmost block in Fig. 4.2) and determine the Boolean expression relating its output variables to its input variables.

EXERCISE

E4.1 Construct the truth table for the traffic-light-control problem of Fig. 4.1*a*, where the city requires that the traffic along the one-way street have priority over the two-way street.

4.1 LOGIC IMPLEMENTATION

In actually designing a system, the designer has usually made, early in the design, some tentative decisions regarding implementation. These relate to selection of the circuit family, degree of integration (i.e., from all SSI to all LSI), and how the ICs will be mounted and interconnected. As usual, these decisions interact, and a change in one can affect the others. Since the method of interconnections and the degree of integration are major factors in determining the cost of a system, we shall touch on these in the next section.

4.1.1 Interconnection and Degree of Integration

A thorough discussion of packaging a digital system and the related question of custom-made LSI circuits are all beyond the scope of this text. However, the methods used for interconnections play such a large role in any system that we cannot ignore them. To introduce these ideas, then, let us consider some of the cost factors that would enter into designing a minicomputer *processor*† which is approximately a 3000-gate digital system. We assume the method of assembly will be to solder IC packages into the holes that are drilled into *printed circuit boards* (PCBs). We further assume that PCB *edge connectors* will be used to contact the printed fingers on the PCBs. The edge connectors will have terminals suitable for *wire wrapping* the wires that will interconnect the various PCBs.

For this relatively standard method of assembly, we have assigned costs‡ for the various items: PCBs, holes, edge connectors, wire wrapping, and ICs. In Table 4.1 we have computed the costs of this system for two different implementations: the first, an all-SSI implementation with an average of three gates per IC package, and the second, a mixed system with 200 SSI, 50 MSI, and 5 LSI packages. The *major assumption* here is that one could find MSI and LSI functions that would permit this second implementation. The likelihood of finding MSI and LSI functions to do almost any subsystem job is increasing continually with time, due to the increasing number of standard types of functions that are commercially available§ in MSI and LSI form.

The major conclusion, as seen on the bottom line of Table 4.1, is that all the parts for such a mixed system could be purchased at approximately one-half the cost of the all-SSI system. Thus we find that it is much more economical to utilize the mixture of ICs (SSI, MSI, and LSI) primarily due to the cost of the PCB and interconnections.

The question then raised is "Why not replace all the remaining 200 SSI in the mixed system with one or two custom-made LSI circuits?" The answer is that this would most likely be uneconomical, since the development and other costs of specialized LSI circuits would usually swamp out the savings in interconnections. Hence the decision to use custom-made LSI is highly dependent upon the projected number of systems to be built and the period of time that one has to build them. Because of this, there is a continuing need to design sizable portions of system circuitry utilizing conventional gates, rather than buying a special

---

† Additional electronics for a complete minicomputer include that of the memory and the peripheral equipment.

‡ These prices are estimates that also assume purchasing material in large volume.

§ Indeed, the Intel 4004 and 8008 ICs are 4- and 8-bit central processor units (CPUs) that are made as a single-chip LSI.

Table 4.1   3000-GATE-SYSTEM COST ANALYSIS

		All SSI (1000 packages)	Mixed (200 SSI, 50 MSI, 5 LSI packages)
1	No. of PCBs (at 100 packages/PCB)	10	3
2	Cost of a 10 in × 14 in PCB without holes	$7.00	$7.00
3	No. of holes/PCB	1800	2000
4	Cost/PCB including cost to drill and plate holes at 1.5¢/hole	34.00	37.00
5	Cost of 80 terminal edge connectors (one edge connector used/PCB)	2.50	2.50
6	Cost of wire wrapping on edge connector/PCB (each wire wrap costs 10¢ with 2 wire wraps/ terminal and 80 terminals/PCB)	16.00	16.00
7	"Interconnection" cost/PCB (items 4 + 5 + 6)	52.50	55.50
8	Total cost of interconnections for system [items (7)(1)]	525.00	166.50
9	Cost of ICs:		
	SSI = $0.20/package	200.00	40.00
	MSI = $1.50/package		75.00
	LSI = $8.00/package		40.00
10	Cost of system power supply (5 V) (20 A all SSI, 10 A mixed)	150.00	100.00
11	Total "component" cost of system (items 8 + 9 + 10)	$825.00	$421.50

(custom-made) IC.† This justifies the amount of time we spend in this chapter on design procedures that utilize simple gates to realize a more complex function.

4.1.2   Selection of Circuit Family

In tackling a practical problem, the selection of the IC family is an important decision. In some cases the requirements are so stringent that only one family is feasible.

As an example where one family type would almost certainly be selected is the use of ECL for implementing the arithmetic and control circuitry of an ultra-high-speed large scientific computer. Here, the small propagation delay would be the overriding factor in the family selection.

As another example of this, consider what kind of logic ICs should go into an unmanned Arctic weather station that has low-speed electronic processing that has to operate for several years before the station is refueled. Clearly power drain is the most vital consideration. Complementary MOS circuits would be a likely first choice because of their extremely low power drain at low operating frequencies.

---

† For certain logic applications a special form of custom-made LSI circuits, read-only memories (ROM), are attractive from the viewpoint of low cost and ease of manufacture.

Table 4.2 FAMILY-SELECTION CRITERIA

1 Circuit performance:
  (*a*) Gain (fanout)
  (*b*) Speed (propagation delay time, loading effects of capacitive loads and transmission lines)
  (*c*) Noise margin
  (*d*) High voltage and/or high current capability outputs
  (*e*) Power drain
2 Logic flexibility:
  (*a*) Types of gates: NAND, NOR, AND, OR (available as quad 2-input, triple 3-input, dual 4-input, etc.) inverters, buffers, AND-OR-NOT, etc.
  (*b*) Availability of expander and wired logic versions of gates
  (*c*) Compatible sequential SSI circuits (flip-flops and monostable multivibrators) available
  (*d*) Compatible MSI and LSI circuits (counters, decoders, arithmetic units, read-only memory, random-access memory) available
  (*e*) Compatible with other SSI families (logic levels and loading)
  (*f*) Compatible *interface* circuits available such as *line* receivers and transmitters, voltage comparators, analog switches, analog-to-digital and digital-to-analog converters

For most applications there are two or more families that can satisfy the most stringent of the specifications, and thus the selection must be based on a combination of factors. These factors† relate to (1) circuit electrical performance and (2) logic flexibility. Table 4.2 lists these factors and also introduces some additional factors that we will discuss in Chaps. 5 and 6.

Table 4.3 lists some of the major IC families available and indicates some

Table 4.3 COMMON IC FAMILIES WITH SOME TYPICAL CHARACTERISTICS

TTL:	
standard	10 mW & 10 ns /gate
high-speed (H series)	22 mW & 6 ns /gate
low-power (L series)	1 mW & 33 ns /gate
Schottky high-speed (S series)	19 mW & 3 ns /gate
low-power Schottky (LS series)	2 mW & 10 ns /gate
DTL:	
standard	8 mW & 25 ns /gate
high-level	40 mW & 125 ns /gate
low-power	1 mW & 60 ns /gate
ECL:	
standard	31 mW & 6.5 ns /gate
high-speed	29 mW & 4 ns /gate
ultra-high-speed	55 mW & 1 ns /gate
MOS:	
p-channel	6-V or 1.5-V threshold; static or clocked
n-channel	1-V threshold; static or clocked
CMOS	(1.5-V threshold)

† Cost, reliability requirements, and maintenance features would also be considerations.

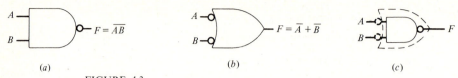

FIGURE 4.3
Representations for the positive-logic NAND gate. (*a*) The output is the negation of ANDed inputs; (*b*) the inputs are negated, then ORed together to form the output; (*c*) superimposed symbols of parts (*a*) and (*b*).

pertinent values for the characteristics of subclasses within a family. Because of the widespread use of TTL, we find that most of the other circuit families have tended to become compatible with TTL, or interface circuits are available to do this.

EXERCISE

E4.2   As a system designer of the 3000-gate system of Table 4.1, you are extremely lucky and find that two LSI packages (which go on one 85-hole PCB) do everything required for your job. Each package has 40 pins, and an 80-pin edge connector will be needed with two wire wraps per terminal. Use $25 as the power-supply cost. What is the average price per LSI circuit that you could pay and still match the cost of the mixed system of Table 4.1?

4.2   ANALYSIS OF LOGIC SCHEMATICS

Before we can begin the task of designing the network (i.e., gates and the wiring between them), we will look at some techniques for analyzing some simple connections of logic blocks. We will find that these *analysis* techniques will be of great use in the later sections on *synthesis*.

4.2.1   Analysis of All-NAND or All-NOR Logic Schematics

In converting schematics to equations, we shall make use of the relationships shown in Fig. 4.3. Figure 4.3*a* shows a NAND gate. Now, recalling DeMorgan's theorem (Theorem 9B in Chap. 2), we can write the NAND-gate output as the ORing of the inverted (i.e., negated) inputs. This can be shown schematically, as in Fig. 4.3*b*, where we show the same two inputs going into an OR gate but with small circles in front of the inputs to indicate inversion. Hence, as in Fig. 4.3*c*, we can show a NAND gate in two ways: (1) in the conventional manner, as a solid line, and (2) in dotted-line form surrounding the first version. We can use either form to represent the circuit, depending upon what is most convenient for us. Figure 4.3*c* is really DeMorgan's theorem for the NAND gate cast in a graphic form.

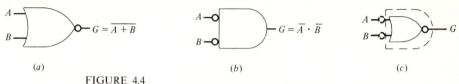

$G = \overline{A + B}$ $G = \overline{A} \cdot \overline{B}$ G

(a)   (b)   (c)

FIGURE 4.4
Representations for the positive-logic NOR gate. (a) The output is the negation of ORed inputs; (b) the inputs are negated, then ANDed together to form the output; (c) superimposed symbols of parts (a) and (b).

In Fig. 4.4 we treat a NOR gate in an analogous manner; thus in Fig. 4.4a we show the conventional form for the NOR and an alternate form in Fig. 4.4b after applying DeMorgan's theorem (Theorem 9A in Chap. 2). In Fig. 4.4c we show both renditions for the NOR gate with the negated-input AND gate in dotted form.

EXAMPLE 4.1   To apply the above ideas, we consider the three-NAND-gate circuit connected as shown in Fig. 4.5. This is referred to as two-level NAND logic or as NAND-NAND logic. In lieu of writing out the expressions and making all the conversions, we simply replace the NAND gate on the extreme right of Fig. 4.5a by its equivalent negated-input OR gate, as shown in Fig. 4.5b. The result

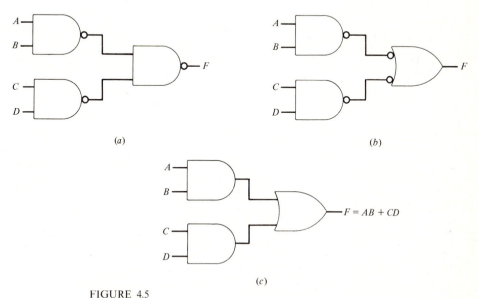

(a)   (b)

$F = AB + CD$

(c)

FIGURE 4.5
NAND-NAND transformed to AND-OR. (a) All gates shown as in Fig. 4.3a; (b) input gates shown as in Fig. 4.3a, output gate as in Fig. 4.3b; (c) the result of ignoring the double negation in part (b).

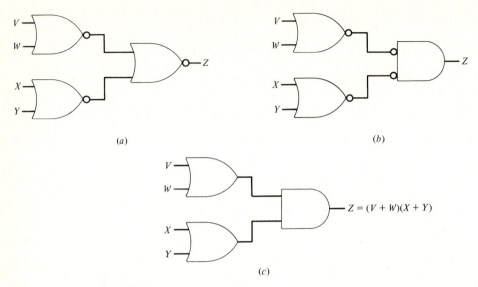

(a)

(b)

(c)

FIGURE 4.6
NOR-NOR transformed to OR-AND. (a) All gates shown as in Fig. 4.4a;
(b) input gates shown as in Fig. 4.4a, output gates as in Fig. 4.4b; (c) the result
of ignoring the double negation in part (b).

of the little circles on both ends of the signal lines (inside the circuit complex) is
that we have inverted twice (a double negation) and hence, for purposes of looking
at the output signal alone, we can eliminate these circles, giving us Fig. 4.5c. In this
we can readily see that the output F is equal to $AB + CD$, as shown in the figure.
Notice that Fig. 4.5c is not an exact representation of the *internal* circuitry in
Fig. 4.5a and b, since the signal in the internal portion of the circuitry has been
inverted. However, for purposes of evaluating the output function, it is equivalent.

We can do the same thing with NOR logic, as shown in Fig. 4.6a and b,
where in the part (b) version we have replaced the rightmost NOR with its
equivalent negated-input AND, and consequently we can write the output as
$Z = (V + W)(X + Y)$. Again, the same considerations hold in that Fig. 4.6c is an
exact representation except for the fact that the signal on the internal line connecting the first level and second-level gate has had the inversion removed. If one
looked in the actual circuit at the internal signal lines with an oscilloscope, one
would see on those lines the inverted signals that would be obtained as the outputs
of the OR gates of Fig. 4.6c.

EXAMPLE 4.2 In another circuit shown in Fig. 4.7a, we have shown another
input E applied as the third input of the second-level NAND gate. On making our
conversion in Fig. 4.7b, we convert the second-level NAND gate to a negated-
input OR gate and further remove the circles when they occur at both ends of a
line. We are left with an unresolved situation for the input E. This we can take care

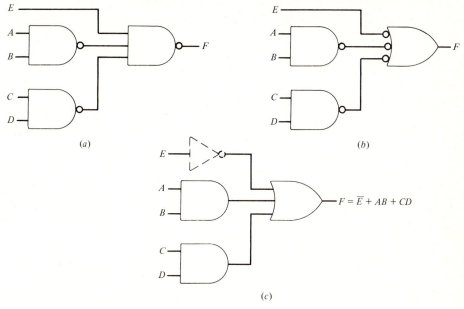

FIGURE 4.7

Using a phantom inverter. (*a*) NAND-NAND logic; (*b*) viewing the input gates as in Fig. 4.3*a* and the output gate as in Fig. 4.3*b*; (*c*) ignoring the double negation requires the addition of the phantom inverter on the line for *E*.

of in Fig. 4.7*c* by drawing in a *phantom* inverter shown in the dotted line which, in effect, takes care of the inversion property. The phantom inverter gives us an inverted input into the conventional OR gate, and the output function again is the proper form; that is,

$$F = \bar{E} + AB + CD$$

EXAMPLE 4.3 Let us now apply the results of the preceding examples to the analysis of various forms of logic gates. Figure 4.8*a* shows a *three-level* NAND-gate connection. This is called three level since one pair of logic signals (*A* and *C*) must travel through three NAND gates before getting to the output.

On making the decision that we want a sum-of-products expression for the output, we convert the output NAND to the negated-input OR, leave the intermediate NAND as is, and convert the leftmost NAND to a negated-input OR, as shown in Fig. 4.8*b*. The resulting expression for the output is

$$F = A + B(\bar{A} + \bar{C})$$

Let us now see if we can obtain a simpler version of this circuit that produces the same result. In Fig. 4.8*c*, we have drawn the Karnaugh map for this three-variable logic expression. Utilizing the results of Chap. 2, we can cover the 1s to

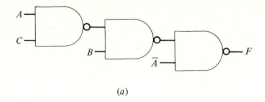

(a)

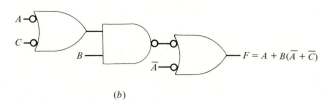

(b)

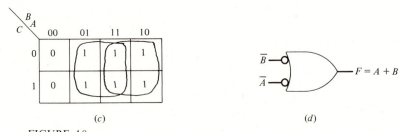

(c)   (d)

FIGURE 4.8
An illustrative example of logic analysis and logic reduction. (a) A given logic
network of NAND gates; (b) one way to view the logic network of part (a);
(c) the Karnaugh map for the output F; (d) a simpler circuit that produces the
same output as part (a).

obtain an expression with the largest possible groupings by utilizing two groups of
four squares each. This gives us the reduced function F as being $A + B$. As shown
in Fig. 4.8d, this can be obtained from a NAND gate with inputs $\bar{A}$ and $\bar{B}$. In terms
of Fig. 4.2, we have progressed along the bottom from a given circuit to a minimized circuit.

EXAMPLE 4.4   Let us further practice our newfound techniques on the three-
level NOR logic shown in Fig. 4.9a. In this case, on making the assumption that
we want the output to be written in the product-of-sums form, we will consider the
output NOR as converted into a negated-input AND gate. We then let the middle
gate stay as it is shown, and we then convert the leftmost gate to a negated-input

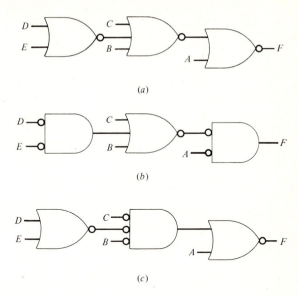

(a)

(b)

(c)

FIGURE 4.9
Interpreting a logic network made from NOR gates. (a) The original network;
(b) a different view of the network in part (a); (c) yet another view of the network
in part (a).

AND gate, as shown in Fig. 4.9b. Having done this, we can, *without any inter-mediate steps*, write the result at the output terminal as

$$F = \bar{A}(C + B + \bar{D} \cdot \bar{E})$$

We could just as well have decided that we would keep the output gate as a NOR function and therefore modify the middle gate to be a negated-input AND and keep the input gate the same. This approach is shown in Fig. 4.9c, where we have expressed the output function F as

$$\bar{F} = A + \bar{C}\bar{B}(D + E)$$

The reader can prove by means of Boolean algebra that this expression is the equivalent to that in the expression in Fig. 4.9b. This proof depends on the repeated application of DeMorgan's two theorems.

EXERCISES

E4.3 Determine the expression for F as a sum-of-products for the circuit of Fig. E4.3.

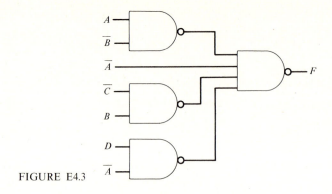

FIGURE E4.3

E4.4   Determine the expression for F as a product-of-sums form for the circuit in Fig. E4.4.

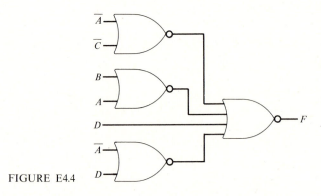

FIGURE E4.4

E4.5   Determine the expression for F in the form of a sum-of-products for the circuit in Fig. E4.5. Eliminate by inspection the extra terms that are redundant.

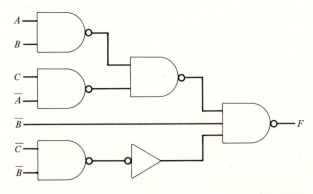

FIGURE E4.5

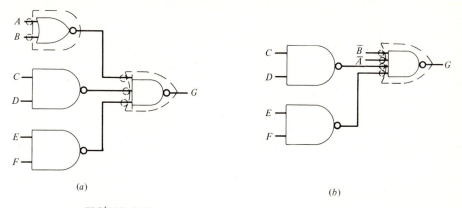

(a)   (b)

FIGURE 4.10
Hybrids of NAND and NOR gates. $G = A + B + CD + EF$. (a) The original
network; (b) a possibly simpler network to replace the one in part (a).

4.2.2 Hybrids of NAND and NOR Logic Blocks

Sometimes we will find circuitry that is not all NAND or all NOR, but is a hybrid
of both NAND and NOR gates. An example of such a circuit is shown in
Fig. 4.10a. If we elect to look at the output as the sum-of-the-products form, that
is, considering the output gate as a negated-input OR gate, we can write the
expression for the output very simply as

$$G = A + B + CD + EF$$

It is worthwhile pointing out that an equivalent logic function to that of
Fig. 4.10a could be generated by the circuitry shown in Fig. 4.10b, where we have
$\bar{A}$ and $\bar{B}$ as inputs to the output NAND gate. The simplification might not be
desired in some applications where the NOR of A and B is desired as a secondary
output. Moreover, there might be excess loading on the $\bar{B}$ and $\bar{A}$ lines where they
are inputs to other gates, so that it would not be feasible to use the circuit shown
in Fig. 4.10b. If the signal lines A and B were lightly loaded and a 3-input NAND
gate was more available than a 4-input NAND gate, the configuration in
Fig. 4.10a could be a more practical realization.

As a further example where we would want to treat an individual gate in
either its normal form or its DeMorgan equivalent form, let us consider Fig. 4.11a.
Here we have a NOR gate with inputs B and C, whose output drives both the
output gate and an intermediate gate that has another input A. We can, as shown
in Fig. 4.11b, show the lower NOR gate as a negated-input AND gate. If we desire
the output written in a form of the product-of-sums, we readily obtain the following expressions:

$$F = (B + C) \cdot (A + \bar{B} \cdot \bar{C})$$
$$= \overline{\bar{B} \cdot \bar{C}} \cdot \overline{[\bar{A} \cdot (B + C)]}$$
$$= \overline{\bar{B} \cdot \bar{C}} \cdot \overline{[\bar{A}B + \bar{A}C]}$$

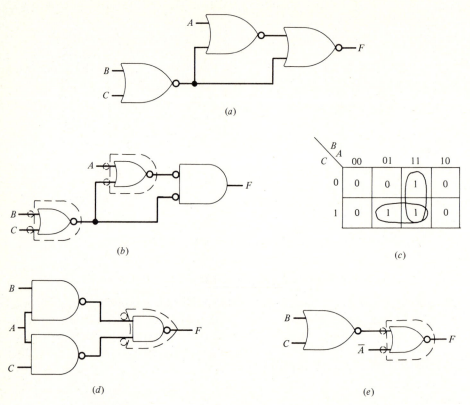

(a)

(b)

(c)

(d)

(e)

FIGURE 4.11
An illustrative example where we view a gate differently so as to get the simplest expression. (a) The original network of NORs; (b) the network in part (a) with alternate representations shown; (c) the Karnaugh map for the function F; (d) a minimum NAND network to realize F; (e) a minimum NOR network to realize F.

Obviously, the first one is the simplest because there are no multiple negations shown in it.

We can learn some important points about logic implementation by further study of the logic generated by the circuit in Fig. 4.11a. First, let us construct a Karnaugh map for this function, as shown in Fig. 4.11c. We have made use of the minimized expression in the sum-of-products notations to obtain $F = BA + CA$, which is best obtained "graphically" from the Karnaugh map. This can also be shown algebraically by the expansion of the factors in the first of the previous expressions:

$$F = (B + C) \cdot (A + \bar{B}\bar{C})$$

$$= BA + B\bar{B}\bar{C} + CA + C\bar{B}\bar{C}$$

$$= BA + CA$$

$$= A(B + C)$$

To realize this by NAND-NAND logic, we can make use of the results shown in the third of the above expressions for F to get the circuit shown in Fig. 4.11d. We can also rewrite the expression as a product of factors, as in the fourth expression, and realize it with NOR-NOR logic, as shown in Fig. 4.11e. If the complemented inputs and NOR gates were available, Fig. 4.11e would be the most efficient way of implementing this logic function.

4.3 SYNTHESIS OF A COMBINATIONAL FUNCTION

In the preceding section we have, by employing the mapping technique of Chap. 2, done parts of a logic synthesis problem. In this section, we will give a more detailed approach to the general problem of how to synthesize a given logic function.

4.3.1 Solving a Synthesis Problem: The Vote Indicator

Let us consider the following example as a representative logic synthesis problem, starting from a word statement of the problem. Three people named A, B, and C are the total population of a desert island. They want an electronic vote-taking system for their parliament that will give them the result of their three votes. They desire an output line labeled M that will be a logic 1 when a majority vote "yes" on a given proposition; otherwise the line output will be logic 0. We will consider an individual's "yes" vote as a logic 1 input to the system. They also desire a second output line labeled U which will have an output level logic 1 when they are unanimous in voting "yes." In order to state this problem in logical form, we can construct a truth table, as shown in Table 4.4. Here we have listed all eight possible combinations of votes, giving us the rows numbered 0 to 7, and in the right-hand columns we have listed the outputs M and U.

Table 4.4 TRUTH TABLE FOR THE VOTE-TAKING SYSTEM

Row no.	C	B	A	M	U
0	0	0	0	0	0
1	0	0	1	0	0
2	0	1	0	0	0
3	0	1	1	1	0
4	1	0	0	0	0
5	1	0	1	1	0
6	1	1	0	1	0
7	1	1	1	1	1

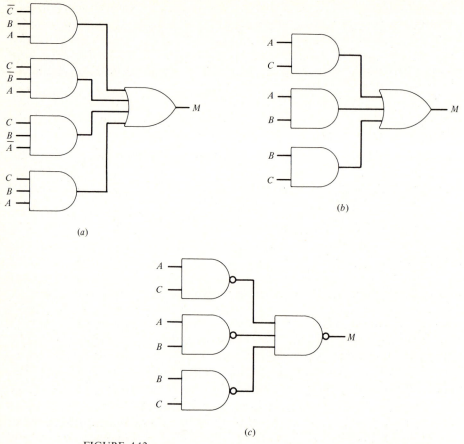

FIGURE 4.12
Logic networks to generate the majority function M. (a) The AND-OR arrangement that generates the standard sum-of-products form; (b) a minimized form of AND-OR realization of part (a); (c) the NAND-NAND version of part (b).

One way of generating logic functions is to write down what is called the *standard sum-of-products* form, which is guaranteed to always give the right result, although rarely is this the most "efficient" circuit. To generate this for the function M, we look at the entries in column M of Table 4.4. For every line where the output is 1, we will write a term in the expression for M. This term will be made up of a *product* of all three input variables; this requires a 3-input AND gate. Each variable (A, B, and C) or its complement ($\bar{A}$, $\bar{B}$, and $\bar{C}$) must be in every such term. The resultant of all these individual terms will now be summed in a single OR gate. The number of inputs to the OR gate is equal to the number of terms (i.e., the number of times that a 1 appears in the M column, which is 4 in our example). Thus, we generate in the circuit of Fig. 4.12a the expression for M given by

$$M = \bar{C}BA + C\bar{B}A + CB\bar{A} + CBA \qquad (4.1a)$$

The three input variables can take on one of two values (either a 0 or a 1); hence we have eight (2^3) possible combinations of input variables. Since in each row of our truth table only one combination of these inputs exists, if the expression for this row (say, row 3) is to be included in the sum, the term $\bar{C}BA$ produces the net result that the output will be a 1 at the time the input variable takes on the values they had in row 3. For the seven other combinations of input variables, this term is 0. Thus each of the four terms will be a 1 only once for the eight possible combinations of input signals. Hence, for the other four combinations, the output expression will be 0.

We can similarly write an expression for U. In this case, there is only a single term:

$$U = CBA \qquad (4.1b)$$

Equation (4.1a) is not necessarily the most efficient way of implementing M, but it is a good beginning for doing that job. Our problem now is to reduce the number of terms necessary to reproduce M, and within these reduced terms to have a lowest possible number of variables involved. That is, we will get the minimum number of OR-gate inputs, and we will have the minimum number of inputs required for the AND gates.

All that we have said so far assumes that we are only interested in implementing the function M and U separately. If there are possible common terms between the two, one would not necessarily independently create two separate functions, but would look for such common terms and attempt to get a most economic solution for implementing both functions. This is a much more difficult task than optimizing each independently and requires a greater degree of art, i.e., practice and experience. Computer techniques are also available for helping to solve these kinds of problems.

In proceeding to minimize this expression, we make use of the Karnaugh map, as shown in Fig. 4.13. We see that we can indeed minimize the number of terms by using the expression

$$M = CA + BA + CB \qquad (4.2)$$

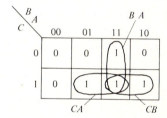

FIGURE 4.13
The Karnaugh map for the function M.

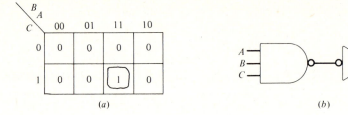

FIGURE 4.14
Realizing the function U. (a) Karnaugh map of U; (b) realization of U using NAND-NAND logic.

With this expression, we reduce the circuit requirement considerably over that of Eq. (4.1a). In Fig. 4.12a is shown the circuit for the standard sum-of-products form which requires one 4-input OR gate and four 3-input AND gates. Utilizing the results of the minimization by the Karnaugh map, we obtain the result shown in Fig. 4.12b, where we now have one 3-input OR gate with three 2-input AND gates. Using our technique for converting this to NAND-NAND circuitry, we have the circuit of Fig. 4.12c.

In Fig. 4.14a, we show the Karnaugh map for our functions for the unanimous "yes" vote U, which is not reducible and thus required all three inputs for the single term. The realization with NAND-NAND logic is shown in Fig. 4.14b.

$$U = \overline{\overline{ABC}} = ABC \qquad (4.3)$$

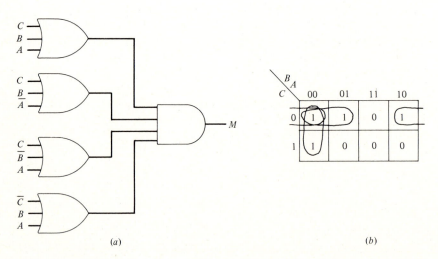

FIGURE 4.15
Generating the product-of-sums form. (a) The standard product-of-sums form for M; (b) the Karnaugh map for $\overline{M}$.

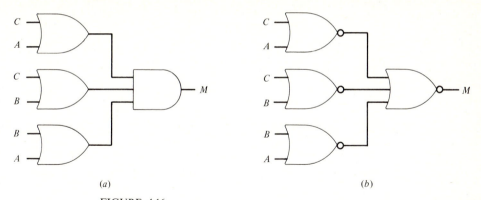

(a)   (b)

FIGURE 4.16

Obtaining M by using NOR gates. (a) The OR-AND minimized circuit to obtain M; (b) the NOR-NOR version of part (a).

Another way to get the logic for our island parliament is to use what is termed the *standard product-of-sums* approach. The standard product-of-sums form of M is shown in Fig. 4.15a, where each 0 entry for M in Table 4.4 has to be generated by an OR gate. These OR-gate outputs are ANDed to form M. Since each OR-gate output is 0 only when all its inputs are 0, we see that the circuit in Fig. 4.15a gives all the required 0s. An alternate way of generating the product-of-sums network is to make use of DeMorgan's theorem and work on the 0s in M, which are the 1s in $\bar{M}$. We begin by minimizing the expression for $\bar{M}$ in the sum-of-products form, as shown in Fig. 4.15b. Again grouping together the largest group of 1s that we possibly can, we obtain the following result:

$$\bar{M} = \bar{C}\,\bar{A} + \bar{C}\,\bar{B} + \bar{B}\,\bar{A} \qquad (4.4a)$$

Applying DeMorgan's theorem to this result, we obtain

$$M = (C + A)(C + B)(B + A) \qquad (4.4b)$$

This gives us the desired result, since M is now written as a product of sums.

The realization of Eq. (4.4b) is shown in Fig. 4.16a, using three OR gates with their outputs ANDed together. The equivalent NOR-NOR logic is shown in Fig. 4.16b.

Let us again look at the realization of the expression of Eq. (4.2) for M. If it were desired to have the OR of any of the two input variables for some other function in the parliament building, we could make use of that output in another set of circuits that generate M, as shown in Fig. 4.17a through c. Here we have shown all possible renditions, where we have made use of two of the three variables ORed together, and have shown that if this input were available, we would require only three 2-input NAND gates for generating the majority function M.

Returning to our function U, if we again consider $\bar{U}$ and minimize it by the sum-of-products form, we would get the grouping as shown in the Karnaugh map

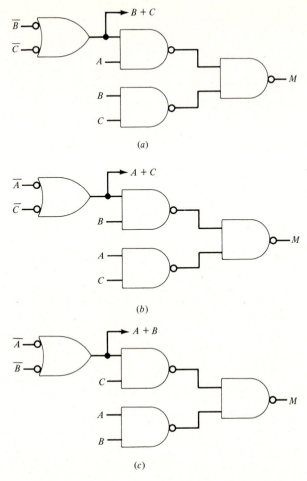

FIGURE 4.17

Alternate ways to generate M by using three-level NAND logic. (a) A version where $B + C$ is desired; (b) a version where $A + C$ is desired; (c) a version where $A + B$ is desired.

in Fig. 4.18a. The resulting expression written in the sum-of-products form for $\bar{U}$ is given in Eq. (4.5a), and consequently applying DeMorgan's theorem to U, we have Eq. (4.5b), which is identical to our previous result.

$$\bar{U} = \bar{C} + \bar{B} + \bar{A} \qquad (4.5a)$$

$$U = \overline{\bar{C} + \bar{B} + \bar{A}} = CBA \qquad (4.5b)$$

Figure 4.18b shows this circuit function implemented with a NOR gate.

Let us now assume that we only have the true inputs A, B, and C and that we want to generate both U and M. One way of doing this efficiently if both NAND and NOR gates are available is shown in Fig. 4.19. In Fig. 4.19a we have redrawn

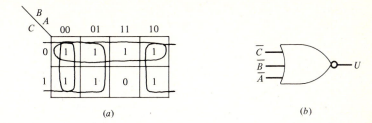

(a) (b)

FIGURE 4.18
Generating the function U. (a) The Karnaugh map of $\overline{U}$; (b) realizing U with one NOR gate.

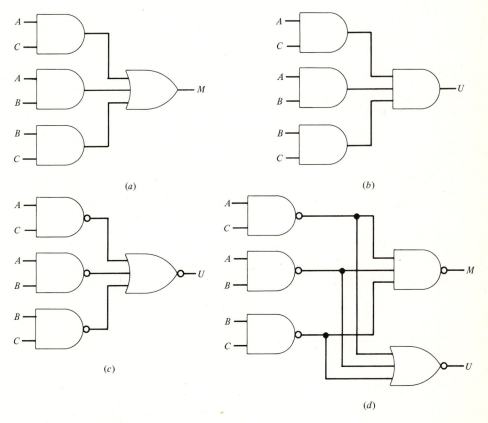

(a)

(b)

(c)

(d)

FIGURE 4.19
Generating both M and U with true inputs only. (a) Generating M by using AND-OR gates; (b) generating $U = ABC$ with the input gates of part (a); (c) the equivalent to part (b) by using NAND and NOR gates; (d) the complete circuit to generate both M and U by using the hybrid of NAND and NOR gates.

Fig. 4.12*b* to show how *M* is generated. These input AND gates can also be used to generate *U* by using the circuit of Fig. 4.19*b*. This can be verified by the use of Theorem 3B from Chap. 2. Hence, with the use of a NOR gate as in Fig. 4.19*c* (treated as a negated-input AND gate), we can generate *U*. Putting the total system together, we have Fig. 4.19*d*, which shows a total of five gates to generate the whole function with only the true inputs. Table 4.5 summarizes the results and also shows the loading on the input signal lines. The hybrid solution is obviously superior on the basis of the least number of gates and the lowest loading on the input lines.

Table 4.5   SUMMARY   OF   CIRCUITS   FOR   VOTE-TAKING SYSTEM

	NAND-NAND	NOR-NOR	Hybrid
M	3 2-input, 1 3-input	3 2-input, 1 3-input	
U	1 3-input, 1 inverter	3 inverters, 1 3-input	
Total	2 3-input	2 3-input	2 3-input
	3 2-input	3 2-input	3 2-input
	1 inverter	3 inverters	
Loading on line			
A	3	3	2
B	3	3	2
C	3	3	2

EXERCISES

E4.6   Use the truth table of Fig. 4.1*b*, with $X = 0$ and $Y = 1$, to show the standard sum-of-products circuit that generates G_1. Use NAND gates.

E4.7   Show the standard product-of-sums circuit that generates the function shown in the Karnaugh map of Fig. 4.8*c*. Use NOR gates.

E4.8   Repeat Exercise E4.6, but now determine the minimum sum-of-products circuit. Realize it with NAND gates and inverters.

E4.9   Repeat Exercise E4.7, but now determine the minimum product-of-sums circuit. Realize it with NOR gates and inverters.

4.3.2   A Further Example of Logic Synthesis: The Binary Adder

For our next example, we consider the single-bit full binary adder. A binary adder permits one to perform the arithmetic operation of integer addition. To do this,

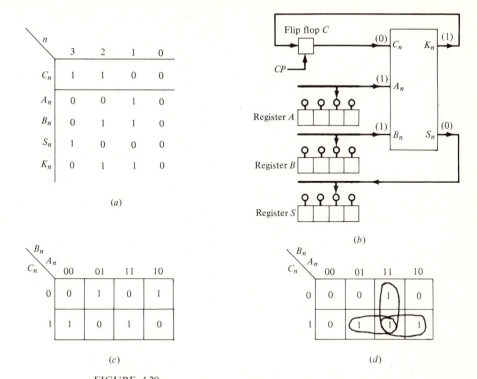

n	3	2	1	0
C_n	1	1	0	0
A_n	0	0	1	0
B_n	0	1	1	0
S_n	1	0	0	0
K_n	0	1	1	0

(a)

(b)

(c)

(d)

FIGURE 4.20

The single-bit full binary adder. (a) The results of adding two 4-bit numbers. (b) The block diagram of a full adder and associated storage circuits shown for the condition $n = 1$. Whenever the three switches are moved to the left 1 position a pulse is applied on the CP line so that the *flip-flop* C will take on the value of K. (c) The Karnaugh map for S_n. (d) The Karnaugh map for K_n.

Boolean variables are assigned to represent the variable values in the standard positional representation of binary numbers.

The truth table for a binary adder is shown in Table 4.6. The inputs to this one-position adder are the three inputs from the nth position in a binary number: A_n, B_n, and the input carry C_n, which is the output carry K_{n-1} from the previous position. The output consists of the sum S_n, which stays at the nth position, and the carry K_n, which is used as C_{n+1}, the input carry of the next higher-order position. The student should verify that this table does, indeed, represent the results of binary addition. Figure 4.20a shows the result for adding A ($= 0010_2$) to B ($= 0110_2$) and obtaining S ($= 1000_2$). The carry input and output for each column are shown in the rows labeled C_n and K_n.

Figure 4.20b shows a 1-bit binary adder, where the inputs are obtained from *registers* that store in cells the binary values used to represent A and B. The A, B, and S registers each have four cells, as in Fig. 4.20b. For our purposes we can

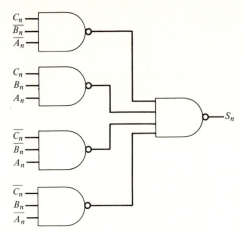

FIGURE 4.21
NAND-NAND realization for S_n.

consider addition as being done 1 bit at a time, with the inputs being switched along the top of the registers from right to left as n increases. The input to the S register is also stepped along concurrently with the switching of the A and B registers. The carry register C only has to be 1 bit long. As we shall see in the next chapter, this cell is called a *flip-flop*. At the time the switch steps, the information at the carry output K_n is transferred into the input of the C cell.

In Fig. 4.20c we have shown Table 4.6 as the Karnaugh map for S_n, while in Fig. 4.20d we have shown a Karnaugh map for K_n. Looking at the Karnaugh map for S_n, we cannot find any groupings which would reduce the complexity of this expression, and consequently we write out all the terms. Therefore, the minimized form is the standard sum-of-products form.

$$S_n = C_n \bar{B}_n \bar{A}_n + C_n B_n A_n + \bar{C}_n \bar{B}_n A_n + \bar{C}_n B_n \bar{A}_n \qquad (4.6a)$$

Table 4.6 TRUTH TABLE FOR BINARY ADDER

C_n	B_n	A_n	S_n	K_n
0	0	0	0	0
0	0	1	1	0
0	1	0	1	0
0	1	1	0	1
1	0	0	1	0
1	0	1	0	1
1	1	0	0	1
1	1	1	1	1

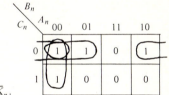

FIGURE 4.22
The Karnaugh map for $\bar{K}_n$.

For the output carry K_n we again see our friend, the majority function, from the vote-taking example, and again have its minimized sum-of-products form as

$$K_n = C_n A_n + B_n A_n + C_n B_n \qquad (4.6b)$$

We can realize the S_n function utilizing NAND-NAND logic, as shown in Fig. 4.21. The K_n function could be realized separately, as done in the previous section. However, we want both these functions simultaneously. Hence we would be interested in solutions that would make use of shared† common or intermediate products. One way of doing this is to recast the expression for K_n by finding the simplest expression for its complement $\bar{K}_n$. Doing this by means of the Karnaugh map of Fig. 4.22, we obtain

$$\bar{K}_n = \bar{B}_n \bar{A}_n + \bar{C}_n \bar{B}_n + \bar{C}_n \bar{A}_n \qquad (4.7)$$

We can now look at Eq. (4.6a), and with a considerable amount of insight and/or luck, we can obtain the following expression for S_n (as a function of $\bar{K}_n$):

$$S_n = C_n(\bar{K}_n) + B_n(\bar{K}_n) + A_n(\bar{K}_n) + A_n B_n C_n \qquad (4.8)$$

To realize the expression in Eq. (4.8), we first generate K_n, as shown in the lower half of Fig. 4.23a. Using this as an input to an inverter, we have $\bar{K}_n$, which is then used in the expression for S_n. This is done in the upper five gates of Fig. 4.23a to form the S_n function at the upper output. The net result of this not-so-obvious NAND circuit is to give us the sum function with five gate delays and the carry function with two gate delays.

Table 4.7 shows the component count for the two full-adder realizations. Assume we want to realize them by utilizing NAND gates that are available in a conventional 14-lead dual-in-line package. We see that the straightforward NAND-NAND logic requires all the gates from four packages.‡ Our not-so-obvious NAND circuit has a longer delay in its S_n output (five units) delays, but only two units of delay (the same as in the NAND-NAND logic) for the carry. Both renditions require four packages, but the second has five inverters that were unused in the adder but are available for use elsewhere in the system.

---

† This concept is discussed more fully in Sec. 4.4.1.
‡ We make use of the fact that a 4-input NAND gate can be used as a 3-input NAND gate, as will be described in Sec. 4.5.1. Thus we always start with the gate with the most inputs (4) and if any gates of the same kind are left over from the package, these are used for the gates that require less inputs (3).

Table 4.7 FULL ADDER REALIZED WITH ONLY TRUE INPUTS

Type of gate	Straightforward NAND-NAND version		Version utilizing $\bar{K}_n$ in S_n	
	No. required	ICs used	No. required	ICs used
4-input	1	→ 1 Dual 4-input	1	→ 1 dual 4-input
3-input	5	→ 2 triple 3-input	2	→ 1 triple 3-input
2-input	3	→ 1 quad 2-input	6	→ 1 quad 2-input
1-input	3		1	→ 1 hex inverter
	Left over: 0		Left over 5	
	Total: 4 ICs		Total: 4 ICs	
Loading on input lines				
A	5		4	
B	5		4	
C	5		4	

We have chosen the above example of constructing an adder to illustrate some of the common problems in many logic designs. In actual practice one would not build an adder from SSI gates but would instead use a commercially available MSI package. A logic diagram of a standard commercial unit† that contains two such binary adders is shown in Fig. 4.23b. This circuit is designed so that it is to be used where the bit positions are adjacent, and so the carry generated by the low-order bit is not connected to an output pin.

† TTL type 7482.

FIGURE 4.23
Binary full adders made by using $\bar{K}_n$. (a) NAND circuitry to generate S_n and K_n in a 1-bit full adder. (b) The logic diagram for a 2-bit binary full adder (TTL type 7482). C_1 is the input-carry bit, while K_2 is the output-carry bit.

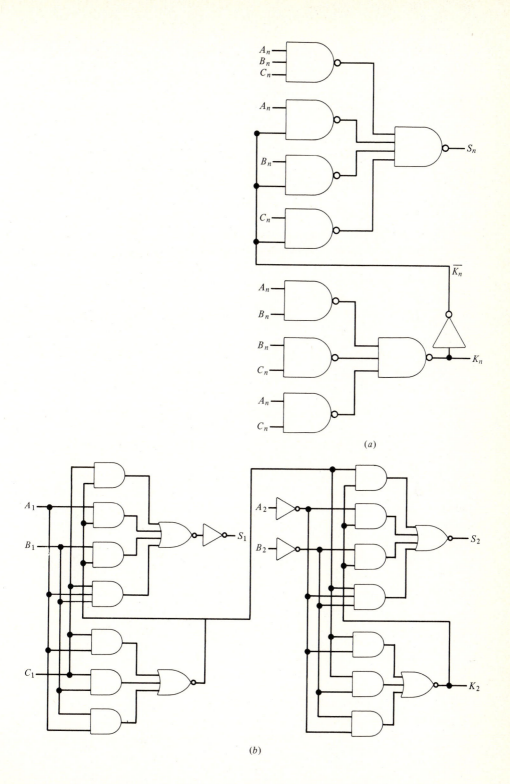

(a)

(b)

EXERCISE

E4.10   In reference to Fig. 4.20a, if $A = 1011_2$ and $B = 0110_2$, find the value of S. This is an *overflow* answer since register S does not have the capacity to be able to store this value.

4.3.3   The EXCLUSIVE-OR and the Digital Comparator

A large and growing number of functions are commercially available in MSI form. Since the EXCLUSIVE-OR function is widely used in digital systems, it is found in many of the families discussed in Chap. 3.

The EXCLUSIVE-OR truth table is given in Table 4.8, with inputs A and B and output X. The Karnaugh map given in Fig. 4.24a shows that no simplification is possible; hence we have the standard sum-of-products function given by

$$X = \bar{B}A + B\bar{A} = A \oplus B \qquad (4.9a)$$

The circled plus sign is a convention used to indicate the EXCLUSIVE-OR operator.†

We can realize the EXCLUSIVE-OR as shown in Fig. 4.24b, with NAND-NAND logic with two gate delays. Note that we need both true and complement inputs of the variables A and B.

Another realization is generated by adding in two redundant terms that are logical 0s in Eq. (4.9a). We do this by rewriting Eq. (4.9a) as

$$X = \bar{B}A + A\bar{A} + B\bar{A} + B\bar{B} \qquad (4.9b)$$

We then can factor this, giving

$$X = A(\bar{B} + \bar{A}) + B(\bar{B} + \bar{A}) \qquad (4.9c)$$

This expression is realized in Fig. 4.24c. We can replace the leftmost OR gate by a NAND gate, as shown in Fig. 4.24d. In this version we only need the two true inputs, but the overall circuit delay is three gate delays. Figure 4.24e shows a logic

Table 4.8   TRUTH   TABLE FOR EXCLUSIVE-OR

A	B	X
0	0	0
0	1	1
1	0	1
1	1	0

† The EXCLUSIVE-OR operation is commutative; i.e., it can be proven (by truth table comparisons) that for three variables C, B, and A,

$$C \oplus (B \oplus A) = (C \oplus B) \oplus A = (C \oplus A) \oplus B$$

so that we can write the above result as $C \oplus B \oplus A$.

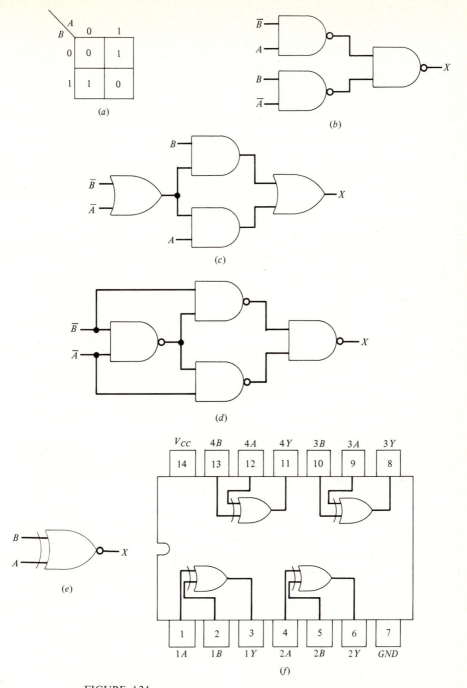

FIGURE 4.24
The EXCLUSIVE-OR function. (*a*) Karnaugh map for the EXCLUSIVE-OR function; (*b*) NAND-NAND implementation for the EXCLUSIVE-OR; (*c*) another realization for the EXCLUSIVE-OR; (*d*) NAND circuit realization for the circuit in part (*c*); (*e*) logic symbol for the EXCLUSIVE-OR gate; (*f*) top view of a quad of 2-input EXCLUSIVE-OR gates.

symbol that is widely used for the EXCLUSIVE-OR gate. Four of these gates in one chip are commercially available as the MSI circuits† shown in Fig. 4.24*f*. Demonstration D4.1 applies these gates to a simple problem (P4.28).

The digital comparator is another widely used MSI circuit. The output F of a single-bit comparator with inputs A and B is given as

$$F = AB + \bar{A}\bar{B} \qquad (4.10)$$

so that if $A = B = 0$ or $A = B = 1$, then $F = 1$. This is related to the EXCLUSIVE-OR of A and B by

$$F = \bar{X} = \overline{A \oplus B} \qquad (4.11)$$

Because of this relationship this function is sometimes called the EXCLUSIVE-NOR.

Commercial digital comparators are available‡ that compare two 4-bit words, W_A and W_B, and give one output that indicates equality $(W_A = W_B)$ and also two other outputs that indicate greater than $(W_A > W_B)$ and less than $(W_A < W_B)$.

EXERCISES

E4.11   A half-adder satisfies the first four rows of the truth table given in Table 4.6, where the C_n input variable is not used. Construct a half-adder using the minimum number of gates. You can use inverters, NAND gates, and EXCLUSIVE-OR gates.

E4.12   Given the function F in the Karnaugh map of Fig. E4.12, generate it by using only two EXCLUSIVE-OR gates. [*Hint:* Write out the expression and gather like terms together. Also be on the lookout for the complement of the EXCLUSIVE-OR expression, Eq. (4.10).]

C \ $^{B}_{\ \ A}$	00	01	11	10
0	0	1	0	1
1	1	0	1	0

FIGURE E4.12

4.4   THE DON'T CARE CONDITION

In many problems, entries in the truth table or Karnaugh map are not given for all possible combinations of input variables. This occurred in our traffic-light example at the beginning of this chapter. When this occurs in a problem where one is attempting to synthesize a "minimum" circuit, one can then insert values for the missing entries which will give the most economical or efficient solution to the problem.

To illustrate this, assume that we are given a function in Karnaugh map

† TTL type 7486.
‡ TTL type 74L85.

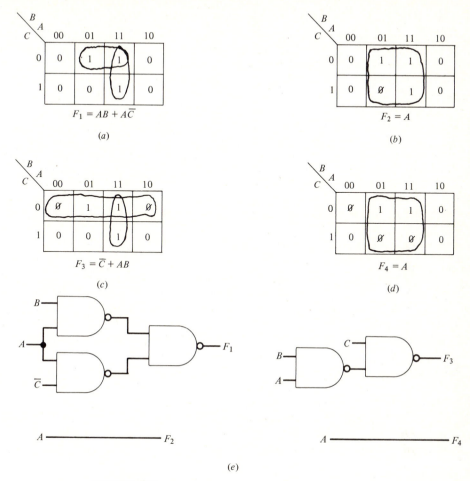

FIGURE 4.25
An illustrative example of the effects of don't care conditions. (*a*) The original function with all entries specified; (*b*) with one don't care condition specified in part (*a*); (*c*) with two don't care conditions specified in part (*a*); (*d*) with three don't care conditions specified in part (*a*); (*e*) realizations of parts (*a*) to (*d*) by using NAND-NAND logic.

form, as in Fig. 4.25*a*. Here we are given eight specific entries: three entries where a 1 is required and the other five entries where 0s are required. The minimized expression is shown underneath the map. Now, looking at the Karnaugh map in Fig. 4.25*b*, we are still given the same 1s requirement as in Fig. 4.25*a*, but now under the condition that when $C = 1$, $B = 0$, and $A = 1$, the function is not specified so it can be either 1 or 0; that is, we have a don't care condition. Hence, we place the symbol $\emptyset$ in this location, which can stand either for a 0 or a 1, depending upon how you, the designer, decide to utilize this extra degree of freedom to obtain a "minimum" circuit. For the version in Fig. 4.25*b*, we obtain on minimization a new function F_2 that generates the required 1s and 0s.

However, here we have elected to evaluate $\emptyset$ as a 1, and we get the simple result that $F_2 = A$.

In Fig. 4.25c, we again have the same functional requirements on the 1s as in Fig. 4.25a, but now we are given in two other places that the entry is a don't care condition. This generates another function F_3, which is the minimum form. In Fig. 4.25d we have don't care conditions in three places, and now we have again the simplest expression, $F_4 = A$. Note that it is not necessary to assign $\emptyset$ the same value throughout the truth table; i.e., $\emptyset$ is not a variable that takes on the value of 0 or 1, rather it is an indication of don't care and can be set to 0 in one entry of the truth table and then set to 1 in another entry of the same truth table.

Using NAND-NAND logic synthesis, we have shown in Fig. 4.25e realizations for the four different functions discussed. Obviously, if we do have the degree of freedom given by the $\emptyset$s, we can see that F_2 and F_4 are the simplest variations since no gates are required.

In practical work, one often is not minimizing for a single function but is attempting to get multiple outputs. In that situation, the use of the don't care condition becomes much more complicated. Then it is possible to share circuitry by the use of terms that are intermediate results obtained in generating one of the output functions, i.e., the binary-adder example. In doing this, a considerable amount of experience and skill is necessary to be able to take full advantage of the extra design freedom given by don't care conditions.

EXERCISE

E4.13 Given the Karnaugh map of F in Fig. E4.13:
  (a) Determine the minimum sum-of-products expression for F.
  (b) Determine the minimum product-of-sums expression for F.

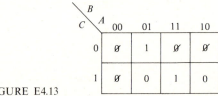

FIGURE E4.13

4.4.1 Code Conversion

A common problem in digital systems is to convert a number or a character from one code to another. For example, consider that a two-position binary number, which could be the last two significant bits in an 8421 BCD code, is to be displayed to a machine operator as a decimal number. For this application we would desire a logic circuit that has four outputs, S_0 through S_3. Each output would drive a light bulb that would indicate the numbers 0, 1, 2, and 3. Thus, when the inputs $B = 1$ and $A = 0$, the integer value is 10_2 and the operator sees the decimal number 2 light up. Our problem is to construct a multi-output logic

$$S_0 = \overline{\overline{B} + A} = \overline{B}\overline{A}$$

$$S_1 = \overline{B + \overline{A}} = \overline{B}A$$

$$S_2 = \overline{\overline{B} + A} = B\overline{A}$$

$$S_3 = \overline{\overline{B} + \overline{A}} = BA$$

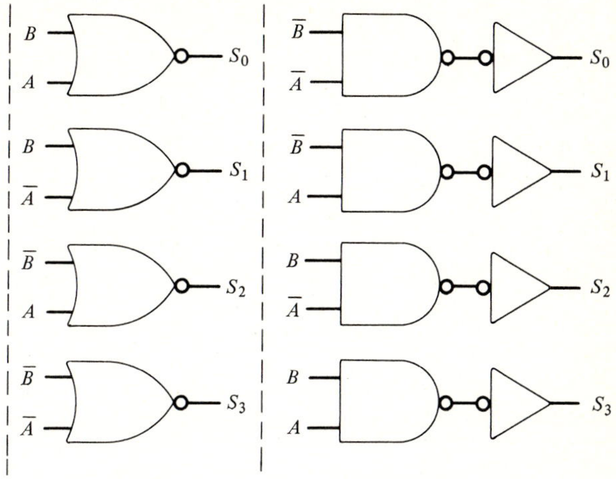

(a)   (b)   (c)

FIGURE 4.26

Code-conversion example: BCD to Decimal. (a) The minimized expressions for the output lines; (b) realizations for the expressions in part (a) by using NOR gates; (c) realizations for the expressions in part (a) using NAND gates.

network that satisfies the truth table given in Table 4.9. Here the inputs are the binary code for the number given in columns B and A, and the outputs are the signal lines shown in the columns S_3, S_2, S_1, and S_0. We show in Fig. 4.26a the minimized Boolean expressions for these terms; in Fig. 4.26b the circuitry that generates the outputs using NOR logic; while in Fig. 4.26c we show the circuitry that generates the outputs using NAND logic. If both true and complement values of the input variables are available, then the NOR logic is a simpler one to implement. However, if only the true values of the variables are to be used, then one would have to carefully analyze the situation and possibly consider a mix of NOR and NAND logic.

Let us now consider the other side of the problem, where the operator is to press only one of four switches labeled as a decimal number which will generate its binary-coded equivalent numbers for the machine to use. Now we have to make a code conversion from decimal to binary. We can use our previous truth table again, except in this case we have as our inputs the signals S_0, S_1, S_2, and S_3, and the outputs will now be variables B and A. An important point to remember is

Table 4.9   TRUTH   TABLE   FOR
DECODER

B	A	S_3	S_2	S_1	S_0
0	0	0	0	0	1
0	1	0	0	1	0
1	0	0	1	0	0
1	1	1	0	0	0

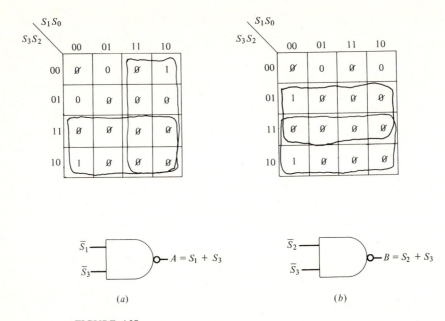

FIGURE 4.27

Code-conversion Example: Decimal to BCD. (*a*) Karnaugh map and solution for
A; (*b*) Karnaugh map and solution for *B*.

that with four input variables the number of possible combinations of input
variables is $16(2^4 = 16)$. Thus Table 4.9 should have 16 lines if it were a complete
truth table. Even though there are 16 possible combinations, we are only required
to handle four of these exactly; that is, there are only four lines where we have
specific assigned values to *B* and *A*. For the other 12 combinations, we will have
don't care as our entry into the value of *B* or *A*. In Fig. 4.27*a*, we have the
Karnaugh map for *A* with the don't care conditions indicated by the entry Ø. The
minimum expression of *A* is written next to the gate where we have decided that
all but two of the Øs take on the value 1. In Fig. 4.27*b* we have the Karnaugh map
for *A*. Here again, all but two of the Ø terms have been assigned the value of 1 to
minimize the expression.

In most logic design problems it is necessary to generate multiple outputs for
a set of input variables rather than a single output, as we have generally discussed.
The previous example of code conversion is an example of design with multiple
outputs in mind. The example of the binary adder with sum and carry outputs is
another.

4.4.2 Multiple Outputs

With multiple outputs, it is sometimes feasible to make use of techniques other
than the classical one of minimizing on an individual function to get the most

efficient design. This is best explained by means of an example, such as that shown in Fig. 4.28a and b. Here we are given two functions shown in the Karnaugh maps with the function F_1, that in *a classical sense* in applying the conventional method would have been minimized with the expression under the Karnaugh map, and implemented by the three-gate NAND circuit shown in Fig. 4.26c.

However, if we want to generate this function F_1 and at the same time the function F_2 shown in the Karnaugh map of Fig. 4.28b, we should keep in mind that we can make use of possible joint terms. For instance, if considered individually, F_2 normally would have been generated by the expression shown in Fig. 4.28b, and the circuit would have been as shown in Fig. 4.28d. However, let us look now at these two Karnaugh maps jointly and realize that we can generate a common term in the lower left corner of both maps which can be shared between these functions. After doing this, the expression for F_1 is as shown in Fig. 4.28e, and for F_2 as shown in Fig. 4.28f.

Implementing this result with NAND gates, as shown in Fig. 4.28g, we now would require five NAND gates, where previously with the techniques we used we required six NAND gates. Because of our simple example, the savings are not very spectacular. However, these considerations can become of considerable importance in a very complex system.

EXERCISES

E4.14   Design the two code converters for converting "both ways" similar to that in the first example of Sec. 4.4.1, except that the machine code, instead of being simple binary, is in Gray code, i.e.,

$$00_2 \leftrightarrow 0_{10} \qquad 01_2 \leftrightarrow 1_{10} \qquad 11_2 \leftrightarrow 2_{10} \qquad 10_2 \leftrightarrow 3_{10}$$

E4.15

(a) Given the Karnaugh map of F in Fig. E4.15a, design a minimum NAND logic circuit that generates F by using the minimum number of gates (3-input, 2-input, and inverters). Only the true of each variable is available to you.

(b) Repeat part (a) for the function G in Fig. E4.15b.

(c) With the circuit for G available to you, now redesign part (a) for minimum gate count.

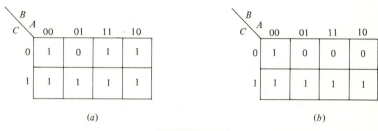

(a)   (b)

FIGURE E4.15
(a) Karnaugh map of F; (b) Karnaugh map of G.

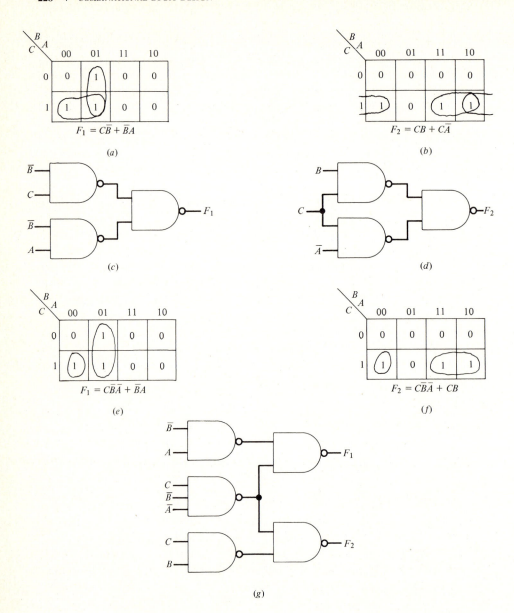

FIGURE 4.28
An example with multiple outputs. (a) The Karnaugh map for F_1; (b) the Karnaugh map for F_2; (c) NAND-NAND realization to the solution in part (a); (d) NAND-NAND realization to the solution in part (b); (e) the Karnaugh map and alternate solution of F_1; (f) the Karnaugh map and alternate solution of F_2; (g) NAND-NAND realization of F_1 and F_2 by using the solutions in parts (e) and (f).

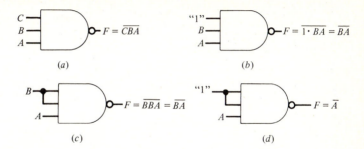

FIGURE 4.29
Ways to handle unused inputs to NAND gates. (*a*) A 3-input NAND gate; (*b*) connecting one input to a 1 makes it into a 2-input NAND gate; (*c*) tying two inputs to one logic line also converts it into a 2-input NAND; (*d*) making it into an inverter.

4.5 CIRCUIT CONSIDERATIONS

By properly using some circuit considerations, improved usage of logic circuits can be obtained when actually implementing a logic circuit. These circuit ideas are related to such things as unused inputs, the difference between the fanout with a high state in the output and the fanout with a low state in the output, and a technique which we have previously referred to as *collector logic*.

4.5.1   Unused Inputs

The question of unused inputs can be best explained by means of an example. Consider Fig. 4.29*a* as a 3-input NAND gate. Now let us assume that we are only interested in using this as a 2-input NAND gate. What to do with the unused input? As shown in Fig. 4.29*b*, if we place a 1 on the input that was called *C* previously, the new output is that of a 2-input NAND gate.

Alternatively, we could have tied both the *B* and the former *C* line together, as shown in Fig. 4.29*c*, and obtained the same output, as shown in Fig. 4.29*b*. However in this case it is possible, but not always necessarily so, that we will have now overloaded the circuit that drives the signal line that has the multiple inputs on it. As another example of unused inputs, Fig. 4.29*d* shows what happens when 1s are applied to the inputs previously labeled *C* and *B*. The output is now simply the inverted version of *A*, so that the 3-input gate acts as an inverter.

We can apply similar concepts to NOR gates. Figure 4.30*a* shows a 3-input NOR gate that we want to use as a 2-input NOR gate. Again we have two ways of doing this: (1) by returning the unused input to the 0 level, as shown in Fig. 4.30*b*, or (2) by tying the unused input to another driven input, as shown in Fig. 4.30*c*. Whether this increased loading on the *B* line overloads the circuit driving the *B* line depends on the number of loads on it and the type of circuit used for the driven gate.

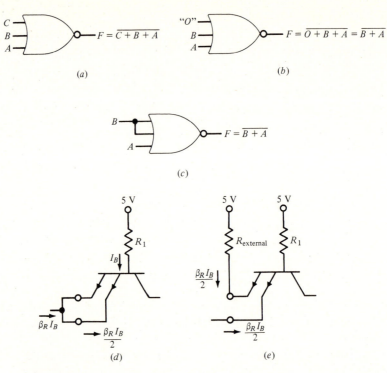

FIGURE 4.30
Unused inputs of NOR gates. (a) A 3-input NOR gate; (b) converting it into a 2-input NOR; (c) an alternate conversion into a 2-input NOR; (d) a 2-input TTL input with both inputs tied together; (e) a 2-input TTL input with one input returned to HIGH reduces the sourcing current required from the circuit driving it.

In the case of DTL and TTL, if multiple inputs into a gate are tied together, the driving circuit does not have to sink any additional current when the multiple inputs are low. However, for the case where the signal input into the "strapped" inputs is high, more sourcing current is required by strapping, as in Fig. 4.30d, as compared to returning the extra input to a fixed high level, as in Fig. 4.30e. In many newer TTL circuits with totem-pole outputs that have Darlingtom active pullups (Figs. P4.22 through P4.24), the high-level fanout N_H is much larger than N_L, so that strapping rarely causes a problem.

EXERCISE

E4.16   A low-power TTL gate is to be used to drive a number of regular Series 74 2-input gates, each of which is connected as an inverter. When its output is HIGH, the low-power gate can source 100 μA, and when LOW, it can sink 2 mA. The maximum current into the input of the regular TTL gate when HIGH is 40 μA.

The maximum current out of the input of the regular TTL gate when LOW is 1.6 mA. Referring to the output of the low-power gate, what are its HIGH and LOW condition fanouts?

4.5.2 Wired Logic

We discussed collector logic in Chap. 3 in regard to TTL circuitry, where we said that with the totem-pole output circuit, which is generally† used in TTL, we are prohibited from doing collector logic. Other logic families besides the standard TTL use the totem-pole output circuits, and there, also, we cannot do collector logic. The term *collector logic* is a description of the ability to create an additional logic function by the tying together of outputs from transistor-inverter stages. This is also called wired-AND, implied-AND, or dot-AND. If any of the logic gates had its output level LOW, then the output of all the gates that are tied at the common-collector point is LOW. This, in effect, for positive logic, is a wired-AND.‡ We will usually refer to it as collector logic, since this tells us specifically how the circuit functions. For positive logic, collector logic produces the AND function. This can be seen in Fig. 4.31a, since if any of the individual outputs "unwired" would go to 0, the composite "wired" output goes to 0 with collector logic. Only when *all* outputs individually would be 1, or in their HIGH state, will the resultant "wired output" be at its HIGH state.

With collector logic, we often can save logic gates compared to other methods. As an example, consider Fig. 4.31b. Here we have two NAND gates, one with inputs A and B, the other with inputs C and D, producing outputs X and Y, respectively. We then consider the effect of collector logic (wired-AND), which we show in Fig. 4.31c as an AND-gate symbol but with wires entering completely into the gate and a solid dot showing where they join the output wire. This is not actually a gate but is a symbol to show that we are going to make use of the collector logic of the wired-AND type.

On performing this in Fig. 4.31c with the above two gates of Fig. 4.31b, we now have the output F, which is X AND Y, since if either X or Y is zero, the output is 0. On applying DeMorgan's theorem, we have at the output, as shown in Fig. 4.31d, the ANDed inputs ORed together and the resulting sum negated. This produces the function AND-OR-INVERT. It should be pointed out that with collector logic the output lines of the individual gates do not take on the values X and Y, as they were previously shown in Fig. 4.31b, but now *all* the actual signal lines at the output of the gates that are bused or wired together take on the value F. Hence in collector logic the intermediate term such as $\overline{AB}$, which was the output of the upper gate *by itself,* is not available for use as a term in another function. Thus collector logic is *not* applicable in places where one specifically

---

† TTL gates suitable for collector logic have *open-collector* outputs. The circuit in Fig. 3.19 would be an open-collector version on removing R_3, Q_4, and D_1, which gives us the circuit in Fig. P4.37.

‡ Many pre-IC logic circuits used *p-n-p* transistors, where this did a wired-OR function with positive logic. This terminology has been used in many references, where this technique for *n-p-n* outputs is incorrectly referred to as "wired-OR."

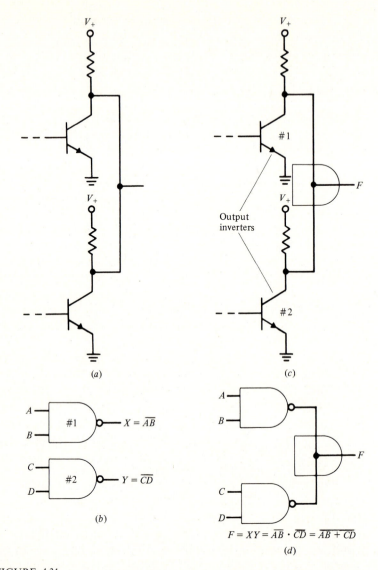

FIGURE 4.31
Collector logic. (a) Gate outputs that can be tied together; (b) two 2-input NAND gates each produce their own outputs X and Y; (c) with the outputs of the circuits in part (b) tied together, we obtain the new output F; (d) logic symbol and expression for NAND gates that permit collector logic.

wants to make use of ANDed terms, as we have discussed previously, for generating multiple-output functions.

To illustrate this difference, let us reconsider the example as discussed in Fig. 4.28, using collector logic. In this problem, we want as outputs the functions F_1 and F_2. Let us realize F_1 first, using NANDs with collector logic. Since the use of collector logic gives us a negation after the ORing of the ANDed terms, it is

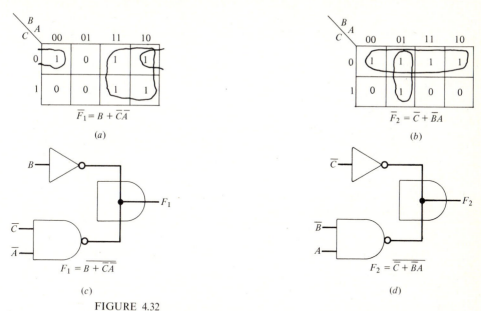

FIGURE 4.32

An illustrative example using collector logic. (a) Karnaugh map and minimized expression for $\bar{F}_1$; (b) Karnaugh map and minimized expression for $\bar{F}_2$; (c) collector logic with NAND gates realization of F_1; (d) collector logic with NAND gates realization of F_2.

more convenient to work with $\bar{F}_1$ and realize $\bar{F}_1$ as a minimized sum-of-products form. The minimized expression for the Karnaugh map of $\bar{F}_1$ is shown in Fig. 4.32a. We will realize this with NAND gates tied together with collector logic, as shown in Fig. 4.32c. Similarly, the minimum sum-of-products term for F_2, shown in Fig. 4.32b, would be given by collector-logic NAND circuitry in Fig. 4.32d. Note that we have had to add inverters to obtain $\bar{B}$ and C from B and $\bar{C}$, respectively, rather than tying them directly to the signal lines $\bar{B}$ and C. This is because if the signal line is not *buffered* in this manner, the signal line B would have taken on the function value F_1 in the circuit of Fig. 4.32c, and in Fig. 4.32d, signal line $\bar{C}$ would have been changed to F_2 due to the collector-logic interaction. Hence, all inputs have to be buffered by an inverter if they are to be used elsewhere. If $\bar{B}$ was only to be used in generating F_1 and it was available from a gate that permitted collector logic, we could have dispensed with the buffering inverter in Fig. 4.32c.

The circuits in Fig. 4.32c and d now require only a total of four gates, with two 2-input gates and two inverters. This is a considerable saving over the circuit in Fig. 4.28g, which required five gates, most of which were of greater complexity. It should be pointed out that it is not always possible to get this improvement by utilizing collector logic. Each problem has to be looked at independently, since at times it may be more efficient to use regular NAND-NAND logic. (However, see Prob. P4.21 for a generalization.) In general, one has to try both approaches if one is looking for the minimum cost and has the time to do alternate designs.

Another form of wired logic is possible with circuits that have emitter-follower outputs such as in ECL. In Fig. 4.33a we show the emitter-follower outputs of two gates whose output logic signals are individually X and Y. If these two outputs are tied together, then whichever output is highest will set the level at the tied output. This is the same as the diode OR gate in Fig. 2.23. For the ECL voltage levels in Fig. 4.33a, the emitter at Q_X is at a higher potential than the emitter of Q_Y. After they are tied together, Q_X will still be ON but Q_Y will be OFF and the output level will be HIGH. Only if both bases are at -0.62 V will the output be LOW. For positive logic this gives the OR function, as shown in Fig. 4.33b. The schematic convention for a *wired*-OR is shown in Fig. 4.33c, where the two NOR gates are "wired-ORable."

A further variation on wired logic with totem-pole outputs (Tri-State Logic†) is useful in many system applications (see Prob. P4.44).

The techniques used for wired logic in bipolar circuits are also applied to MOS ICs.

For wired logic and other uses (see Prob. P4.37), some ICs are made with *open collectors* (in TTL and DTL) and open emitters in ECL. A discrete resistor can then be selected to give optimum performance (see Probs. P4.37 to P4.39).

EXERCISE

E4.17   Realize the expression for the Karnaugh map of F_1 in Fig. 4.25a by using NAND logic that permits collector logic (wired-AND). Do this two ways:

(a) Both true and complements of all variables are available.
(b) Only the true are available.

4.5.3   BCD-to-Seven-Segment Decoder/Driver

In this section we will discuss a commercial MSI circuit‡ that has four input lines (D, C, B, and A) which represent a decimal digit in BCD. Seven of the output lines are used to drive a seven-segment display that makes visible this decimal digit.

We discuss this circuit for two reasons. First, it brings together in one application most of the techniques that we have covered so far. Second, we shall use both this decoder and seven-segment displays in the digital clock that we will design in Chap. 6.

One form§ of a seven-segment display is shown in Fig. 4.34a. This is made up of seven long and narrow incandescent 5-V bulbs that are arranged to form the numeral 8. Consider what happens if the two output transistors (open collector) of the decoder that are tied to segments b and c are saturated and the other five output transistors (not shown) are cut off. Then the right-hand side of the figure lights up to produce the numeral 1.

---

† Trademark, National Semiconductor.
‡ TTL type 7447.
§ Other forms of seven-segment displays are gas discharge tubes, light-emitting diodes (LEDs), and liquid crystals.

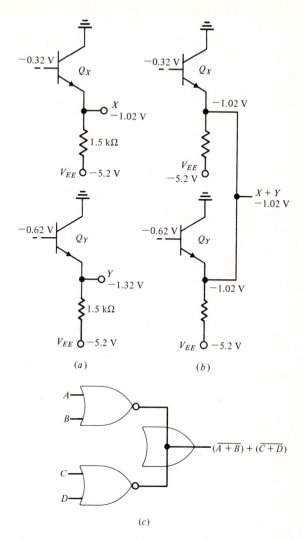

FIGURE 4.33

Wired logic with emitter-follower outputs. (a) Individual ECL gate-output emitter followers produce functions X and Y; (b) tying these two gate outputs together gives the function $X + Y$; (c) NOR gates shown with the wired-OR schematic convention.

Complete segment identification and the resulting displays for all 16 combinations of inputs are shown in Fig. 4.34b. Since this decoder is designed to *sink*
current when the display segment is ON, we require the output to be LOW when
we want the segment ON. Thus when the segment *b* is active, its terminal at the
decoder is LOW. This is referred to as an *active-low* output. Thus we see in the
first row of the decoder truth table (Fig. 4.34c) that for the BCD value of 0_{10} as
the input, all outputs except *g* are LOW.

The truth table for the first 10 rows is generated by the 0 to 9 requirements
shown in Fig. 4.34c. We add the additional requirement that for all 1s as inputs
the visual output be blank; i.e., all decoder output lines are HIGH. For rows 10
through 14, we have don't care conditions as outputs. After these don't care
conditions are used to minimize the network (chip size), we obtain the resultant
displays shown for 10 through 14.

Other inputs and outputs are provided that make it possible to test the
"lamps" (which are the indicator segments) as well as blank out superfluous zeros
in the display.

The logic diagram for doing all this is shown in Fig. 4.35, and the circuit
schematic is shown in Fig. 4.36a. Note that common terms (that is, *BD* in *a* and *b*)
are generated twice. This is done to avoid the current-hogging problem. The logic
stage before the output stage uses a very simplified form of TTL (Fig. 3.17b
without D_1 and R_2). This is possible since within the package we do not have the
same loading and noise-margin considerations that we must consider for a
general-purpose gate. Note that collector logic is used at the base of the output
transistor.

The top view of the decoder package showing pin connections is given in
Fig. 4.36b. The small circle at the output terminals on the upper edge indicate that
these are *active-low* outputs. (In Demonstration D4.2 we make a limited version of
such a decoder.)

FIGURE 4.34
Seven-segment displays and decoders to drive them. (*a*) An incandescent bulb-type
seven-segment display will produce the number 1 if the two transistors shown are
saturated; (*b*) segment identification and displayed characters for the 16 possible
inputs; (*c*) truth table for the TTL type 7447 decoder. *Notes*: (1) BI/RBO is wire-OR
logic serving as blanking input (BI) and/or ripple-blanking output (RBO). The BI
must be open or held at a logical 1 when output functions 0 through 15 are
desired, and ripple-blanking input (RBI) must be open or at a logical 1 during the
decimal 0 input. X = input may be high or low. (2) When a logical 0 is applied
to the BI (forced condition), all segment outputs go to a logical 1 regardless of
the state of any other input condition. (3) When RBI is at a logical 0 and
$A = B = C = D =$ logical 0, all segment outputs go to a logical 1 and the RBO
goes to a logical 0 (response condition). (4) When BI/RBO is open or held at a
logical 1, and a logical 0 is applied to lamp-test input, all segment outputs go to
a logical 0.

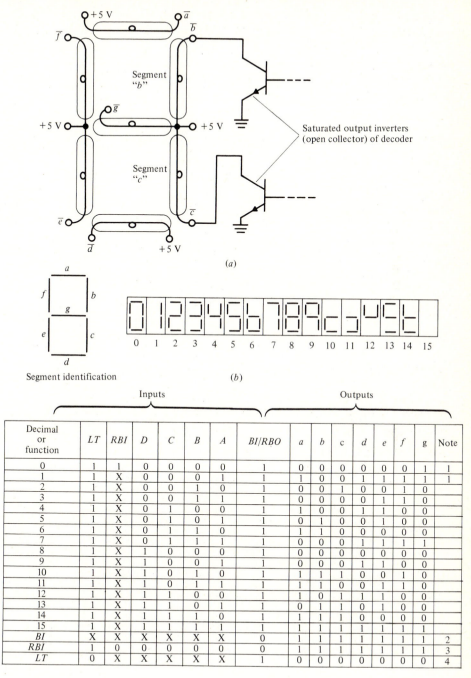

(a)

Saturated output inverters
(open collector) of decoder

Segment identification

(b)

		Inputs						Outputs							
Decimal or function	LT	RBI	D	C	B	A	BI/RBO	a	b	c	d	e	f	g	Note
0	1	1	0	0	0	0	1	0	0	0	0	0	0	1	1
1	1	X	0	0	0	1	1	1	0	0	1	1	1	1	1
2	1	X	0	0	1	0	1	0	0	1	0	0	1	0	
3	1	X	0	0	1	1	1	0	0	0	0	1	1	0	
4	1	X	0	1	0	0	1	1	0	0	1	1	0	0	
5	1	X	0	1	0	1	1	0	1	0	0	1	0	0	
6	1	X	0	1	1	0	1	1	1	0	0	0	0	0	
7	1	X	0	1	1	1	1	0	0	0	1	1	1	1	
8	1	X	1	0	0	0	1	0	0	0	0	0	0	0	
9	1	X	1	0	0	1	1	0	0	0	1	1	0	0	
10	1	X	1	0	1	0	1	1	1	1	0	0	1	0	
11	1	X	1	0	1	1	1	1	1	0	0	1	1	0	
12	1	X	1	1	0	0	1	1	0	1	1	0	0	0	
13	1	X	1	1	0	1	1	0	1	1	0	1	0	0	
14	1	X	1	1	1	0	1	1	1	1	0	0	0	0	
15	1	X	1	1	1	1	1	1	1	1	1	1	1	1	
BI	X	X	X	X	X	X	0	1	1	1	1	1	1	1	2
RBI	1	0	0	0	0	0	0	1	1	1	1	1	1	1	3
LT	0	X	X	X	X	X	1	0	0	0	0	0	0	0	4

(c)

238   4   COMBINATIONAL LOGIC DESIGN

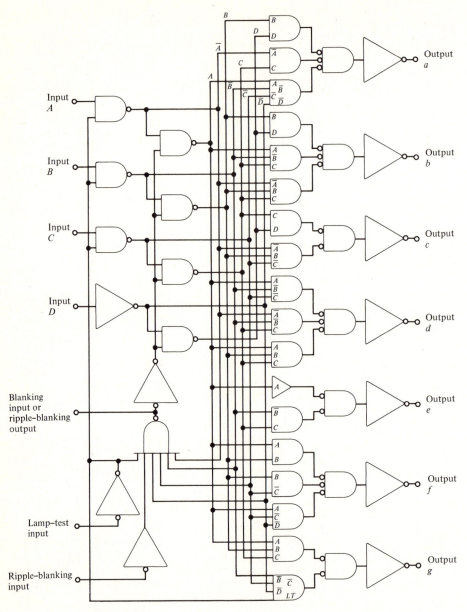

FIGURE 4.35
Logic diagram for the TTL type 7447 seven-segment decoder.

FIGURE 4.36
Circuit schematic for the 7447 decoder. (a) Complete schematic diagram of the 7447
decoder. Component values shown are nominal. (b) Pin connections for the 7447
decoder. For positive logic, see truth tables.

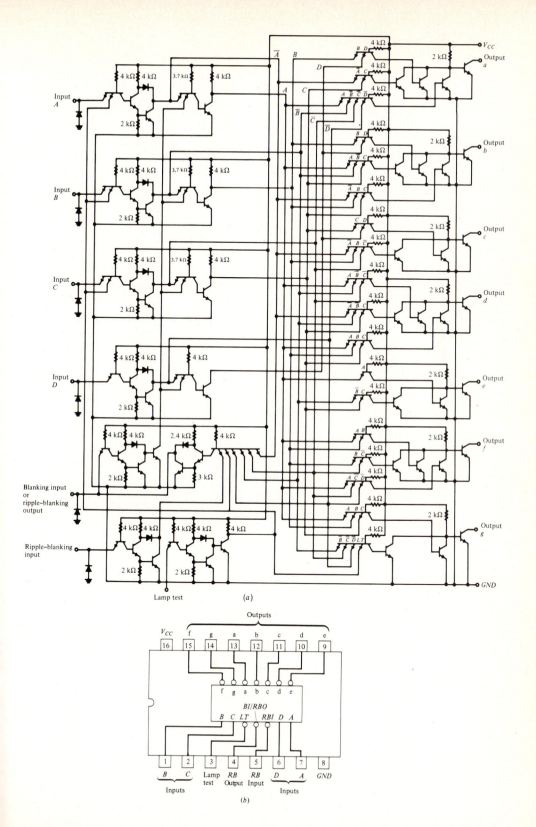

(a)

(b)

EXERCISE

E4.18 For practice with segmented displays, design the following unlikely system. A four-segment display (arranged in a square) is to be used as a floor-status indicator for a special passenger elevator that stops at four floors. The binary value assigned to the floor and the display symbol follow in ascending order: (00; cellar, ⊏), (01; lobby, L), (10; first floor, |) (11; seventh floor, ⌐). The display elements are active LOW, and only the true of the BCD values are available. Design the logic that drives the four segments by using a minimum number of NAND gates and inverters.

This completes our discussion on combinational circuits. In Chap. 5 we will show how interconnecting simple gates can provide *sequential* circuits with memory. These circuits are the building blocks of the IC counters and registers described in Chap. 6. We will find that in designing and using these sequential circuits our newfound knowledge of combinational logic design will be vitally needed.

4.6 SUMMARY

In this chapter we discussed some of the engineering problems that have to be solved in an actual implementation of ICs to a digital system.

We saw that system considerations, the IC family used, and the degree of integration possible are important interrelated factors in the final design.

We also discussed minimization methods that are useful tools in designing with ICs and applied these to several examples.

Examples were presented of the use of don't care conditions for circuit minimization.

Methods and limitations of wired logic were described.

DEMONSTRATIONS

D4.1 Implement Prob. P4.28. If EXCLUSIVE-OR gates are not available, make them by using the circuit in Fig. 4.24d. Use a single LED as the hallway "light." If a standard TTL (7400 Series) IC is used, the 130-Ω current-limiting resistor in the totem-pole circuit will protect the LED.

D4.2 Implement the solution to Prob. P4.33, which is the "special" BCD-to-seven-segment decoder that is "good" for CBA of 000, 001, 010, 011, and 100 (D is always 0). This is to give visual outputs of 0, 1, 2, 3, and 4, respectively. If a seven-segment LED display is used† that has one LED per segment (where the anodes of several of the segments are common), the circuit of Fig. D4.2 can be

† Such as the Monsanto MAN 3, Litronix DATA-LIT 8, Fairchild FND-70, or equivalent.

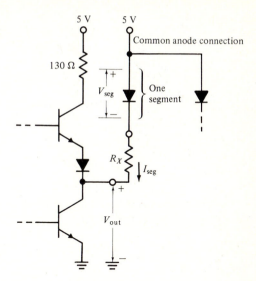

FIGURE D4.2
Circuit for seven-segment display made
from LEDs.

used, where a current-limiting resistor R_X must be added in series between each output and the cathode of its associated LED. Find R_X nominal when the maximum $I_{LED} = 10$ mA and $V_{CC} = 5$ V $\pm$ 5 percent, R_X is a 5 percent resistor, and the following values for the gate output and LED on voltages are given:

	Minimum	Typical	Maximum
V_{OL} (V)	0.1	0.2	0.4
$V_{LED(on)}$ (V)	1.55	1.65	2.0

Calculate what the corresponding minimum and typical I_{LED} are. Measure the actual LED currents that are controlled by the gates.

Record what the displays are for 101, 110, and 111, and verify that the logic relationships produce this.

D4.3 Implement with standard TTL a binary full adder (truth table of Fig. 4.20c and d) by using two 2-input EXCLUSIVE-OR gates and three 2-input NAND gates, as in Fig. D4.3a. To prove this gives the correct result, make use of the results of Exercise E4.12 and Prob. P4.3. If EXCLUSIVE-ORs are not available, use the circuit in Fig. D4.3b. Use LEDs as status indicators, as shown in Fig. D4.3c. For displaying the inputs, the double negation provided by the inverters decouples the LEDs so they do not load down the signal lines and also sets the current limit in the LED. This is an example of the power that an IC has in replacing discrete components (resistors, in this case).

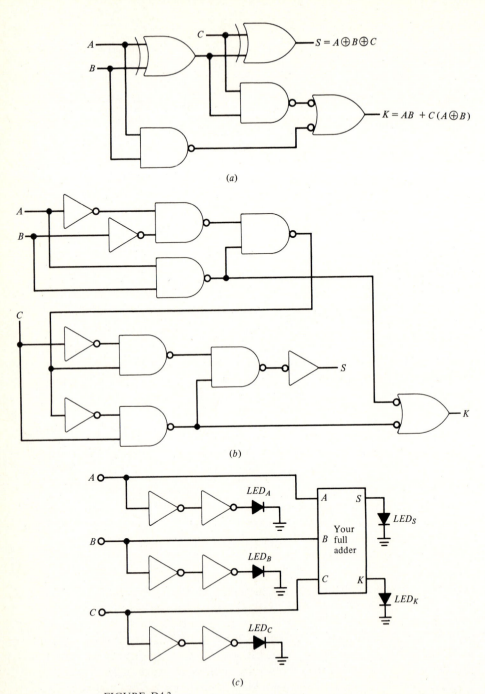

FIGURE D4.3
Full-adder circuits. (a) Full adder made using EXCLUSIVE-ORs and NANDS; (b) full adder made using NANDs and inverters; (c) full-adder display circuitry with LEDs.

REFERENCES

For alternate treatments of the material in this chapter, see either WICKES, W. E.: "Logic Design with Integrated Circuits," Chaps. 3 and 5, Wiley, New York, 1968; or OLDHAM, W. G. AND S. E. SCHWARTZ: "An Introduction to Electronics," chap. 9, Holt, New York, 1972.

For further in depth study, see PEATMAN, J. B.: "The Design of Digital Systems," chaps. 2, 3, and 8, McGraw Hill, New York, 1972.

PROBLEMS

P4.1 Obtain an expression for the output F of Fig. P4.1.

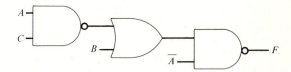

FIGURE P4.1

P4.2 (*a*) Determine the expression for F in the circuit of Fig. P4.2.
(*b*) Convert the circuit of part (*a*) to NAND-NAND logic.
(*c*) Realize the circuit of part (*a*) with three NOR gates and an inverter.

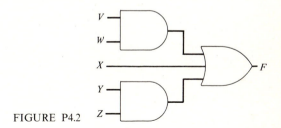

FIGURE P4.2

P4.3 Determine the expressions for S and K in Fig. P4.3. Make use of the EXCLUSIVE-OR operator to simplify the expressions.

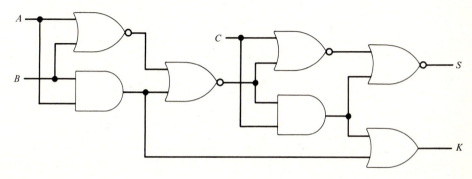

FIGURE P4.3

P4.4   Given the function in the Karnaugh map of Fig. P4.4:
(a) Determine the simplest expression in sum-of-products form.
(b) Determine the simplest expression in product-of-sums form.
(*Hint:* In this simple example the answers should be the same.)
(c) Realize the expression by using an OR gate and inverters as necessary. Only the true of a variable is available.
(d) Repeat part (c) using a NAND gate and inverters.
(e) Repeat part (c) with NOR gates and inverters.

$D\ \backslash^{B}_{A}$ C	00	01	11	10
00	1	1	1	1
01	0	0	1	1
11	0	0	1	1
10	1	1	1	1

FIGURE P4.4

P4.5   Given $F = \bar{A}BCD + ABC + DC + D\bar{C}B + \bar{A}BC$:
(a) Fill in the Karnaugh map.
(b) Minimize F in the sum-of-products form.

P4.6   Given the following truth table:

D	C	B	A	F
0	0	0	0	0
0	0	0	1	0
0	0	1	0	1
0	0	1	1	1
0	1	0	0	0
0	1	0	1	1
0	1	1	0	0
0	1	1	1	1
1	0	0	0	0
1	0	0	1	1
1	0	1	0	1
1	0	1	1	1
1	1	0	0	0
1	1	0	1	0
1	1	1	0	0
1	1	1	1	0

construct its Karnaugh map, then find:
(a) The (or one) minimum sum-of-products expression for *F*.
(b) Repeat part (a) for product-of-sums.

P4.7   Given the Karnaugh map of Fig. P4.7, find the lowest cost two-level circuit that can generate the mapped function. You can use either NAND or NOR gates. Both the true and complement of all variables are available. For costs we use 1¢ per signal terminal of a gate (i.e., a 2-input gate costs 3¢).

$D\!\diagdown\!C$ $\ \ B\!\diagdown\!A$	00	01	11	10
00	1	1	1	1
01	1	1	0	0
11	1	1	1	0
10	1	1	1	0

FIGURE P4.7

P4.8   Repeat Prob. P4.7 for the Karnaugh map of Fig. P4.8.

$D\!\diagdown\!C$ $\ \ B\!\diagdown\!A$	00	01	11	10
00	1	1	0	0
01	1	0	0	0
11	1	0	0	0
10	1	1	0	0

FIGURE P4.8

P4.9   Repeat Prob. P4.7 for the Karnaugh map of Fig. P4.9.

$D\!\diagdown\!C$ $\ \ B\!\diagdown\!A$	00	01	11	10
00	1	0	1	1
01	1	0	1	1
11	0	0	1	0
10	0	0	0	0

FIGURE P4.9

P4.10 Repeat Prob. P4.7 for the Karnaugh map of Fig. P4.10.

$D\,{}^{B}_{C}\!\!\diagdown^{A}$	00	01	11	10
00	0	1	0	0
01	0	0	0	0
11	1	1	0	1
10	1	1	0	1

FIGURE P4.10

P4.11 Repeat Prob. P4.7 for the Karnaugh map of Fig. P4.11.

$D\,{}^{B}_{C}\!\!\diagdown^{A}$	00	01	11	10
00	0	0	1	1
01	0	0	0	1
11	0	0	0	1
10	0	0	0	1

FIGURE P4.11

P4.12 Given the expression for *F* as

$$A\overline{D}\overline{C} + \overline{A}\overline{C}D + \overline{A}CD + \overline{A}BC\overline{D} + \overline{A}BCD + \overline{A}\overline{B}D$$

(*a*) Find the minimum sum-of-products form for *F*.
(*b*) Find the minimum product-of-sums form for *F*.
(*Hint:* There are more than one.)

P4.13 Repeat Prob. P4.7 for the Karnaugh map of Fig. P4.13.

$C\,{}^{B}_{D}\!\!\diagdown^{A}$	00	01	11	10
00	0	0	0	0
01	1	0	0	0
11	1	1	1	0
10	1	0	1	0

FIGURE P4.13

P4.14 For the circuit in Fig. P4.14:
    (*a*) Determine the minimum sum-of-products expression for its output.
    (*b*) Realize this expression with NAND logic. The true and complement of all variables are available.

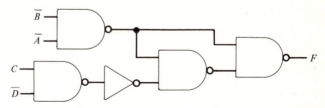

FIGURE P4.14

P4.15 Repeat Prob. P4.14 for the circuit in Fig. P4.15.

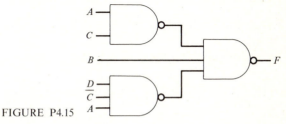

FIGURE P4.15

P4.16 Repeat Prob. P4.14 for the circuit in Fig. P4.16.

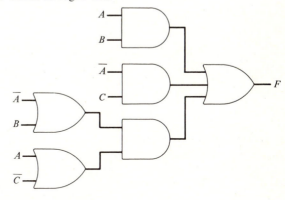

FIGURE P4.16

P4.17 Repeat Prob. P4.14 for the circuit in Fig. P4.17.

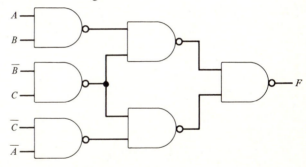

FIGURE P4.17

P4.18 Repeat Prob. P4.14 for the circuit in Fig. P4.18.

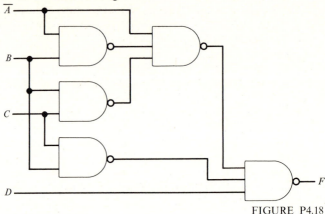

FIGURE P4.18

P4.19 Given the Karnaugh map of Fig. P4.19, using 1¢ per signal pin for costing:
  (a) Realize this Karnaugh map with a minimum-cost NAND logic circuit.
  (b) Realize this Karnaugh map with a minimum-cost NOR logic circuit.

P4.20 Given four input signals whose values are always a valid BCD code, determine the NAND logic required that gives a 1 output if the BCD value is equal or greater than 6_{10} and gives a 0 if the BCD value is less than 6_{10}. The four true inputs are all that you have available.

$D\,C$ \ $B\,A$	00	01	11	10
00	0	1	Ø	Ø
01	1	1	Ø	1
11	Ø	0	0	Ø
10	0	Ø	1	Ø

FIGURE P4.19

P4.21 (a) Show that a wired-AND followed by an inverter gives the same logic output (AND-OR) as conventional NAND-NAND logic. If one has no need for sharing common terms, show that the guaranteed savings in cents of this method over NAND-NAND logic (if one is dealing with M terms in an expression) is $(M - 1)$¢. Gate costs are based on the data in Prob. P4.7.
  (b) Figure P4.21 gives the Karnaugh map of a function F. Implement it with minimum cost by using ICs that permit collector logic (wired-AND). Use 1¢ per signal pin costs.

$D\,C$ \ $B\,A$	00	01	11	10
00	0	1	Ø	Ø
01	1	Ø	Ø	1
11	0	Ø	Ø	1
10	1	0	0	Ø

P4.22 Redo Prob. P4.7, with wired-AND permissible.
P4.23 Redo Prob. P4.8, with wired-AND permissible.
P4.24 Redo Prob. P4.9, with wired-AND permissible.
P4.25 Redo Prob. P4.10, with wired-AND permissible.
P4.26 Redo Prob. P4.13, with wired-AND permissible.

FIGURE P4.21

P4.27 The traffic problem discussed in the introduction of this chapter has four possible Karnaugh maps for G_1 and G_2, depending on the requirements.

(*a*) Determine the minimum expression for all eight maps by using both totem-pole NAND gates and open-collector NAND gates.

*(*b*) If NOR gates were available, show an alternate solution that only uses one IC of the 14-pin dual in-line type.

P4.28 A light in a long hallway is to be controlled by three switches C, B, and A. If the switch is UP, its logical value is 1. If an odd number of switches are UP, the light is ON; if an even number, OFF (remember 0 is an even number). Generate the logic signal F ($F = 1$ for light ON) from C, B, and A by using EXCLUSIVE-OR gates only. Only the true of the inputs is available. (*Hint:* Look for a checkerboard-pattern arrangement of 1s and 0s on the Karnaugh map that indicates the EXCLUSIVE-OR can be used effectively.)

P4.29 Prove algebraically that the MSI-TTL circuit of Fig. P4.29*a* produces the outputs shown in Fig. P4.29*b*.

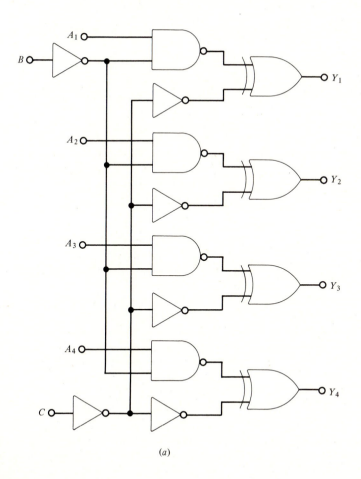

4–bit true/complement, zero/one element

B	C	Y_n
0	0	$\overline{A}_n$
0	1	A_n
1	0	1
1	1	0

(*b*)

(*a*)

FIGURE P4.29

P4.30 Given the two Karnaugh maps of F_1 and F_2 in Fig. P4.30 (because of speed requirements the wired-AND is not permissible), realize the functions with NAND-NAND logic:
(*a*) Separately optimized.
(*b*) Making use of common terms.
(*c*) Compare the number of signal terminals used in part (*a*) to that used in part (*b*).

$DC \backslash BA$	00	01	11	10
00	1	0	0	0
01	0	1	0	0
11	0	1	1	1
10	1	0	1	1

(*a*)

$DC \backslash BA$	00	01	11	10
00	1	0	1	1
01	0	1	0	0
11	0	0	0	0
10	1	0	1	1

(*b*)

FIGURE P4.30

P4.31 A 12-h digital clock is to be designed that has one seven-segment character such as in Fig. 4.34*a* which is to be driven so that in the A.M. it reads A and then reads P during the P.M. Given a logic signal Z which is 0 during the A.M. and 1 during the P.M., show the wiring to the seven-segment character. The decoders output are to be active LOW.

P4.32 Redo the problem in Exercise E4.18; but now the outputs are L, I, ˥, □ corresponding to 00, 01, 10, 11 respectively.

P4.33 Design a decoder by using Series 74 NAND gates to drive an incandescent-type (Series 90 Minitron) seven-segment display, as shown in Fig. 4.34*a*. Each segment requires 8 mA when ON. The display is to show the integers 0, 1, 2, 3, and 4 when the corresponding BCD values are inputs. Make the 1 on the right-hand side of the character. We "don't care" what the display looks like for any other possible inputs. The decoder outputs are to be active LOW.

Realize your decoder with the minimum number of packages. Package types available: standard Series 74 TTL hex inverters, quad 2-input NAND, triple 3-input NAND, and dual 4-input NAND. All outputs are guaranteed to be able to sink 16 mA.

P4.34 Design a one-out-of-ten decoder that has four input lines (8421 BCD) that select the appropriate one out of the ten outputs. When the selected output is active, it is LOW. For nonvalid BCD we "don't care" what the outputs are.

P4.35 Find N_L, N_H, NM_L and NM_H by using the data in Table P4.35 for the following conditions:
(*a*) Standard driving low-power.
(*b*) Low-power driving standard.
(*c*) High-speed driving standard.
(*d*) High-speed driving low-power.

Table P4.35

	I_{OL}	I_{IL}	V_{OL}	V_{IL}	I_{OH}	I_{IH}	V_{OH}	V_{IH}
Standard Series 54/74	16	1.6	0.4	0.8	0.4	0.04	2.4	2.0
High-speed (Series 74H)	20	2.0	0.4	0.8	0.5	0.05	2.4	2.0
Low-power (Series 74L)	2	0.18	0.3	0.7	0.1	0.01	2.4	2.0

NOTE: Voltages in volts, currents in milliamperes.

P4.36 (*a*) A low-power TTL unit is to drive a mixture of standard and low-power TTL. What are the possible loads it can handle? Use the data in Table P4.35.

(*b*) Repeat Exercise E4.16, but now determine how many low-power TTL "inverters" (formed by tying together both inputs of a 2-input gate) can be driven by a standard Series 54/74 gate.

P4.37 An open-collector TTL gate is to be used as a driver for the light-emitting diode (LED), as shown in Fig. P4.37. Given that V_{CC} is 5 V $\pm$ 10 percent and R is to be a 5 percent resistor, the minimum saturation voltage of the output transistor of the gate is 0.1 V, while its maximum is 0.4 V. $V_{D(on)}$ can vary from 1.5 to 2 V.

(*a*) Find R_{nom} so that the lowest current through the LED is 5 mA.

(*b*) What is the maximum current that the gate has to sink?

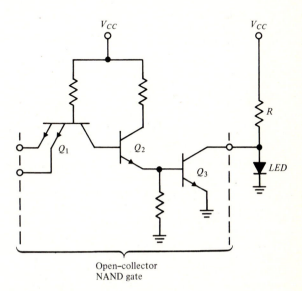

FIGURE P4.37

P4.38 A Series 54/74 gate with an open-collector output, as in Fig. P4.38, is to drive a Series 74 gate (data in Table P4.35). With $V_{CC} = 5$ V and $I_{leakage} = 0$ mA, it is desired to have $N_H = 10$ with $NM_H = 0.4$ V. What is the maximum value of the pullup resistance R_L that can be used?

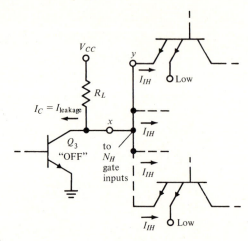

FIGURE P4.38

P4.39 The open-collector TTL Series 54/74 gate circuit is to be used as in Fig. P4.39, where it is desired that it have an $N_L = 5$. For $V_{CC} = 5$ V, find the minimum value of the pullup resistance R_L.

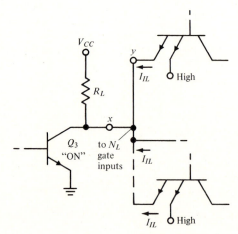

FIGURE P4.39

P4.40 Collector logic is to be used in the circuit of Fig. P4.40, where m open-collector outputs share one pullup resistor R_L.

(*a*) Prove that

$$\bar{R}_L = \frac{V_{CC} - (V_{IH} + NM_H)}{mI_{\text{leakage}} + N_H I_{IH}}$$

$$\underline{R}_L = \frac{\bar{V}_{CC} - (V_{IL} - NM_L)}{I_{OL} - N_L I_{IL}}$$

Assume $I_{\text{leakage}} = 0$ when $V_0 < V_{IL}$.

(*b*) If it is desired to have $NM_L = 0.5$ V with Series 54 TTL, to what value would the guaranteed value of V_{OL} have to be changed?

(*c*) For $m = 6$, $N_H = 15$, $N_L = 6$, $\bar{V}_{CC} = 5.25$ V, $\underline{V}_{CC} = 4.75$ V, and $I_{\text{leakage}} = 250 \ \mu\text{A}$, find:
  (1) $\bar{R}_L$.
  (2) $\underline{R}_L$.
  (3) The nominal value of R_L that satisfies parts (1) and (2) which uses the loosest tolerance resistor. Standard tolerances are 1, 2, 5, 10, and 20 percent.

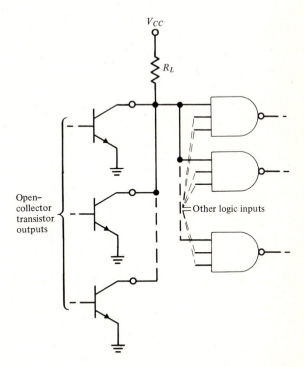

FIGURE P4.40

P4.41 Figure P4.41 shows an *expander* (made up of Q_7, Q_8, and R_5), whose two output terminals are tied to the expander nodes on a separate expandable AND-OR-INVERT gate. As shown, phase-inverter transistors Q_2, Q_6, and Q_8 share the same collector and emitter leads.

(*a*) Verify that the output expression is as given.

(*b*) If all inputs are HIGH and each is drawing 40 μA, what is the base overdrive factor on Q_3 when it is sinking its maximum load of 16 mA? Use $V_{CC} = 5$ V, $V_{BE(sat)} = V_{BE(on)} = V_{D(on)} = 0.7$ V, $V_{CE(sat)} = 0.2$ V, $V_{BC(on)} = 0.7$ V (in the inverted active region), $\beta_F = 50$.

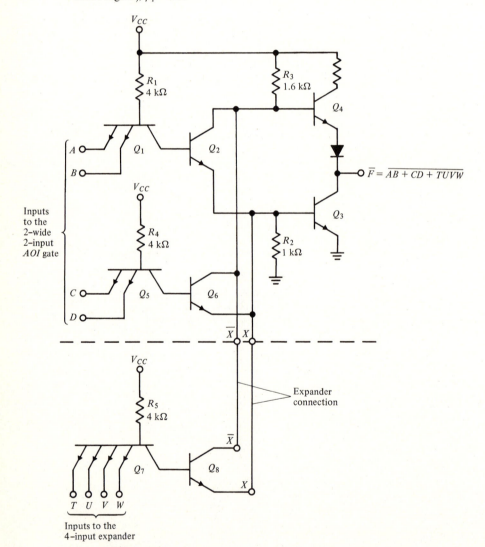

FIGURE P4.41

An expandable 2-wide 2-input AND-OR-INVERT gate with a 4-input expander connected to it.

*P4.42 The TTL-NAND buffer in Fig. P4.42 is used to provide more output drive than a standard 7400 Series active pullup of Fig. 3.19. In this circuit the Darlington-connected pair (Q_4 and Q_5) acting as an emitter follower provides 1.2 mA, which is greater than the 0.4-mA sourcing current from the single transistor of the circuit in Fig. 3.19. The configuration in Fig. P4.42 is used in the 74H and 74S Series.

Determine the typical V_0 for both circuits, sourcing their maximum current (0.4 mA and 1.2 mA) with $V_{CC} = 5$ V. Use all inputs at $V_I = 0.8$ V. Transistor data are $V_{BE(sat)} = V_{BE(on)} = V_{D(on)} = 0.7$ V, $\beta_F = 50$, $V_{CE(sat)} = 0.2$ V, except for Q_1, where where $V_{CE(sat)} = 0.1$ V.

(*Hint:* For all inputs at 0.8 V, Q_1 is in saturation with current out of the collector of Q_1 into the base of Q_2. Q_2 is in the active region.)

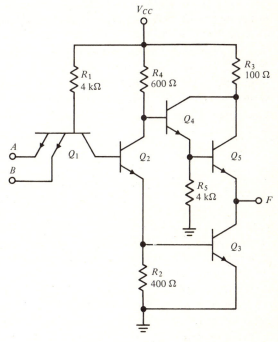

FIGURE P4.42
TTL NAND buffer.

*P4.43 Another variation of the TTL circuit is shown in Fig. P4.43. In addition to having a Darlington in the upper portion of the totem pole, the emitter load on Q_2 has been modified. This gives a squarer transfer characteristic than the Series 54/74 type and is used in the 74S and 74LS Series.

Assume: $V_{BE(on)} = V_{BE(sat)} = 0.6$ V (except for $Q_6 = 0.55$ V), $V_{CE(sat)} = 0.1$ V, $\beta_F = 20$, $\beta_R = 0.2$.

(a) Carefully determine the voltage-transfer characteristic for the Series 74 circuit shown in Fig. 3.19, noting particularly the first breakpoint on the curve.

(b) Repeat part (a) for the circuit shown in Fig. P4.43.

(c) Show that with all inputs HIGH, the base current to Q_3 is the same in both circuits.

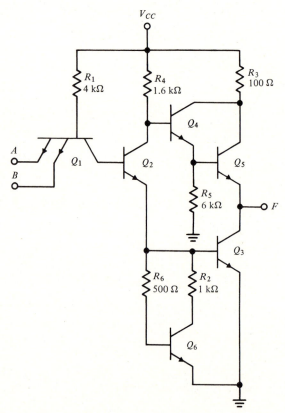

FIGURE P4.43
Modified TTL gate, where R_2 of Series 54/74 is replaced by R_6, Q_6, and R_2.

P4.44 This problem analyzes Tri-State Logic (TSL), which is a *busable* form of TTL. Figure P4.44 shows the basic gate circuit, which is similar to that in Fig. P4.42, with the addition of the CONTROL logic shown enclosed in the dotted lines. With CONTROL = LOW, the gate behaves as a normal TTL gate since Q_8 is OFF and no current flows in the diode D_1. The gate is disabled with CONTROL = HIGH. Now with Q_8 and D_1 conducting, the voltage at the base of Q_4 is about 0.9 V and both Q_4 and Q_5 are cut off. Also, Q_2 will be cut off since the third emitter of Q_1 will sink the current flowing down R_1. As a consequence, Q_3 is also cut off. The net result is that only leakage currents flow at both input and output terminals. Hence when the gate is disabled, it is effectively "cut out" of the circuit.

Determine the maximum number of TSL gates whose outputs can be wired together, as in Fig. P4.44, if the number of load gates is 3. In the disabled state the maximum leakage current at the output of a TSL gate is 40 μA.

$$I_{OH} = 5.2 \text{ mA at } V_{OH} = 2.4 \text{ V} \qquad I_{IH} = 40 \text{ }\mu\text{A}$$

$$I_{OL} = 16 \text{ mA at } V_{OL} = 0.2 \text{ V} \qquad I_{IL} = 1.1 \text{ mA}$$

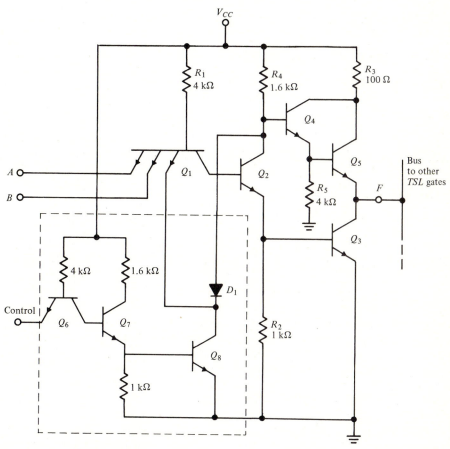

FIGURE P4.44
Basic Tri-State Logic (TSL) gate.

5

LATCHES AND FLIP-FLOPS

In the circuits discussed in Chaps. 3 and 4, the desired output signals existed only when the input signals were present. That is, with combinational logic circuits the output logic is directly related to the input logic, so that any change at the input causes an immediate change at the output (except for some propagation delay time). As a class, these circuits are described as *nonregenerative*. In this class of circuits no path is provided for the output signal to connect back to the input.

Digital circuits in general, however, have need of a *memory element*, where digital information is stored either temporarily or permanently. This is a circuit whose output is related to the input logic signal at a specified time, and then "remembers" the state of the input at that time even though the input logic signal may subsequently change. This introduces us to *regenerative* circuits and a simple memory element, the *bistable circuit* or *flip-flop*, which in the simplest form we call a *latch*. In these circuits a connection is deliberately made from the output terminal back to the input.

Storage circuits, such as those we have just mentioned, are also given the class name *sequential circuits*. This name arises from the fact that the information available from the generalized circuit depends not only on the input signal at a given time but also on the sequence of signals that have led to the existing state. In

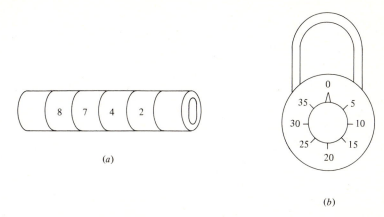

FIGURE 5.1
(a) A *combinational* bicycle chain lock; (b) *sequential* padlock.

this chapter, we will deal with very simple sequential circuits that consist of gate circuits that interconnect outputs to inputs to provide *positive feedback*, another name for regeneration. The circuits to be described are latches, flip-flops, and binary counters. These basic circuits are very useful in digital systems, but their greatest use is as building blocks for making higher-order functions like *counters* and *registers*, which we describe in Chap. 6.

The differences between combinational and sequential circuits can be brought out by analogy to two prevalent types of bicycle locks. Figure 5.1a shows a simple combination lock, in which the lock opens when the proper number (as displayed) is dialed. Figure 5.1b shows a second type of bicycle lock. The requirement for opening this lock is that the knob is turned to one number at a time in the proper sequence. The lock in Fig. 5.1a is the analog of a simple combination circuit, while the lock in Fig. 5.1b is analogous to a sequential circuit.

5.1 LATCHES

The basic ingredient of a sequential circuit is the *latch*. A latch circuit can also be called a bistable or flip-flop. However, in ICs we reserve the term flip-flop for a more complex function that includes latch circuits in it. A latch can be constructed by connecting two 2-input NOR gates in cascade, as shown in Fig. 5.2a. The output of gate 1 is connected to one input of gate 2, and the output of gate 2 is connected to one of the inputs of gate 1. If both the unused inputs are set to 0 (for positive logic this is LOW), we have the equivalent circuit of Fig. 5.2b. This shows two inverters connected in cascade, with the output of the second inverter tied into the input of the first inverter. If each individual inverter has the familiar voltage

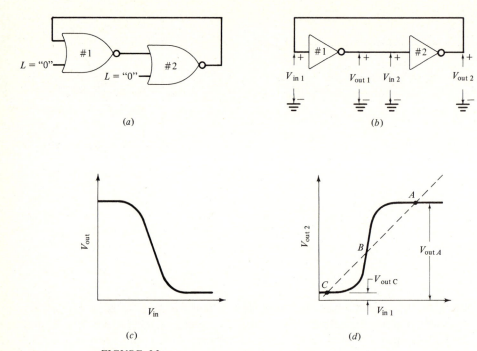

FIGURE 5.2
(a) Latch circuit made with two 2-input NOR gates; (b) equivalent circuit of part
(a) using two inverters; (c) voltage transfer characteristic of the inverters; (d) voltage
transfer characteristic of the latch.

transfer characteristic, shown in Fig. 5.2c, the composite transfer function will be
the solid S-shaped line shown in Fig. 5.2d. Notice that when $V_{in, 1}$ is LOW, $V_{out, 2}$ is
LOW; and when $V_{in, 1}$ is HIGH, $V_{out, 2}$ is HIGH. We can then close the *feedback*
path by connecting the output of gate 2 to the input of gate 1 without disturbing
the circuit. Thus the voltage $V_{out, 2} = V_{in, 1}$, and this gives the dotted line in
Fig. 5.2d. The intersections of the dotted line and the S-shaped line give us three
points at which all circuit conditions are satisfied.

   Let us now consider the *stability* of the points at which the circuit conditions
are satisfied. We begin with the point labeled A. In Fig. 5.3a, we have redrawn the
basic circuit with a small voltage δ inserted in series with the input to gate 1. This
small voltage could be due to "noise," generated either internally or externally to
the circuit.

   If the circuit is operating at point A, and if we increase the input voltage $V_{in, 1}$
by an amount δ over and above $V_{out, 2}$, the output voltage will remain at the
HIGH value, $V_{out, A}$. Thus the small voltage δ does not produce a change in the
output state. A similar argument holds at point C, where the output voltage is at a
LOW level, $V_{out, C}$. We say that points A and C are stable operating points because

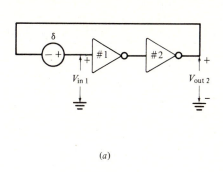

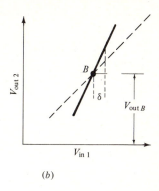

(a)   (b)

FIGURE 5.3
Stability of latch illustrated with (a) a noise source at the input to gate 1, (b) Magnified view of the voltage transfer characteristic in the transition region.

the output voltage is unaffected by a small perturbation (δ) in the input voltage. A quite different situation occurs for operation at point B. Figure 5.3b shows a magnified plot of the solid line, which is the voltage transfer characteristic, and the dotted line, which is the line expressing equality of the input and output voltage. The intersection is at point B, and so we would have, without a voltage δ, an output voltage $V_{out,\,B}$. If we now insert a small perturbing signal δ in series with the input voltage, the output voltage will increase above the dotted line to a new value proportional to the slope of the solid line. A further increase in input voltage then results as this effect propagates through the two inverters. We therefore have a regenerative effect that causes the output voltage to move from $V_{out,\,B}$ to point $V_{out,\,A}$. If the perturbation had been in the other direction (i.e., if δ had a negative value in Fig. 5.3a), the circuit would have moved down to point C. Consequently, point B is an unstable equilibrium point, and the circuit will never rest† at that point. Notice in Fig. 5.3b, a requirement for regeneration is that the slope of the solid line, which is the ratio of $V_{out,\,2}$ and $V_{in,\,1}$ (i.e., the voltage gain of the circuit), must be greater than the slope of the dotted line (where the voltage gain is equal to unity). In other words, for regeneration to occur in this circuit, the voltage gain ($V_{out,\,2}/V_{in,\,1}$) must be greater than 1.

    In summary, the feedback connection of the two inverters has yielded a *bistable* circuit (i.e., a circuit with two stable states), $V_{out,\,A}$ and $V_{out,\,C}$. Additionally we have seen that the circuit will regeneratively move toward one of these two stable states. It will *flip* toward one and *flop* toward the other. However, we do need a method of determining a particular state for the circuit.

---

† If an ideal voltage source of magnitude $V_{out,\,B}$ is applied at the input to gate 1, we would have a stable condition at point B. However, we never use circuits with ideal precision voltage sources, so we can ignore this.

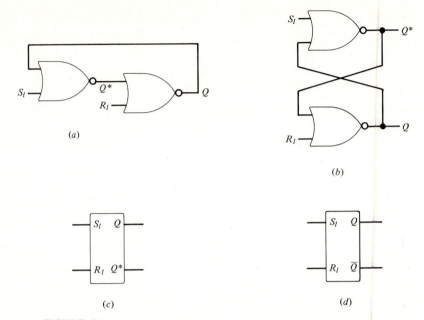

(a)

(b)

(c)

(d)

FIGURE 5.4

(a) Latch circuit made with two 2-input NOR gates, showing S_1 and R_1; (b) re-drawing of part (a) showing symmetry and cross-coupling of the latch; (c) generalized block-diagram symbol for a latch; (d) The logic symbol for a latch.

5.1.1   The Characteristics of a Latch Made from NOR Gates

In Fig. 5.4a we have redrawn the latch circuit made with two NOR gates. The inputs which previously were set at 0 are now labeled S_ℓ and R_ℓ. To determine the effects of signals at the S_ℓ and R_ℓ terminals, we must find the combinations of S_ℓ, R_ℓ, Q, and Q^* that will satisfy the NOR relationship

$$Q^* = \overline{Q + S_\ell}$$
$$Q = \overline{Q^* + R_\ell}$$

To help find these combinations, we have prepared Table 5.1, where we have listed all four variables S_ℓ, R_ℓ, Q, and Q^*, with all their possible combinations. We then determine which relationships have to be satisfied because of the NOR function and draw a line through the rows in which those relationships are *not* satisfied. Of the 16 possible combinations, we find 5 that satisfy the NOR relationship. These five are shown in Table 5.2a. We note one further feature of Table 5.2a; namely, that if the condition $S_\ell = 1$ and $R_\ell = 1$ simultaneously is disallowed, then in all other rows the value of Q^* is equal to $\bar{Q}$. Hence, if we are assured that the $S_\ell = 1$, $R_\ell = 1$ combination of input signals into the latch can never occur, we can then relabel Q^* everywhere as $\bar{Q}$.

Table 5.1 LATCH TRUTH TABLE

S_ℓ	R_ℓ	Q	Q^*
~~0~~	~~0~~	~~0~~	~~0~~
0	0	0	1
0	0	1	0
~~0~~	~~0~~	~~1~~	~~1~~
~~0~~	~~1~~	~~0~~	~~0~~
0	1	0	1
~~0~~	~~1~~	~~1~~	~~0~~
~~0~~	~~1~~	~~1~~	~~1~~
~~1~~	~~0~~	~~0~~	~~0~~
~~1~~	~~0~~	~~0~~	~~1~~
1	0	1	0
~~1~~	~~0~~	~~1~~	~~1~~
1	1	0	0
~~1~~	~~1~~	~~0~~	~~1~~
~~1~~	~~1~~	~~1~~	~~0~~
~~1~~	~~1~~	~~1~~	~~1~~

A more compact way of presenting the information given in Table 5.2a is the *characteristic table* shown in Table 5.2b. The input S_ℓ is the SET input, while the R_ℓ input is the RESET input. As seen in Table 5.2b with both SET and RESET equal to 0, the latch keeps its previous state. There is no change at either Q or Q^*. The state of the latch is that given by the Q output. If the SET signal is 1 and the RESET is 0, the Q output goes to 1, and we say the latch is *set*. Conversely, for SET equal to 0 and RESET equal to 1, the Q output goes to 0, and the latch is *reset*. For both inputs 1, we show an ambiguous output: both Q and Q^* would be 0.

Table 5.2a LATCH TRUTH TABLE: VALID COMBINATIONS (FOR LATCH MADE FROM NOR GATES)

S_ℓ	R_ℓ	Q	Q^*
0	0	0	1
0	0	1	0
0	1	0	1
1	0	1	0
1	1	0	0

Table 5.2b CHARACTERISTIC TABLE FOR LATCH MADE FROM NOR GATES

S_ℓ	R_ℓ	Q after inputs are applied
0	0	Q before inputs are applied
0	1	0
1	0	1
1	1	(not valid inputs)

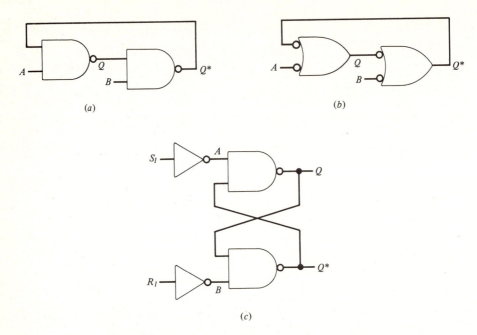

FIGURE 5.5
(*a*) Latch circuit made with two 2-input NAND gates; (*b*) the logic equivalent of
part (*a*); (*c*) the logic symbol of Fig. 5.4*d* can represent the NAND latch if inverters are added at S_1 and R_1.

The circuit in Fig. 5.4*b* is a redrawing of Fig. 5.4*a* to show the symmetry of
the circuit and the cross-coupling of the two NOR gates. The circuit in Fig. 5.4*c* is
a generalized block-diagram symbol for a latch in which we have indicated that
the two outputs are Q and Q^*, while in Fig. 5.4*d* we have replaced Q^* with $\bar{Q}$, on
the assumption that S_ℓ and R_ℓ are never both equal to 1 at the same time.

5.1.2   A Latch Made from NAND Gates

Instead of using two 2-input NOR gates, we could just as well have developed the
latch by interconnecting two NAND gates, as shown in Fig. 5.5*a*. Using our
graphical DeMorgan theorem conversion, we can then proceed to Fig. 5.5*b*,
which, in effect, is similar to Fig. 5.4*a*. We have labeled the inputs A and B for
convenience. For the NAND realization of the latch, one can go through a similar
argument, as that in Table 5.1, and find that everywhere Q^* would equal $\bar{Q}$
except for the combination $A = 0$ and $B = 0$. In that situation, the outputs would
both equal 1. Hence, if we avoid having both inputs simultaneously 0, we can treat
the outputs as being complementary.

In Fig. 5.5*c* we have shown inverters in front of the A and B inputs; therefore the logic symbol of Fig. 5.4*d* would now represent the circuit shown in Fig. 5.5*c*. Again, we will be able to use this representation only as long as S_ℓ and R_ℓ are not equal to 1 simultaneously. If they were equal to 1 when NAND gates made up the latch, then both outputs would be 1.

Since in logic diagrams one does not generally know whether the internal circuitry of a latch is made up of NOR or NAND gates, one would then show for the characteristic table of Table 5.2*b* that when both inputs to a latch were 1, the output is indeterminate.

The characteristics of the latch are further explored in Demonstration D5.1, where the latch function is formed with:

1   NOR gates, made using discrete components
2   A quad 2-input NAND-gate package of the TTL family

EXERCISES

E5.1   A grounded-emitter RTL inverter circuit has the following parameters:

$$R_C = 1 \text{ k}\Omega \qquad V_{CC} \quad = 4 \text{ V}$$

$$R_B = 3 \text{ k}\Omega \qquad V_{BE(\text{sat})} = 0.7 \text{ V}$$

$$\beta_F = 20 \qquad V_{CE(\text{sat})} = 0.2 \text{ V}$$

(*a*) With two such inverters connected as a latch, find the HIGH and LOW values of voltage at each collector.

(*b*) Compute the overdrive factor k for each inverter in the latch connection.

E5.2   A latch, similar to that shown in Fig. 5.5*c*, is constructed using a type 7400, which is a quadruple 2-input TTL-NAND gate. For each gate the nominal fanout is 10 and the maximum propagation delay time is 18 ns.

(*a*) What is the nominal fanout of the latch circuit?

(*b*) Determine the minimum pulse width required at the S_ℓ input to set the latch.

5.1.3   An Example of Latch Usage: Debouncing Switches

In many systems utilizing digital circuits, it is necessary to have an operator initiate a sequence of events by means of a push switch. Unfortunately, most switches exhibit a bounce phenomenon, which makes it impossible to reliably use a simple push switch by itself to generate a single pulse. The memory property of a latch can be used to "debounce" a switch. A circuit to do this is shown in Fig. 5.6*a*; it consists of a latch, two resistors, and the switch. Normally the switch rests at the position *a* and has its center position returned to ground. The switch is a push switch, as opposed to a toggle switch, which means that it will be in the *b* position

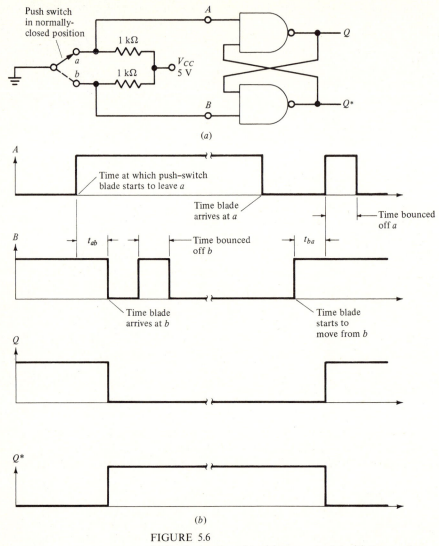

FIGURE 5.6
(a) Latch circuit used to debounce a switch; (b) the waveforms.

only as long as the operator pushes down on it. As soon as the operator releases the button, a spring inside the switch restores it to position *a*.

Let us examine the waveforms that will be generated in a sequence of switching events. In Fig. 5.6b the upper waveform shows the voltage at point *A* as a function of time. Initially, the switch is in position *a*, which means that point *A* is grounded. The operator pushes the button down and the blade of the switch starts moving across to *b*. As soon as contact is broken, the voltage of point *A* rises very rapidly toward 5 V and reaches a HIGH level set by the input current of the gate.

With negligible input current, A would go to 5 V. The voltage at point B, however, does not change instantaneously; we have to wait a time t_{ab}, which represents the time required for the blade to travel from point a to b. Once the switch reaches point b, the voltage of B goes to ground level. However, since the switch blade can vibrate, the switch will bounce off contact b for a few milliseconds. In a properly functioning switch, it will not go all the way back to a but will spend some time in between before finally settling down, resting on b again, as is shown in the middle part of the B waveform. We have assumed here that only a single bounce occurs.

If we look at the output waveform now, we note that, since A was initially at ground level, Q was HIGH and therefore the output Q^* is LOW. At the time that the blade touches contact b, the input to B goes to ground; therefore, Q^* goes HIGH immediately (i.e., in nanoseconds) and through the feedback connection the output Q is forced to LOW, since the input A is already HIGH. Now that Q^* is HIGH, and as long as A remains HIGH (remember we assume that the switch did not bounce all the way back to a), the output Q will stay LOW ensuring that Q^* stays HIGH. Hence, the switch bounce off of contact b does not affect the waveform at Q or Q^*.

In the opposite sequence of operations, the operator first releases the switch, causing the B waveform to rise to its HIGH level while the blade starts to move from b to a. There is again a finite length of time t_{ba} required for the spring to move the blade back to position a. At the end of this time interval, the input A goes LOW; as soon as it goes LOW, the Q output goes HIGH, and the regeneration action on the gate is such that Q^* will then go LOW. Again, if the switch bounces after it returns to a, as shown in the upper diagram, this will have no effect since the B input stays HIGH, and the Q output stays HIGH, which therefore holds the Q^* LOW. Hence again the switch bounce at the a point has no effect on either Q or Q^*.

The desired output pulse shown in the waveform of Q^* is a single pulse of duration that is approximately the length of time that the operator held the switch button down. This example of the use of a latch is part of Demonstration D5.2.

5.2 CLOCKED FLIP-FLOPS

Where many latch circuits are used in a digital system, it is generally necessary that the output waveforms of the various latches be synchronized in time irrespective of the arrival of the S and R signals to the individual latches. This requires additional gating to the latch circuit. For purposes of this text, a latch with additional gating, where one of the inputs to the gating circuit affects both inputs to the latch, will be referred to as a *flip-flop*. This additional input is called the *clock input*.

5.2.1 The *R-S* Flip-Flop

We now consider the useful circuit found in ICs, namely, the *R-S flip-flop*. As shown in Fig. 5.7a, we have made it from a NOR latch to which we have added two AND gates. Note that the inputs S and R go to two AND gates that also have a common input C_p, which is the clock input. The inputs into the latch now can only be activated when the clock is 1. Figure 5.7b shows an all-NAND-gate version of this type of flip-flop. The commonly accepted logic symbol for the R-S flip-flop is shown in Fig. 5.7c.

Figure 5.8 shows a series of waveforms from an R-S flip-flop, with the sequence of events where initially the flip-flop is reset; that is, $Q = 0$ and both the set (S) and reset (R) inputs are 0. The result of these initial conditions is that the first clock pulse ($n = 1$) has no effect on the state of the flip-flop, (that is, Q remains unchanged, which is shown as a $0 \rightarrow 0$, after the clock pulse activates the input to the latch). This is confirmed in Fig. 5.7a, where with S and R both 0, S_ℓ and R_ℓ are also 0 even with $C_p = 1$. Hence there is no change at the output of the latch with clock pulse 1.

The next sequence of events in Fig. 5.8 is that $S = 1$ before clock pulse 2 arrives; R is still 0. Hence upon the arrival of the clock pulse, the latch input (S_ℓ in Fig. 5.7a) goes HIGH, Q^* goes LOW, and consequently the Q output goes to 1. This sequence of events continues through all valid combinations of S, R, and Q in Fig. 5.8.

Table 5.3 shows the same sequence of events, where we have numbered the clock pulse in sequence from 1 to 6. The Q_{n+} column in the table stands for the value of Q immediately after the effect of the nth clock pulse.

In Table 5.3 we also show the inputs which give an indeterminate output for the case of the R-S flip-flop. Notice in Fig. 5.7a that when C_p goes HIGH, with 1 at both S and R inputs, the latch inputs are both 1 and then both outputs are 0. Now after the clock pulse, when C_p goes LOW, both latch inputs are 0 and the output of the flip-flop is left in an indeterminate state; either Q will finish up LOW and Q^* HIGH, or vice versa. Small unpredictable variations in the circuit parameters will determine the final state of the flip-flop.

Table 5.3 R-S **FLIP-FLOP TRUTH TABLE**

n	S	R	Q_n	Q_{n+}
1	0	0	0	0
2	1	0	0	1
3	1	0	1	1
4	0	0	1	1
5	0	1	1	0
6	0	1	0	0
...	1	1	0	?
...	1	1	1	?

5.2.2 Excitation and Characteristic Tables for Flip-Flops

The same information as given in Table 5.3 can be given in another form called the *excitation table*, as shown in Table 5.4. Here we show on the left the value of the Q output (Q_n) before the clock pulse is applied, and in the next column the

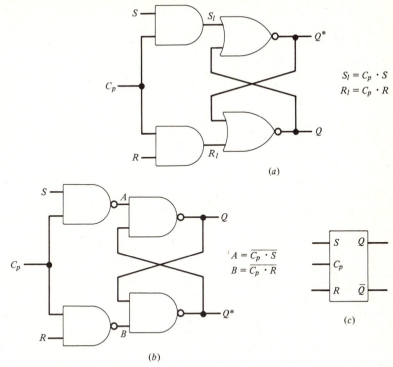

$$S_l = C_p \cdot S$$
$$R_l = C_p \cdot R$$

(a)

$$A = \overline{C_p \cdot S}$$
$$B = \overline{C_p \cdot R}$$

(b)

(c)

FIGURE 5.7
(a) R-S flip-flop made from NOR latch and two AND gates; (b) NAND-NAND
version of R-S flip-flop; (c) logic symbol for an R-S flip-flop.

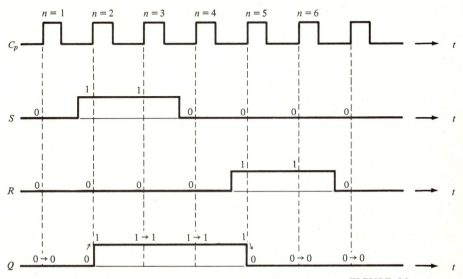

FIGURE 5.8
R-S flip-flop waveforms.

value of Q output after the clock pulse is applied, Q_{n+}. In the next two columns, the set and reset values that will cause the indicated transition to take place are shown. Note the use of the don't care symbol $\emptyset$, which allows for more compactness in the table.

Table 5.4 R-S FLIP-FLOP EXCITA- TION TABLE

Q_n	Q_{n+}	S_n	R_n
0	0	0	$\emptyset$
0	1	1	0
1	0	0	1
1	1	$\emptyset$	0

Table 5.5 FLIP-FLOP CHARAC- TERISTIC TABLE

S_n	R_n	Q_{n+}
0	0	Q_n
0	1	0
1	0	1
1	1	?

The same information, cast in a different form, is given in Table 5.5, which is called the *characteristic table.* Here we show in the left two columns the set and reset inputs, and in the rightmost column, the value taken by Q after the clock pulse is applied. The excitation table indicates the required input excitation for a desired logic state at the output. The characteristic table, on the other hand, shows the result at the output due to a particular logic state at the input.

Because of the ambiguity, described in the previous section, the excitation table does not show provisions for both $S = 1$ and $R = 1$, which is shown in the characteristic table. Since we plan to avoid this set of input signals, the excitation table carries all the information that we need for properly using this flip-flop.

Inputs that are effective only during the presence of a clock pulse are referred to as the *synchronous inputs.* Later on we will discuss inputs that are *asynchronous,* which means that they are effective without the clock.

5.2.3 Data Storage Latch

Another common application for the latch is shown in Fig. 5.9. This has the usual cross-coupled internal circuitry with a single data input called DATA 1 and a line called the *clock* line that *enables* the data to be entered into the latch. Note the use of the inverter to ensure that the data inputs into the inner latch are truly complementary. Another name for the clock line is the *enable* line, as it enables the data to enter into the latch.

Commercial MSI circuits are available that consists of four† or eight such circuits of Fig. 5.9 in a single package. These 4- and 8-bit latches are useful for temporary storage, such as taking a "snapshot" of information that exists in some other circuit for only a short length of time and holding it for further operations or display at a time in the future.

† See, for example, the TTL type 7475.

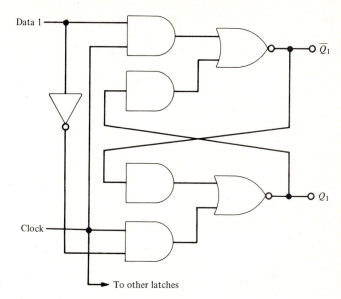

FIGURE 5.9
Data storage latch.

To other latches

EXERCISES

E5.3   Given the different characteristic table for an *R-S* flip-flop shown below:
(*a*) Derive the excitation table (similar to Table 5.4) for this flip-flop.
(*b*) Determine the state of Q_{n+} for the eight logic states similar to that shown in Table 5.3.

S_n	R_n	Q_{n+}
1	1	Q_n
1	0	1
0	1	0
0	0	?

E5.4   Use the excitation table for the *R-S* flip-flop given in Table 5.4 and the circuit of Fig. 5.7*b* to complete the waveform at the *Q* output in Fig. E5.4.

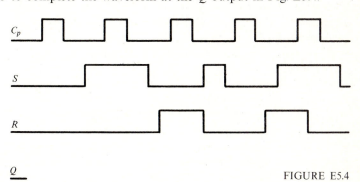

FIGURE E5.4

5.3 FLIP-FLOPS WITH MEMORY

Beside the set and reset functions, in many applications of sequential circuits it is desired to have a *toggle* function. This circuit function is also called a *binary counter*, *divide-by-two*, *scale-of-two*, or *trigger* flip-flop. We shall see later that a clocked form of flip-flop, the toggle or T flip-flop, is a circuit which can be used to give a single output pulse for every two input pulses.

The digital clock (which we briefly described in the Introduction and which we will complete in Chap. 6) will require many such flip-flops to divide the 60-pps pulse train to 10 pps and then successively to 1 pps, 1 ppm, etc. In Chap. 6 we describe how to connect a series of flip-flops to divide by six or ten, but first we must determine how with successive pulses at the T input the flip-flop will change state, irrespective of its original condition. We find that we need some method of *steering* the input pulses so that the flip-flop will set on the first input pulse, reset on the second, etc.

5.3.1 Toggle Flip-Flop Containing a Delay Element

In Fig. 5.10*a* we show an example of a binary counter or toggle flip-flop implemented with NAND gates. It shows a clocked R-S flip-flop with a pulse T applied to the input that we have previously referred to as the clock input. In addition, we show a cross-coupled feedback connection from the output terminals back to the input NAND gates through a *delay* network, which provides a time delay Δ with no signal attenuation.

For toggle operation both R and S are connected HIGH. Now notice that when Q is HIGH the T pulse is steered to the reset side of the flip-flop, causing Q to go LOW. Then with $\bar{Q}$ now HIGH the next pulse on the T line is steered to the set side, causing another change of state for the flip-flop. Hence we have the waveform at the Q output as shown in Fig. 5.10*b*, where for two pulses at the T input we generate one output pulse. Hence this is a binary counter.

However, also note that if the pulse at T is wider than Δ, T will still be HIGH after the switching has taken place; now the T pulse is directed to the other latch input and the flip-flop will change state twice for one input pulse. Indeed we would find that the output would alternate between 1 and 0 for the duration of the input pulse. We must therefore control the width of the pulse at T to be less than Δ. The problem of pulse-width control with this simple toggle flip-flop is shown in Demonstration D5.3.

The circuit shown in Fig. 5.10*a* is impractical for an IC since it requires precise pulse-width control as well as an element (pure delay) which is not amenable to IC technology. In the next section we will discuss a circuit that permits us to obtain the binary-counter function with IC technology.

5.3.2 Master/Slave Flip-Flop

In order to eliminate the need for physical delay elements, the *master/slave* principle is now utilized in most IC flip-flops.

In the master/slave method, the delay is supplied by a second latch and gate

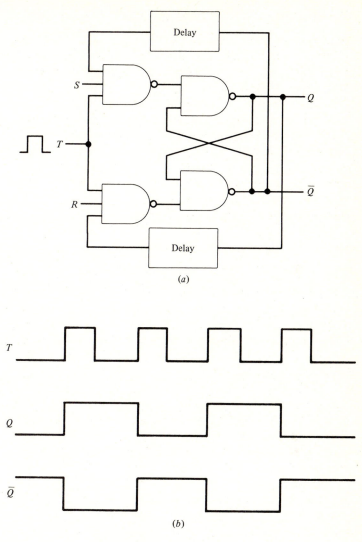

FIGURE 5.10
(a) T flip-flop made with NAND gates; (b) the waveforms.

circuit. In Fig. 5.11a, we have shown the logic diagram for an R-S master/slave flip-flop. Here we have two latches; the one to the left, i.e., nearest to the S and R inputs, termed the *master* latch, and the one where the outputs Q and Q̄ are taken from, called the *slave* latch. We have really just two R-S flip-flops with one additional feature, namely, the inversion of the clock pulse going to the slave input gates.

The operation of this circuit can be best described by looking at the effect of the clock pulse and the inverted clock pulse on the inputs into the AND gates, as

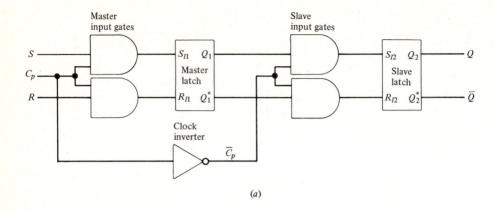

(a)

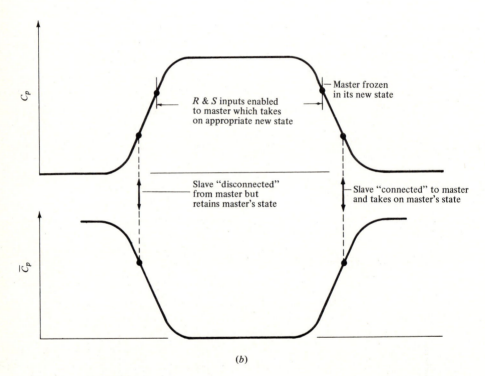

(b)

FIGURE 5.11

(a) R-S master/slave flip-flop; (b) data transfer in master/slave flip-flop during clock pulse.

shown in Fig. 5.11*b*. Here we initially show the clock pulse as being LOW. Then because $\bar{C}_p$ is HIGH, the slave latch is connected to the master latch, through the slave input gates, and the output of the slave is the same as that of the master. As the clock-pulse voltage increases with time, we find that the inverted clock-pulse magnitude decreases below the threshold level of the slave input gates so, in effect, the slave latch is disconnected from the master latch. On increasing the magnitude of the clock-pulse signal, we come to the threshold of the master input gates. When we increase the clock pulse past this amplitude, the *S* and the *R* inputs are enabled to the master, and the master now takes on the state which the input data dictate. This new state is a function both of *S* and *R* inputs, with the constraint that both cannot be 1 if we want a determinate state. During all this time, the slave latch has been disconnected from the master. Now when the clock pulse starts decreasing, we come to a point where the clock-pulse input to the master input gates decreases to a level where the gates no longer respond to the input variables *S* and *R*. Thus the state of the master latch is frozen to that which was determined during the time the inputs to the master latch were enabled. As the clock-pulse amplitude decreases further, we finally come to the point where the inverted clock-pulse magnitude becomes large enough to enable the slave input gates. At this point, the slave latch is connected to the master latch, and the slave takes on the status corresponding to the new state of the master.

Thus the output of the slave latch is not able to change during the time the *S* and *R* inputs are able to affect the master latch. Also, when the slave latch is changing, the *S* and *R* inputs to the master latch are disabled.

R-S master/slave flip-flops are useful by themselves and are available in integrated form. We find them particularly useful in a cascade arrangement such as in the 8-bit shift register.† However, one of the more popular flip-flops utilizing the master/slave technique is the *J-K* flip-flop, in which the ambiguity associated with 1 at both *S* and *R* inputs is avoided.

5.3.3 *J-K* Master/Slave Flip-Flop

Let us now see how to realize a *J-K* master/slave flip-flop from an *R-S* master/slave flip-flop. This is shown in Fig. 5.12, where the only difference from Fig. 5.11*a* is the use of 3-input AND gates at the master latch input and the cross-coupled feedback connections to these extra inputs from the outputs of the slave latch.

The *J* and *K* are the synchronous inputs; the *J* input is the clocked-set input, and the *K* input is the clocked-reset input. With these inputs, data are entered into the master latch when the clock (C_p) is HIGH. When the clock is LOW, these inputs are disabled, and the master dictates the condition of the slave latch through the slave input gates. Now when the clock goes HIGH, the slave is locked in its present condition isolated from the master. The master input gates are now enabled, but we find that either the *J* input or the *K* input is disabled by the feedback line from the output terminals. Thus when $J = 1$ and $K = 1$ with the

† Available in TTL as the 7491.

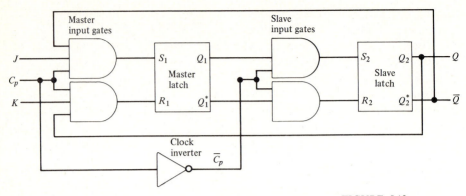

FIGURE 5.12
J-K master/slave flip-flop.

clock HIGH, the only master latch input permitted is for the master to take up a condition opposite to that of the slave. For example, with C_p LOW and the master latch in the SET state, the slave latch is also SET. Consequently Q is 1 and $\bar{Q}$ is 0. Now when C_p goes HIGH, only the K side of the master input gates are enabled. Hence the only input permitted is for the master latch to RESET. Of course with both J and K LOW there is no change of state in the master latch.

With both J and K made HIGH, the clock line becomes a toggle input, but now there is no width dependence on the pulse input. That is, when the clock line (the T input in Fig. 5.10a) goes HIGH, the pulse is directed to change the state of the master latch under the influence of the J or K feedback line. But it is not until the clock line goes LOW (i.e., at the end of the T pulse), after the master input gates are disabled, is there any transfer of information into the slave and hence any change at the output terminals.

Needless to say this is a rather complex circuit; a typical J-K master/slave flip-flop has a total of about 40 transistors, diodes, and resistors. However, this is all contained in a 60-mils-sq chip, so that they are quite economical. Indeed two such independent flip-flops in one package are a common IC product.

Let us next discuss the J-K flip-flop from a truth table point of view. The purpose of the J-K flip-flop is to eliminate the uncertainty which arises in the R-S flip-flop characteristic, when both S and R inputs are 1. The J-K flip-flop gives a definitive response under these conditions. This can be seen in the J-K flip-flop truth table in Table 5.6. In particular, under clock-pulse entries 7 and 8, we show that for 1 input at both J and K the output changes to its opposite state after being triggered by the clock pulse.

Table 5.7 shows the same results cast in the form of an excitation table, where the desired transitions of the output are shown in the two left columns and the necessary excitations are shown in the J and K columns. Notice the increased number of Øs available with the J-K as opposed to the R-S flip-flop excitation table shown in Table 5.4.

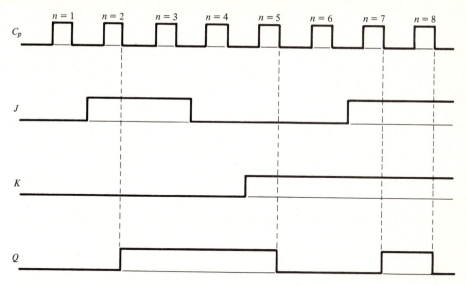

FIGURE 5.13
Waveforms of the J-K master/slave flip-flop.

Table 5.8 shows the J-K flip-flop characteristics, where we have the same result for the first three entries as in the R-S flip-flop. Then at the fourth entry, with 1 at both inputs, the flip-flop toggles (i.e., it changes state), as shown by the fact that the state of the flip-flop after the nth pulse (Q_{n+}) is given as $\bar{Q}_n$.

Figure 5.13 shows the sequence of events for the eight possible combinations of the J, K, and Q states of Table 5.6. We see that the transitions at the Q output (and similarly for the $\bar{Q}$ output) take place at the trailing or negative edge of the clock pulse, as shown at the end of clock pulse 2, where the state of Q changes from 0 to 1. In the figure we also see that the Q changes state back to 0 after clock pulse 5, since K had been 1 during clock pulse time, while J was 0. Finally, we see that at the negative edge of clock pulse 7, Q goes to 1 because both J and K inputs

Table 5.6 J-K FLIP-FLOP TRUTH TABLE

n	J_n	K_n	Q_n	Q_{n+}
1	0	0	0	0
2	1	0	0	1
3	1	0	1	1
4	0	0	1	1
5	0	1	1	0
6	0	1	0	0
7	1	1	0	1
8	1	1	1	0

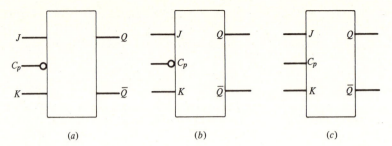

FIGURE 5.14
(*a*) Generalized block-diagram symbol for a *J-K* flip-flop; (*b*) logic symbol for *J-K* flip-flop, showing outputs change on negative edge of clock pulse; (*c*) logic symbol for positive edge-triggered *J-K* flip-flop.

were HIGH. Similarly, at the negative edge of clock pulse 8, we see that Q returns to 0, again because both J and K inputs were 1.

The logic symbol for a *J-K* master/slave flip-flop of the type that we have described is shown in Fig. 5.14*a*, where the small circle on the clock-pulse line indicates that the output of the flip-flop changes state on the negative edge of the clock-pulse waveform. Conventionally, we label the terminals of the flip-flop on the inside of the square to save space in drawings (see Fig. 5.14*b*). It should be understood that this does not mean inversion (i.e., the signal labeled is still applied external to the inverting circle).

It is possible to have *J-K* flip-flops which respond on the leading or positive edge of the clock pulse, in which case the block diagram would be that given in Fig. 5.14*c*. This type is often referred to as edge-triggered, with the word *positive* left out. Figure 5.15 compares these two types of *J-K* flip-flops, with a clock-pulse waveform shown in Fig. 5.15*a*. We show the output for both *J-K* flip-flops with both J and K equal to 1. For the negative edge-triggering, we see that the change in Q takes place at the trailing edge, as shown in Fig. 5.15*b*. In Fig. 5.15*c*, we show the result of a flip-flop that is triggered on the positive edge of the clock pulse. Here the change of state occurs immediately upon application of the clock pulse. Both types of flip-flops are available in integrated form, and the user has to look carefully at his own application to determine which is the most appropriate for his

Table 5.7 *J-K* FLIP-FLOP EXCITATION TABLE	**Table 5.8** *J-K* FLIP-FLOP CHARACTERISTIC TABLE

Q_n	Q_{n+}	J_n	K_n
0	0	0	Ø
0	1	1	Ø
1	0	Ø	1
1	1	Ø	0

J_n	K_n	Q_{n+}
0	0	Q_n
0	1	0
1	0	1
1	1	$\bar{Q}_n$

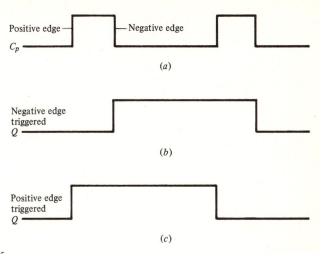

FIGURE 5.15
Waveforms for the two types of triggering for J-K flip-flop. (a) Clock-pulse input.
(b) Q output with negative edge-trigger. $J = K = 1$. (c) Q output with positive edge-
trigger. $J = K = 1$.

needs. Thus we have two TTL J-K flip-flops, the 7470 and the 7476. The 7470 is
positive edge-triggered, while in the 7476, the change of state at the output takes
place on the negative or trailing edge of the clock pulse.

5.3.4 Initializing Flip-Flops

So far we have not discussed what the initial state of the flip-flop is, and the only
way that we could affect its state is through the synchronous inputs that are
functionally tied to the clock pulse. However, in many situations, it is desirable to
preset the flip-flop (force $Q = 1$) or clear it (force $Q = 0$). To be specific, we often
desire another set of inputs, P and C, which can be actuated independently of the
clock pulse in order to PRESET a 1 into a flip-flop, or conversely, to CLEAR the
flip-flop so it has 0 value. Table 5.9 shows a typical excitation table for the
asynchronous inputs. Asynchronous inputs are also called *direct* inputs.

Table 5.9 J-K FLIP-FLOP ASYNCHRONOUS-
INPUT EXCITATION TABLE

P	C	Q (after asynchronous inputs)
0	0	Q (before asynchronous inputs)
0	1	0
1	0	1
1	1	(not valid inputs)

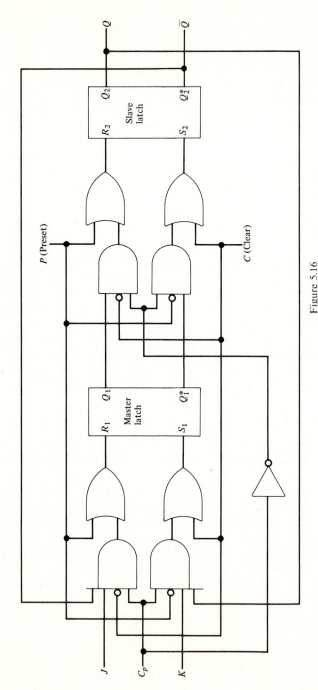

Figure 5.16
J-K master/slave flip-flop with asynchronous present and clear.

Figure 5.16 shows the modification necessary to a *J-K* master/slave flip-flop in order to have a preset and clear capability. The little circles in front of the AND gates simply represent inversion. This circuit assures us that the preset or clear signal will affect both the master and the slave latch. Notation for such a circuit is shown in Fig. 5.17a, corresponding to our Fig. 5.16. If inverters were used on the chip ahead of the preset and clear inputs of Fig. 5.16, then the logic symbol of Fig. 5.17b would apply.

The data sheet for a *dual J-K* master/slave flip-flop is shown in Fig. 5.17c. A schematic diagram of each flip-flop, as well as the block diagram, are found on this data sheet. Included is a truth table for each flip-flop, along with a description of the operation of the clock pulse. The pin connections for this 16-pin dual in-line package are also included.

EXERCISES

E5.5 Given the different characteristic table for a *J-K* flip-flop shown below:
(*a*) Derive the excitation table (similar to Table 5.7) for this flip-flop.
(*b*) Determine the state of Q_{n+} for the eight logic states similar to that shown in Table 5.6.

J_n	K_n	Q_{n+}
0	0	$\bar{Q}_n$
0	1	0
1	0	1
1	1	Q_n

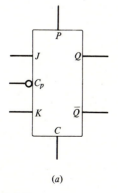

(a)

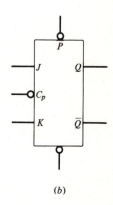

(b)

FIGURE 5.17
(a) Logic symbol for *J-K* flip-flop of Fig. 5.16; (b) logic symbol for *J-K* flip-flop with direct entry made to the flip-flop with logic LOW; (c) data sheet for TTL type 5476/7476, dual *J-K* master/slave flip-flops (overleaf).

**DUAL J-K MASTER-SLAVE FLIP-FLOPS
WITH PRESET AND CLEAR**

logic

TRUTH TABLE (Each Flip-Flop)		
t_n		t_{n+1}
J	K	Q
0	0	Q_0
0	1	0
1	0	1
1	1	$\overline{Q}_0$

NOTES: 1. t_n = Bit time before clock pulse.
  2. t_{n+1} = Bit time after clock pulse.

description

The SN7476 J-K flip-flop is based on the master-slave principle. Inputs to the master section are controlled by the clock pulse. The clock pulse also regulates the state of the coupling transistors which connect the master and slave sections. The sequence of operation is as follows:

1. Isolate slave from master
2. Enter information from J and K inputs to master
3. Disable J and K inputs
4. Transfer information from master to slave.

recommended operating conditions

	MIN	NOM	MAX	UNIT
Supply Voltage V_{CC}: SN5476 Circuits	4.5	5	5.5	V
SN7476 Circuits	4.75	5	5.25	V
Operating Free-Air Temperature Range, T_A: SN5476 Circuits	-55	25	125	°C
SN7476 Circuits	0	25	70	°C
Normalized Fan-Out From Each Output, N			10	
Width of Clock Pulse, $t_{p(clock)}$ (See figure 69)	20			ns
Width of Preset Pulse, $t_{p(preset)}$ (See figure 70)	25			ns
Width of Clear Pulse, $t_{p(clear)}$ (See figure 69)	25			ns
Input Setup Time, t_{setup} (See figure 69)	$\geq t_{p(clock)}$			ns
Input Hold Time, t_{hold}	0			

6-58

---

functional block diagram (each flip-flop)

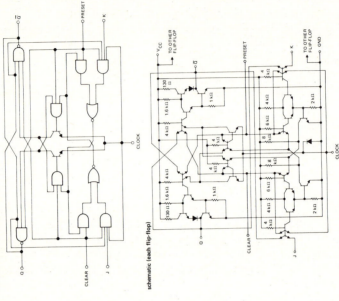

schematic (each flip-flop)

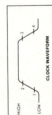

NOTE: Component values shown are nominal.

**J OR N DUAL-IN-LINE OR
W FLAT PACKAGES (TOP VIEW)†**

positive logic: Low input to preset sets Q to logical 1
Low input to clear sets Q to logical 0
Clear and preset are independent of clock

†Pin assignments for these circuits are the same
for all packages.

CLOCK WAVEFORM

FIGURE 5.17(c)

E5.6 The excitation table of a *J-K* flip-flop is as given in Table 5.7. Complete the timing diagram in Fig. E5.6 for the *Q* output. The flip-flop responds on the trailing edge of the clock pulse and to the *P* and *C* inputs independent of the clock pulse.

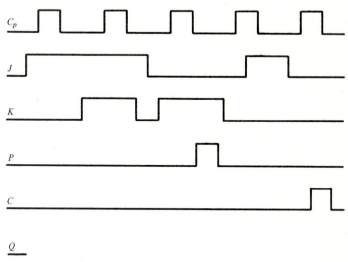

FIGURE E5.6

5.4 *D* FLIP-FLOP

In addition to the *R-S* and *J-K* flip-flops, another useful flip-flop available as an IC is the *D* or DELAY flip-flop. We will find in Chap. 6 that *D*-type flip-flops are most useful in the design of shift registers as well as in simple counters. The logic symbol for the *D* flip-flop is shown in Fig. 5.18, and the excitation table is given in Table 5.10.

Notice that the characteristics of the *D* flip-flop are very similar to the data latch (Fig. 5.9) in that the *Q* output follows the logic state of the *D* input. However, there is a difference. In the case of the latch, information present at the data input is transferred to the *Q* output when the clock goes HIGH. Then while the clock continues HIGH, the *Q* output will directly follow the logic state of the data input. Whereas in the case of the *D* flip-flop, information at the input is again transferred to the output when the clock goes HIGH, but immediately thereafter the *D* input is inhibited while the clock remains HIGH. Hence information may be stored in

Table 5.10 *D* FLIP-FLOP EXCITA-
TION TABLE

Q_n	Q_{n+}	D_n
0	0	0
0	1	1
1	0	0
1	1	1

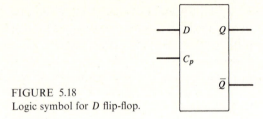

FIGURE 5.18
Logic symbol for *D* flip-flop.

the *D* flip-flop until cleared by a pulse at the direct clear input or replaced with new information at the *D* input during another clock pulse.

The type of *D* flip-flop made in IC form† is most commonly triggered on the positive edge of the clock pulse.

5.5 CONVERSION BETWEEN FLIP-FLOPS

In the use of ICs we sometimes need a particular type of flip-flop which may frequently be unavailable. The conversion from one type of flip-flop to another is possible by the use of additional combinational logic gates called *control logic*. Note that this conversion applies to the function of the flip-flop and does not affect whether the triggering is on the positive or negative edge of the clock pulse.

To convert from one type of flip-flop to another (of the types *D*, *T*, and *J-K*), we use the simple technique as shown in the following example. Consider the problem of converting a *D* flip-flop to a *J-K* flip-flop. This problem is shown in Fig. 5.19*a*, where we are given a *D* flip-flop and we desire to find a combinational circuit shown to the left with inputs *J*, *K*, *Q*, and $\bar{Q}$ that will drive the *D* input and give us the equivalent function shown in Fig. 5.19*b*, that is, the *J-K* flip-flop.

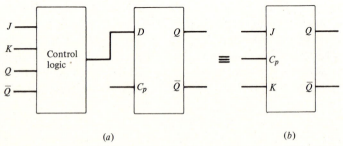

(*a*)   (*b*)

FIGURE 5.19
Conversion of *D* flip-flop to *J-K* flip-flop, showing need of some combinational control logic.

† For example, the 7474 made with TTL or the 4013A fabricated with CMOS.

A systematic method of doing this is to list all possible combinations of the input variables to the control logic, as shown in Table 5.11. This is a special form of the truth table and is referred to as a *state table* for the J-K flip-flop. We have shown the three input variables to the left. The desired output is Q_{n+}, which is the next state of the flip-flop, that is, after the clock pulse has come along. But note in particular that each value for Q_n corresponds to the immediate previous value for Q_{n+}. This ordering, while not important here, will be found especially useful in dealing with sequential logic circuits. We have arbitrarily assigned state numbers to the sequence of events. Thus we show eight state numbers, and in Table 5.12 we can determine what the value of D_n should be to obtain this state. This value of D_n is obtained from the excitation table in Table 5.10.

As an example of filling in these tables, at state number $n = 1$, with $J_n = 0$ and $K_n = 1$ we note from the J-K characteristic table in Table 5.8 that $Q_{n+} = 0$. That is, at this state number there is no change in the Q output. Now from the D excitation table (Table 5.10) we see that for this condition $D_n = 0$, which is entered in Table 5.12 at $n = 1$.

At state number $n = 2$, we refer to Table 5.8 and see that with $J_n = 1$ and $K_n = 0$ that $Q_{n+} = 1$. At this state number there is a change at the Q output from 0 to 1. From Table 5.10 we see that this requires $D_n = 1$, which is so entered at $n = 2$ on Table 5.12. We then progressively complete the requirements of D_n in Table 5.12 from Table 5.11 with reference to Tables 5.8 and 5.10.

In Fig. 5.20a we have made a Karnaugh map where, instead of the binary entries, we have now entered in the state numbers that are consistent with the three input variables J_n, K_n and Q_n from Table 5.11. This is our *reference map* and is our means of keeping track of numbers. We will find this technique of considerable help in the following chapter on counter design.

In Fig. 5.20b we now enter the values of D_n into another Karnaugh map called a *control map*. This is done by taking the values from Table 5.12 and putting them in the appropriate state-numbered location. A simple reduction, in sum-of-products form, from the control map gives $D_n = J_n \bar{Q}_n + \bar{K}_n Q_n$. This is our con-

Table 5.11 STATE TABLE FOR A J-K FLIP-FLOP

State no.	Present state			Next state
n	J_n	K_n	Q_n	Q_{n+}
0	0	0	0	0
1	0	1	0	0
2	1	0	0	1
3	0	0	1	1
4	1	0	1	1
5	0	1	1	0
6	1	1	0	1
7	1	1	1	0
0	0	0	0	0

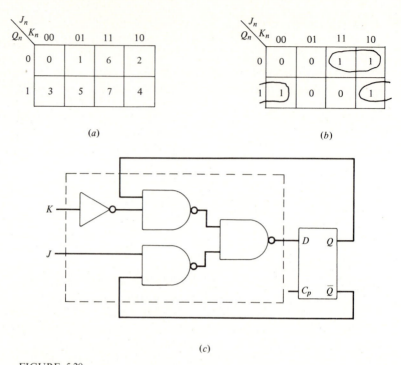

(a)

(b)

(c)

FIGURE 5.20
(a) Reference map for control logic of Fig. 5.19; (b) Karnaugh map of control logic; (c) NAND-NAND solution to control logic.

trol logic, which is realized in the NAND-NAND circuit shown in the dotted square in Fig. 5.20c. Thus with three 2-input NAND gates and one inverter we can readily convert a D flip-flop to a J-K flip-flop. This is the subject of Demonstration D5.4. The technique of state number and present-to-next state listings described in this conversion process will also apply in the design of counters, as we shall see in the next chapter.

Table 5.12 REQUIRED D_n AS A FUNCTION OF n

State no.	
n	D_n
0	0
1	0
2	1
3	1
4	1
5	0
6	1
7	0

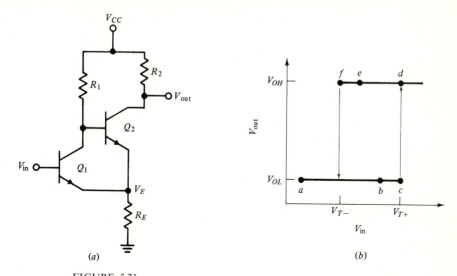

FIGURE 5.21
(*a*) Basic emitter-coupled Schmitt trigger circuit; (*b*) voltage transfer characteristic.

EXERCISE

E5.7   Use the methods described in this section to convert a positive edge-triggered *J-K* flip-flop to a similarly triggered *D* flip-flop. The characteristics of the two flip-flops are shown in Tables 5.8 and 5.10, respectively.

5.6   SCHMITT TRIGGER

Another useful regenerative circuit found in many digital systems is the *Schmitt trigger*. This circuit is especially useful when it is desired to steepen the leading and trailing edges of a slowly rising and falling pulse.

A simplified diagram of an emitter-coupled version of the Schmitt circuit is shown in Fig. 5.21*a*. The desirable feature of this trigger circuit is that the voltage transfer characteristic, shown in Fig. 5.21*b*, has different input threshold levels for positive- and negative-going signals. The difference between the threshold levels is known as *hysteresis*.

That such a characteristic is useful is best shown by considering the circuit in the digital clock (Fig. I.4*a*) which is to generate pulses from a voltage derived from the power line. In Fig. 5.22*a* the solid line shows a pure sine-wave input voltage to the Schmitt trigger. In Fig. 5.22*b* the solid line shows the output of the trigger circuit with $V_{T+} = 1.7$ V and $V_{T-} = 0.9$ V; that is, the hysteresis is equal to 0.8 V. The solid line in Fig. 5.22*c* is the output of the circuit with the hysteresis equal to 0.1 V; here $V_{T+} = 1.7$ V and $V_{T-} = 1.6$ V. Now if the line voltage is not a pure

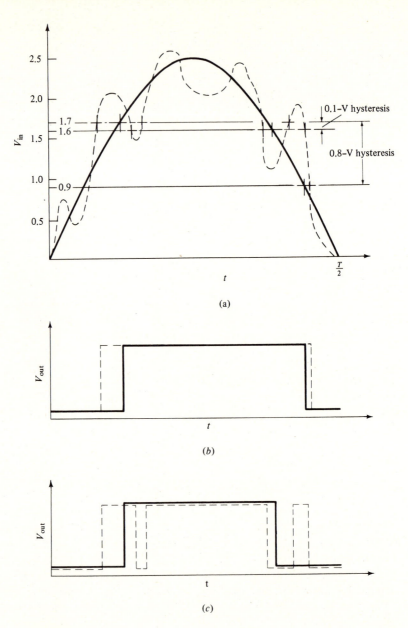

(a)

(b)

(c)

FIGURE 5.22
(a) Input voltage to the Schmitt circuit, with pure sine wave shown in solid line and typical input shown dotted; (b) output voltage for the two cases of part (a), with hystersis equal to 0.8 V; (c) output voltage for the two cases of part (a), with hystersis equal to 0.1 V.

sine wave (due to load transients, etc.), as shown by the dotted waveform in Fig. 5.22a, the output pulse with the circuit having 0.8-V hysteresis will be shifted slightly in time, but we will still get only *one* output pulse for this input. However, as shown in Fig. 5.22c, with only 0.1-V hysteresis we would obtain *three* pulses, which would make our clock very much in error.

This operation of the circuit in Fig. 5.21a is as follows. Assume V_{in} is low enough so that Q_1 is cut off (this is point a in Fig. 5.21b). Then with β_F large, Q_2 will be saturated. This establishes some value for V_E (the common-emitter voltage of Q_1 and Q_2). Now increasing V_{in} we come to the point b, where Q_1 becomes active but Q_2 is still saturated. For β_F large, V_E will remain unchanged. Further increases in V_{in} to V_{T+} bring us to the point c, where Q_2 is at the edge of saturation. Now any minute increase in V_{in} will cause I_{C1} to increase, which causes I_{C2} to decrease. The result is that Q_2 turns off rapidly, due to the regenerative action of the voltage at the base of Q_2 falling, as I_{C1} increases, and the voltage at the emitter of Q_2 rising, as V_{in} increases. The collector current of Q_1 keeps on increasing until Q_2 is cut off and Q_1 is saturated. This gets us to point d on the transfer function. Further increases in V_{in} have no effect on V_{out}.

If now we start to decrease V_{in} to a value less than V_{T+}, we find the interesting result that Q_1 stays saturated and Q_2 cut off until we get to point e, at which time Q_1 goes into the active region while Q_2 is still cut off. Decreasing V_{in} further to V_{T-} (point f) brings Q_2 to the edge of conduction while Q_1 is still active. Now any minute decrease in V_{in} will cause I_{C1} to decrease, which will cause I_{C2} to increase. The regenerative action is now due to the voltage at the base of Q_2 rising, as I_{C1} decreases, and the voltage at the emitter of Q_2 falling, as V_{in} decreases. The result is that Q_2 rapidly turns on and saturates as Q_1 cuts off. Further reductions in V_{in} have no effect other than to make Q_1 go further into cutoff.

This circuit with a 4-input AND gate at the input and a totem-pole circuit at the output is used in the TTL type 7413.

5.7 MONOSTABLE MULTIVIBRATOR

We now briefly turn our attention to another member of the regenerative circuit family, namely, the *monostable multivibrator*. The bistable circuit, it may be recalled from our discussion of latches and flip-flops, had two stable states—in either one of which it may permanently remain. The monostable circuit, on the other hand, has only one stable state and one so-called *quasi-stable* state. That is, a trigger signal is required for the monostable to make the transition from the stable state to the quasi-stable state. The circuit will remain in its quasi-stable condition for a predetermined time set by the circuit parameters, usually a simple choice of value for a resistor and a capacitor. However, at the end of this time the circuit will make the transition back to its stable state, where it will remain until again triggered. The transition from one state to the other is a regenerative process and so can be very fast. The output of the monostable multivibrator is therefore a voltage pulse with a width of some predetermined time and relatively fast transition times. The monostable circuit (also called a *one-shot*) finds application in

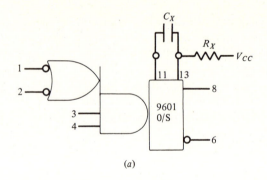

(a)

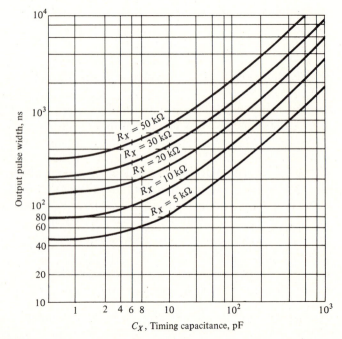

FIGURE 5.23
(a) Block diagram of IC monostable multivibrator (type 9601); (b) output pulse
width vs. timing resistance (R_x) and capacitance (C_x).

pulse regeneration and in *time delay*. It can also be used as a convenient *clock-pulse
generator*, as described in Sec. 10.1.9.

A simple monostable multivibrator can be made from either NOR or
NAND logic gates, with the addition of a resistor and a capacitor (see Prob.
P5.16). However, the general use of this circuit has resulted in the development
of easy-to-use monostable multivibrator ICs.

An example is the TTL type 9601, for which the block diagram is shown in
Fig. 5.23a. The input and output voltage levels of this IC are compatible with TTL

circuits. The combinational logic at the input allows for some logic control of the triggering of the monostable. Also, because of this, the circuit may be triggered with a positive-going input (leading edge-triggering) or a negative-going input (trailing edge-triggering). The triggering occurs at a particular input voltage level and is not directly related to the transition time of the input signal. This is due to a Schmitt circuit (similar to that described in Sec. 5.6) at the input of the IC.

The duration of the pulse at the complementary outputs is determined by the value of the timing components R_X and C_X. The value of R_X must be between 5 and 50 kΩ. For $C_X \geq 1000$ pF, the expression for the pulse width is

$$t = 0.32 R_X C_X \left(1 + \frac{0.7}{R_X}\right) \qquad (5.1)$$

where t = nanoseconds (ns)
R_X = kilohms (kΩ)
C_X = picofarads (pF)

For values of $C_X < 1000$ pF, the graphical relationship shown plotted in Fig. 5.23b must be used. Hence if we desire an output pulse width of 1000 ns, we have an ample choice of component values, say, $R_X = 10$ kΩ and $C_X = 220$ pF. The above relationships are accurate to within 15 percent.

5.8 SUMMARY

\# A digital circuit can generally be characterized as either *nonregenerative* or *regenerative*.

\# Representative of nonregenerative circuits are logic *gates* used in *combinational logic*.

\# *Sequential logic* makes use of regenerative circuits, in particular the *bistable* circuits, like *latches* and *flip-flops*.

\# A simple latch may be formed by *cross-coupling* either two NOR gates or two NAND gates.

\# The logic of a latch or flip-flop is contained in the *characteristic table* or *excitation table* for the device.

The characteristic table shows the result at the output due to a particular logic state at the input.

The excitation table indicates the required excitation at the input for a desired output logic state.

\# It is necessary to check the characteristic table of a latch or flip-flop to determine the possibility of ambiguous outputs.

\# The *R-S flip-flop* is made with a latch circuit and additional gates that control the clock input, so that changes at the output of the flip-flop are synchronized with the timing of the clock pulse.

\# The *J-K flip-flop* is the most versatile IC flip-flop, performing both the clocked SET and RESET functions, as well as having no ambiguity of output state.

\#   Most *J-K* flip-flops operate on the *master/slave* principle.

\#   A very useful logic element is the *T flip-flop*, or ÷ 2 circuit. While not available as a separate IC device, this logic function is easily formed with the *J-K* flip-flop.

\#   The *D flip-flop* is an IC bistable element used extensively in digital systems.

\#   One type of flip-flop can readily be converted to another type by the use of *control logic*.

\#   Flip-flop inputs that are effective only with a clock pulse are referred to as *synchronous* inputs.

\#   *Asynchronous* or *direct* inputs act directly upon the latches within the flip-flop.

\#   The *Schmitt trigger* is a regenerative circuit, though not a true bistable since in general use the output follows the input up and down, but with fast transition times. *Hysteresis* prevents false triggering of the circuit.

\#   The *monostable multivibrator* is a regenerative circuit that has one stable and one *quasi-stable* state. Useful in pulse generation and time delay.

DEMONSTRATIONS

D5.1   Latches

(*a*) Make a NOR latch by using the discrete component RTL circuit, as in Fig. D5.1*a*. Test it with the waveforms of Fig. D5.1*b*. Dc levels will do. Check (for the various input combinations) the outputs with dc voltmeters or use the visual display of LEDs (with TTL inverters as LED drivers, as in Fig. 4.3).

(*b*) Make the latch of Fig. 5.5*c* by using *one* IC package that contains a quad of 2-input NAND gates (TTL 7400), then re-do part (*a*) above.

D5.2   Latch as switch debouncer
Construct the circuit of Fig. 5.6*a*. Using an oscilloscope (either a dual beam or a dual trace in the chopped mode), view the signals at point *a* and point *b*, triggering the oscilloscope from the signal at point *b*. Note the bounce seen at point *b*. Now view *Q*, then *Q** again, triggering the oscilloscope off of *b*.

Now trigger the scope at point *a* and notice the "noise" caused by the "unclean" break at *a*. Explain why this does not affect the outputs *Q* and *Q**.

D5.3   Binary counter made from a single latch

(*a*) Construct the circuit of Fig. 5.10*a* except that the input gates can be made from 2-input NAND gates (effectively *S* and *R* are both set to 1). This is shown in Fig. D5.3, where the delay is obtained using an even number of inverters (at least two) along with the 0.1-μF capacitors.

Using a variable pulse-width pulse generator, set the pulse width at 20 ns and verify that the circuit "toggles." Also measure the delay Δ through the inverter

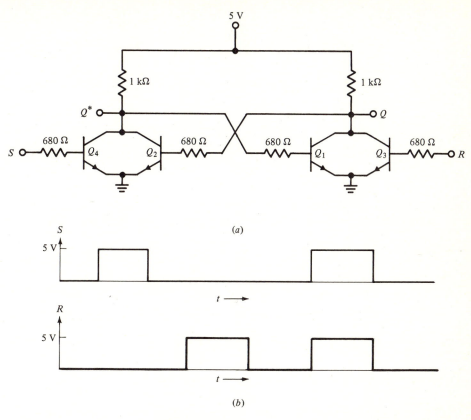

FIGURE D5.1
(a) NOR latch circuit; (b) waveforms for testing circuit in part (a).

pairs. Decrease the pulse width until toggling does not take place. Measure this width and explain why toggling stopped.

Now, starting at 20 ns, keep increasing the pulse width until toggling ceases. Determine the input pulse width and compare to your measurement of Δ.

(b) Construct a master/slave binary counter by using the circuit of Fig. 5.12, where in effect we have set J and K to be 1. Thus the circuit can be constructed using eight 2-input NAND gates and one inverter. Verify the circuits operation similar to part (a). What restrictions must be placed on the propagation delay of the clock inverter to have this circuit function properly?

(c) Using a dual J-K flip-flop,† construct two binary counters. Cascade them to obtain a $\div 4$ circuit. With inverters and LEDs, make a visual binary display of the counters' contents. Initially preset all zero with the CLEAR inputs. Use a debounced switch to enter pulses so that the counter counts through its full cycle.

† TTL 74107

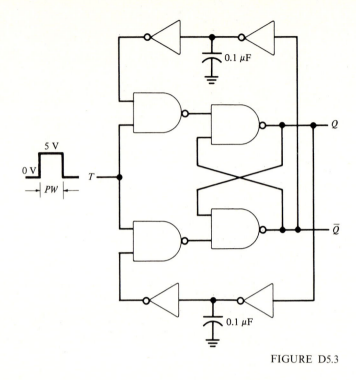

FIGURE D5.3

D5.4 Conversion between flip-flops Construct a *J-K* edge-triggered flip-flop by using a *D* flip-flop† and NAND gates. With a scope and pulse generator (or a debounced switch), verify that it does change state on the positive-going (leading) edge of the clock pulse.

REFERENCES

Chapter 7, Flip-Flops, in MORRIS, R. L., and J. R. MILLER: "Designing with TTL Integrated Circuits," McGraw-Hill, New York, 1971, contains a description of the basic flip-flop as well as the characteristics of all flip-flops in the 54/74 logic family.

MILLMAN, J., and C. C. HALKIAS: "Integrated Electronics," McGraw-Hill, New York, 1971, briefly discuss each of the general type of IC flip-flops in a chapter on Digital Systems, chap. 17.

A short description of the logic properties of IC flip-flops is found in WICKES, W. E.: "Logic Design with Integrated Circuits," chap. 8, Wiley, New York, 1968, though most of the chapter is devoted to the use of state tables and control maps.

† TTL 7474

For a more extensive treatment of the characteristics of IC latches and flip-flops, the reader is advised to consult the many application notes from the IC manufacturers. Especially note,

> Texas Instrument Incorporated: for TTL devices
> Motorola Semiconductor Products Incorporated: for ECL devices
> Radio Corporation of America: for CMOS devices

The data sheets for a particular latch or flip-flop are also a very useful source of information.

PROBLEMS

P5.1   For the circuit of Fig. D5.1, use $V_{BE(on)} = V_{BE(sat)} = 0.7$ V, $V_{CE(sat)} = 0.2$ V, $\beta_F = 50$.
   (a) Determine the value of the overdrive factor k_2 on Q_2 when there is no external load on Q_1.
   (b) If the input to a RTL inverter (of the same type used in making the latch) is loaded to the collector of Q_1, find k_2.

P5.2   For the circuit of Fig. D5.1 assume Q_2 is saturated and Q_1 is cut off. Find the value of the voltage at the RESET terminal that will start the latch to switch states.

P5.3   For the circuit of Fig. D5.1 estimate how long the voltage you found in Prob. P5.2 must be applied in order for the latch to "latch-up" in its new state. Assume the propagation delay of an individual inverter used in making the circuit is 100 ns.

P5.4   It is desired to display the contents of a latch made from a TTL-NAND gate by using a LED, as shown in Fig. P5.4. We desire that the latch be able to drive five standard TTL loads as well as the cross-coupled input. Use a V_{CC} of 5 V $\pm$ 5 percent and a 5 percent tolerance on R_X. The $V_{D(on)}$ of the LED can vary from 1.5 to 1.9 V.
   (a) Find the value of the current in the LED under worst-case conditions for standard series 54/74 shown in Table P4.35.

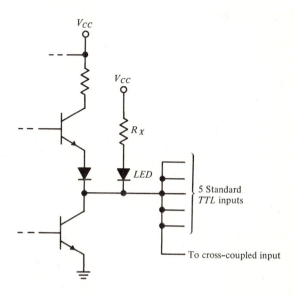

FIGURE P5.4

(b) Find the nominal value of R_X.

(c) Use the value determined in part (b) to find the minimum worst-case value of the current through the LED.

(d) Now find the maximum value of current in the LED. Assume $V_{CE(sat)} = 0.1$ V.

P5.5 A "power" latch (made from discrete components) is to drive a LED and regular TTL gate, as shown in Fig. P5.5. The value of $V_{LED(on)}$ ranges from 1.5 to 2.0 V. Assume $V_{BE(sat)} = 0.7$ V, $V_{CE(sat)} = 0.2$ V, and $\beta_F = 50$.

(a) Find the worst-case minimum value of I_{LED} (assume that I_{IH} of the TTL gates is negligible).

(b) For the overdrive factor of Q_2, k_2 set equal to 3, determine the maximum number of standard TTL gates that can be loaded on the collector of Q_2.

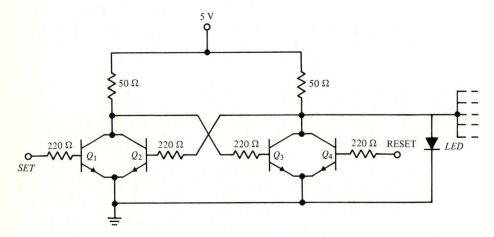

FIGURE P5.5

P5.6 Use a J-K flip-flop and NAND gates to construct a D flip-flop.

P5.7 For the input waveforms given in Fig. 5.13, show the output waveform for a J-K flip-flop that is edge-triggered (not a master/slave).

P5.8 A J-K master/slave flip-flop of the type in Fig. 5.12 has applied the clock J and K waveforms shown in Fig. P5.8. With the flip-flop initially reset show the resulting waveform for the Q output of the master and slave latches. This effect, called "1s catching" or "0s catching," must be accounted for in using such flip-flops.

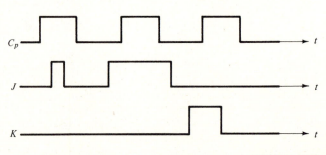

FIGURE P5.8

P5.9 A T flip-flop schematic is shown in Fig. P5.9. It is another form of clocked flip-flop, which has the excitation table shown in Table P5.9.

Table P5.9

Q_n	Q_{n+}	T
0	0	0
0	1	1
1	0	1
1	1	0

(a) What input must be applied to the T in order to have a binary counter that counts the clock pulses?
(b) Show the control logic that must be added to a J-K flip-flop to make a T flip-flop.
(c) Use NAND and EXCLUSIVE-OR gates in the control logic to convert a D to a T flip-flop.

Q_n	Q_{n+}	M	N
0	0	∅	1
0	1	∅	0
1	0	0	∅
1	1	1	∅

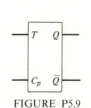

FIGURE P5.9   FIGURE P5.10

P5.10 In your position as Chief Designer of flip-flops at Flip-Flops, Amalgamated, you have to design all kinds of special flip-flops, one of which is the following: Given the excitation table of Fig. P5.10 for an M-N flip-flop with two inputs M and N. Design the two logic networks whose possible inputs are M, N, and Q that have outputs which drive the J-K flip-flop, whose excitation table is given in Table 5.7, so that the composite satisfies Fig. P5.10. The logic networks should be of NAND gates and inverters and of minimum cost, where cost is proportional to the number of gate-signal terminals used (equal weight to all terminals whether it be an input or output).

*P5.11 The circuit in Fig. P5.11a is a D flip-flop (edge-triggered) as could be made with three NOR R-S latches, two inverters, and one OR gate. For the waveforms given in Fig. P5.11b, determine for the nine intervals of time the waveforms at S_1, R_1, Q_1, S_2, R_2, Q_2, Q_2^*, Q_3 and Q_3^*.

Hints: (1) When both S and R inputs of this type of latch are 1, both outputs of the latch are 0; (2) when $C_p = 0$, both S_3 and R_3 can be shown to always be 0; (3) Q_2 should always take on the value of D; (4) finite propagation delays need to be considered for proper operation.

P5.12 A D flip-flop by its nature only produces a storage of data for one clock period; hence its name, delay flip-flop. Therefore, if one wants to store data in a D flip-flop for more than one period, several schemes are possible. Show that the circuits in Fig. P5.12a and b are logically equivalent in doing this. What must be the logic signal on the ENABLE line so that new data can enter the flip-flop in Fig. P5.12a and b?

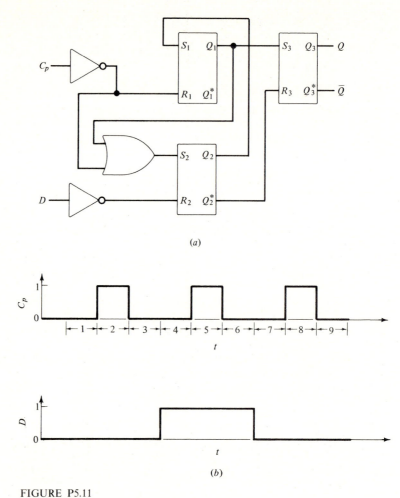

(a)

(b)

FIGURE P5.11
(a) A D flip-flop made with R-S latches (NOR gate type); (b) waveforms to test the edge-triggered D flip-flops of part (a).

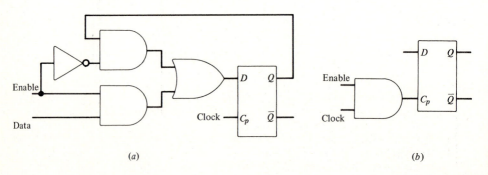

(a)

(b)

FIGURE P5.12

P5.13 From the results of Prob. P5.9a and c:

(a) Show how to make a $\div 2$ counter by using a single D flip-flop.

(b) With the aid of two D flip-flops, show how to make a $\div 4$ counter.

P5.14 For the basic emitter-coupled Schmitt trigger circuit shown in Fig. 5.21a, find values for R_1 and R_2 in terms of R_E so that $V_{T+} = 2.45$ V and $V_{T-} = 1.55$ V. Assume $V_{CC} = 5$ V, and that $V_{BE(EOC)} = 0.5$ V and $V_{BE(on)} = V_{BE(sat)} = 0.7$ V. $V_{CE(sat)} = 0.2$ V, and β_F is infinite.

P5.15 For a monostable multivibrator:

(a) Use Eq. (5.1) to determine the value of C_X to give an output pulse width of 10,000 ns. Use a type 9601 with the values of R_X (1) 5 kΩ, (2) 27 kΩ, and (3) 50 kΩ.

(b) For $C_X = 2000$ pF, what should R_X be for a 9601 so that the output pulse width is 4000 ns?

(c) Use the curves in Fig. 5.23b to determine what C_X should be so that the output pulse width is 200 ns when R_X is 20 kΩ.

(d) Making use of the curves in Fig. 5.23b, what is the smallest R_X and its associated C_X that would give a pulse width of 100 ns? What is the largest R_X and its associated C_X that would produce a 100-ns pulse width?

*P5.16 A simple monostable multivibrator, which is useful over a limited temperature range, made with one-half of a TTL type 7400 is shown in Fig. P5.16. Assume for these gates that V_{OL} and V_{OH} is equal to 0 and 3.6 V, respectively, with $V_{IL} = V_{IH} = 1.5$ V.

(a) Show the complete timing waveforms at the outputs Q and $\bar{Q}$ and at the input V_{i2} after the trigger pulse has been applied to the input.

(b) Derive a simple equation suitable for the approximate determination of the output pulse width.

(c) What is the minimum width of the trigger pulse?

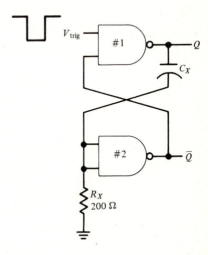

FIGURE P5.16

*P5.17 Design a simple monostable multivibrator made with two TTL-type NOR gates. Use a positive-going input trigger pulse.

6

COUNTERS AND REGISTERS

Electronic counters are found in many digital applications, from counting clock pulses and program steps in a computer to counting bottles on a filling line. Counters are also found in diverse places: dividing down a quartz-crystal oscillation frequency to obtain the tones in a mighty electronic organ or dividing down frequency to 1 pps in a tiny electronic wristwatch. Indeed, in the demonstrations at the end of this chapter we include some suggestions and guidelines for building the digital clock, to which we have often made reference along the path of our description of digital ICs.

There are also many variations in the design of electronic counters. In an electronic organ, where a simple division of the input frequency is required, a *ripple* counter may be used. But in a digital computer the timing of the output from the counter is important, leading to a *synchronous* counter. Some counters, like those within a computer, count in the *binary code*. Other counters whose output must be intelligible to the human being, like in the digital clock, generally use the *BCD code*. A variety of counters, as well as *shift registers* and *memory cells*, will be described in this chapter.

Basic to all counters is the bistable circuit or the flip-flop. Several types of flip-flops have already been described in Chap. 5. The *J-K* master/slave flip-flop is

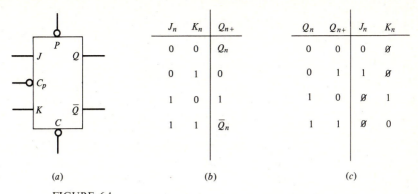

J_n	K_n	Q_{n+}
0	0	Q_n
0	1	0
1	0	1
1	1	$\bar{Q}_n$

Q_n	Q_{n+}	J_n	K_n
0	0	0	∅
0	1	1	∅
1	0	∅	1
1	1	∅	0

(a) (b) (c)

FIGURE 6.1
The *J-K* flip-flop. (*a*) Logic symbol; (*b*) characteristic table; (*c*) excitation table.

most generally used in counter designs because its output states are completely determinate. Shown in Fig. 6.1*a* is the generalized block-diagram symbol of a *J-K* flip-flop. The characteristic table and excitation table for this flip-flop are shown in Fig. 6.1*b* and *c*, respectively.

Unless otherwise noted all flip-flops in this chapter will be of the *J-K* master/slave type, although applications for the *D*-type flip-flop will also be described.

6.1 ASYNCHRONOUS COUNTERS

From the characteristic table of the *J-K* flip-flop (Fig. 6.1*b*), we see that with 1s at both the *J* and *K* inputs the flip-flop will change state with each clock pulse. This is used to advantage in the simplest form of counter, as shown in Fig. 6.2*a*. This is a *cascade* of flip-flops, with the *J-K* inputs connected to a positive voltage (for positive logic) and each *Q* output connected to the C_p input of the succeeding flip-flop. This is known as an *asynchronous* or *ripple* counter. The effect of the trigger pulse rippling down the counter is to set and reset the flip-flops with each alternate input pulse. The ripple counter is widely used as a *frequency counter*, where it is used to divide down the input frequency, or as a *random-event counter*, where the total number of events needs to be recorded.

6.1.1 Binary Ripple Counter

Four *J-K* flip-flops connected in series form a 4-bit binary ripple counter, as shown in Fig. 6.2*a*.†

The waveform timing diagram (Fig. 6.2*b*) shows a representation of the C_p line and the output of the four flip-flops Q_A, Q_B, Q_C, and Q_D. The C_p line shows a

---

† The block diagram and timing waveform, as shown in this figure, as well as subsequent counters and registers, form the basis of Demonstration D6.1.

series of pulses. For a frequency counter these pulses would be uniformly spaced in time, while for a random-event counter they would be irregularly spaced. As we would expect with this type of flip-flop, there is a change of state of flip-flop A (FF-A) following the negative edge of each C_p. Similarly, the output of each flip-flop in the cascade changes state when the output of the preceding flip-flop changes from a HIGH to a LOW. Notice in the asynchronous counter that, due to the inherent propagation delay time through each flip-flop, there is a cumulative delay in the time at which switching occurs at the output of each flip-flop. This delay is illustrated in Fig. 6.2c, where the time scale has been expanded so that the time delay of each flip-flop changing state at the eighth clock pulse can be seen to be cumulative as the effect of the C_p ripples down the counter.

A truth table may be written for this counter, as shown in Fig. 6.3. The code at the output of each flip-flop, following the C_p, is shown. This may be obtained directly from Fig. 6.2b by assigning a 1 to a HIGH level and a 0 to a LOW level.

Notice that a simple cascade of flip-flops yields a binary counter. Each flip-flop effectively divides its input by 2, so that each flip-flop has a numerical weighting of the next higher power of 2 than the one before it. This weighting is shown in the truth table. For example, after six clock pulses the output of the counter is

$$\begin{array}{cccc} D & C & B & A \\ (8) & (4) & (2) & (1) \\ 0 & 1 & 1 & 0 \end{array}$$

hence $\bar{D}CB\bar{A}$ represents the binary equivalent for a decimal 6 (that is, 0110_2).

To avoid problems of notation, the A flip-flop will always represent the *least significant bit* (LSB), and we will proceed consecutively through the alphabet to the *most significant bit* (MSB). It is the usual practice to *draw* logic block diagrams of counters with the input and the LSB flip-flop on the left and the MSB flip-flop on the right; but to *write* out binary numbers with the MSB to the left and the LSB to the right. This practice will be followed in this text.

With four flip-flops connected as a binary ripple counter, all four have returned to the 0 count after 16 clock pulses. That is, there are 16 unique states with a 4-bit binary counter. The addition of another flip-flop to the cascade yields a 5-bit binary counter with 2^5 or 32 unique states. For N flip-flops we have 2^N states. Binary counters of 12 and 16 bits are common.† Notice that, starting from 0, the maximum decimal number uniquely counted or stored by a binary ripple counter is given by $2^N - 1$, where N equals the number of flip-flops.

EXERCISES

E6.1 Using binary ripple counters, how many flip-flops are required to count from 0 up to the following decimal numbers:

(*a*) 12; (*b*) 60; (*c*) 144.

---

† An example of a 12-bit counter is the TTL type 7492 or the CMOS type 4040. The TTL type 7493 is a 16-bit counter.

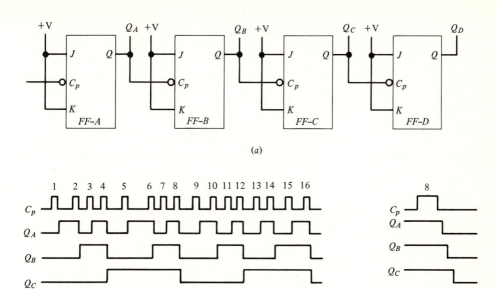

(a)

(b)   (c)

FIGURE 6.2

A 4-bit binary ripple counter. (a) Block diagram; (b) timing waveforms; (c) expanded time scale showing ripple effect of clock pulse.

State number ↓ C_p	D (8) Q_{D+}	C (4) Q_{C+}	B (2) Q_{B+}	A (1) Q_{A+}	← Flip-flop ← Weighting
0	0	0	0	0	
1	0	0	0	1	
2	0	0	1	0	
3	0	0	1	1	
4	0	1	0	0	
5	0	1	0	1	
6	0	1	1	0	
7	0	1	1	1	
8	1	0	0	0	
9	1	0	0	1	
10	1	0	1	0	
11	1	0	1	1	
12	1	1	0	0	
13	1	1	0	1	
14	1	1	1	0	
15	1	1	1	1	
16 or 0	0	0	0	0	

FIGURE 6.3

Truth table for 4-bit binary counter.

E6.2 How many states can be represented by a binary ripple counter having the following number of flip-flops:

(a) 8; (b) 12; (c) 16.

6.1.2 Countdown Counter

The counter that has been described above is of the *countup* type. The number that is stored in the counter indicates the total or sum of pulses or events that have occurred at the input. However, it is sometimes required to initially store a number in a counter and then, as pulses are applied to the counter, have it decrease the value stored in the counter and then give an indication when the counter is brought back to zero. This *countdown* type of counter may be found counting down the seconds remaining to play on an electronic scoreboard. It is also used in the control unit of a computer to indicate the end of a predetermined number of program steps.

By changing the trigger connection from the Q to the $\bar{Q}$ output of each flip-flop, a countup type of counter can be converted to a countdown type. A 4-bit countdown type of binary ripple counter is shown in Fig. 6.4a. Notice that the encoded output is still taken from the Q output of each flip-flop. The operation of the countdown counter can be followed by referring to the waveform timing diagram of Fig. 6.4b and the truth table shown in Fig. 6.4c.

A number is initially stored in the counter by using the direct preset and clear inputs. For this example, flip-flops A and B have been cleared and flip-flops C and D have been preset so that 1100, the binary equivalent of the decimal number 12, is stored in the counter. This is indicated on the truth table as the *state number* 0, or the initial state of the counter. With the first C_p, flip-flop A (FF-A) changes state, Q_A changes from LOW to HIGH, but the trigger to FF-B is taken from $\bar{Q}_A$, which has made a HIGH to LOW change. With the output of the J-K flip-flop changing when its C_p input makes this transition, FF-B also changes state. Again, Q_B changes from LOW to HIGH, but with $\bar{Q}_B$ changing from HIGH to LOW, FF-C also changes state. Q_C changes from HIGH to LOW, but $\bar{Q}_C$ changes from LOW to HIGH so there is no change at the output of FF-D. The result, as shown in Fig. 6.4b or c, is that the binary equivalent of the decimal number 11 is now stored in the counter. With each succeeding C_p the number stored in the counter is reduced by 1. When the number reaches 0 a LOW logic signal is obtained at F, indicating the end of the operation.

EXERCISE

E6.3 Design a binary ripple counter that is able to count either up to or down from 7. A single control line determines the direction of the count. With the control line LOW the count is DOWN. Use flip-flops similar to that shown in Fig. 6.1a. Implement all control logic, including the SET 7 logic, with NAND gates.

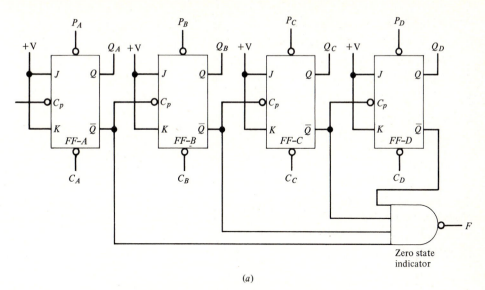

(a)

(b)

State number	Decimal number	D (8)	C (4)	B (2)	A (1)
0	12	1	1	0	0
1	11	1	0	1	1
2	10	1	0	1	0
3	9	1	0	0	1
4	8	1	0	0	0
5	7	0	1	1	1
6	6	0	1	1	0
7	5	0	1	0	1
8	4	0	1	0	0
9	3	0	0	1	1
10	2	0	0	1	0
11	1	0	0	0	1
12	0	0	0	0	0

(c)

FIGURE 6.4
A 4-bit binary *countdown* counter. (a) Block diagram; (b) timing waveforms;
(c) truth table.

6.1.3   Decade Ripple Counter

So far the discussion has concerned only binary counters. However, a number of applications require a binary-coded-decimal (BCD) counter. Where the number stored in the counter needs to be displayed, such as in a frequency counter or an event counter, a BCD counter will generally be used. This requires coding and connecting the flip-flops of the counter to count in decades of the decimal number 0 through 9.

We have seen that the maximum natural, or binary, count of three flip-flops is decimal 7.† With four flip-flops, it is possible to count from 0 to decimal 15. The decade counter will, therefore, use four flip-flops, and the design must eliminate the last six states to obtain a maximum count of 9.

As mentioned in Chap. 2, the most common code used today for a BCD counter is the 8421 code. Again, this refers to the numerical weighting given to each flip-flop of the counter. The most significant bit (MSB) is the first number on the left and controls the 8 value. The least significant bit (LSB) is the last number to the right and controls the 1 value. Other BCD codes are 2421 and 4221.

Decade counters may be implemented as ripple counters or synchronous counters. An 8421 BCD ripple counter is shown in Fig. 6.5, along with a waveform diagram and truth table.

The circuit makes use of four flip-flops and one AND gate. Reference to the waveform diagram shows that, starting with the four flip-flops in the *reset state* (i.e., all Q's are 0), the count follows the normal binary mode up to and including the count of 8. This may also be confirmed by checking the truth table, where the binary count is in agreement with the state number up to and including state number 8. The schematic diagram does show a direct connection from Q_A to the C_p of FF-D. However, FF-D is prevented from changing to the *set state* (that is, Q_D cannot go to 1) until the AND gate output (at the J input of FF-D) is 1. This first occurs at the count of 6, when both FF-B and FF-C are set. (See state number 6 of the truth table.) The way is now prepared so that at the next down-going transition of Q_A, at the count of 8, FF-D is set. There is a feedback connection from $\bar{Q}_D$ to the J input of FF-B. Therefore, with FF-D in the set state, $\bar{Q}_D$ is LOW, and FF-B is prevented from changing to the set state. That is, FF-B can only set when FF-D is in the reset state. Notice that there is no control logic to prevent the four flip-flops from changing to the reset state. The K input of all four flip-flops is connected to a 1, a positive voltage. Therefore, following the count of 8, the next time FF-A resets, at the count of 10, FF-D also resets. With FF-B and FF-C already in the reset state, all flip-flops are now reset and one complete counting cycle has been made with 10 input pulses.

With this design, the master/slave feature of the J-K flip-flop is used to advantage. Referring to the waveform diagram of Fig. 6.5b, at the tenth clock pulse the output of FF-A is used as the clock input to reset FF-D. With FF-D reset, the FF-B is now permitted to set. With simple flip-flops and edge-type

---

† Including zero as a state, we have $2^3 = 8$ numbers.

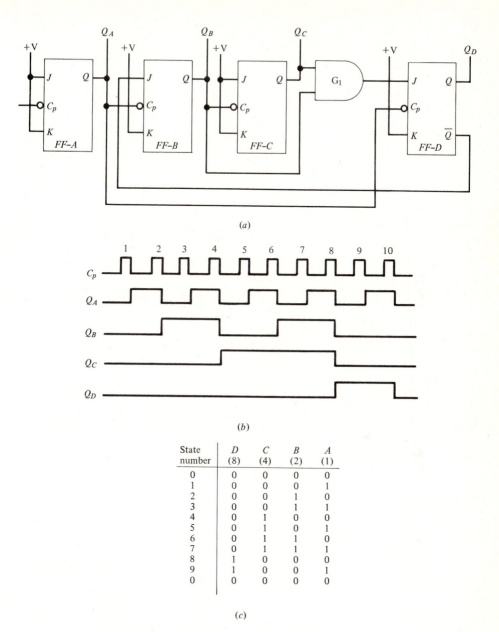

FIGURE 6.5
8421 BCD ripple counter. (*a*) Block diagram; (*b*) timing waveforms; (*c*) truth table.

triggering, delay must be introduced between $\bar{Q}_D$ and the J input of FF-B to ensure that the C_p input to FF-B is gone before the feedback signal from $\bar{Q}_D$ is applied to FF-B, which otherwise would be set. With the master/slave type of flip-flop, entry to the master can only be made when the C_p is HIGH, but the output level of the flip-flop only changes when the C_p is LOW. The output of FF-D changes when Q_A goes LOW, but entry into FF-B can only be made with Q_A at a HIGH, that is, with the next clock pulse. The decade counter shown in Fig. 6.5a, made with the four flip-flops and gate circuits fabricated in one chip, was among the first commercially available MSI circuits.†

The design of BCD ripple counters is generally dependent upon some special design feature either of the individual flip-flops or of the counter function itself. There is, therefore, room for the designer to use his ingenuity in the design of these counters.

6.2 SYNCHRONOUS COUNTERS

In many computer applications close control is required of the clocking operation of counters to prevent the condition called *racing*. This problem is particularly evident in ripple counters, where the cumulative delay in the change of state of the flip-flops permits the generation of false codes in the counting cycle. For the example shown in Fig. 6.2c, during the interval from the transition of the count of 7 to the count of 8, binary numbers corresponding to 6 and 4 are also momentarily produced. This is a race condition and is unacceptable. To eliminate this problem we require that all flip-flops change state at the same time.

This concept leads to the *synchronous* counter. However, before describing a synchronous counter, an additional feature of J-K flip-flops must be mentioned. Most IC J-K flip-flops have more than one J and K input. There may be as many as three J and three K inputs.‡ Generally these are arranged as inputs to an AND gate, as shown in Fig. 6.6, where $J = J_1 J_2 J_3$ and $K = K_1 K_2 K_3$. These AND gates are integrated with the flip-flop in a single chip.

To avoid problems of notation we shall use a consistent set of symbols and block diagrams. Notice in Fig. 6.6 the placement of the synchronous set, the direct preset, and the set output. These are mirrored by the synchronous reset, the direct clear, and the reset output. The clock-pulse input appears between the synchronous inputs.

6.2.1 Binary Synchronous Counter

An example of a 4-bit binary synchronous counter is shown in Fig. 6.7. Use is made of AND gates (inside the IC) at the J and K inputs of the flip-flop, of the type described in Fig. 6.6. Flip-flop C need only have a 2-input AND gate at each J and K input, or unused inputs of a 3-input AND gate can be connected HIGH.

† TTL type 7490.
‡ TTL type 7472.

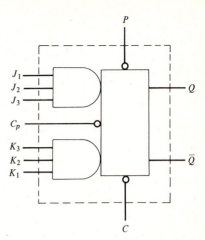

FIGURE 6.6
J-K flip-flop with AND inputs.

The truth table for the 4-bit synchronous binary counter is the same as for the 4-bit binary ripple counter (see Fig. 6.3). Logically, the counter is still a simple cascade of *J-K* flip-flops. The timing diagram is also similar, except now all the outputs change directly on the command of the C_p. The propagation delay time for each flip-flop is just that of the flip-flop and its associated AND gate. Using similar flip-flops, each stage of the counter will have similar delay with no cumulation of delay times.

With the clock pulse now being fed directly to the C_p terminals of all flip-flops, the need for a *buffer/driver* should be apparent for a counter of any significant size. That is, a large fanout is required of the circuit that provides C_p. Also notice that the number of loads (fanout) on each flip-flop increases with each bit added to the counter, as does the number of inputs (fan-in) for the *J* and *K* gates. This is shown in Table 6.1, where *N* is the number of flip-flops.

The design of synchronous counters is easily implemented with the aid of *state tables* and *control maps*. The meaning and use of these two aids are best described in a design problem. The same 4-bit binary synchronous counter will now be analyzed by using state tables and control maps.

The state table of Fig. 6.8a is a truth table which shows the state of each of the four flip-flops making up the counter, as well as the equivalent decimal number and the assigned state number. This information is shown for the present state and for the next state of the counter, that is, the counter state one clock pulse

Table 6.1 LOADING OF A BINARY SYNCHRONOUS COUNTER

		$N = 4$	$N = 12$
Fanout	FF-*A*	$2(N - 1) = 6$	$2(N - 1) = 22$
Fan-in	FF-*N*	$N - 1 = 3$	$N - 1 = 11$

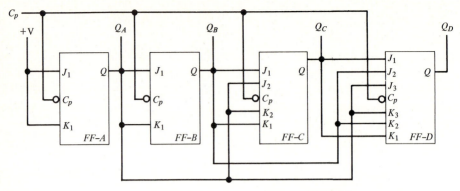

FIGURE 6.7
A 4-bit binary synchronous counter.

later. With the counter consisting of four flip-flops, there are 2^4 or 16 rows in the truth table; again notice the systematic listing and that A is assigned the LSB.

As an aid in setting up the control maps, a *reference map* is derived from the state table. The reference map (Fig. 6.8*b*) locates the state number in a Karnaugh-like map by using the present state of the four flip-flops for the location within the reference map.

With the help of the excitation table (Fig. 6.8*c*) for the particular type of flip-flop being used, the control maps for the counter are now compiled. This is done by considering the logic necessary at the J and K inputs of each flip-flop, at each state number, for the flip-flop to make the required transition to the next state number.

As an example, at the state number 0, FF-A is required to make the transition 0 to 1. Reference to the excitation table shows that the J input of FF-A (J_A) requires a 1 condition, while the K input of the same flip-flop (K_A) is a $\emptyset$ condition. These requirements (1 and $\emptyset$) are placed in the state 0 square of the A control map. Two control maps could be used for each flip-flop, one for the J input and one for the K input. However, it is generally simpler to locate both J and K input requirements on the one control map. Continuing, at state 1, FF-A is to make a change from 1 to 0. The J-K excitation for this are $\emptyset$ and 1; this then is placed on the state 1 square of the A control map. In similar fashion the control conditions of FF-A for each state number are recorded in the control map. With the control map for FF-A complete, a similar procedure is used to compile the control maps of FF-B, FF-C, and FF-D.

With the maps complete, consideration is now given to minimizing the control logic. The technique for minimization is the same for minimizing any logic function with a Karnaugh map. To reduce the number of variables, as many as possible of the adjacent squares containing 1s are grouped together. It may sometimes be more convenient to group the 0s leading to a NOT statement. Remember, within each square of the control map the left value is for the J input and the right is for the K input.

State number	Present state Decimal number	D (8)	C (4)	B (2)	A (1)	Next state Decimal number	D (8)	C (4)	B (2)	A (1)
0	0	0	0	0	0	1	0	0	0	1
1	1	0	0	0	1	2	0	0	1	0
2	2	0	0	1	0	3	0	0	1	1
3	3	0	0	1	1	4	0	1	0	0
4	4	0	1	0	0	5	0	1	0	1
5	5	0	1	0	1	6	0	1	1	0
6	6	0	1	1	0	7	0	1	1	1
7	7	0	1	1	1	8	1	0	0	0
8	8	1	0	0	0	9	1	0	0	1
9	9	1	0	0	1	10	1	0	1	0
10	10	1	0	1	0	11	1	0	1	1
11	11	1	0	1	1	12	1	1	0	0
12	12	1	1	0	0	13	1	1	0	1
13	13	1	1	0	1	14	1	1	1	0
14	14	1	1	1	0	15	1	1	1	1
15	15	1	1	1	1	0	0	0	0	0
0	0	0	0	0	0	1	0	0	0	1

(*a*)

DC \ BA	00	01	11	
00	0	1	3	2
01	4	5	7	6
11	12	13	15	14
10	8	9	11	10

(*b*)

DC \ BA	00	01	11	10
00	1Ø	Ø1	Ø1	1Ø
01	1Ø	Ø1	Ø1	1Ø
11	1Ø	Ø1	Ø1	1Ø
10	1Ø	Ø1	Ø1	1Ø

J_A - K_A map

$J_A = 1$
$K_A = 1$

DC \ BA	00	01	11	10
00	0Ø	1Ø	Ø1	Ø0
01	0Ø	1Ø	Ø1	Ø0
11	0Ø	1Ø	Ø1	Ø0
10	0Ø	1Ø	Ø1	Ø0

J_B - K map

$J_B = A$
$K_B = A$

Q_n	Q_{n+}	J_n	K_n
0	0	0	Ø
0	1	1	Ø
1	0	Ø	1
1	1	Ø	0

(*c*)

DC \ BA	00	01	11	10
00	0Ø	0Ø	1Ø	0Ø
01	Ø0	Ø0	Ø1	Ø0
11	Ø0	Ø0	Ø1	Ø0
10	0Ø	0Ø	1Ø	0Ø

J_C - K_C map

$J_C = AB$
$K_C = AB$

DC \ BA	00	01	11	10
00	0Ø	0Ø	0Ø	0Ø
01	0Ø	0Ø	1Ø	0Ø
11	Ø0	Ø0	Ø1	Ø0
10	Ø0	Ø0	Ø0	Ø0

J_D - K_D map

$J_D = ABC$
$K_D = ABC$

(*d*)

FIGURE 6.8

A 4-bit binary synchronous counter. (*a*) State table; (*b*) reference map; (*c*) excitation table; (*d*) control maps.

From the control map for FF-A, we note that the control logic for the J_A input is always either 1 or $\emptyset$. This can be minimized to a simple 1. That is, a 1 will always satisfy the requirements for the J_A input. Notice that control logic for the K_A input is also always either 1 or $\emptyset$. Again, this can be reduced to a simple 1, which will always satisfy the control logic for the K_A input. The control map indicates that FF-A must change state with each clock pulse. The logic conditions are met by connecting the J_A and K_A inputs to a positive voltage, as shown in Fig. 6.7.

Checking the control map for the B flip-flop, we find the J_B requirements are all 0s in the 00 column, all 1s in the 01 column, and all $\emptyset$s in both the 11 and 10 columns. By considering these $\emptyset$s in the 11 column to be 1s, we can group 1s, as shown on the map, and the logic for the J_B input can be reduced to a simple A. This requirement is met by connecting the J_B input to the Q_A output. The K_B requirements are all $\emptyset$s in both the 00 and 01 columns of the control map. There are all 1s in the 11 column and all 0s in the 10 column. To obtain a minimized function for the K_B input, we consider the $\emptyset$s in the 01 column to be 1s and make the grouping shown on the map. The control logic for the K_B input is minimized to a simple A. The K_B and J_B inputs can, therefore, be connected together, both going to the Q_A output, as shown in Fig. 6.7.

For the C flip-flop, the J_C requirements are two 1s and two $\emptyset$s in the 11 column. All the other columns contain either 0s or $\emptyset$s for the J_C logic. At the J_C input an AND function is therefore required of the Q_A and Q_B outputs. The K_C logic is also two 1s and two $\emptyset$s in the BA column, with 0s or $\emptyset$s in all the other columns. The function at the K_C input is then also the AND of the Q_A and Q_B outputs. If one used separate AND gates at the input to the flip-flop, the output of one AND gate, having Q_A and Q_B at its input, could be connected directly to both the J_C and K_C inputs. However, with the AND gates for the J_C and K_C inputs of the J-K flip-flop inside the package, this is not possible.

Finally, the control map of FF-D shows that an AND function of Q_A, Q_B, and Q_C is required for both the J_D and K_D inputs. These connections are shown in Fig. 6.7.

EXERCISE

E6.4 Use flip-flops similar to that shown in Fig. 6.1 to design a synchronous binary counter that counts from 0 to 6 and then repeats the cycle. NAND gates are available for additional control logic.

6.2.2 Decade Synchronous Counter

The design of BCD synchronous counters follows the same systematic approach. An 8421 BCD synchronous counter is shown in Fig. 6.9. The waveform diagram and truth table are the same as for the decade ripple counter and are shown in Fig. 6.5.

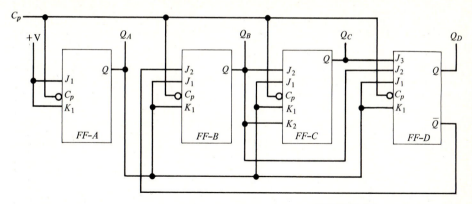

FIGURE 6.9
8421 BCD synchronous counter.

Like the binary synchronous counter, the C_p is connected directly to each flip-flop, and use is made of the AND gates at the J-K inputs of the master/slave–type flip-flop. In similar fashion, also, the design starts with compiling the control map for each flip-flop from the state table of the counter, using the reference map as an aid. These are shown in Fig. 6.10 with the excitation table for the flip-flop. The open squares of the control maps represent don't care conditions for both the J and K inputs.

From the state table it can be noted that the flip-flop of the LSB, FF-A, changes state with each clock pulse. The J_A and K_A inputs can then be connected to a positive voltage representing a 1 for positive logic. This control logic can also be derived from the control map of FF-A. Both the J_A and K_A input requirements are always either a 1 or a $\emptyset$ condition. This can be minimized to a simple 1, or a positive voltage for positive logic.

The control map of FF-B shows, for the J input, 1s in the first two rows of the 01 column. There are 0s in the first two rows of the 00 column, so that is no help. The 1s of the 01 column will be grouped with the $\emptyset$s of the first two rows of the 11 column. The 0 located at $D\bar{C}\bar{B}A$ prevents a larger grouping. For the four-square grouping, the chosen control logic for the J_B input is $A\bar{D}$. For the K input of this flip-flop, the 1s of the first two rows of the 11 column may be grouped with the $\emptyset$s of the same column and the $\emptyset$s of the 01 column. A grouping of eight squares here leads to three redundant variables, and thus the K_B input is just A. The schematic, in Fig. 6.9, shows the connection to J_1 and J_2 of FF-B from Q_A and $\bar{Q}_D$, respectively. Also made is the connection to K_B (K_1 of FF-B) from Q_A.

In determining the control logic for FF-C, the $\emptyset$ conditions of column 11 are considered to be 1s for both the J and K inputs. The resulting minimization leads to a control logic of AB for both inputs. In the schematic, connection is made from Q_A to J_1 and K_1 of FF-C and from Q_B to J_2 and K_2 of FF-C.

In the control map of FF-D the only 1 for the J input is surrounded by 0s, except for the $\emptyset$ condition at $DCBA$. This $\emptyset$ condition is grouped with the 1, and

State number	Decimal number	Present state D (8)	C (4)	B (2)	A (1)	Decimal number	Next state D (8)	C (4)	B (2)	A (1)
0	0	0	0	0	0	1	0	0	0	1
1	1	0	0	0	1	2	0	0	1	0
2	2	0	0	1	0	3	0	0	1	1
3	3	0	0	1	1	4	0	1	0	0
4	4	0	1	0	0	5	0	1	0	1
5	5	0	1	0	1	6	0	1	1	0
6	6	0	1	1	0	7	0	1	1	1
7	7	0	1	1	1	8	1	0	0	0
8	8	1	0	0	0	9	1	0	0	1
9	9	1	0	0	1	0	0	0	0	0
0	0	0	0	0	0	1	0	0	0	1

(a)

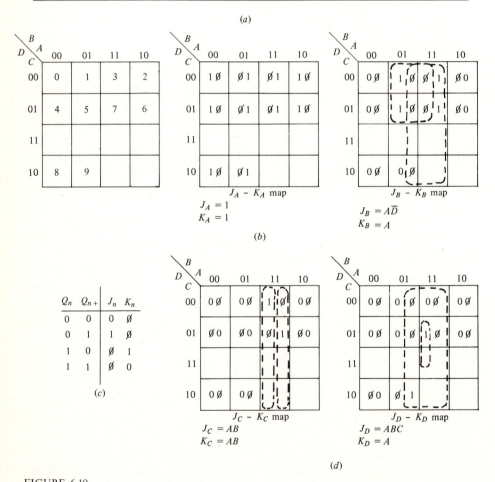

$J_A = 1$
$K_A = 1$

$J_B = A\overline{D}$
$K_B = A$

(b)

Q_n	Q_{n+}	J_n	K_n
0	0	0	Ø
0	1	1	Ø
1	0	Ø	1
1	1	Ø	0

(c)

$J_C = AB$
$K_C = AB$

$J_D = ABC$
$K_D = A$

(d)

FIGURE 6.10
8421 BCD synchronous counter. (a) State table; (b) reference map; (c) excitation table; (d) control maps.

the minimum control logic for the J_D input is an AND function of Q_A, Q_B, and Q_C. For the K_D input eight squares are able to be grouped, and the minimization leads to a simple Q_A. These connections are shown in Fig. 6.9.

EXERCISE

E6.5 Use flip-flops similar to that shown in Fig. 6.1 to design a synchronous BCD decade counter that counts in the 2421 code. (See Table C.2 in Appendix C). NAND gates are available for additional control logic.

6.2.3 General BCD Counters

A synchronous counter of any coding can be designed in a similar manner to that given above. The counters analyzed so far have all started from zero and have had an increment of 1 for each clock pulse. Because of this, the state number has always agreed with the decimal number of the state of the counter. However, these are not necessary conditions for the design of a counter. Suppose a counter is desired to count in a 1326401 sequence. The design procedure is exactly as has been described.

A decision must first be made on how many flip-flops are required for the counter. For this example, with 6 the highest number to be counted, three flip-flops are adequate ($2^3 - 1 = 7$). The state table is written, and with the aid of the excitation table and reference map, the control maps are compiled, as shown in Fig. 6.11. As mentioned earlier the excitation table for all flip-flops is not the same. For this example the flip-flop excitation table is *different* from the one previously used. The open squares in the control map again represent don't care conditions.

In minimizing the logic for the K input of FF-A, notice that a grouping of 0s has been made to yield $\overline{B}\overline{C}$ as a minimum function. A grouping of all the 1s into a single AND function is impossible. A solution could be obtained by grouping the 1s in the 10 and 11 columns and by grouping the 1s in the C row, leading to $B + C$; but using DeMorgan's theorem, this is equivalent to $\overline{B}\overline{C}$. Assuming that NAND gates were preferred over NOR gates, we would implement this as shown in the block diagram of Fig. 6.12a.

A similar situation exists in determining the logic for the K input of FF-C. A grouping of 0s leads to $\overline{A}B$ as a minimum function. A grouping of 1s reduces to the equivalent, $A + \overline{B}$. The waveform diagram for this counter is shown in Fig. 6.12b.

The design of synchronous counters of any coding is therefore a straightforward affair, and simplification of the control logic is greatly helped by the use of Karnaugh maps.

Synchronous 4-bit counters, both binary and decade, are available as MSI products implemented with either TTL or CMOS devices.† Many of these count-

---

† See, for example, the TTL types 74161 and 74160, and the 4520 and 4518 in CMOS.

State number	Present state: Decimal number	Present state: C (4)	Present state: B (2)	Present state: A (1)	Next state: Decimal number	Next state: C (4)	Next state: B (2)	Next state: A (1)
0	1	0	0	1	3	0	1	1
1	3	0	1	1	2	0	1	0
2	2	0	1	0	6	1	1	0
3	6	1	1	0	4	1	0	0
4	4	1	0	0	0	0	0	0
5	0	0	0	0	1	0	0	1
0	1	0	0	1	3	0	1	1

(*a*)

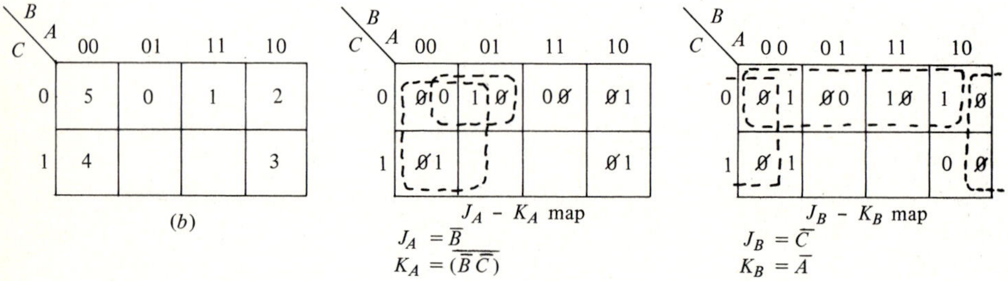
(b)
J_A - K_A map
$J_A = \overline{B}$
$K_A = (\overline{\overline{B}\,\overline{C}})$
J_B - K_B map
$J_B = \overline{C}$
$K_B = \overline{A}$

Q_n	Q_{n+}	J_n	K_n
0	0	Ø	1
0	1	Ø	0
1	0	0	Ø
1	1	1	Ø

(*c*)

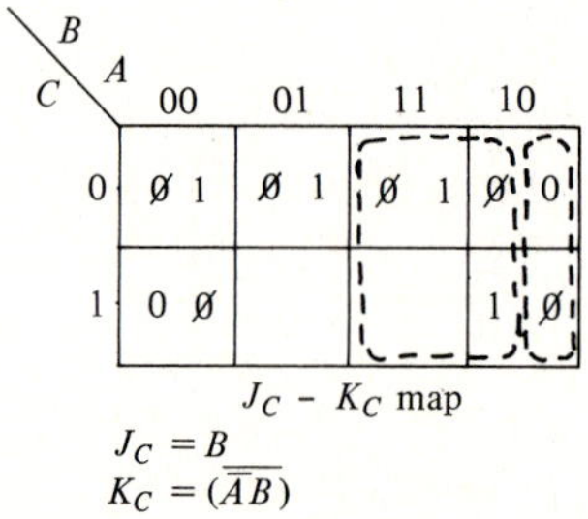
J_C - K_C map
$J_C = B$
$K_C = (\overline{\overline{A}B})$

(*d*)

FIGURE 6.11
General BCD synchronous counter. (*a*) State table; (*b*) reference map; (*c*) excitation table; (*d*) control maps.

ers may be operated in either the countup or countdown mode.† The preset connection to each of the flip-flops in these counters allows for initially storing a number into the counter. The clear connection is generally common to all four flip-flops.

Programmable counters are also available as MSI functions.‡ These circuits provide counting by almost any number up to 16 with no extra control logic circuits being required. These counters are also referred to as variable MODULO counters.

† In TTL, types 74193 and 74192, and in CMOS, types 4516 and 4510.

‡ TTL type 9305.

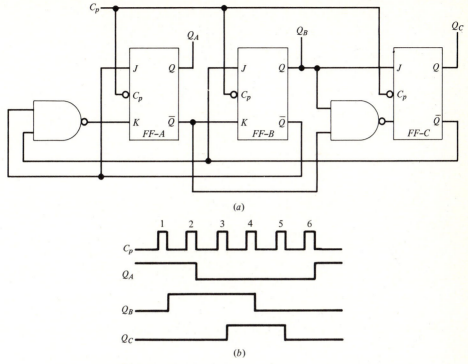

FIGURE 6.12
General BCD synchronous counter. (*a*) Block diagram; (*b*) timing waveforms.

Applied to counter circuits, MODULO (or MOD) indicates the number of inputs pulses required before the count cycle is repeated. Thus a decade counter can be described as a MOD 10 counter. A flip-flop, with its two states, counts in MOD 2.

EXERCISE

E6.6   Use flip-flops similar to that shown in Fig. 6.1 to design a synchronous BCD counter that counts in the sequence described in this section, that is, 1326401. NOR gates are available for additional control logic.

6.3   SHIFT REGISTERS

When performing binary arithmetic in a computer, it is usual to store the binary numbers as data words in registers, as was indicated in Fig. 4.20*b*. The bit storage element is a simple flip-flop, so that we have as many flip-flops in the register as we have bits in the data word. However, means must be provided for shifting the data bits into and out of the register. A shift register is a simple synchronous counter which shifts binary data under the control of a clock pulse. The flip-flops of the counter are connected such that data shifts from one flip-flop to the next in synchronism with a command clock pulse. The use of shift registers in the arithmetic unit of a computer is described in Demonstration D6.4.

Typically, data is applied in *serial* form, on a single wire (for example, as DATAIN in Fig. 6.13*a*), to the synchronous input of the MSB flip-flop (FF-*D*) in the register. That is, one bit follows another bit in time sequence, with the LSB coming first. The information is shifted through the register under the control of the clock pulse. By using additional control gates, the data in the register can be shifted either to the right or to the left.

In a more versatile form, the shift register acts as a *serial-to-parallel* converter. A data word may be entered into the register in serial form, then after N clock pulses for a word of N bits, the word now appears in parallel form, with one bit at the output of each flip-flop in the register. We can now connect an indicator, or decoder, to each flip-flop output and display the contents of the register as one complete data word.

Notice, the parallel transmission of a data word requires as many wires as there are bits to the word. However, with a *parallel-to-serial* converter we can transmit the same information on a single wire as a timed sequence of pulses. Data is entered into the register in parallel form, via the direct inputs of the flip-flop. Then later, on command of the clock pulse, the data is shifted out in serial form on to the transmission line. We could use a serial-to-parallel converter at the end of the transmission line to convert the data back to a parallel form.

6.3.1 Serial-to-Parallel Converters

A 4-bit shift register of the serial-to-parallel type is shown in Fig. 6.13. Also shown is the waveform diagram and the truth table. This shift register uses J-K master/ slave flip-flops, one for each bit required to be stored or shifted in the register. The inverter at the input ensures that the signal at K_D is opposite in polarity to that at J_D. The common clock-pulse line to all flip-flops ensures synchronism of the shifting operation. Shift registers are a variation of synchronous counters that require very simple control logic. On the single entry line, DATA IN, data enters in serial form. With each clock pulse, each bit of data moves down the shift register. In this example all flip-flops have been initialized to 0, and the data word 0101 is to be entered into the register. Then, as shown, the first entry into the shift register is the LSB, which is the 1 at the J_D input. This causes the master of FF-*D* to be set when the clock pulse is HIGH; this is transferred to the slave and output when the clock pulse goes LOW. The other flip-flops would have a 0 at the J input and a 1 at the K input. They would, therefore, continue in the reset state after the first clock pulse. However, after the first clock pulse, there is now a 1 at the J_C input and a 0 at the J_D input. Now with the second clock pulse, FF-*C* is set and

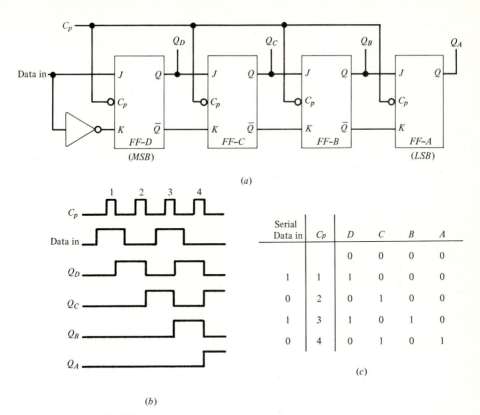

(a)

(b)

Serial Data in	C_p	D	C	B	A
		0	0	0	0
1	1	1	0	0	0
0	2	0	1	0	0
1	3	1	0	1	0
0	4	0	1	0	1

(c)

FIGURE 6.13
A 4-bit shift register (serial-to-parallel). (a) Block diagram; (b) timing waveforms; (c) truth table.

FF-*D* is reset. With a 4-bit shift register, 4 clock pulses are required to store a 4-bit word. The serial input data then appears in parallel form at the output of the flip-flops, with the LSB at FF-*A*. This is also shown in the truth table.

6.3.2   Parallel-to-Serial Converter

With parallel-to-serial conversion, data is preset into each flip-flop of the register. Use is made here of the direct inputs (preset and clear) of the flip-flop. Each bit of data then moves down the shift register under command of the clock pulse. The data then appears in serial form at the output of the register; this is the output of the last flip-flop in the shift register. An example of a parallel-to-serial 4-bit shift register is shown in Fig. 6.14.

For this example, the direct inputs are disabled after loading the initial states into the flip-flops. Therefore, even before the first clock pulse, the first serial data bit, the LSB, appears at the DATA OUT terminal. At the time of the negative edge

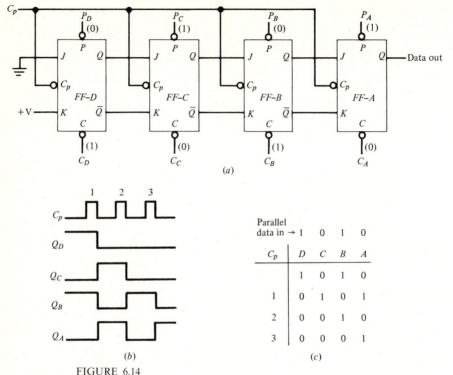

(a)

(b) (c)

FIGURE 6.14
A 4-bit shift register (parallel-to-serial). (a) Block diagram; (b) timing waveforms; (c) truth table.

of each clock pulse, each flip-flop of the register has transferred into it the contents of the flip-flop to its left. The MSB flip-flop (FF-D) has 0s entered into it, which 0s then propagate down the register.

These shift registers have been implemented by using J-K master/slave flip-flops. It should be noted that for the control logic requirements of a simple shift register the D-type flip-flop, as described in Chap. 5, can be used to advantage. The Q output follows the D input when the flip-flop is clocked. A simple shift register would consist of a cascade of D flip-flops with the Q output of one driving the D input of the next.

Present-day shift registers of IC design (primarily TTL and CMOS) are available that operate in any or all modes of operation, either serial (serial in–serial out), or serial-to-parallel, or parallel-to-serial conversion. Some IC shift registers contain additional gating at the input of the individual flip-flops that allows the data to be shifted to the right or to the left on command of a shift control signal.†

† Manufacturers' data books contain detailed descriptions of these IC devices.

Extended shift registers, using MOS transistors, of up to 1024 bits are now available. With this number of bits, the capacitive loading (> 100 pf) on the clock-pulse line can become a problem. As MOS transistors are essentially high-input resistance devices, the loading effect only lowers the operating speed of the register.

EXERCISE

E6.7 Use flip-flops similar to that shown in Fig. 6.1 to draw a block diagram of a 4-bit shift register which receives serial information from the DATA line during clock pulses 1 through 4. At C_{p5} the output is delivered in parallel on four output lines. At C_{p6} and C_{p7} all flip-flops in the register are reset to zero. After C_{p8} the shift register will start receiving serial information again and the cycle repeats. You may use additional flip-flops and NAND gates to make any required counters, etc.

6.4 RING COUNTERS

So far, the counters that have been considered have consisted of a cascade of flip-flops. This has necessitated one flip-flop for each binary bit to be handled. Decoding of the binary information has also been necessary to realize the data in decimal form. Decoders have already been described in Chap. 4.

Another type of counter is the *ring counter*. This counts decimal numbers directly, and no decoding is necessary. It does, however, require one flip-flop for each decimal number. That is, a decade counter contains 10 flip-flops for the units column, 10 for the tens column, etc. (Recall that the BCD decade counter required only four flip-flops per column.) As shown in Fig. 6.15 the ring counter consists of a cascade of flip-flops closed back on itself directly. This is an example of a ring of five, or quinary counter, implemented with D-type flip-flops. (For clarity in the figure, the direct preset and clear have been omitted).

The counter is initially reset using the direct preset of FF-A and the direct clear of all the other flip-flops. The result is that all the flip-flops are reset (0), except FF-A, which is set (1). Therefore, the D input of FF-B is at a 1, and all the other D inputs are at 0. Now on the first clock pulse, FF-B is set, FF-A is reset, and all the other flip-flops continue in the reset state. A 1 is now at the D input of FF-C, and all other D inputs are at a 0. With the second clock pulse, FF-C is set and FF-B is reset. Thus the 1 moves down the counter with each clock pulse. After the fourth clock pulse, FF-E is set, and due to the connection from FF-E to FF-A there is a 1 at the D input of FF-A. Now, with the fifth clock pulse, FF-A is set, and it has taken five clock pulses to complete one counting cycle.

Again, the master/slave type of flip-flop ensures correct counting. A 0 appears at the Q output of FF-A when the first clock pulse goes LOW, but entry

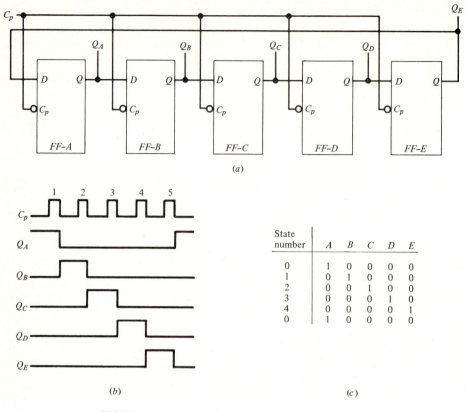

(a)

(b)   (c)

State number	A	B	C	D	E
0	1	0	0	0	0
1	0	1	0	0	0
2	0	0	1	0	0
3	0	0	0	1	0
4	0	0	0	0	1
0	1	0	0	0	0

FIGURE 6.15
Quinary ring counter. (a) Block diagram; (b) timing waveforms; (c) truth table.

into the master of FF-B is only permitted when the clock pulse is HIGH, that is, with the next clock pulse. Therefore the counter is prevented from making two counts with only one clock pulse.

Notice that the simple control logic of the ring counter is similar to that of a shift register. The ring counter is another special case of the synchronous counter.

An advantage of this type of counter, besides requiring no decoding, is that it is very fast. The only delay is in setting just one flip-flop. Also, each flip-flop has to switch at $1/N$ ($= 1/5$ for the quinary counter) of the maximum input repetition rate. That is, by using five D-type flip-flops that individually have a maximum toggle frequency of 20 MHz, in a quinary counter we would have a maximum input repetition rate of 100 MHz. Compare this with a cascade of flip-flops, either binary or BCD, where the maximum repetition rate of the counter is limited to the maximum switching rate of the first flip-flop. The disadvantage of the ring counter is that it is expensive. If made from individual flip-flops the cost of the flip-flops in the decade ring counter is 2.5 times that of a decade BCD counter.

6.5   SHIFT COUNTERS

An example of the *shift counter* is shown in Fig. 6.16a. The shift counter is similar to the ring counter of Fig. 6.15a, but now the ring is closed with the $\bar{Q}_E$ going to J_A and Q_E going to K_A. Hence this counter is also referred to as a *twisted-ring counter*; another name is a Johnson counter. A MOD 10 shift counter is shown in Fig. 6.16. Now with five flip-flops we have a MOD 10 counter as compared to the 10 flip-flops required with a ring counter. Some decoding is now necessary for displaying the result.†

To begin with, all flip-flops are reset by means of a pulse at the direct clear input (not shown). With the counter initially reset, there is a 1 at the $\bar{Q}$ outputs and a 0 at the Q outputs. There is therefore a 1 at the J_A input with all other J inputs being at 0. With the first clock pulse, FF-A is set, with no change in the state of the other flip-flops. The 1 therefore continues at the J_A input, but now there is also a 1 at the J input of FF-B. A second clock pulse now causes FF-B to set. After two clock pulses, both FF-A and FF-B are set, and there is now a 1 at the J input of FF-C. A third clock pulse causes FF-C to set. Therefore, with each successive clock pulse, the flip-flop on the right is set. For the example shown in Fig. 6.16, after five clock pulses all five flip-flops are in the set state. With FF-E in the set state, the Q output is at a 1 as is the K input of FF-A. The J input of FF-A is at a 0. Now, with the sixth clock pulse, FF-A is reset. With every additional clock pulse another flip-flop to the right is reset in turn. After 10 clock pulses, one complete counting cycle has been made, and all flip-flops have returned to the reset state.

The natural modulus of a shift counter is an even number. Two clock pulses are counted by each flip-flop. One clock pulse is required to set a flip-flop and another is used to reset the flip-flop. Five flip-flops lead to a MOD 10 counter. Three flip-flops lead to a MOD 6 counter. All flip-flops are set in the first half of the counting cycle, and they are all reset in the second half of the cycle. However, a simple change allows for odd-numbered moduli. This is shown in Fig. 6.17.

Here, the J input of the first flip-flop is connected to the $\bar{Q}$ output of the last flip-flop, as with a regular shift counter. However the K input of the first flip-flop is connected to the Q output of the next-to-last flip-flop instead of coming from the Q output of the last flip-flop.

For the example shown, after the count of 2, both FF-A and FF-B are set and only FF-C is in the reset state. The J input of FF-C is at a 1, the normal case, but the K input of FF-A is also at a 1. The result is, at the count of 3, FF-C sets and FF-A resets. The way is now prepared for FF-B to reset on the next clock pulse, the fourth, and FF-C resets on the fifth. Therefore, an odd number of clock pulses are required to complete one counting cycle; that is, the counter has an odd-numbered modulus. Note, that while a shift counter with an even-numbered modulus may be designed using D-type flip-flops, a J-K type is required for an odd-numbered modulus counter.

---

† A decade shift counter with 10 decoded decimal outputs is the CMOS type 4017.

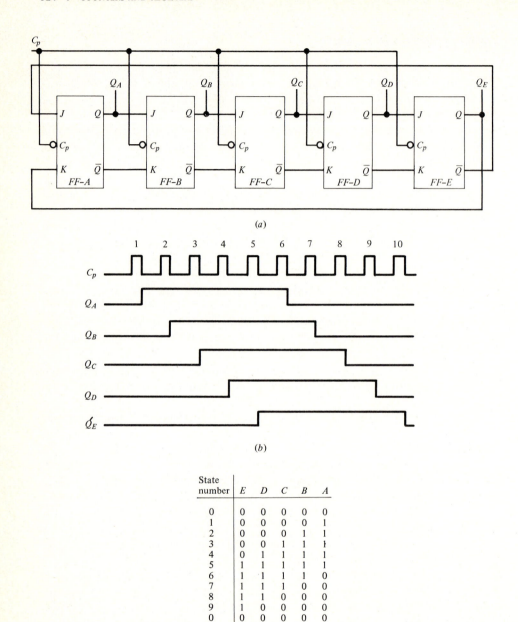

(a)

(b)

State number	E	D	C	B	A
0	0	0	0	0	0
1	0	0	0	0	1
2	0	0	0	1	1
3	0	0	1	1	1
4	0	1	1	1	1
5	1	1	1	1	1
6	1	1	1	1	0
7	1	1	1	0	0
8	1	1	0	0	0
9	1	0	0	0	0
0	0	0	0	0	0

(c)

FIGURE 6.16
MOD 10 shift counter. (a) Block diagram; (b) timing waveforms; (c) truth table.

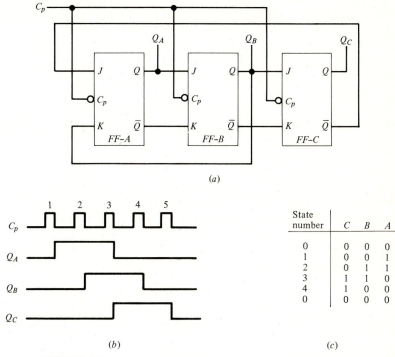

$$(a)$$

State number	C	B	A
0	0	0	0
1	0	0	1
2	0	1	1
3	1	1	0
4	1	0	0
0	0	0	0

$$(b) \qquad\qquad\qquad (c)$$

FIGURE 6.17
MOD 5 shift counter. (*a*) Block diagram; (*b*) timing waveforms; (*c*) truth table.

EXERCISE

E6.8 Show the waveform diagrams of a MOD 9 counter made, using master-slave flip-flops:

  (*a*) Implemented as a ring counter.
  (*b*) Implemented as a shift counter.

6.6   MEMORY CELL

The large-scale use of the flip-flop circuit is not limited to counters. Another common usage for these circuits, or more generally, the simple latch, is in the memory portion of a digital system. Historically, the mainframe memory in a computer has made use of magnetic cores. However, with the advent of large-scale integration (LSI), the low cost and small size of the monolithic IC latch has made this circuit very attractive for use as the memory element. In addition, by using bipolar or MOS transistors, the latch can be set and reset much faster than the magnetic core.

  In the memory array, the latches are arranged as a matrix of memory cells. The row of the matrix is selected as the binary word, and the columns represent

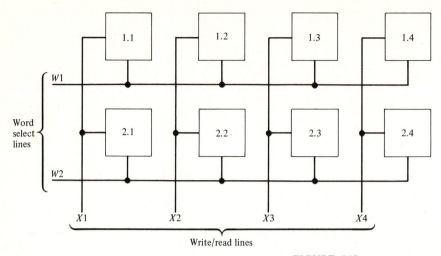

FIGURE 6.18
Two-word 4-bit memory array.

the bits of the word. A two-word 4-bit memory array is shown in Fig. 6.18. Of course, this could be extended vertically to increase the number of words or horizontally to increase the number of bits in the word. A common arrangement for a bipolar or CMOS IC memory chip is 256 words by 1 bit. For a MOS memory array, 1024 words by 1 bit is a popular IC package.

The typical bipolar memory cell makes use of two multi-emitter transistors and two resistors, forming the latch circuit shown in Fig. 6.19.

With the WORD line at V_1, current flow, in either Q_1 or Q_2, is out of these emitters. No current flows in either the SENSE line or the REFERENCE line because the WORD-line emitter voltage is the lowest of all the control voltage levels. A WORD row is selected by raising the WORD line to V_4, the highest of all the voltage levels.

A gate (WRITE ENABLE) on the SENSE line is used to control whether data is to be written into the cell or data read out. In the WRITE ENABLE condition, with the WORD line at V_4, a 1 is written into the cell by lowering the emitter of Q_1 below the REFERENCE-line voltage. Current then flows in Q_1, which saturates, ensuring that Q_2 is OFF. The latch is now set, with Q_1 ON and Q_2 OFF. To store a 0, the latch is reset, with the SENSE line and WORD line at V_4. The voltage of the REFERENCE line is lower than that on the SENSE line so that current now flows in Q_2, which saturates and holds Q_1 off.

Reading of data is accomplished by sensing the current in Q_1. With the cell selected by the WORD line at V_4, current flowing on the SENSE line connected to the emitter of Q_1 indicates that Q_1 is ON. That is, the latch is in the set state; thus a 1 is read out of the memory cell. The absence of current will indicate that the latch is in the reset or zero state.

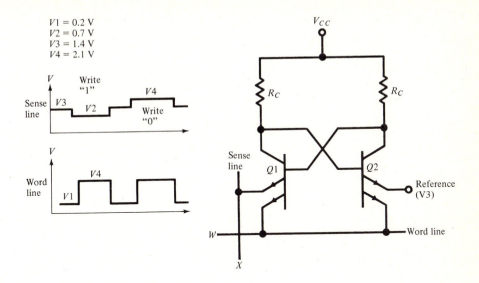

FIGURE 6.19
Bipolar transistor memory cell, including waveform for selecting a cell and writing the 1s or 0s, ($V1 = 0.2$ V, $V2 = 0.7$ V, $V3 = 1.4$ V and $V4 = 2.1$ V).

6.7 SUMMARY

\#   Counters and registers are interconnections of the basic *flip-flops* and *gates.*

\#   Electronic counters are generally classified as being either *synchronous* or *asynchronous.* Another name for the asynchronous counter is the *ripple* counter.

In a synchronous counter the clock-pulse input is connected directly to each flip-flop in the counter. The output from each flip-flop is then synchronized in time with the clock pulse.

In a ripple counter the clock-pulse input connects directly only to the first of a cascade of flip-flops. Each succeeding flip-flop changes state only as the immediate previous flip-flop makes the necessary triggering transition.

\#   Counters usually count in either the *binary* or *BCD* code. One flip-flop is required for each binary digit to be stored in the counter.

\#   Counters may also operate in either the *countup* or *countdown* mode.

\#   The design of synchronous counters is greatly aided by the use of *state tables* and *control maps.*

\#   The *shift register* is a simple synchronous counter, used for the temporary storage of binary data.

\#   Two particularly useful forms of the shift register are the *serial-to-parallel* and *parallel-to-serial* converter.

\#   With additional *shift-right* and *shift-left* gating, the data in a shift register may be moved either up or down the register.

The *ring counter* can be used as a decimal counter that requires no decoding, since there is one flip-flop for each number stored in the counter. The ring counter is a form of synchronous counting.

The *shift*, or *twisted-ring*, counter is similar to the ring counter but requires only half the number of flip-flops for a given decimal storage.

Many different types of counters and registers are available as MSI or LSI functions, using either bipolar or MOS transistors.

DEMONSTRATIONS

D6.1   Ripple counters

(*a*) Use dual *J-K* flip-flops† to design the 4-bit binary ripple counter shown in Fig. 6.2*a*. With input clock pulses from a pulse generator and the Q_D output connected to the external trigger input of an oscilloscope, observe the waveforms shown in Fig. 6.2*b* and note the time relationships between the flip-flop outputs and the clock pulses.

(*b*) With the addition of an AND-gate function, convert the 4-bit binary counter to the BCD ripple counter of Fig. 6.5*a*. Display the state of the counter by using LEDs (as in Demonstration D4.3) or a seven-segment display and an appropriate BCD-to-seven-segment decoder/driver.‡ The clock-pulse input may be obtained from a "debounced" switch.

(*c*) Make use of the flip-flop clear inputs plus an additional NAND gate to convert the 4-bit binary counter to a ÷ 12 ripple counter. Test as in part (*b*).

D6.2   Synchronous counters

(*a*) Use *J-K* flip-flops§ to design the 4-bit binary synchronous counter shown in Fig. 6.7. With the oscilloscope connections as in Demonstration D6.1*a*, again observe the time relationships between the flip-flop outputs and the clock pulses.

(*b*) Convert the 4-bit binary counter to the BCD synchronous counter of Fig. 6.9. Test and display the counter outputs as in Demonstration D6.1*b*.

(*c*) Make three BCD synchronous counters: one ÷ 6 and two ÷ 10. Use a 60-Hz sine wave (stepped-down ac line voltage) as the input into a Schmitt trigger∥ to get 60 pps. Use the *D* output of the ÷ 6 counter as the input to the first ÷ 10 counter. The *D* output of the ÷ 10 counter will have 1 pps on it. By using the 1-pps signal as the input to the second ÷ 10 counter, its BCD outputs can be used for the seconds display (of a digital clock) when coupled with the appropriate decoder and display element, such as in part (*b*).

---

† TTL type 74107.
‡ TTL type 7447.
§ TTL type 7472.
∥ TTL type 7413.

D6.3 Shift-left shift-right shift register   Use four D flip-flops† and the necessary NAND control logic to design, construct, and test a 4-bit shift-right shift-left shift register that has a serial-in and parallel-out capability. Use a signal (MODE) which if 1 causes data on the input line DATAR to be entered at the "near end" of the register and shifted right; when MODE is LOW, data on the input line DATAL will be entered at the "far end" of the register and shifted left. Use the input and display suggested in Demonstration D6.1b. Enter in 1001_2, 0110_2, and 1101_2 for both values of MODE. Use the circuit of Fig. 5.6 to make a manually operated clock-pulse source. Observe outputs after each shift pulse.

D6.4 Serial binary addition employing two registers and one full binary adder   Use the circuit of Fig. D4.3b to make the full binary adder that is incorporated in Fig. D6.4. Use D flip flops to make two 4-bit shift-right shift registers (X and Y) and for the flip-flop Z. For generating pulses and displaying contents of the registers, use the methods discussed in Demonstration D6.1b. In place of a third register for the sum, as shown in Fig. 4.20b, we use register X as an *accumulator* where the sum (of contents of Y and original value in X) is stored. Again use the circuit of Fig. 5.6 to make a manually operated clock-pulse source.

Use the accumulator to add 4_{10} to 5_{10}. The steps are:

1 Clear all registers by setting CLEAR to LOW. This sets 0 as the initial carry value into the carry flip-flop Z. Then return CLEAR to HIGH.
2 Set LOAD to HIGH then enter into Y_{in} the sequence HIGH, LOW, HIGH, LOW, where the value of Y_{in} is changed between CLOCK pulses. Thus the contents of the Y register read 0101_2 ($= 5_{10}$) after four clock pulses.
3 Set LOAD to LOW.
4 Apply four CLOCK pulses, which transfers 0101 into the contents of register X through the SUM output of the full adder.
5 Repeat step (2) but with the proper sequence of levels at Y_{in} so that at the end of four CLOCK pulses the contents of Y are 0100_2.
6 Set LOAD to LOW.
7 Apply four CLOCK pulses and check that register X now has accumulated 4_{10} to its previous value 5_{10} so that it now holds 9_{10}. Now add 3_{10} and 5_{10} in the accumulator.

D6.5 Digital-clock project   Build the 24-h digital clock of Fig. I.4a by using as many MSI circuits as practical. For an economy‡ version, display only minutes and hours. Use the circuit in Fig. D6.5a to generate the 5-V supply voltage and the waveform v_y (shown in Fig. D6.5b), which is the input for generating the time base. To eliminate counting "spikes" generated on the ac power line, use the RC filter and Schmitt trigger shown in Fig. D6.5c.

---

† TTL type 7474.
‡ The lowest-cost clock would be made by using an LSI circuit (Mostek MK5017P or National MM5314N as in Fig. I.6b) that drives displays that are *multiplexed*.

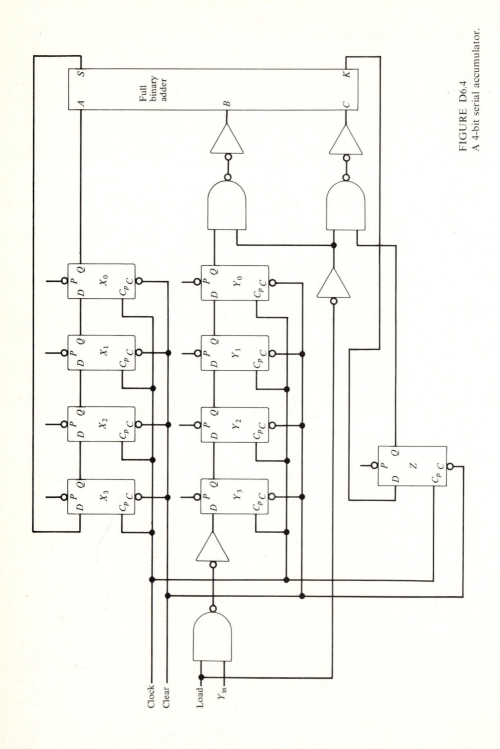

FIGURE D6.4
A 4-bit serial accumulator.

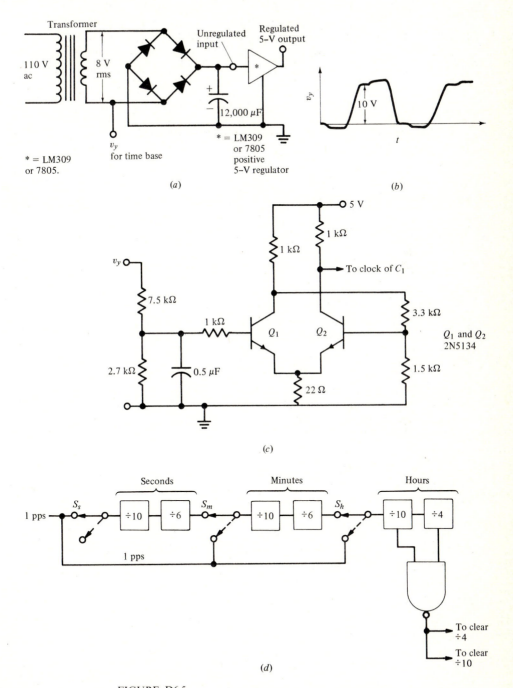

FIGURE D6.5

(a) Power supply and time base for the digital clock; (b) waveform of v_y; (c) RC filter and Schmitt trigger for generating 60-pps signal for first counter input; (d) Block diagram of 24-h digital clock.

For setting the time use the switch circuit shown in Fig. D6.5d in the following manner: (1) To set hours, push S_h down so that the hours input receives 1 pps. Allow this to continue until the desired reading is in the display, then return S_h to its normal (solid-line) position. (2) In the same fashion by means of S_m set the minutes. Then hold S_s down until the actual time coincides with the reading of your clock.

For the counters use TTL type 7492 for C_1 and C_8 in Fig. I.4a and TTL type 7490 everywhere else. For a $\div$ 6 with BCD output, use the 7490 with appropriate logic to set 0000 into the counter at the instant when the count reaches 0110_2. Seven-segment displays, such as discussed in Demonstration D4.2, can be directly driven from the TTL type 7447 decoder/driver. All logic can be done with one TTL type 7400 quad 2-input NAND gate and one TTL type 7404 hex inverter.

REFERENCES

The use of state tables and control maps in the design of counters and registers is well presented in a chapter on Flip-Flop Counters by WICKES, W. E.: "Logic Design with Integrated Circuits," chap. 9, Wiley, New York, 1968.

The operation of IC counters and registers is described in a chapter on Digital IC Modules by FITCHEN, F. S.: "Electronic Integrated Circuits and Systems," chap. 11, Van Nostrand Reinhold, New York, 1970; and in the chapter on Digital Systems by MILLMAN, J., and C. C. HALKIAS: "Integrated Electronics," chap. 17, McGraw-Hill, New York, 1972.

Another practical reference is MORRIS, R. L., and J. R. MILLER: "Designing with TTL Integrated Circuits," McGraw-Hill, New York, 1971, chap. 10 on Counters and chap. 11 on Shift Registers.

Further topics on the use of counters and registers is covered in a chapter on The Design of Sequential Circuitry by PEATMAN, J. B.: "The Design of Digital Systems," chap. 5, McGraw-Hill, New York, 1972.

PROBLEMS

P6.1 Design a 4-bit binary ripple counter that counts up or down depending on the level of a control signal. When the control signal is high, the counter should count up. Use SN7473 flip-flops and NAND gates.

P6.2 Design a binary ripple counter with a 60-Hz input to produce a pulse every second. Use SN7473 flip-flops.

P6.3 Design a binary ripple counter that counts down after the binary equivalent to the number 13 is set into the counter by the use of logic gates. A HIGH-control signal should be produced as an output when the counter reaches zero. Use SN7476 flip-flops and NOR gates.

P6.4 Design a 4-bit binary ripple counter that counts up using positive edge-triggered flip-flops (for example, the SN7470). Sketch the waveforms at Q_A, Q_B, Q_C, and Q_D in relationship to the clock pulse waveform C_p.

P6.5 Design a 4-bit binary ripple counter that counts up by using *J-K* flip-flops similar to that shown in Fig. 6.1a but with the excitation table of Fig. 6.11c.

P6.6   Given a 1-Hz pulse, design a synchronous counter that outputs a pulse every minute and every hour. Implement it as a MOD 10 followed by a MOD 6 counter. Use SN7472 flip-flops.

P6.7   For the counter shown in Fig. P6.7, make a truth table for the first five input pulses, assuming the SN7473 flip-flops are reset in the initial state.

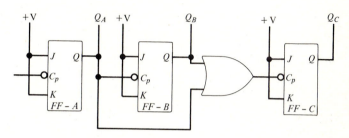

FIGURE P6.7

P6.8   Solve Prob. P6.1 by using a synchronous counter in lieu of the ripple counter.

P6.9   Solve Prob. P6.3 by using a synchronous counter in lieu of the ripple counter.

P6.10   Design a 4-bit synchronous binary counter that counts up from the zero state. At each state, the content of the counter is compared to the content of a 4-bit register, and a HIGH-level control signal is produced as an output when the contents of the two devices are the same. Use the SN7472 flip-flops.

P6.11   Design a synchronous BCD decade counter to count in the 2421 code by using *J-K* flip-flops similar to that shown in Fig. 6.1*a* but with the excitation table of Fig. 6.11*c*.

P6.12   A synchronous counter is required to count the number of iterations of a certain process and output a signal at the end of the process. The total number of iterations required for the process can be any number between 1 and 13, available in parallel 8421 BCD code. Design the gating and control lines. (*Note:* One possible design for this problem uses the results of P6.10.)

P6.13   A parking garage has space for 12 cars. Design a synchronous counter using SN7472 flip-flops and NAND gates that counts the cars entering and leaving the garage. The counter should output signals to light either the "VACANCY" or "NO VACANCY" sign. Assume two input signal lines to your system: when a pulse occurs on one line, it indicates a car entering the garage; the other line has a pulse on it when a car leaves the garage. The system is so structured that the pulses never overlap in time.

P6.14   Design a MOD 6 synchronous counter to count in 531420 binary sequence by using SN7473 flip-flops.

P6.15   Design a MOD 4 synchronous counter to count in 3572 binary sequence by using SN7473 flip-flops.

P6.16   Design a MOD 13 synchronous counter to count in 1-4-7-10-13-3-6-9-12-2-5-8-11 binary sequence by using SN7473 flip-flops.

P6.17   Design a MOD 11 synchronous counter to count in 15-8-10-3-6-12-0-14-5-2-4 sequence by using *J-K* flip-flops similar to that shown in Fig. 6.1*a* but with the excitation table of Fig. 6.11*c*.

*P6.18   Two 4-bit binary numbers are to be logically ORed together. The numbers, available in parallel form, are transferred into two 4-bit right-shift registers by a control pulse. The contents of the registers are then shifted right every clock pulse, with the rightmost bits (one from each register) sent to the OR gate. The output of the OR gate is placed back into the leftmost bit position of one of the shift registers. Assuming that a second control pulse initiates the shifting process, design a synchronous system by using SN7476 flip-flops such that a PROCESS COMPLETE signal is produced after all the bits have been ORed.

*P6.19   Modify Prob. P6.18 to include a full adder in lieu of the OR gate so that the two numbers are added together with the results left in one of the registers. The operation should include provisions for indicating a carry overflow.

*P6.20   Recirculating shift registers are sometimes used in computers as memory in addition to core memory or memory cells. Assume a recirculating shift register memory (made with J-K flip-flops) contains 256 bits in which individual bits can be stored or retrieved from memory as the result of control signals. The store-and-retrieve control signal must be synchronized with the clock pulse that drives the shift register. Show the logic diagram including the control lines. Only the first two and last two flip-flops of the shift register need be shown. Use SN7476 flip-flops.

P6.21   Each bit in the recirculating shift register of Prob. P6.20 is assigned an address number. A counter is used to designate the address number of the bit presently at the output of the shift register. Another register, called the *address* register, is used to contain the address number of the bit to be entered or retrieved from the memory. Assuming the address number exists in the address register in binary form, draw the logic diagram showing the gates and control lines. Show only the first two and last two flip-flops of the shift register, counter, and address register.

P6.22   Given the bipolar-transistor memory cell shown in Fig. 6.19, calculate the current in the SENSE line when the WORD line is HIGH, for a WRITE 0 and 1 and READ 0 and 1. Assume $V_{CC} = 5.0$ V, $V_{BE(sat)} = 0.7$ V $V_{CE(sat)} = 0.2$ V, $\beta_F = 100$, and $R_C = 10$ kΩ.

SMALL-SIGNAL AMPLIFIERS

So far in this text we have concerned ourselves with digital circuits where there was a nonlinear relationship between the input and output signals. In particular, recall the transfer characteristic of the transistor inverter used in a saturating logic circuit. The output voltage can take only one of two discrete levels, not withstanding small variations of the input voltage. However, as the circuit is switched from one discrete level to another, it necessarily passes through the *transition region*. In digital circuits we avoid staying in this region, but it is this area that interests us now, where the transistor is operating in the *active region*. In normal use of this class of circuits there is a *linear* relationship between the input and output signals. That is, the output voltage is essentially a scaled-up replica of the small variations of the input voltage. A simple example of such a linear circuit is an audio amplifier.

More specifically, consider a *preamplifier* connected to the tone arm of a phonograph. The nominal output voltage of a magnetic cartridge (contained in the tone arm) is about 5 millivolts (mV). A typical phonograph preamplifier will have sufficient voltage gain that this small signal will be amplified, without distortion, by a factor of about 100. The output signal of the preamplifier will therefore be a replica of the input signal but with an amplitude of 0.5 V. In some circuits the output signal may not be required to be an exact replica of the input signal; the general class of these circuits is sometimes described as *analog* circuits.

As with digital ICs, we will find that linear ICs use many designs that are variations of earlier circuits with discrete transistors and resistors. However, advantage is taken of the close matching and tracking of components implicit in ICs. Moreover, ingenuity is used to cope with the disadvantages as well, namely, the lack of large-valued capacitors in the IC chip.

It will also be useful to observe many parallels in the analysis of digital and analog systems. We have seen that a digital system may be simply regarded as an interconnection of various logic blocks. For instance, the basic logic circuits of our digital clock were gates and flip-flops. These circuits, in turn, were interconnections of stages of transistor circuits. By developing a model for the transistor used in a digital application, we were able to derive useful expressions for the input, output, and transfer characteristics of these logic blocks. We take a similar approach in dealing with analog systems. With many such systems we will find a popular functional block to be the *operational amplifier*. Now a basic element of the operational amplifier is the differential amplifier circuit, and in its most simplest form this circuit is composed of just two transistors and three resistors. If we now develop a model for the transistor operating in the active region, we will be able, in a manner somewhat analogous to the digital circuit, to obtain expressions for the input, output, and transfer characteristics of the amplifier or *gain block*.

The ideal gain block would have the following characteristics:

1 Infinite input resistance
2 Infinite voltage and current gain
3 Zero output resistance

The infinite input resistance will ensure that the gain block will have no loading effect on the source of the input signal voltage. In a practical system we will require a finite voltage or current gain, but this can easily be obtained by *negative feedback*, even if the original gain is extremely large. The zero output resistance means the signal from the gain block will be unaffected by any loading at the output terminal. Of course, no circuit can exhibit all, or any, of these characteristics. However, there are ICs which for many applications can be considered an ideal gain block. In this chapter we will examine the basic transistor circuit configurations and relate their properties to the ideal amplifier.

7.1 TRANSISTOR CIRCUIT MODELS

A single-stage transistor amplifier circuit is shown in Fig. 7.1a. The capacitor, shown at the input to the amplifier, presents an almost infinite resistance to the flow of direct current (dc). However, providing they have large enough capacitance, capacitors are almost a short circuit to alternating current (ac). We will assume in this analysis that the signal voltage (v_s) is a sine-wave generator of such a frequency, and the capacitors have sufficient capacitance that they only block dc but have no ac voltage developed across them.

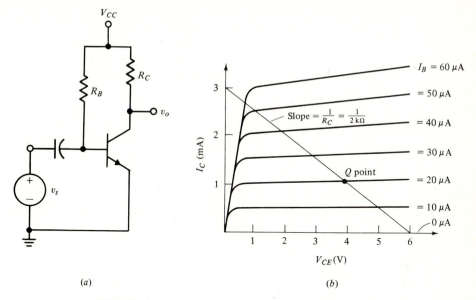

FIGURE 7.1

(a) Single-stage transistor amplifier circuit; (b) transistor collector characteristics.

Suppose we use our old friend the 2N4275 transistor in this circuit. The collector characteristics of this transistor, taken from Chap. 1, are shown in Fig. 7.1b. While this figure shows the dc collector characteristics (I_C versus V_{CE}), in the absence of any signal voltage this plot also represents the relationship between the instantaneous values i_C and v_{CE}. The *operating bias point*, called the Q point, is chosen so that the transistor is operating in the normal-mode active region. Also shown on the collector characteristics is the load line due to a 2-kΩ collector load resistor, R_C.

Now small variations of the signal voltage v_s will cause small changes of the instantaneous base current i_B. The Q point will move along the load line in accordance with these changes of base current. The result is a change of the collector current i_C, which is much larger than the incremental change of i_B. Of course with a sufficiently large increase of i_B, the transistor enters the saturation region; or with a large decrease of i_B, the transistor cuts off.†

One problem we face is to determine the input resistance of the amplifier stage so that we will know the effect of this load on the output of the sine-wave generator. We also want to know the voltage or current gain (amplification) of the stage. To determine the capability of this stage to drive other circuits, we need to know the output resistance of the stage. For this kind of analysis of the *small-signal* behavior of the amplifier, we will find it useful to develop a *small-signal circuit model* for the transistor.

† The bias circuit shown in Fig. 7.1a is for simplification only and is not recommended for practical design. A more practical method of biasing a discrete transistor is shown in Demonstration D7.2.

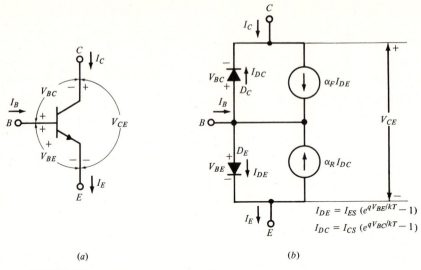

$$I_{DE} = I_{ES} (e^{qV_{BE}/kT} - 1)$$
$$I_{DC} = I_{CS} (e^{qV_{BC}/kT} - 1)$$

(a)   (b)

FIGURE 7.2
(a) Reference directions for *n-p-n* transistor; (b) Ebers-Moll model of bipolar transistor.

7.1.1   The Ebers-Moll Model

We begin with the Ebers-Moll model for an *n-p-n* transistor taken from Chap. 1 and shown again in Fig. 7.2.

Now with the transistor operating in the normal-mode active region, we have shown in Chap. 1 that we have the following very good approximations:

$$I_C = \alpha_F I_E \qquad (7.1a)$$

and

$$I_B = (1 - \alpha_F)I_E \qquad (7.1b)$$

Further, we have by definition that

$$\beta_F = \frac{\alpha_F}{1 - \alpha_F} = \frac{I_C}{I_B} \qquad (7.2)$$

To complete the expression for the operating currents of the transistor, we note from Fig. 7.2 and Eq. (7.1) that

$$I_E = I_B + I_C$$

We can now draw the simple circuit model applicable for the dc operating point of the transistor. We have reproduced Fig. 1.19c as Fig. 7.3. Note that in referring to the dc or quiescent currents in the transistor we are using capital letters and subscripts.

7.1.2   Small-signal Transistor Model

Now that we have a model for the transistor at the dc operating point, our next interest is to take into account the effects of small variations about this operating

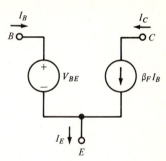

FIGURE 7.3
Normal-mode active-region model of transistor.

point. To do this we need to derive a value for the *small-signal* or *incremental* conductance of the base-emitter junction diode.

The VI characteristic of a silicon diode is shown in Fig. 7.4. This can be the base-emitter junction of a transistor. In the figure, we will represent the change in the diode current from the operating-point current I_0 as

$$\Delta I_D = i_d \qquad (7.3a)$$

and the change in the diode voltage as

$$\Delta V_D = v_d \qquad (7.3b)$$

Again from Chap. 1, we recall that with a diode forward-biased such that $V_D > 5V_T$, then I_D and V_D are related by the diode equation; namely,

$$I_D = I_S e^{V_D/V_T} \qquad (7.4)$$

Now the incremental conductance of the diode is given by the ratio of i_d and v_d. Then

$$g_d = \frac{i_d}{v_d} = \frac{\Delta I_D}{\Delta V_D} \qquad (7.5a)$$

By taking an infinitesimal increment of I_D, we have

$$\frac{\Delta I_D}{\Delta V_D} = \frac{dI_D}{dV_D} \qquad (7.5b)$$

That is, differentiating I_D with respect to V_D at the operating-point current I_0, and replacing I_0 with the expression in Eq. (7.4),

$$\frac{dI_D}{dV_D} = \frac{d(I_S e^{V_D/V_T})}{dV_D} = \frac{(I_S e^{V_D/V_T})}{V_T} \qquad (7.5c)$$

Then again making use of the diode equation (7.4) in Eq. (7.5c), we have the simple relation for g_d; namely,

$$g_d = \frac{I_D}{V_T} \qquad (7.5d)$$

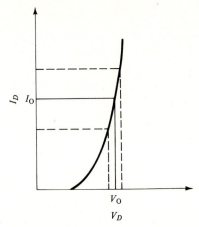

FIGURE 7.4
VI characteristic of a silicon diode.

At normal room temperature therefore, where $V_T = 26$ mV, the small-signal conductance of a junction diode at a given operating current is simply given by the ratio of the diode current and 26 mV. That is, at 27°C,

$$g_d = \frac{I_D}{26 \text{ mV}}$$

We have a direct analogous case for the incremental conductance of the base-emitter junction of the transistor,† where, viewed from the base side, we have

$$g_{be} = \frac{\Delta I_B}{\Delta V_{BE}} = \frac{I_B}{V_T} \qquad (7.6a)$$

Generally we find it easier to make use of the incremental resistance of the base-emitter junction. This is the reciprocal of the incremental conductance and is commonly labeled r_π. Then

$$r_\pi = \frac{1}{g_{be}} = \frac{V_T}{I_B} \qquad (7.6b)$$

An alternative form of writing the equation for r_π is by substituting I_C/β_0 for I_B; then

$$r_\pi = \frac{\beta_0 V_T}{I_C} \qquad (7.6c)$$

We can now draw a circuit model which is applicable for the small-signal operation of the transistor. This is shown in Fig. 7.5a, where we have replaced the voltage source, V_{BE} of Fig. 7.3, with the small-signal resistance r_π; and the collector-current generator is now $\beta_0 i_b$. Note in Fig. 7.5a that the small-signal

---

† Equation (7.6a) can be proven rigorously by the use of Eq. (1.12b).

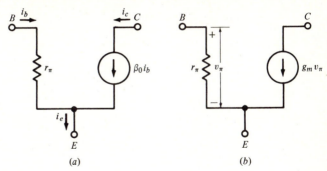

FIGURE 7.5
Small-signal circuit model of transistor showing (a) collector-current generator as $\beta_0 i_b$ and (b) collector-current generator as $g_m v_\pi$.

currents are shown with lower-case letters and lower-case subscripts. We also indicate the small-signal current gain of the transistor as β_0 in contrast with the dc parameter β_F. (Transistor data sheets generally refer to these two parameters as h_{fe} and h_{FE}, respectively.) In the broad region of collector current where β_F is approximately constant, these two parameters will usually have about the same value. This fact should be confirmed in Demonstration D7.1.

7.1.3   Transconductance

From the circuit model shown in Fig. 7.5a, we see that a small change of current at the input to the transistor (the base current i_b) will result in a change of the current at the output of the transistor (the collector current i_c). There two currents are related by the current gain of the transistor, β_0. Then

$$i_c = \beta_0 i_b \qquad (7.7)$$

We can also see from Fig. 7.5a that a small change of voltage at the input to the transistor will also cause a change in base current and hence cause a change in the collector current at the output of the transistor. This introduces us to the *transconductance* parameter, or *transfer conductance*, that relates the change of output (collector) current to the change of input (base) voltage v_π, which is the voltage across r_π due to i_b. That is,

$$g_m = \frac{\Delta I_C}{\Delta V_{BE}} = \frac{i_c}{v_\pi} \qquad (7.8a)$$

(This parameter is also referred to as the *mutual* conductance, hence the term g_m.)
     By rearrangement of Eq. (7.8a), we have

$$i_c = g_m v_\pi \qquad (7.8b)$$

This alternative representation for the collector-current generator is shown in the transistor small-signal circuit model of Fig. 7.5b. Both forms of circuit model shown in Fig. 7.5 are commonly used.

We can substitute $\beta_0 i_b$ for i_c in Eq. (7.8a) and obtain another useful relationship for g_m. That is,

$$g_m = \frac{\beta_0 i_b}{v_\pi} \qquad (7.9a)$$

We should realize that this is equivalent to

$$g_m = \frac{\beta_0 \, \Delta I_B}{\Delta V_{BE}} \qquad (7.9b)$$

A simplification of the expression for the transconductance is now possible because we can use the latter part of Eq. (7.6a) to substitute in Eq. (7.9b) and show that

$$g_m = \frac{\beta_0 I_B}{V_T} \qquad (7.10a)$$

or

$$g_m = \frac{I_C}{V_T} \qquad (7.10b)$$

This equation may be rigorously derived as in Prob. P7.2.

Finally, by including Eq. (7.6b) in Eq. (7.10a), we can show that the product of the transconductance (g_m) and the base-emitter small-signal resistance (r_π) of a transistor is equal to the current gain (β_0). That is,

$$\beta_0 = g_m r_\pi \qquad (7.11)$$

It is important to note, from Eqs. (7.6c) and (7.10b), respectively, that both the small-signal parameters r_π and g_m are dependent upon the quiescent or operating collector current of the transistor. We will find this very useful in the design of transistor amplifiers.

EXAMPLE 7.1   For an ambient temperature of 27°C, calculate the g_m and r_π of a transistor that has $\beta_0 = 50$, when operating in the active region, at a collector current of 5 mA.

SOLUTION

$$g_m = \frac{I_C}{V_T}$$

$$= \frac{5 \text{ mA}}{26 \text{ mV}} = 0.192 \text{ mho or } 192 \text{ mmhos}$$

$$r_\pi = \frac{\beta_0}{g_m}$$

$$= \frac{50}{0.192} = 260 \text{ } \Omega$$

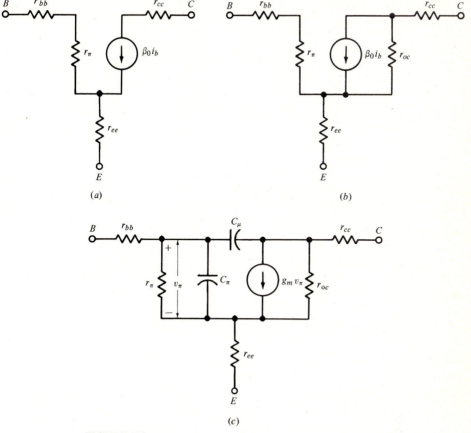

FIGURE 7.6
(a) Circuit model of transistor, including ohmic bulk resistances; (b) the circuit model with the collector output resistance included; (c) the basic hybrid-π model for the transistor.

Alternatively,

$$r_\pi = \frac{\beta_0 V_T}{I_C}$$

$$= \frac{(50)(26\ \text{mV})}{5\ \text{mA}} = 260\ \Omega \qquad \qquad ////$$

7.1.4 Ohmic Resistance

The model shown in Fig. 7.5a or b is generally adequate for most low-frequency transistor circuit applications. However, other elements must be added to obtain a more accurate model. Since the emitter, base, and collector regions are made from semiconductor materials, there are ohmic resistances in each region corresponding to conduction through the bulk semiconductor. These additional elements are shown in the circuit model of Fig. 7.6a. Typical values for these bulk resistances are given below:

$$r_{bb} = 20 \text{ to } 200 \; \Omega$$

$$r_{cc} = 10 \text{ to } 100 \; \Omega$$

$$r_{ee} = 0.5 \text{ to } 5 \; \Omega$$

7.1.5 Base-width Modulation

So far our small-signal model of the collector of the transistor has consisted of just an ideal current generator $\beta_0 \, i_b$. There is a resistance element we can add to the collector circuit to account for voltage variations at the collector of the transistor.

In particular notice from the collector characteristic of the 2N4275 in Fig. 7.1b that with a fixed base current, an increase of the collector voltage (V_{CE}) causes a small increase in the collector current. For example, at $V_{CE} = 4.0$ V and $I_C = 1.0$ mA, the base current is 20 μA. Keeping the base current constant at 20 μA but increasing V_{CE} to 12 V, we would find the collector current has increased to 1.2 mA.

The explanation of the increase of collector current is as follows. With an increase of V_{CE}, though more particularly V_{CB}, the region of the reverse-biased junction between the collector and the base is widened. Hence the base-width region narrows. The width of the forward-biased base-emitter junction is little affected. Now the narrower base region results in less recombination of the minority carriers (electrons in the case of an n-p-n transistor) in the base, resulting in an increase of collector current. Thus the effective base width of the transistor changes inversely with the collector-to-emitter voltage. This base-width-modulation effect accounts for the finite slope of the $V_{CE} I_C$ characteristic shown in Fig. 7.1b.

Since this slope of the characteristic is intimately related to the collector voltage and the collector current, we can model the base-width-modulation effect by a collector output resistance r_{oc}. However, also notice from Fig. 7.1b that the slope of the characteristic increases as both I_C and V_{CE} are increased. Hence we must restrict any value for r_{oc} to a particular operating point. It is therefore another small-signal parameter.

$$r_{oc} = \frac{dV_{CE}}{dI_C}$$

For the ideal transistor, the slope of the $V_{CE} I_C$ characteristic is zero and the collector-output resistance is infinite. For the 2N4275 we can measure the slope at $I_C = 1$ mA and $V_{CE} = 4$ V in Fig. 7.1b and obtain a value for the collector-output resistance of about 40 kΩ. For the typical transistor, operating at 1 mA,

$$r_{oc} = 10 \text{ to } 100 \text{ k}\Omega$$

Now we see that the collector of the transistor is modeled with a finite resistance in parallel with the ideal current generator, as shown in Fig. 7.6b. Generally, however, the value of r_{oc} is very much greater than the value of any other resistor in the collector circuit so that the effect of r_{oc} can be ignored.

7.1.6 Hybrid-π Model

A complete circuit model for the transistor is not established until charge-storage effects are taken into account. A consideration of the charge-storage effects in the transistor ultimately leads to a study of the frequency limitations on the operation of the transistor. In this text we are predominantly concerned only with the low-frequency operation of the transistor, where the effects of charge storage in the transistor may be ignored.

However, when we later (in Chap. 9) consider the application of transistors in feedback amplifiers, it will be necessary to realize that there is a frequency limitation in the use of a practical transistor. We therefore show in Fig. 7.6c the small-signal model for the transistor, often referred to as the basic *hybrid-π model*. The charge-storage effects in the base region of the transistor, due to the forward-biased base-emitter junction, are modeled by the capacitor C_π. The charge stored between the base and collector regions with the junction reverse-biased is modeled by the capacitor C_μ.

In summary of the circuit model for the transistor it should be pointed out the transistor action is a phenomenon taking place within the semiconductor device. This action is modeled with circuit elements, like resistors and current generators, only for ease of analysis and design of electronic circuits containing transistors. Again, we repeat, the circuit model shown in Fig. 7.5a or b is generally adequate for most low-frequency applications. However, the model is to be used only for consideration of small-signal effects. Small signal in this context generally means voltage variations at the input to the transistor much less than the thermal voltage V_T. This is because V_T is intimately related to two important parameters of the small-signal circuit model that we do not want the signal level to change, namely, r_π and g_m. Hence, at room temperature, the voltage signal to the input of the transistor (across the base-emitter junction) should be less than 5 mV. With much larger signals at the input, nonlinearities may be introduced with consequent inaccuracies of the model.

EXERCISE

E7.1 For an ambient temperature of 27°C, calculate the g_m and r_π of a transistor operating in the active region at a collector current of 1 mA. Also determine these small-signal parameters at operating temperatures of 0 and 70°C. Assume β_0 is invariant with temperature and is equal to 50.

7.2   BASIC COMMON-EMITTER AMPLIFIER

We now return to the single-stage transistor amplifier circuit shown in Fig. 7.1a. The transistor connection shown is referred to as *common emitter* (CE). The input signal to the transistor is applied between the base and the emitter. The output signal is taken between the collector and the emitter. The transistor emitter is therefore a common terminal for both the input and output signals. We will later see that the CE connection is the only single transistor configuration which yields both voltage and current gain from the input to the output. It is therefore the basis of many transistor circuits. Our main purpose in this section is to develop circuit equations so that we may determine the voltage gain of the stage, the current gain, and the input and output resistance presented by the stage.

The following circuit and bias data apply to Fig. 7.1a,

$$V_{CC} = 6\text{ V} \qquad V_{CE} = 4\text{ V} \qquad T_A = 27°\text{C}$$
$$I_C = 1\text{ mA} \qquad V_{BE(on)} = 0.7\text{ V} \qquad V_T = 26\text{ mV}$$
$$\beta_F = \beta_0 = 50$$

7.2.1   DC Conditions

The first task is to determine values for the base-biasing resistor R_B and the collector-load resistor R_C such that the transistor is biased at the correct operating point with $I_C = 1$ mA and $V_{CE} = 4$ V.

Using Kirchhoff's voltage law (KVL) in the base circuit of the transistor, we know

$$V_{BE(on)} + I_B R_B = V_{CC} \qquad (7.12a)$$

but also

$$I_B = \frac{I_C}{\beta_F}$$

We may therefore rewrite Eq. (7.12a) as

$$V_{BE(on)} + \frac{I_C R_B}{\beta_F} = V_{CC} \qquad (7.12b)$$

we now have an equation for R_B as

$$R_B = \frac{\beta_F(V_{CC} - V_{BE(on)})}{I_C} \qquad (7.13)$$

Then, using the data we have been given above,

$$R_B = \frac{50(6 - 0.7)}{1} = 265\text{ k}\Omega$$

We may also use KVL in the collector circuit to calculate a value for R_C, since

$$V_{CE} + I_C R_C = V_{CC} \qquad (7.14)$$

that is,

$$R_C = \frac{V_{CC} - V_{CE}}{I_C}$$

$$= \frac{6 - 4}{1} = 2 \text{ k}\Omega$$

7.2.2 Small-signal Characteristics

Since at this time we are neglecting all frequency effects in the small-signal model, we can assume that the capacitor in Fig. 7.1a can be replaced by a short circuit. We can also assume that the voltage supply V_{CC} has zero internal resistance. We further assume an ideal transistor with $r_{oc} = \infty$. The small-signal model for the circuit is then as shown in Fig. 7.7.

To complete the values for the circuit model elements, we calculate the transistor parameter r_π from Eq. (7.6c),

$$r_\pi = \frac{50(26 \text{ mV})}{1 \text{ mA}} = 1.3 \text{ k}\Omega \qquad (7.15)$$

Input resistance (r_i) The small-signal input resistance is simply determined by including R_B in parallel with r_π. Then

$$r_i = \frac{v_s}{i_s} = \frac{R_B r_\pi}{R_B + r_\pi} \qquad (7.16)$$

Output resistance (r_o) Since the collector-current generator is an ideal current source with infinite internal resistance, the output resistance is just R_C. (The output resistance is measured with v_s set to zero.)

$$r_o = \frac{v_o}{i_o} = R_C \qquad (7.17)$$

Voltage gain (a_v) To determine the voltage gain v_o/v_s of the stage, we note from Fig. 7.7 that the output signal voltage is given as

$$v_o = -i_c r_o = -\beta_0 i_b R_C \qquad (7.18)$$

and that the input signal voltage is equal to

$$v_s = i_b r_\pi \qquad (7.19)$$

Therefore, using the latter part of Eq. (7.18), the following expression can be written for the voltage gain:

$$a_v = \frac{v_o}{v_s} = \frac{-\beta_0 R_C}{r_\pi} \qquad (7.20a)$$

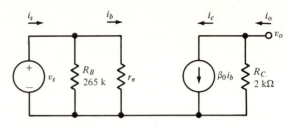

FIGURE 7.7
The small-signal model for the amplifier circuit of Fig. 7.1a.

Notice in Eq. (7.16) that with $R_B \gg r_\pi$, $r_i = r_\pi$. Then by using the first part of Eq. (7.18), the voltage gain may simply be written as

$$a_v = \frac{v_o}{v_s} = \frac{-\beta_0 r_o}{r_i} \qquad (7.20b)$$

The voltage gain is equal to β_0, multiplied by the ratio of the output and input resistances. The minus sign indicates the polarity inversion of the CE amplifier. The output voltage signal at the collector is inverted from the input voltage signal at the base.

Current gain (a_i) The current gain for the circuit in Fig. 7.7 is i_c/i_s. With $R_B \gg r_\pi$, this reduces to i_c/i_b, that is, β_0.

EXAMPLE 7.2   As a numerical example we will compute the small-signal input resistance, output resistance, and voltage gain for the amplifier circuit shown in Fig. 7.1a.

SOLUTION   Referring to Fig. 7.7, we have, from Eq. (7.15), $r_\pi = 1.3$ kΩ. Then using Eq. (7.16) to determine the input resistance,

$$r_i = \frac{(265)(1.3)}{265 + 1.3} \approx 1.29 \text{ k}\Omega$$

Because $R_{BB} \gg r_\pi$, the effect of R_B on r_i is negligible. The output resistance of the amplifier stage is, from Eq. (7.17), simply

$$r_o = 2 \text{ k}\Omega$$

The voltage gain is calculated with the aid of Eq. (7.20); namely,

$$a_v = \frac{-(50)(2)}{1.3} = -77 \qquad ////$$

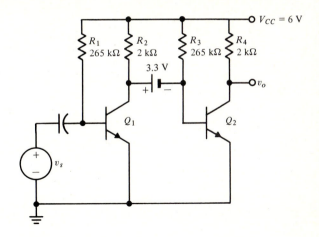

FIGURE 7.8
Cascade of two CE stages coupled with a level-shifting "battery."

A very practical method for experimentally determining these small-signal characteristics of the CE transistor configuration is included in Demonstration D7.2.1.

EXERCISE

E7.2 An amplifier circuit is similar to that shown in Fig. 7.1a, but $V_{CC} = 15$ V, $R_B = 360$ kΩ, and $R_C = 3$ kΩ. The transistor has the same characteristics as that shown. That is, $\beta_F = \beta_0 = 50$; $V_{BE(on)} = 0.7$ V. Compute I_C, V_{CE}, r_i, r_o, and a_v.

7.3 CASCADE OF CE STAGES

In the introduction to this chapter we cited the need of a phonograph preamplifier to have a voltage gain of about 100. The vertical deflection amplifier in an oscilloscope might require a voltage gain of more than 1000. A simple method of obtaining a larger factor of amplification is to follow one gain stage with another. An example of a cascade of two CE stages is shown in Fig. 7.8.

The circuit shown in Fig. 7.8 will be recognized as two stages of amplification, each one similar to the circuit shown in Fig. 7.1. Of course, this duplication is not necessary; indeed, the larger voltage swing at the output of the second collector will generally require a different operating point for this transistor. Or perhaps, a different type of transistor. We show a 3.3-V battery connecting the two gain stages. This is necessary because the quiescent voltage level at the collector of Q_1 is 4 V, and at the base of Q_2 it is 0.7 V. With discrete components (transistors, resistors, and capacitors) the two gain stages would be connected

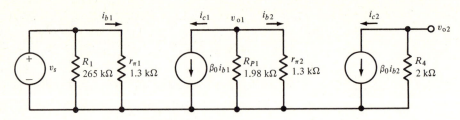

FIGURE 7.9

Small-signal circuit model for the amplifier circuit shown in Fig. 7.8.

with a coupling capacitor of some large value. However, we have seen that with ICs, capacitors are generally to be avoided. Therefore in Chap. 8 we will see how to readily shift voltage levels without using capacitors and without any loss of signal.

With both transistors having similar characteristics as given for Example 7.2 in the previous section, we have, for each transistor, that $I_C = 1$ mA and $V_{CE} = 4$ V. In this section we will only be concerned with the small-signal characteristics of the amplifier. We then draw the small-signal circuit model for the complete amplifier as shown in Fig. 7.9. Notice for the small-signal model that R_2 and R_3 effectively appear in parallel and have so been combined as R_{P1} in Fig. 7.9.

To determine the small-signal voltage gain of the complete amplifier, we initially compute the signal base current of the first transistor as

$$i_{b1} = \frac{v_s}{r_{\pi 1}} \qquad (7.21)$$

The signal current in the collector of the first transistor is then

$$i_{c1} = \beta_0 i_{b1} = \frac{\beta_0 v_s}{r_{\pi 1}} \qquad (7.22)$$

We may now solve for the output voltage of the first stage as

$$v_{o1} = -i_{c1} R_{L1} = \frac{-\beta_0 R_{L1} v_s}{r_{\pi 1}} \qquad (7.23)$$

where R_{L1} is equal to the parallel combination of R_{P1} and $r_{\pi 2}$. That is,

$$R_{L1} = \frac{R_{P1} r_{\pi 2}}{R_{P1} + r_{\pi 2}} \qquad (7.24)$$

Continuing to the second stage, we have for the signal base current,

$$i_{b2} = \frac{v_{o1}}{r_{\pi 2}} \qquad (7.25)$$

and for the collector current,

$$i_{c2} = \beta_0 i_{b2} = \frac{\beta_0 v_{o1}}{r_{\pi 2}} \qquad (7.26)$$

The output voltage of the second stage is therefore

$$v_{o2} = -i_{c2}R_4 = \frac{-\beta_0 R_4 v_{o1}}{r_{\pi2}} \qquad (7.27)$$

Combining Eq. (7.23) with Eq. (7.27), we have

$$v_{o2} = \left(\frac{-\beta_0 R_4}{r_{\pi2}}\right)\left(\frac{-\beta_0 R_{L1}}{r_{\pi1}}\right)v_s \qquad (7.28)$$

We have now determined the small-signal voltage gain for the complete amplifier circuit shown in Fig. 7.8. That is,

$$a_v = \frac{v_{o2}}{v_s} = \left(\frac{\beta_0 R_4}{r_{\pi2}}\right)\left(\frac{\beta_0 R_{L1}}{r_{\pi1}}\right) \qquad (7.29)$$

For a numerical solution we may use the values given in Fig. 7.8. But first we calculate a value for R_{L1} from Eq. (7.24).

$$R_{L1} = \frac{(1.98)(1.3)}{1.98 + 1.3} = 0.785 \text{ k}\Omega$$

Then the voltage gain from Eq. (7.29) is

$$a_v = \left[\frac{(50)(2)}{1.3}\right]\left[\frac{(50)(0.785)}{1.3}\right] = 2320$$

Notice the overall gain of the amplifier is the product of the gains for the individual stages. But we must include the loading effects of the input of the second stage upon the output of the first stage. From Eq. (7.23), the voltage gain for the first stage may be written as

$$\frac{v_{o1}}{v_s} = \frac{-(50)(0.785)}{1.3} = -30.2$$

and for the second stage, from Eq. (7.27),

$$\frac{v_{o2}}{v_{o1}} = \frac{-(50)(2)}{1.3} = -77$$

therefore the voltage gain v_{o2} from v_s is

$$a_v = \left(\frac{v_{o2}}{v_{o1}}\right)\left(\frac{v_{o1}}{v_s}\right) = (-77)(-30.2) = 2320$$

The two-stage cascaded amplifier has a voltage gain of over 2000. Therefore with a 1-mV-rms sinewave signal at the input of the amplifier we would calculate that we have a 2.3-V-rms signal at the output of the amplifier. This corresponds to a 6.5-V peak-to-peak voltage across R_4. However, since our supply voltage is only 6 V, we would get "clipping" on the sine wave. Assuming zero collector saturation voltage for Q_2, we would obtain a maximum undistorted output signal of 6-V peak to peak (2.12 V rms) with an input of 0.91 mV.

Notice, however, that with this maximum unclipped output signal the power delivered at the output of the amplifier is given by

$$\frac{(v_{o2})^2}{R_4} = \frac{(2.12)^2}{2 \text{ k}\Omega} = 2.25 \text{ mW}$$

With a pair of high-resistance earphones connected across the load resistor R_4, this amplifier could serve as the audio amplifier for the phonograph cited in the introduction to this chapter. But generally we would want the output of an audio amplifier to drive a loudspeaker. The typical resistance of a loudspeaker is about 8 Ω. With such a speaker capacitively coupled to the output of our amplifier and a 1-mV input voltage, the voltage across the speaker is 9.2 mV rms, and the power into the loudspeaker is

$$\frac{[(9.2)(10^{-3})]^2}{8} = 10.6 \text{ } \mu\text{W}$$

This power level is far below any listening level. If we could get the 2.3 V across the 8-Ω speaker, we would have a power level of 0.66 W, which is pleasantly audible. Thus we require an electronic circuit which will allow us to drive a low-resistance (8-Ω) load from a high-resistance (2-kΩ) source without any loss of signal voltage. How we do this is the subject of the next section.

EXERCISE

E7.3 Determine the small-signal voltage gain of an amplifier consisting of two cascaded stages of the circuit described in Exercise E7.2. Determine the value of an appropriate level-shifting "battery," as shown in Fig. 7.8. Also determine the power delivered to an 8-Ω load with a 0.3-mV-rms signal at the input.

7.4 BASIC EMITTER FOLLOWER

A transistor circuit which helps provide a match between a high resistance source and a lower resistance load with almost no loss of signal voltage is the *emitter follower*. The basic circuit is shown in Fig. 7.10. The circuit is also referred to as a common-collector (CC) amplifier. The input signal is applied between the base and ground, and the output signal is taken between the emitter and ground. For the small-signal variations we can usually consider V_{CC} as having zero resistance to ground. Then with the transistor collector connected directly to V_{CC}, the collector is therefore the common terminal for both the input and output signals. The name *emitter follower* is used because the output signal voltage is taken from the emitter, which follows or tracks the voltage at the base to which the input signal has been applied.

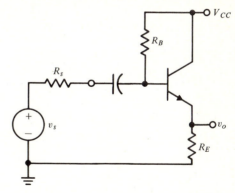

FIGURE 7.10
Basic emitter-follower circuit.

The following circuit and bias data apply to the circuit shown in Fig. 7.10.

$$V_{CC} = 6 \text{ V} \qquad V_{BE(on)} = 0.7 \text{ V}$$

$$V_{CE} = 2.7 \text{ V} \qquad \beta_F = 50$$

$$I_C = 1 \text{ mA} \qquad R_s = 2 \text{ k}\Omega$$

7.4.1   DC Conditions

For the basic circuit shown, the bias calculations reduce to simply finding values for R_B and R_E. To determine a value for R_E, by KVL we have

$$I_E R_E + V_{CE} = V_{CC} \qquad (7.30a)$$

but

$$I_E = \frac{(\beta_F + 1)I_C}{\beta_F}$$

therefore

$$R_E = (V_{CC} - V_{CE}) \frac{\beta_F}{(\beta_F + 1)I_C} \qquad (7.30b)$$

$$= \frac{(6 - 2.7)(50)}{(51)(1)} = 3.24 \text{ k}\Omega$$

Again, we can use KVL to compute a value for R_B. Thus

$$I_E R_E + V_{BE(on)} + I_B R_B = V_{CC} \qquad (7.31a)$$

but

$$I_B = \frac{I_C}{\beta_F}$$

and

$$I_E = \frac{(\beta_F + 1)I_C}{\beta_F}$$

therefore

$$R_B = \left[V_{CC} - V_{BE(on)} - \frac{(\beta_F + 1)(I_C)(R_E)}{\beta_F} \right] \frac{\beta_F}{I_C} \qquad (7.31b)$$

$$= \left[6 - 0.7 - \frac{51}{50}(1)(3.24) \right] \frac{50}{1} = 100 \text{ k}\Omega$$

7.4.2   Small-signal Characteristics

In either using or designing an emitter follower, the usual small-signal parameters are of interest, namely, the input and output resistance r_i and r_o and the current and voltage gain a_i and a_v. In order to derive these quantities we again assume that at the frequencies of interest the capacitors and voltage sources have zero resistance. We can then draw the small-signal circuit model shown in Fig. 7.11a. The circuit has been redrawn in Fig. 7.11b to show clearly that the dependent current generator $\beta_0 i_b$ appears in parallel with R_E. We have also reduced the input circuit to the Thévenin equivalent, where

$$v_{sT} = \frac{R_B}{R_s + R_B} v_s \qquad (7.32a)$$

$$R_{sT} = \frac{(R_s)(R_B)}{R_s + R_B} \qquad (7.32b)$$

Generally we will find (as is the case here) that $R_B \gg R_s$, then

$$v_{sT} \approx v_s \qquad (7.33a)$$

and

$$R_{sT} \approx R_s \qquad (7.32b)$$

Current gain (a_i) From Fig. 7.11b, the current gain is given by the ratio of the output current i_e and the input current i_b. Then

$$a_i = \frac{i_e}{i_b} \qquad (7.34a)$$

Substituting $i_b (\beta_0 + 1)$ for i_e, we have

$$a_i = \frac{i_b(\beta_0 + 1)}{i_b} = \beta_0 + 1 \qquad (7.34b)$$

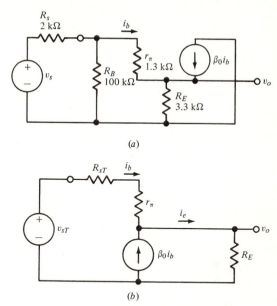

FIGURE 7.11
(a) Small-signal model of the emitter-follower circuit; (b) a simplified version.

The emitter follower therefore exhibits a current gain equal to the current gain of the transistor (β_0) plus 1 (i.e., the ratio of i_e to i_b).

Voltage gain (a_v) We also use Fig. 7.11b to obtain the voltage gain of the stage. From the figure we take the ratio of v_o and v_{sT} and by KVL show that

$$\frac{v_o}{v_{sT}} = \frac{i_e R_E}{i_b(R_{sT} + r_\pi) + i_e R_E} \qquad (7.35a)$$

but again substituting $i_b(\beta_0 + 1)$ for i_e and noting the approximation for v_{sT} and R_{sT} in Eq. (7.33), we have

$$a_v = \frac{v_o}{v_s} = \frac{(\beta_0 + 1)R_E}{R_s + r_\pi + (\beta_0 + 1)R_E} \qquad (7.35b)$$

Hence the voltage gain is always less than 1, but since usually $(\beta_0 + 1)R_E \gg R_s + r_\pi$, there is little attenuation of the voltage signal from input to output. Neither is there any polarity inversion from input to output. The output voltage is almost an exact replica of the input signal voltage.

Input resistance (r_i) To obtain the input resistance let us use Fig. 7.11a but set R_s equal to $0\,\Omega$, so that v_s now appears at the input terminal of the transistor. Consequently, by KVL,

$$v_s = i_b[r_\pi + (\beta_0 + 1)R_E] \qquad (7.36)$$

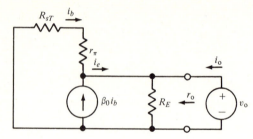

FIGURE 7.12
Modified circuit model of the emitter follower to determine the output resistance.

and the input resistance to the transistor is

$$\frac{v_s}{i_b} = r_\pi + (\beta_0 + 1)R_E \qquad (7.37)$$

This expression is the same as for the input resistance of a CE stage with any resistance between the emitter and ground increased by a factor of $\beta_0 + 1$.

The small-signal input resistance of the complete circuit in Fig. 7.10, neglecting the source resistance R_s, is the parallel combination of R_B and the input resistance to the transistor. That is,

$$r_i = \frac{R_B[r_\pi + (\beta_0 + 1)R_E]}{R_B + r_\pi + (\beta_0 + 1)R_E} \qquad (7.38)$$

Output resistance (r_o) To determine the output resistance r_o, we modify our circuit model to that shown in Fig. 7.12. The input signal is set to zero, and then we effectively apply a small-signal voltage source across the output terminals of the circuit and measure the current drawn from this voltage source. Then

$$r_o = \frac{v_o}{i_o} \qquad (7.39)$$

The derivation is simplified if we notice that R_E also appears across the output terminals. We will then initially remove R_E from the circuit and solve for r_o. That is, r_o is the small-signal output resistance in the absence of R_E. Or, R_E appears as a load resistance across the small-signal output resistance of the circuit.

Neglecting, for now, R_E in Fig. 7.12, we have by Kirchhoff's current law (KCL) that

$$i_o = -i_e = -(\beta_0 + 1)i_b \qquad (7.40a)$$

Again neglecting R_E, we see that the voltage v_o causes a current $-i_b$ to flow through the series combination of r_π and R_{sT}. That is, by KVL,

$$v_o = -i_b(R_{sT} + r_\pi) \qquad (7.40b)$$

then

$$r_o = \frac{v_o}{i_o} = \frac{(R_{sT} + r_\pi)}{\beta_0 + 1} \qquad (7.40c)$$

At the output of the emitter follower any source resistance (seen between the base and ground), as well as r_π, appears reduced in value by the factor $\beta_0 + 1$. Values obtained for a_v, r_i, and r_o by these analytical methods can be compared with experimental results in the practical demonstration, D7.2.2.

EXAMPLE 7.3   For a numerical example we will calculate the small-signal parameters r_i, r_o, a_i, and a_v for the emitter-follower circuit shown in Fig. 7.10.

SOLUTION   The small-signal circuit model is shown in Fig. 7.11a. Then, by using Eq. (7.32), the Thévenin equivalent for the input circuit,

$$v_{sT} = \frac{100 v_s}{2 + 100} = 0.98 v_s \quad \text{and} \quad R_{sT} = \frac{(2)(100)}{2 + 100} = 1.96 \text{ k}\Omega$$

We see that R_B is so much greater than R_s that we may assume, as in Eq. (7.33), that $v_{sT} = v_s$ and $R_{sT} = R_s$. Calculating for the small-signal parameter r_π,

$$r_\pi = \frac{(50)(26)}{1} = 1.3 \text{ k}\Omega$$

Now the input resistance r_i is given by Eq. (7.38) as

$$r_i = \frac{100[1.3 + (51)(3.3)]}{100 + 1.3 + (51)(3.3)} = 63 \text{ k}\Omega$$

To obtain the input resistance seen by v_s in Fig. 7.10, we would have to include R_s in series with r_i. We can obtain the output resistance from Eq. (7.40c), where we have included the effect of R_s. Then

$$r_o = \frac{2 + 1.3}{51} = 0.065 \text{ k}\Omega = 65 \ \Omega$$

With the assumption of $R_B \gg R_s$, the current gain is given by Eq. (7.34b) as

$$a_i = \beta_0 + 1 = 51$$

Using the same assumption, we can obtain the voltage gain directly from Eq. (7.35b),

$$a_v = \frac{(51)(3.3)}{2 + 1.3 + (51)(3.3)} = 0.98 \qquad \text{////}$$

With R_E ($= 3.3 \text{ k}\Omega$) appearing as a load across the output, the voltage gain of the emitter follower is approximately 1, and the source resistance ($R_s = 2 \text{ k}\Omega$) has been effectively reduced to 65 Ω. While such a value for the output resistance is quite tolerable in many applications, there are times when an even lower output resistance is required. With an 8-Ω loudspeaker capacitively coupled to the output

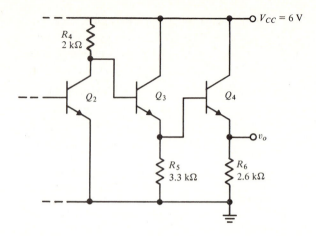

FIGURE 7.13
Cascade of two emitter-follower circuits directly coupled to the collector transistor Q_2.

of our emitter-follower circuit, the signal developed across the voice coil of the speaker will only be a fraction of the input signal to the emitter follower. In particular, we have an 8-Ω load being driven from a voltage source of 65-Ω resistance. The fraction of the source signal appearing across the load is therefore

$$\frac{8}{8 + 65} = 11 \text{ percent}$$

which is still too low. In the next section we will improve on this.

EXERCISE

E7.4 An emitter-follower circuit is similar to that shown in Fig. 7.10, but $V_{CC} = 15$ V, $R_s = 3$ kΩ, $R_B = 22$ kΩ, and $R_E = 1$ kΩ. Compute the small-signal input resistance seen by v_s, the output resistance r_o, and the voltage gain v_o/v_s. $\beta_F = 50$; $V_{BE(on)} = 0.7$ V.

7.5 CASCADE OF CC STAGES

To obtain a lower output resistance than that attainable with one emitter-follower stage, a general solution is to cascade two emitter-follower circuits. Such a circuit directly connected to the collector of Q_2 in our phonograph amplifier is shown in Fig. 7.13. The transistor connection that results in a cascade of two emitter followers is popularly referred to as a *Darlington connection*.

Before we determine the dc operating conditions, we take note that where β_F of the transistors is a reasonably large value (that is, ≥ 50) it is often acceptable to ignore the effect of the base current on the bias calculations. We then use the approximation that $\beta_F \to \infty$, then

$$I_B = 0 \quad \text{and} \quad I_C = I_E$$

We may then assume in Fig. 7.13 that the only current in R_4 is due to the collector current of Q_2. Then from Sec. 7.3 the voltage to ground at the collector of Q_2 is 4 V. With $V_{BE(on)} = 0.7$ V, the voltage at the emitter of Q_3 is 3.3 V. If we now assume that the collector current and emitter current of Q_3 are equal, we have

$$I_{C3} = \frac{V_{E3}}{R_5} \tag{7.41a}$$

$$= \frac{3.3}{3.3} = 1 \text{ mA}$$

Following this approximate but practical method for determining the operating point, we calculate the collector current of Q_4 as

$$I_{C4} = \frac{V_{E4}}{R_6} = \frac{V_{E3} - V_{BE(on)}}{R_6} \tag{7.41b}$$

$$= \frac{V_{E4}}{R_6} = \frac{3.3 - 0.7}{2.6} = 1 \text{ mA}$$

Our assumption of infinite β_F in this example has introduced an acceptable error of less than 2 percent in our values for I_{C3} and I_{C4} (see Prob. P7.22). Following the method outlined in the previous section, we may use Eq. (7.40c) to calculate the small-signal output resistance at the emitter of Q_3 and then at the emitter of Q_4. That is,

$$r_{o3} = \frac{R_4 + r_{\pi3}}{\beta_0 + 1}$$

$$= \frac{2 + 1.3}{51} = 0.065 \text{ k}\Omega = 65 \text{ }\Omega$$

Now R_5 appears in parallel with r_{o3}, but the effect of R_5 is negligible (~ 2 percent), so we may calculate r_{o4} as

$$r_{o4} = \frac{r_{o3} + r_{\pi4}}{\beta_0 + 1}$$

$$= \frac{0.065 + 1.3}{51} = 0.027 \text{ k}\Omega = 27 \text{ }\Omega$$

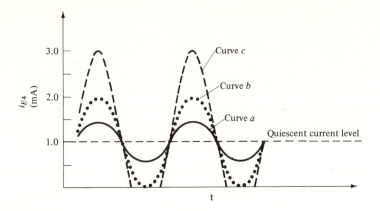

FIGURE 7.14
A plot showing the effects of signal overdrive in a simple emitter-follower circuit.

With an output resistance of 27 Ω, the fraction of the signal appearing across an 8-Ω load, compared to that of an open-circuit load, is still only

$$\frac{8}{8 + 27} = 23 \text{ percent}$$

However, notice the output resistance r_{o4} is the sum of two components. The output resistance of Q_3 contributes $r_{o3}/(\beta_0 + 1)$, or about 1.3 Ω. From transistor Q_4 we have $r_{\pi4}/(\beta_0 + 1)$, or nearly 26 Ω. Obviously to reduce the output resistance we have to decrease $r_{\pi4}$. That is, we would increase the collector current of Q_4, provided there was no decrease in β_0. This could be done by reducing the value of R_6, or better still, by returning R_6 to a negative voltage supply line rather than to ground.

We may also have to increase the collector current of transistor Q_4 for another reason. Notice that the maximum undistorted power output from an emitter follower may be limited by the quiescent current in the transistor. With a 1-mA quiescent current in Q_4 and an output excursion of 0.5 mA there is no problem, as shown by curve a in Fig. 7.14. However, with 1-mA output current we note that Q_4 is just at the point of cutoff on the negative excursion of the output signal. With increased output currents there is distortion ("clipping") of the output signal, as shown by curve c. The problem of obtaining reasonable amounts of undistorted power from a simple emitter follower is illustrated in the output stage of the audio amplifier described in Demonstration D7.3.

EXERCISE

E7.5 The amplifier described in Exercise E7.3 is now followed by a cascaded emitter-follower circuit similar to that shown in Fig. 7.13. $V_{CC} = 15$ V; $R_4 = 3$ kΩ; $R_5 = 1$ kΩ; $R_6 = 1$ kΩ. With $\beta_0 = 50$, $V_{BE(on)} = 0.7$ V and $I_{C2} = 2$ mA, determine the small-signal input resistance looking into the base of Q_3 and the output resistance looking into the emitter of Q_4. Also determine the maximum undistorted power available into a load resistance of 8 Ω that is capacitively coupled to the output.

7.6   BASIC COMMON-BASE AMPLIFIER

The third and final basic transistor circuit configuration is the *common-base* (CB) *amplifier*. A typical circuit is shown in Fig. 7.15. The input signal is developed between the emitter and the base, and the output signal is taken from between the collector and the base. Hence the term common-base configuration. The CB amplifier is a voltage amplifier only; as we shall see later, the current gain of this configuration is always less than 1. The CB amplifier finds its most useful application where it is desired to obtain a large voltage gain, without polarity inversion, from a low resistance source.

The following circuit and bias data apply to the amplifier circuit shown in Fig. 7.14.

$$V_{CC} = 6 \text{ V} \qquad V_{BE(\text{on})} = 0.7 \text{ V}$$

$$V_{EE} = -6 \text{ V} \qquad \beta_F = 50$$

$$I_C = 1 \text{ mA} \qquad R_s = 100 \ \Omega$$

$$V_{CB} = 4 \text{ V}$$

7.6.1   DC Conditions

A value for the collector resistor R_C is found by using KVL in the collector circuit, as

$$V_{CC} = I_C R_C + V_{CB} \qquad (7.42a)$$

then

$$R_C = \frac{V_{CC} - V_{CB}}{I_C} \qquad (7.42b)$$

$$= \frac{6 - 4}{1} = 2 \text{ k}\Omega$$

We may also use KVL in the emitter circuit to determine a value for R_E. That is,

$$V_{BE(\text{on})} + I_E R_E = -V_{EE} \qquad (7.43a)$$

Now recall that the transistor collector current is related to the emitter current by the term α_F. That is,

$$I_C = \alpha_F I_E$$

also

$$\alpha_F = \frac{\beta_F}{\beta_F + 1}$$

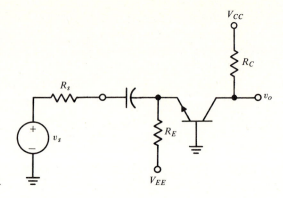

FIGURE 7.15
Basic common-base amplifier.

Hence from Eq. $(7.43a)$ we have

$$R_E = \left(\frac{-V_{EE} - V_{BE(on)}}{I_C}\right)\left(\frac{\beta_F}{\beta_F + 1}\right) \quad (7.43b)$$

$$= \left(\frac{6 - 0.7}{1}\right)\left(\frac{50}{51}\right) = 5.2 \text{ k}\Omega$$

7.6.2   Small-signal Characteristics

In drawing the small-signal circuit model, shown in Fig. 7.16, we have again simplified the input circuit through the use of Thévenin's theorem. Then

$$v_{sT} = \frac{R_E v_s}{R_s + R_E} \quad (7.44a)$$

$$R_{sT} = \frac{R_s R_E}{R_s + R_E} \quad (7.44b)$$

However, with R_E usually much greater than R_s, we can again make the approximation

$$v_{sT} = v_s \quad (7.45a)$$

$$R_{sT} = R_s \quad (7.45b)$$

Input resistance (r_i)   The input resistance at the transistor emitter is given as

$$r_i = \frac{v_i}{i_i}$$

where

$$v_i = -i_b r_\pi$$

and

$$i_i = -(\beta_0 + 1)i_b$$

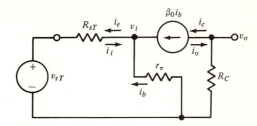

FIGURE 7.16
Small-signal circuit model of the CB amplifier.

therefore

$$r_i = \frac{r_\pi}{\beta_0 + 1} \qquad (7.46a)$$

This relationship could be seen from our results of Eq. (7.40c), since the input resistance of a CB stage is the output resistance of a CC (emitter-follower) stage.

With reference to Fig. 7.15, we would generally be interested in knowing the input resistance at the input terminal of the amplifier. Hence we should include the shunting effect of R_E in our calculations of the input resistance. However the input resistance to a CB transistor is generally so very small that usually $R_E \gg r_i$, then we may make the approximation

$$r_i = \frac{r_\pi}{\beta_0 + 1}$$

Notice, with one further simple approximation, namely, $\beta_0 \gg 1$, we have the following useful relationship for the CB amplifier.

$$r_i = \frac{r_\pi}{\beta_0} = \frac{V_T}{I_C} \qquad (7.46b)$$

This is the reciprocal of the g_m equation, Eq. (7.10b).

Output resistance (r_o) With only two elements connected at the output node of Fig. 7.16, the collector resistor R_C and the ideal current generator $\beta_0 i_b$, the output resistance is simply

$$r_o = R_C \qquad (7.47)$$

Current gain (a_i) The ratio of the output current and the input current is the current gain of the circuit. But, from Fig. 7.16,

$$i_o = -i_c$$

and

$$i_i = -i_e$$

Hence substituting for i_c and i_e their equivalent terms involving i_b, we have

$$a_i = \frac{-\beta_0 i_b}{-(\beta_0 + 1)i_b} = \frac{\beta_0}{\beta_0 + 1} = \alpha_0 \qquad (7.48)$$

Again referring to Fig. 7.15, in determining the current gain for this circuit we should include the shunting effect of R_E. However, with $r_i \ll R_E$, this effect can be ignored, and for the complete circuit,

$$a_i = \alpha_0$$

By definition, α_0, and hence the current gain of the CB amplifier, is always less than 1.

Voltage gain (a_v)   The voltage gain of the circuit shown in Fig. 7.16 is

$$a_v = \frac{v_o}{v_{sT}} = \frac{i_o R_C}{i_i(R_{sT} + r_i)} \qquad (7.49a)$$

but from Eq. (7.48),

$$\frac{i_o}{i_i} = \alpha_0$$

then

$$a_v = \frac{\alpha_0 R_C}{R_{sT} + r_i} \qquad (7.49b)$$

The voltage gain of the CB amplifier can be quite large, since R_s, hence R_{sT}, and r_i are usually very small. Notice also that there is no inversion of the polarity of the signal voltage from input to output.

EXAMPLE 7.4   As a numerical example we will compute the small-signal characteristics for the CB amplifier shown in Fig. 7.15. $R_s = 100\ \Omega$; $R_E = 5.2\ k\Omega$; $R_C = 2\ k\Omega$.

SOLUTION   We can use the small-signal circuit model of Fig. 7.16. Then from Eq. (7.44),

$$v_{sT} = \frac{5.2v_s}{0.1 + 5.2} = 0.98v_s \qquad \text{and} \qquad R_{sT} = \frac{(0.1)(5.2)}{0.1 + 5.2} = 98\ \Omega$$

With $I_C = 1\ mA$ and $\beta_0 = 50$, r_π is equal to 1.3 kΩ. The input resistance of the transistor is, by Eq. (7.46a),

$$r_i = \frac{1.3}{51} = 26\ \Omega$$

The shunting effect of R_E, which is 5.2 kΩ, is therefore insignificant, and we have the small-signal input resistance of the amplifier as

$$r_i = 26 \ \Omega$$

The output resistance of the amplifier is due to R_C only. That is,

$$r_o = 2 \ \text{k}\Omega$$

The current gain of the amplifier is, from Eq. (7.48), equal to the α_o of the transistor. Or

$$a_i = \alpha_0 = \frac{50}{51} = 0.98$$

The voltage gain for the equivalent circuit of Fig. 7.16 is obtained with the use of Eq. (7.49*b*):

$$a_v = \frac{v_o}{v_{sT}} = \frac{(0.98)(2)}{0.098 + 0.026} = 158$$

but

$$v_{sT} = 0.98 v_s$$

therefore

$$a_v = \frac{v_o}{v_s} = 155 \qquad \qquad ////$$

EXERCISE

E7.6 A CB amplifier circuit is similar to that shown in Fig. 7.15, but $V_{CC} = 15$ V, $V_{EE} = -15$ V, $R_s = 50 \ \Omega$, $R_E = 2.7$ kΩ, and $R_C = 1.2$ kΩ. Compute the small-signal input resistance seen by v_s, the output resistance r_o, and the voltage gain v_o/v_s. $\beta_F = 50$; $V_{BE(\text{on})} = 0.7$ V.

7.7 SUMMARY OF SMALL-SIGNAL PARAMETERS

A summary of the small-signal characteristics of the transistor, for each of the basic configurations, has been listed in Table 7.1. These are approximate but practical expressions for the *transistor only*. In a circuit application the effect of loading at the input and output as well as the effect of the signal-source resistance must be included. Particularly note that the signal-source resistance has *not* been included in Table 7.1.

In Table 7.2 values have been obtained for these expressions, considering a typical simple transistor circuit operating at room temperature (27°C). Again note that these numbers do *not* include the effect of the base-biasing circuit or signal-source resistance. For each of the configurations the common terminal is considered to be at ground.

From these tables we see that:

1 The input resistance is different for all three circuits, but the CC has the highest input resistance. For this circuit, the effect of R_E is multiplied by $\beta_0 + 1$. Looking directly into the emitter of the transistor, the input resistance of the CB has a very low value, being approximately equivalent to $1/g_m$ $(= V_T/I_C)$.

2 The output resistance is the same for the CE and CB circuits. This is logical since in both cases, with an ideal transistor, the output resistance is solely due to the collector-load resistor. For the CC, the output is taken from the emitter where (with $R_s = 0$) the output resistance is very low, again approximately equivalent to V_T/I_C.

3 The current gain is about the same for CE and CC circuits. This is the transistor small-signal CE current-gain parameter β_0. The small-signal CB current gain is α_o $[= \beta_0/(\beta_0 + 1)]$, a little less than 1.

4 The voltage gain is high and almost the same for the CE and CB, but there is no polarity inversion for the CB circuit. The voltage gain of the CC (with $R_s = 0$) is approximately equal to 1.

These generalizations can be made another way, by considering the application of the circuit. Then:

1 Where a large input resistance is required to prevent loading of the signal source, we would use a CC stage.

2 Where a low output resistance is needed to drive a low resistance load, a CC stage is again indicated.

3 Either a CE or CC stage would be used if a large current gain is required, dependent upon whether a polarity inversion is also needed.

4 A CE stage would be used for a large voltage-gain requirement if the polarity inversion can be tolerated. Otherwise a CB stage would be used.

Table 7.1 SUMMARY OF SMALL-SIGNAL PARAMETERS FOR TRANSISTOR ONLY

	Common emitter CE	Common collector CC	Common base CB
r_i	r_π	$r_\pi + (\beta_0 + 1)R_E$	$r_\pi/(\beta_0 + 1)$
r_o	R_C	$r_\pi/(\beta_0 + 1)$	R_C
a_i	β_0	$\beta_0 + 1$	α_0
a_v	$-\dfrac{\beta_0 R_C}{r_\pi}$	$\dfrac{(\beta_0 + 1)R_E}{r_\pi + (\beta_0 + 1)R_E}$	$\dfrac{\beta_0 R_C}{r_\pi}$

$$r_\pi = \frac{\beta_0 V_T}{I_C}$$

Notice from Table 7.1 that the CB stage yields a voltage gain only, and the CC only a current gain. The CE stage provides both voltage and current gain.

In this chapter we have dealt almost exclusively with the transistor as a discrete component. This is necessary to present a clear picture of the small-signal characteristics of the three basic transistor configurations. In the next chapter we will consider transistor connections which only have meaning when treated as a composite or as an IC. That is, the devices must be physically located very close to each other and have almost exactly the same characteristics. However, in analyzing or describing these circuits we will find it advantagous to relate them to the basic amplifier circuits we have dealt with in this chapter.

EXERCISE

E7.7 Compile a table similar to Table 7.2 for a transistor circuit having the following characteristics:

$$I_C = 2 \text{ mA} \qquad \beta_0 = 80$$
$$R_C = 5 \text{ k}\Omega \qquad R_E = 1 \text{ k}\Omega$$

DEMONSTRATIONS

D7.1 Using the circuit in Fig. D7.1, we will measure both the small-signal current gain (β_0) and the dc current gain (β_F) of transistor Q_1. The transistor type chosen should be of the kind that typically has a dc gain of 60 at a 1-mA collector current, such as the 2N4275 described in Chap. 1. With the resistance values shown in Fig. D7.1, the transistor would typically be biased with $I_C = 1$ mA and $V_{CE} = 3$ V.

If we measure (with a high-resistance voltmeter) the dc voltages V_{CC}, V_B, and

Table 7.2 TYPICAL SMALL-SIGNAL PARAMETER VALUES

	CE	CC	CB
r_i	260 Ω	61.4 kΩ	5 Ω
r_o	1.2 kΩ	5 Ω	1.2 kΩ
a_i	50	51	0.98
a_v	-230	$\simeq 1$	230

$$r_\pi = 260 \ \Omega$$

$$\begin{array}{ll} I_C = 5 \text{ mA} & \beta_0 = 50 \\ R_C = 1.2 \text{ k}\Omega & R_E = 1.2 \text{ k}\Omega \end{array}$$

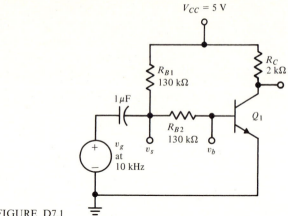

FIGURE D7.1

V_C,† we can calculate the base and collector currents (I_B and I_C). The actual β_F at the actual operating point will be the ratio of these currents.

With a sinusoidal input v_g of 10 kHz we can use an oscilloscope (or a sensitive ac voltmeter) to measure the small-signal output voltage (v_o) at the collector. The output level of the signal generator (v_g) should be adjusted so that the signal at the collector is no greater than 0.2 V peak-to-peak. The input ac signal v_s is measured by using the same metering method as in measuring v_o. We can now calculate the signal currents (i_b, i_c) since $i_b \approx v_s/R_{B2}$ and $i_c \approx v_o/R_C$. Now from the small-signal model for the transistor shown in Fig. 7.7 we can compute the value of β_0. Note that this method of computing β_0 is valid only when R_{B2} is much greater than the input resistance of the transistor. If you have a very sensitive ac voltmeter (or scope), you can measure the ac voltage at the base of the transistor and determine the transistor input resistance r_i.

Heat the transistor by holding a soldering iron against the ceramic portion of the body (of a 2N4275) and monitor the dc collector voltage V_C. What is causing the gradual change noted?

D7.2 In Fig. D7.2 we show an amplifier circuit that employs a biasing scheme commonly used when discrete components make up the amplifier. In this manner the transistor operating point is more precisely established, so that even with transistor parameter and temperature variations the collector current I_C will be very close to 1 mA and V_{CE} will be very close to 2.25 V.

The intent of this demonstration is to determine experimentally values for the three small-signal parameters a_v, r_i, and r_o for both the common-emitter (CE) and common-collector (CC) transistor configurations. After obtaining these values experimentally, they should be compared to those calculated using the

† If V_C is less than 0.5 V, change R_C to 1 kΩ.

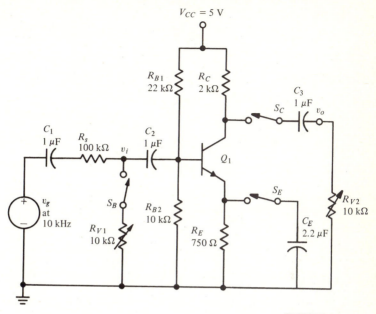

value of β_0 determined in Demonstration D7.1. Why would you not expect an exact match between calculated and experimental results?

D7.2.1 For the CE configuration:

(a) With switches S_B, S_C, and S_E open:

(1) Measure (using an ac voltmeter or oscilloscope) the amplitude of the signal voltage at the collector and at the base. Adjust v_g so that the output signal (v_o) is less than 2 V peak-to-peak (0.7 V rms). Then

$$a_v = \frac{v_o}{v_i}$$

(2) Maintaining v_g as in part (1), monitor the voltage v_i. After closing switch S_B, adjust the variable resistance R_{V1} until the amplitude of v_i is reduced to one-half its original value. Why is the resulting value of R_{V1} equal to the input resistance r_i of the amplifier? Without removing it from the circuit, R_{V1} can be measured with a dc ohmmeter.

(3) Measure the amplitude of the output signal v_o. Then, close S_C to connect the variable resistor R_{V2} as a load. Now determine the output resistance (r_o) of the amplifier by the same procedure used for finding the input resistance in part (2) above.

(b) Repeat these measurements with C_E bypassing R_E (remember to readjust v_g!). Compare these results also to those determined analytically using the β_0 determined in Demonstration D7.1.

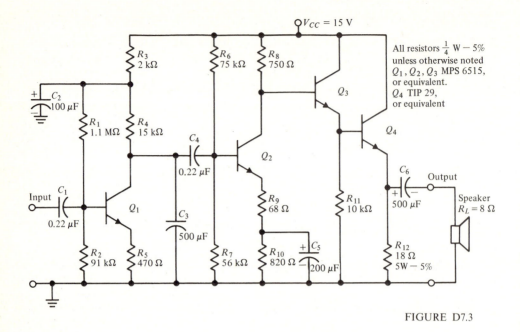

FIGURE D7.3

D7.2.2 For the CC configuration, disconnect C_E and put a short across R_C. Repeat the measurements in Demonstration D7.2.1 for a_v, r_i, and r_o, where the output voltage is now taken from the emitter terminal. Attach C_3 and R_{V2} to the emitter to determine r_o.

D7.3 The final demonstration (lab project) is the audio amplifier shown in Fig. D7.3. This circuit is derived from the principles described in the text with the following differences:

   1 The transistors used for Q_1, Q_2, and Q_3 should be a high-gain type ($\beta_0 > 250$ at 2 mA).
   2 The biasing arrangement for Q_1 and Q_2 uses the method in Fig. D7.2. The emitter resistors are partially bypassed (C_5 for Q_2) or unbypassed (Q_1) so that the small-signal input resistances are increased, permitting the use of small *coupling* capacitors (C_1 and C_4).
   3 The circuit $R_3 C_2$ prevents "motorboating," which is signal feedback from the output stage through the finite power-supply source resistance and thus into the input stages. The design value of the overall voltage gain is 100. The first-stage gain (v_{c_1}/v_{in}) is 10. Measure these values with a 5-mV input signal. With an oscilloscope monitor v_{out} while increasing v_{in} until the onset of clipping. From this value determine the rms power rating of this amplifier working into an 8-Ω load. The power efficiency of this type of output stage is very poor. For a more efficient form of output circuit, see Prob. P7.34.

PROBLEMS

All problems are to be solved assuming a 27°C temperature.

P7.1   Calculate g_m and r_π for the four different transistors at the given operating current:

(a) $\beta_0 = 50$, $I_C = 0.5$ mA

(b) $\beta_0 = 30$, $I_C = 500$ mA

(c) $\beta_0 = 200$, $I_C = 100$ μA

(d) $\beta_0 = 1000$, $I_C = 10$ μA

P7.2   Derive Eq. (7.10b) for $V_{BC} \ll 5V_T$ and $V_{BE} \gg 5V_T$ by making use of Eq. (1.11d).

P7.3   From the active-region expressions $\beta_0 = dI_C/dI_B$ and $\beta_F = I_C/I_B$, prove that

$$\beta_0 = \frac{\beta_F}{1 - \left(\dfrac{I_C}{\beta_F}\right)\left(\dfrac{d\beta_F}{dI_C}\right)}$$

P7.4   Given that $V_{CC} = 15$ V in Fig. 7.1a, find the values of R_B and R_C that will establish the operating currents given in Prob. P7.1. Assume that $\beta_F = \beta_0$ and set $V_{CE} = 7.5$ V.

P7.5   (a) For the four transistor and associated resistor values calculated in Prob. P7.4, determine r_i, r_o, a_v, and a_i.

(b) Show that an alternate expression for the gain expression given in Eq. (7.20a) is

$$a_v = -g_m R_C$$

(c) Obtain a general expression for a_v for the circuit in Fig. 7.1a. Use the results of part (b) to get the expression in terms of the variables V_{CC}, V_{CE}, and V_T.

(d) Use the results of part (c) to find the upper limit to the maximum gain that could be obtained for the amplifier in Fig. 7.1a. Assume it is to be used with extremely small-signal levels at the input so that the output signal is only a few millivolts.

P7.6   The circuit in Fig. P7.6 is similar to that in Fig. 7.1a except that we have added a source resistance R_s at the input and also an external load resistor R_{XL}. Capacitor

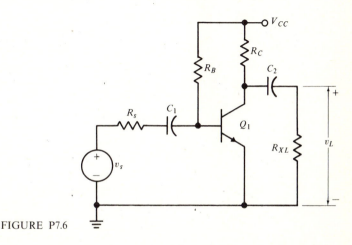

FIGURE P7.6

C_2 blocks dc but is an ac short circuit for our purposes. Determine the ac load R_L at the collector of the transistor, and use this value to determine the voltage gain v_L/v_s. Use the source and external load values listed below for the four different circuits found in Prob. P7.5a.

	R_s	R_{XL}
(a)	5 kΩ	10 kΩ
(b)	3 Ω	8 Ω
(c)	10 kΩ	50 kΩ
(d)	500 kΩ	500 kΩ

P7.7   As shown in Prob. P7.5d, the amplifier in Fig. 7.1a will have a voltage gain independent of β_F if the operating current does not change with β_F. Moreover the available range that the output voltage can swing will be kept constant if the operating point can be made stable against variations in β_F. Various methods for stabilizing the operating point against variations of β_F have been used with discrete transistor circuits. The *constant base bias* circuit of Fig. 7.1a is poor in this respect since a change of β, $\Delta\beta$, will show up as a change in collector current $\Delta I_C = \Delta\beta \cdot I_B$. With I_B fixed, the normal variations in β_F (due to differences in β from unit to unit plus temperature effects) can cause enough change in I_C to reduce the range of output voltage swing, as well as reduce the gain so that the amplifier performance is inadequate for its intended task. One way to reduce the effect of bias sensitivity to β_F is to use the *emitter-biasing circuit* shown in Fig. P7.7a.

By Thévenin's theorem this circuit can be converted into the one shown in Fig. P7.7b. R_B must be kept large enough so that the ac input resistance of the amplifier is not reduced to too low a value for the source resistor used.

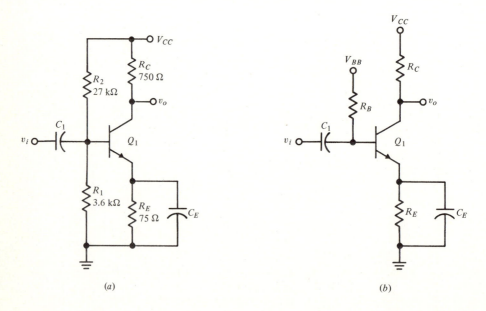

(a)   (b)

FIGURE P7.7

(a) Show that for the circuit in Fig. P7.7b the expression for the emitter current is

$$I_E = \frac{V_{BB} - V_{BE(on)}}{R_E + \dfrac{R_B}{\beta_F + 1}}$$

(b) For the numeric values in Fig. P7.7a, solve for I_E and V_{CE} when $\beta_F = 100$ and $V_{BE(on)} = 0.7$ V.

(c) With the bypass capacitor C_E large enough so that it is a short circuit for our purposes, use the conditions determined in part (b) to solve for (1) r_i; (2) r_o; (3) a_v.

(d) If the β_F of the transistor in part (b) can range from 60 to 190, what are the minimum and maximum values that I_E can have? What percentage changes does this give over that found in part (b)?

(e) If resistors R_1 and R_2 can independently vary ± 10 percent with the other conditions as in part (b), what are the extreme values that I_E can take? What percentage changes are these relative to the value in part (b)?

P7.8   The circuit in Fig. P7.8 is another biasing method used with transistor circuits. This is called the *collector-to-base* biasing method.

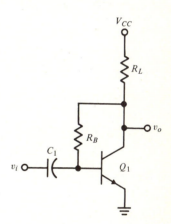

FIGURE P7.8

(a) Show that the following expression holds for the emitter current:

$$I_E = \frac{V_{CC} - V_{BE(on)}}{R_L + \dfrac{R_B}{\beta_F + 1}}$$

(b) For $R_L = 750$ Ω, $V_{CC} = 15$ V, $V_{BE(on)} = 0.7$ V, and $\beta_F = 100$, solve for the value of R_B that gives a 10-mA emitter current.

(c) For this circuit, repeat part (d) of Prob. P7.7.

(d) What are the advantages and disadvantages of the circuit in Fig. P7.8 relative to that in Fig. P7.7a?

(e) As we shall see in Chap. 9 the feedback through R_B affects the input resistance. The result is that

$$r_i = r_\pi \left\| \frac{R_B}{|a_v| + 1} \right.$$

Show that for $V_{CE} = V_{CC}/2$ and $|a_v| \gg 1$, a good approximation for r_i is

$$r_i \approx \frac{r_\pi}{2}$$

P7.9 For the circuit in Fig. 7.8 with $V_{CC} = 15$ V, $I_{C1} = 100$ μA, $V_{CE1} = 5$ V, $I_{C2} = 1$ mA, $V_{CE2} = 7.5$ V, $\beta_F = 200$, and $V_{BE(on)} = 0.7$ V, find the new numeric value for R_1, R_2, R_3, R_4, and the battery voltage so zero dc current flows through the battery.

P7.10 For the amplifier in Prob. P7.9, calculate the gain of the first stage (v_{c1}/v_{b1}), the gain of the second stage (v_{c2}/v_{b2}), the overall gain (v_o/v_i), the input resistance r_i, and the output resistance r_o.

P7.11 A representation of a two-stage CE amplifier is shown in Fig. P7.11. Biasing circuit details are not shown, but we are given that both Q_1 and Q_2 are in the active region, where each has a collector current of 2.6 mA and β_0 of 100. Calculate

(a) R_{L1}.

(b) a_{v1} ($\equiv v_{c1}/v_{b1}$).

(c) a_{v2} ($\equiv v_{c2}/v_{b2}$).

(d) a_v ($\equiv v_{c2}/v_{b1}$).

(e) r_i (the resistance seen looking into the base of Q_1).

(f) a_{vs} ($\equiv v_{c2}/v_s$).

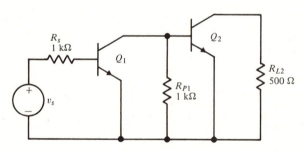

FIGURE P7.11

*P7.12 (a) The three-stage circuit in Fig. P7.12a has a potentiometer R_{BV}, which is adjusted so that $V_{C3} = V_{CC}/2$. For $I_{C1} = 100$ μA, $I_{C2} = 1$ mA, and $I_{C3} = 10$ mA, solve for R_{BV}, R_{C1}, R_{C2}, and R_{C3}. Use $\beta_F = 100$ and $V_{BE(on)} = 0.7$ V.

(b) Determine the value of a_v ($= v_o/v_i$), r_i, and r_o.

(c) For better bias stability the circuit in Fig. P7.12b is used (where the collector-to-base bias method of Prob. P7.8 is extended to this *compound* transistor). In order to not affect the small-signal input resistance the *feedback* resistor is divided up and a capacitor C_2 is used so that the ac performance is not affected by the *dc feedback*. For the values shown in Fig. P7.12b, determine I_{C1}, I_{C2}, and I_{C3} where $V_{CC} = 5$ V and R_2 is adjusted so that $V_{C3} = V_{CC}/2$. Use $\beta_F = 100$ and $V_{BE(on)} = 0.7$ V. (*Hint:* Start at Q_3 and work toward Q_1.)

(d) For the values given in part (c), what would be the adjusted value of R_2?

(e) For the circuit in part (c), calculate a_v, r_i, and r_o.

(f) Assume Q_2 and Q_3 have β_F of 100. What range of β_{F1} can the circuit in Fig. P7.12b handle if we are willing to readjust R_2?

(g) Repeat part (f), but use a fixed resistor for the combination of R_2 and R_3, which is selected so that with $\beta_{F1} = 100$, $V_{C3} = 2.5$ V. Find the range that β_{F1} can cover where V_{C3} stays between 2 and 3 V.

(h) Repeat part (g), except that now allow V_{C3} to vary from 1.5 to 3.5 V.

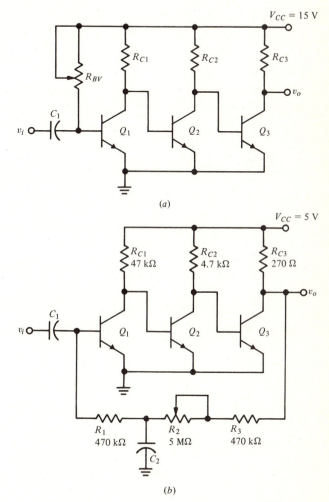

(a)

(b)

FIGURE P7.12

P7.13 The emitter-follower circuit in Fig. P7.13 has a transistor with $\beta_0 = 100$ and $I_C = 1.02$ mA.

(a) For $R_s = 0$ and $R_L = \infty$, find

    (1) r_i (looking to the right of point X).

    (2) r_o (looking to the left of point Y).

    (3) a_v ($\equiv v_L/v_s$).

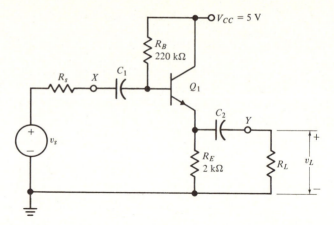

FIGURE P7.13

(b) Repeat part (a) for $R_s = 10$ kΩ and $R_L = 100$ Ω. [*Hint:* The expression for the input resistance of Eq. (7.38) must be modified to account for R_L.]

P7.14 The emitter-follower circuit in Fig. 7.10 has $V_{CC} = 15$ V, $R_B = 68$ kΩ, $R_E = 750$ Ω, $\beta_F = 100$, and $V_{BE(on)} = 0.7$ V.
(a) Find I_C, V_E, and V_{CE}.
(b) For $R_s = 0$, find (1) r_i ($\equiv v_b/i_s$); (2) r_o; (3) a_v ($\equiv v_o/v_s$).
(c) Repeat part (b) but with $R_s = 10$ kΩ.

P7.15 The circuit in Fig. P7.15 has a constant current source for establishing the dc emitter current. For the numeric value with $\beta = 100$ and $V_{BE(on)} = 0.7$ V:
(a) Determine I_C and V_{CE}.
(b) Determine r_o seen looking into the emitter.
(c) For $R_L = \infty$, determine (1) r_i; (2) v_L/v_s.
(d) Repeat part (c) but with $R_L = 75$ Ω.

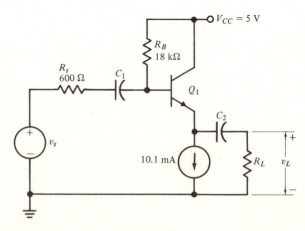

FIGURE P7.15

P7.16 The circuit in Fig. P7.16 is a CE amplifier, where the emitter resistor R_E is not bypassed by a capacitor.

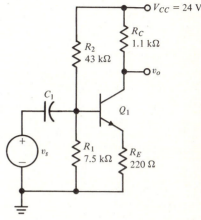

FIGURE P7.16

(a) Use Eqs. (7.36) and (7.18) to prove that the voltage gain v_o/v_s is given as

$$a_v = -\frac{\beta_0 R_C}{r_\pi + (\beta_0 + 1)R_E}$$

(b) For $\beta_0 \gg 1$ and $R_E \gg 1/g_m$, show that the above expressions can be approximated by

$$a_v = -\frac{R_C}{R_E}$$

(c) For the numeric values in Fig. P7.16, determine a_v by both the exact and approximate expressions given above. ($\beta_F = \beta_0 = 100$ and $V_{BE(\text{on})} = 0.7$ V.)

(d) For the numeric values given, solve for the input resistance of the amplifier.

P7.17 Show that a transistor with an unbypassed emitter resistor, as in Fig. P7.17a, can be modeled as the circuit in Fig. P7.17b.

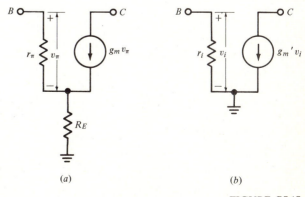

(a)   (b)

FIGURE P7.17

Prove that for $\beta_0 \gg 1$, a good approximation for the circuit elements in P7.17b is that:

(a) The input resistance r_i is

$$r_i \approx r_\pi(1 + g_m R_E)$$

(b) The transconductance g_m' is

$$g_m' = \frac{g_m}{1 + g_m R_E}$$

Note that the same factor $(1 + g_m R_E)$ shows up in both expressions.

P7.18 The two-stage (common-emitter, emitter-follower) circuit in Fig. P7.18 is drawn to indicate ac conditions only. For both transistors the collector currents are 2.6 mA and the current gains are 100. Calculate

(a) R_{L1}.
(b) a_{v2} $(\equiv v_{e2}/v_{b2})$.
(c) a_{v1} $(\equiv v_{c1}/v_{b1})$.
(d) a_v $(\equiv v_{e2}/v_{b1})$.
(e) r_i (the resistance seen looking into the base of Q_1).
(f) r_o (the resistance seen looking into the emitter of Q_2).
(g) a_{vs} $(\equiv v_o/v_s)$.

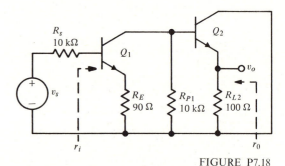

FIGURE P7.18

*P7.19 If the transistor model in Fig. P7.17a is not accurate enough, we can add the output resistance r_{oc}, as shown in Fig. 7.6b. The source resistance between base and ground R_s is usually much less than r_π. For most applications the following relations hold:

$$r_{oc} \gg r_\pi \gg R_E$$

Show that the output resistance of the transistor under these conditions is approximately $r_{oc}(1 + g_m R_E)$.

P7.20 The circuit in Fig. P7.20 has outputs taken at both the collector and emitter terminals. This circuit is called a *phase splitter*, since for ac signals the output voltages are 180° out of phase. With $\beta_F = \beta_0 = 100$ and $V_{BE(on)} = 0.7$ V, determine:

(a) I_C and V_{CE}.
(b) The gain v_1/v_s.
(c) The gain v_2/v_s.
(d) The output resistance r_{o1}.
(e) The output resistance r_{o2}.
(f) The input resistance r_i.

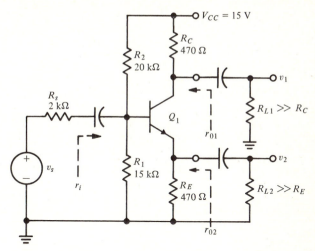

FIGURE P7.20

P7.21 The circuit in Fig. P7.21 shows the ac conditions for a cascade of two emitter followers. Using collector currents of 2.6 mA and current gains of 100, solve for:

(a) a_{v1} $(\equiv v_{e1}/v_{b1})$.

(b) a_{v2} $(\equiv v_{e2}/v_{b2})$.

(c) a_v $(\equiv v_{e2}/v_{b1})$.

(d) r_i.

(e) a_{vs} $(\equiv v_{e2}/v_s)$.

(f) r_o.

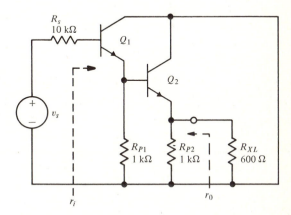

FIGURE P7.21

P7.22 Do the exact numeric work with $\beta_F = 50$ and $V_{BE(on)} = 0.7$ V to solve for the operating points (I_C, V_{CE}) of the circuit in Fig. 7.13. Use $I_{C2} = 1$ mA. (*Hint:* Make use of Thévenin's theorem to replace Q_4 and R_6 as seen looking into the base of Q_4.)

P7.23 The circuit in Fig. P7.23 is an emitter follower driving a CE stage with an unbypassed emitter resistance.
(a) Solve for the operating points, assuming $\beta_F \rightarrow \infty$ and $V_{BE(\text{on})} = 0.7$ V.
(b) Solve for a_{v1}, a_{v2}, a_v, r_i and r_o by using $\beta_0 \rightarrow \infty$.

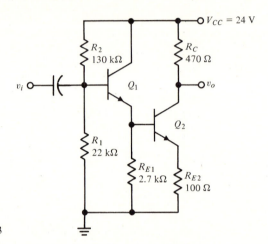

FIGURE P7.23

P7.24 The circuit in Fig. P7.24 has an external load R_{XL} of 50 Ω capacitively coupled to the collector. Assume C_E and C_L are large enough so that for the signal frequencies their voltage does not vary from their dc value. For transistor data use $\beta_F = 100$, $V_{BE(\text{on})} = 0.7$ V, and $V_{CE(\text{sat})} = 0.2$ V.
(a) Solve for the dc operating points of Q_1.
(b) Solve for the maximum positive and negative values of i_{XL} which occur when the transistor is cut off and saturated, respectively.

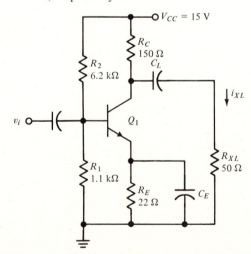

FIGURE P7.24

P7.25 The circuit parameters of Prob. P7.24 are modified in order to get a symmetrical output swing when clipping occurs. Further, we use $R_1 = 0.36$ kΩ, $R_2 = 2$ kΩ, $R_E = 3.9$ Ω, $R_C = 39$ Ω, and $R_{XL} = 50$ Ω. Use the same transistor parameters as in Prob. P7.24 to solve for the same quantities as in Prob. P7.24.

P7.26 The emitter follower in Fig. P7.26 is driven by a CE stage. For the values shown, using $\beta_1 = 100$, $\beta_2 = 30$, $V_{BE(on)} = 0.7$ V:

(a) Solve for the operating points (*Hint:* Solve for I_{C1} first, then get V_{BB2} in order to determine I_{C2}.)

(b) At the operating point determine (1) a_v; (2) r_i; (3) r_o ("seen" by the 8-Ω resistor, R_{XL}).

(c) Solve for the maximum positive and negative values that i_{XL} can assume without clipping. (*Hint:* One value is determined when Q_1 is cut off.)

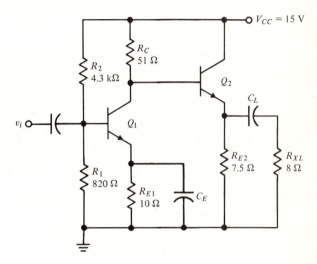

FIGURE P7.26

P7.27 The CB stage circuit in Fig. P7.27 has a constant current source for setting the dc emitter current. Let $\beta_0 = 100$. For $R_s = 0$ and $R_{XL} = \infty$, solve for:

(a) r_i.

(b) a_v ($\equiv v_L/v_s$).

(c) r_o.

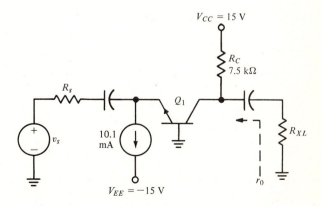

FIGURE P7.27

P7.28 (a) Repeat Prob. P7.27 for $R_s = 75\ \Omega$ and $R_{XL} = \infty$.

(b) Repeat Prob. P7.27 for $R_s = 75\ \Omega$ and $R_{XL} = 600\ \Omega$.

P7.29 For the circuit in Fig. 7.15 with $R_E = 1\ k\Omega$, $R_C = 2.2\ k\Omega$, $V_{EE} = -6\ V$, $V_{CC} = 24\ V$, $\beta = 100$, and $V_{BE(on)} = 0.7\ V$:

(a) Solve for I_C and V_{CE}.

(b) For $R_s = 0$, solve for (1) r_i; (2) a_v ($\equiv v_o/v_s$).

(c) For $R_s = 50\ \Omega$, solve for a_v.

P7.30 The circuit in Fig. P7.30 is the ac circuit for an *emitter-coupled* pair, which is an emitter follower driving a CB stage. Given $\beta_0 = 100$ and $I_{C1} = I_{C2} = 35\ \mu A$, solve for:

(a) a_{v2} ($\equiv v_{c2}/v_{e1}$).

(b) a_{v1} ($\equiv v_{e1}/v_{b1}$).

(c) a_v ($\equiv v_{c2}/v_{b1}$).

(d) r_i.

(e) r_o.

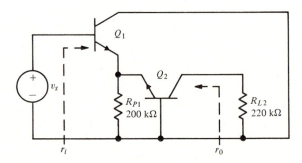

FIGURE P7.30

P7.31 The circuit in Fig. P7.31 is the dc bias circuitry that is the conventional way of showing the emitter-coupled pair in Prob. P7.30. For $v_s = 0$ and identical transistors for Q_1 and Q_2 with $\beta_F = 100$ and $V_{BE(on)} = 0.7\ V$:

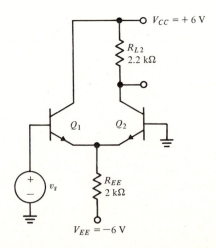

FIGURE P7.31

(a) Solve for I_{C1}, I_{C2}, V_{CE1}, and V_{CE2}.

(b) Solve for the five small-signal quantities as in Prob. P7.30.

P7.32 The circuit in Fig. P7.32a is called a *cascode*, and its small-signal representation as in Fig. P7.32b shows it to be a CE stage driving a CB stage. For $\beta_F = 100$, $V_{BE(on)} = 0.7$ V, and the other given conditions, solve for:

(a) I_{C1}, V_{CE1}, I_{C2}, and V_{CE2} with zero dc volts at the base of Q_1.

(b) a_{v1} ($\equiv v_{c1}/v_{b1}$).

(c) a_{v2} ($\equiv v_{c2}/v_{e2}$).

(d) a_v ($\equiv v_o/v_s$).

(e) r_i of the complete amplifier.

(f) r_o of the complete amplifier.

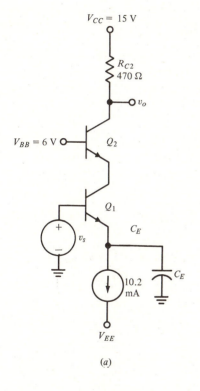

(a)

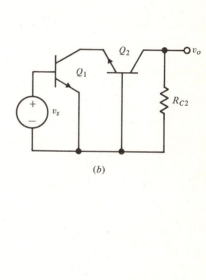

(b)

FIGURE P7.32

P7.33 The circuit in Fig. P7.33a is a *complementary symmetry* version of an emitter-coupled pair. The small-signal version is shown in Fig. P7.33b. For the dc value of v_s set at 0 V and V_{BB2} adjusted so $I_{C1} = 10$ μA, solve:

(a) For V_{CE1}, I_{C2}, and V_{CE2}. Use $\beta_F = 100$ and emitter junction ON voltages of 0.6 V for both Q_1 and Q_2. Solve for the small-signal quantities.

(b) a_{v1}.

(c) a_{v2}.

(d) a_v.

(e) r_i.

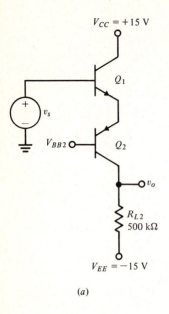

(a)

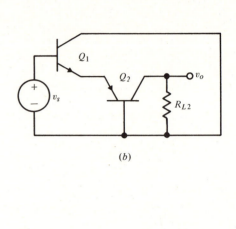

(b)

FIGURE P7.33

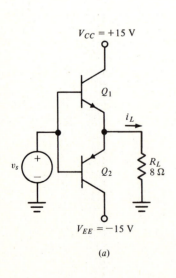

(a)

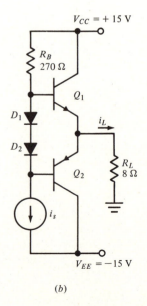

(b)

FIGURE P7.34

P7.34 The circuit in Fig. P7.34a is a complementary symmetry emitter follower that has the advantage of near zero standby power. For a sufficiently positive v_S, v_L "follows" v_S by means of Q_1, and for negative values it follows by means of Q_2.

    (a) Use 0.7 V for the emitter junction ON voltage for both Q_1 and Q_2 to plot v_L versus v_S.

  *(b) The circuit in Fig. P7.34a suffers from *crossover distortion*, which makes it unsuitable for many applications where linear performance is required, such as in an audio amplifier. The circuit in Fig. P7.34b corrects for this by adding a biasing network made up of R_B, D_1, D_2, and the current source i_S. For the numeric values used here, i_S is made up of a dc value of 53 mA and the input signal to the amplifier varies this current sinusoidally about 53 mA from 0 to 106 mA. Assume that the diode drops (when ON) are 0.7 V and that the emitter junctions are also 0.7 V when the collector currents are 53 mA. Both collectors will conduct 53 mA when i_S is 53 mA. Use $\beta_F = 121$ and emitter junction ON voltage of 0.8 V when either transistor is conducting 1.4 A of emitter current.

    (1) When i_S goes to zero, show that i_L becomes 1.4 A. Use the results of Prob. P7.7.

    (2) When i_L is -1.4 A, show that i_S is 106 mA.

    (3) In part (2), what is your estimate for i_{E1} when i_L is -1.4 A?

    (4) What is the maximum rms power delivered into the load?

    (5) What is the rms signal power required to drive the input of the amplifier at the top of the current source i_S?

8
DIFFERENTIAL AMPLIFIERS

Direct-coupled (dc) differential amplifiers, commonly called operational amplifiers (op amps) when ICs, are the fundamental building blocks of many electronic systems. In this chapter we will study some of the circuit techniques used in designing these operational amplifiers in IC form. This knowledge will also be beneficial in understanding and using many other types of linear ICs, including sense amplifiers, voltage comparators, voltage regulators, and analog multipliers.

8.1 SINGLE-ENDED AMPLIFIERS

Before proceeding further, we should review some terms relating to amplifiers. Let us begin by considering the model of an amplifier in Fig. 8.1a. This model, that gives a constant *voltage gain* a_v, only approximates an actual amplifier. However, this approximation can give adequate results over a limited frequency range and range of signal levels.

A key point of this model is that both input and output voltages are determined with reference to the ground terminal. In most practical realizations of

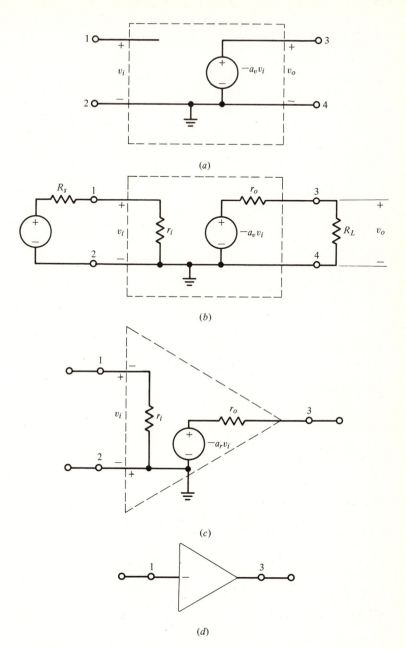

FIGURE 8.1

(a) Model for an inverting amplifier with single-ended input and single-ended output;
(b) more complete model for cases where R_s or r_o are not negligibly small compared
to r_i and R_L, respectively; (c) standard triangular symbol for an amplifier;
(d) simplified symbol for an inverting single-ended amplifier.

amplifiers, this ground point is determined by the power supply and often is one terminal of the power supply.

The amplifier in Fig. 8.1a amplifies and *inverts* the input signal v_i. Thus, if the magnitude of the voltage gain a_v is 100 and $+1$ mV is applied to terminal 1 relative to terminal 2, then terminal 3 will be at -100 mV referred to terminal 4. Terminals 2 and 4 are tied together to the internal ground. We refer to this type of amplifier as having a single-ended input and single-ended output, since all the potentials are determined in relation to a single point, which is the ground terminal here.

If we need a better approximation to account for *loading*, the circuit in Fig. 8.1b can be used. It includes the input resistance r_i, so that the effect of the amplifier-input loading can be accounted for when it is driven by a practical source that has a nonzero source resistance R_s. Similarly, the effect of a resistive load R_L on the amplifier output can be better accounted for with this model because of the inclusion of the output resistance r_o.

Figure 8.1c shows the same circuit model as in Fig. 8.1b, where a triangular outline is used. This symbol is generally used to represent a gain block. For a single-ended input and output amplifier, the symbol in Fig. 8.1d is used, where all reference to the ground terminal is deleted. The minus sign on the input indicates an inversion of the polarity of the signal through the amplifier.

EXERCISE

E8.1

(a) For $R_s = 10$ kΩ, $r_i = 90$ kΩ, $r_o = 100$ Ω, $R_L = 900$ Ω, and $a_v = 10^3$, find v_o/v_s for Fig. 8.1b.

(b) If the amplifier were removed and the source tied directly to the load, what would v_o/v_s be?

8.2 DIFFERENCE AMPLIFIERS

In Fig. 8.2a we show another type of amplifier. This is the *difference* or *differential* amplifier. Here two input terminals can be driven independently with respect to ground. For an ideal differential amplifier, the output voltage v_o (with respect to ground) is proportional to the difference of the input signals ($v_{id} = v_{i1} - v_{i2}$) only. The constant of proportionality is the *differential* voltage gain a_d. The output is single-ended.

We will see later that *bias-current* or *feedback* considerations make it necessary to have differential inputs even in those cases where the signal input is single-ended.

We can add resistive elements r_i and r_o to this amplifier, as we did in the single-ended case; wherein we obtain the model of Fig. 8.2b. We have also indicated a positive and negative power supply connected to the amplifier. The

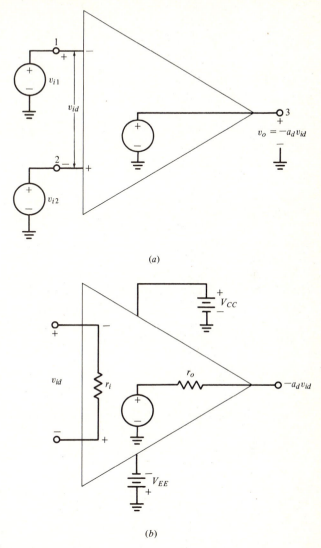

(a)

(b)

FIGURE 8.2
(a) A differential-amplifier model with a single-ended output proportional to the differential input; (b) a differential-amplifier model that includes a differential input resistance r_i and output resistance r_o.

ground terminal for the output is established through the power-supply connections in many practical amplifiers. Thus no ground connection need be made to the amplifier itself.

As an example of where a precision differential amplifier is called for, consider the following instrumentation problem.

Suppose we are given the job of engineering a system that will enable us to continuously monitor the stress in a beam of a bridge as it is subjected to a variety of load conditions due to wind and traffic. One way to do this is to use a strain gage. A strain gage is a resistor made of metal wire or a semiconductor structure which is then cemented to an insulating base, and this base is then cemented to the structure to be monitored, in this case, the beam's surface. Under tension, at the surface of the beam, the resistor lengthens, and the resulting effect is an increase in the resistor value directly proportional to the elongation. Similarly, a decrease in resistance occurs under compression. We want this change in resistance transformed into vertical deflection of a spot on a *cathode-ray tube* (CRT). It is desired to have the information displayed on a large-screen CRT so that for the normal peak load on the beam the vertical deflection of the CRT will be 10 divisions. The deflection sensitivity of the CRT is 10 mV/division, and one end of the input is grounded (single-ended input to CRT). For simplicity, we find that the best way to utilize the strain gage is to connect it in a Wheatstone bridge circuit, as in Fig. 8.3a. Here the excitation to the bridge circuit is 15 V dc with the negative terminal of the battery grounded. The two upper elements of the bridge, R_c and R_d, are precision-fixed resistors of 250 Ω. The lower two elements are identical strain-gage elements whose unstressed values are 250 Ω. The element R_b is not stressed and is used to cancel out temperature effects. Hence $R_b = R_c = R_d$. R_a is the element that is cemented to the beam and is to be stressed. For elongations corresponding to the designed peak-load condition on the bridge structure, the fractional increase in resistance is 8×10^{-4}; that is, the resistance value changes from R_a to $R_a + \Delta R$, where

$$\Delta R = (8 \times 10^{-4})R_a$$

$$= (8)(10^{-4})(250) = 0.2 \ \Omega$$

For this circuit, due to the voltage-divider action of R_b and R_c, v_{i2} is thus fixed at 7.5 V, while under normal full load we have for v_{i1},

$$v_{i1} = \frac{R_a + \Delta R}{R_a + \Delta R + R_d} E = \frac{R_a + \Delta R}{2R_a + \Delta R} E = \frac{R_a\left(1 + \dfrac{\Delta R}{R_a}\right)}{2R_a\left(1 + \dfrac{\Delta R}{2R_a}\right)} E \qquad (8.1)$$

$$v_{i1} \approx \left(1 + \frac{\Delta R}{2R_a}\right)\frac{E}{2} = \left(1 + \frac{0.2}{500}\right)\frac{15}{2}$$

$$= 7.5 + 3 \times 10^{-3} = 7.5 \text{ V} + 3 \text{ mV} = 7.503 \text{ V}$$

By using a difference amplifier, the large fixed portion of v_{i1} (that is, 7.5 V) is canceled out by v_{i2}, and only the difference signal $(v_{i1} - v_{i2})$ of 3 mV is left, which is proportional to the stress in the girder.

If we had tried to amplify v_{i1} itself, the large fixed component would have *overdriven* the amplifier so that it would have not been possible to amplify linearly the desired 3-mV signal.

The gain required for our amplifier is determined from the following argument. For normal full-scale girder stress we obtain a 3-mV signal. The output of

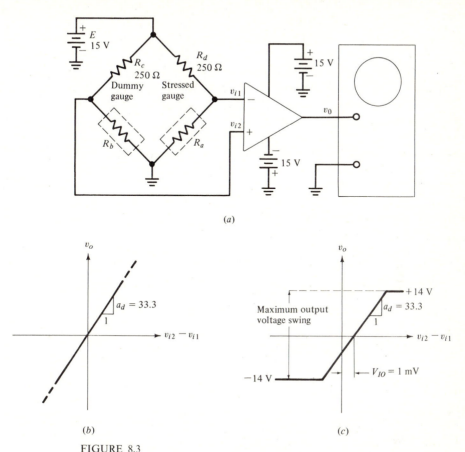

FIGURE 8.3
(*a*) Strain-gage circuit and amplifier driving an oscilloscope display; (*b*) ideal-amplifier transfer characteristic; (*c*) practical-amplifier characteristic showing output voltage limiting and input offset voltage.

the amplifier for this input should drive the CRT 10 divisions, where for our CRT each division of deflection requires 10 mV at the input. Therefore, we need a differential voltage gain a_d,

$$a_d = \frac{\Delta v_o}{\Delta v_{id}} = \frac{(10)(10 \text{ mV})}{3 \text{ mV}} = 33.3 \qquad (8.2)$$

The transfer characteristic (v_o versus v_{id}) for an amplifier that provides the ideal characteristics for this job is shown in Fig. 8.3*b*. This has a straight line through the origin that extends to plus and minus infinity. A practical amplifier has characteristics such as shown in Fig. 8.3*c*. Here the amplifier output swing is limited to ± 14 V when ± 15-V supplies are used. Moreover the straight line does *not* intersect the origin but there is an input *offset voltage* V_{IO} of 1 mV. This offset voltage is due to the fact that even with the utmost care in the design and making of an IC op amp, the final result is not exactly symmetrical. However, with many IC op amps, connections are made available for the user to reduce this input offset voltage to zero. This is known as *offset nulling*.

As we have seen in Chap. 7, the gain we obtain in a voltage amplifier is a function of transistor parameters (primarily β_0 and I_C) and the load resistors. Since these vary with temperature and also from unit to unit, we would not have a well-calibrated system if we attempted to use a simple amplifier to do this job. In order to obtain a precise gain that stays constant with all parameter changes, we would employ an amplifier with *negative feedback*. With the feedback determined by stable, high-accuracy and low-temperature coefficient *external* resistors, we could obtain a *feedback amplifier* that does this job. In Chap. 9 we will discuss the design of the resistive feedback network and the resulting feedback-amplifier characteristic. The final details of this example of a difference amplifier for the strain gage is given in Prob. P9.26.

Now that we have indicated the need for a differential amplifier, we return to see how we can realize it.

8.3   THE DIFFERENTIAL-AMPLIFIER PAIR

A fundamental subcircuit used in operational amplifiers and many other linear ICs is the differential-amplifier pair shown in Fig. 8.4. This circuit is also called a *balanced-amplifier* or an *emitter-coupled pair*. It is very similar to the basic circuit of ECL discussed in Sec. 3.6, but in the case here, both transistors are biased in the normal active region, ideally both having the same operating point. This requires that transistors Q_1 and Q_2 should not only be as well matched as possible but also at the same temperature. This is best accomplished by making them in the same chip and as close together as possible. In other words, the monolithic construction used in ICs is ideal for this circuit.

In the following we will use the same notation as in Chap. 7; that is, the instantaneous voltage v_{I1} at the base of Q_1 is given by

$$v_{I1} = V_{I1} + v_{i1} \qquad (8.3)$$

where V_{I1} = dc quiescent voltage at base of Q_1
$\qquad v_{i1}$ = incremental small-signal voltage at base of Q_1

We first determine the dc operating conditions for the differential amplifier, and then we will develop circuit equations for the small-signal characteristics.

The following circuit and bias data apply to Fig. 8.4. We assume Q_1 and Q_2 have identical electrical characteristics.

$$V_{CC} = +15 \text{ V} \qquad\qquad V_{EE} = -15 \text{ V}$$
$$\beta_F = \beta_0 = 50 \qquad\qquad V_{BE(on)} = 0.7 \text{ V}$$
$$I_{C1} = I_{C2} = 0.1 \text{ mA}$$
$$V_{CB1} = V_{CB2} = \tfrac{1}{2}V_{CC}$$

The last conditions assure us that Q_1 and Q_2 are in the active region.

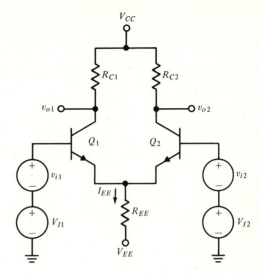

FIGURE 8.4
Differential-amplifier pair.

8.3.1   DC Conditions

We determine values for R_{C1}, R_{C2}, and R_{EE} under the condition that the dc voltage at the base of Q_1 and Q_2 is zero (that is, $V_{I1} = V_{I2} = 0$). Now with $I_{C1} = I_{C2}$ and $V_{C1} = V_{C2}$, then $R_{C1} = R_{C2}$. Hence

$$R_{C1} = R_{C2} = \frac{V_{CC} - V_{C1}}{I_{C1}} \qquad (8.4)$$

where

$$V_{C1} = \tfrac{1}{2}V_{CC} = 7.5 \text{ V}$$

then

$$R_{C1} = R_{C2} = \frac{15 - 7.5}{0.1} = 75 \text{ k}\Omega$$

Also

$$I_{B1} = I_{B2} = \frac{I_{C1}}{\beta_F} = \frac{100 \ \mu\text{A}}{50} = 2 \ \mu\text{A}$$

Since $I_E = I_B + I_C$, we have

$$I_{E1} = I_{E2} = 102 \ \mu\text{A} = 0.102 \text{ mA}$$

By KCL,

$$I_{EE} = I_{E1} + I_{E2} = 0.204 \text{ mA}$$

Then, since $V_{I1} = V_{I2} = 0$ and $V_{BE(\text{on})} = 0.7$ V, we have

$$V_{E1} = V_{E2} = -0.7 \text{ V}$$

Hence

$$R_{EE} = \frac{V_{E1} - V_{EE}}{I_{EE}} \tag{8.5}$$

$$= \frac{-0.7 - (-15)}{0.204} = 70 \text{ k}\Omega$$

This completes the dc design.

EXERCISE

E8.2   For the circuit of Fig. 8.4, find the values of R_{C1}, R_{C2}, and R_{EE} that give the following operating point when the input dc voltages are zero.

$$V_{CC} = +6 \text{ V} \qquad\qquad V_{EE} = -6 \text{ V}$$

$$I_{C1} = I_{C2} = 0.2 \text{ mA}$$

$$V_{C1} = V_{C2} = 3 \text{ V}$$

$$\beta_F = 80 \qquad\qquad V_{BE(\text{on})} = 0.7 \text{ V}$$

8.3.2   Modes of Operation

Before we derive equations for the small-signal characteristics of the differential amplifier, we will describe another mode of operation for this circuit. We have already seen the need for the circuit to amplify the differential input signal. This is the *differential mode* of operation. For this mode we define the differential small-signal input voltage as

$$v_{id} = v_{i1} - v_{i2} \tag{8.6a}$$

and the output differential small-signal voltage as

$$v_{od} = v_{o1} - v_{o2} \tag{8.6b}$$

Then the differential-mode voltage gain is given by

$$a_d = \frac{v_{od}}{v_{id}} = \frac{v_{o1} - v_{o2}}{v_{i1} - v_{i2}} \tag{8.6c}$$

So far we have only dealt with the difference voltage signals of v_{i1} and v_{i2}. We also find in Fig. 8.4 that even with exactly equal signals at v_{i1} and v_{i2} (such as the 7.5 V in our beam-stress problem) we can also obtain an output at v_{o1} and v_{o2}. This is the *common mode* of operation. We define the common-mode input signal as the average of v_{i1} and v_{i2}. That is,

$$v_{ic} = \frac{v_{i1} + v_{i2}}{2} \qquad (8.7a)$$

Similarly we have the common-mode output voltage

$$v_{oc} = \frac{v_{o1} + v_{o2}}{2} \qquad (8.7b)$$

Then the common-mode voltage gain is given by

$$a_c = \frac{v_{oc}}{v_{ic}} = \frac{v_{o1} + v_{o2}}{v_{i1} + v_{i2}} \qquad (8.7c)$$

Of course, we require this mode of voltage gain to be very small, while we maximize the differential-mode voltage gain.

We will find it useful in our analysis of the differential amplifier to relate the voltage signals at the input and output to the differential- and common-mode voltages. Hence by combining Eqs. (8.6a) and (8.7a), we have

$$v_{i1} = v_{ic} + \frac{v_{id}}{2} \qquad (8.8a)$$

and

$$v_{i2} = v_{ic} - \frac{v_{id}}{2} \qquad (8.8b)$$

Likewise, by combining Eqs. (8.6b) and (8.7b), we have for the output voltages

$$v_{o1} = v_{oc} + \frac{v_{od}}{2} \qquad (8.9a)$$

and

$$v_{o2} = v_{oc} - \frac{v_{od}}{2} \qquad (8.9b)$$

We may further develop the last two equations by substituting for v_{oc} from Eq. (8.7c) and for v_{od} from Eq. (8.6c). Then

$$v_{o1} = a_c v_{ic} + \frac{a_d v_{id}}{2} \qquad (8.10a)$$

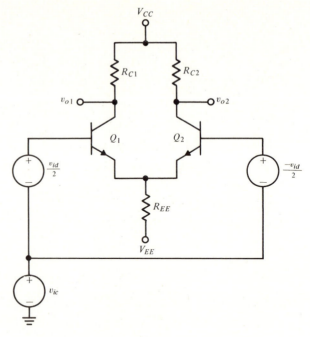

FIGURE 8.5
Differential-amplifier pair showing small-signal difference and common-mode input voltages.

and

$$v_{o2} = a_c v_{ic} - \frac{a_d v_{id}}{2} \qquad (8.10b)$$

These relationships for the differential and common mode of operation are summarized in Table 8.1. We make use of the equations for the input-signal voltages to redraw the circuit for the input signals to the differential amplifier as shown in Fig. 8.5.

8.3.3 Small-signal Characteristics

We now consider the small-signal characteristics of the differential amplifier in two sections. First, we describe the response of the amplifier to difference-mode signals. Second, we investigate the effect of common-mode signals. In each case we develop equations for the small-signal voltage gain and input resistance of the circuit through the use of *half-circuits*.

Difference mode In considering the effects of difference-mode operation on the differential amplifier, we assume that the common-mode input voltage is zero. Hence we have the circuit diagram as shown in Fig. 8.6a. In this figure particularly notice that the two input signals are equal but of the opposite polarity. Now if we also restrict the input signals to a small value so that the operating point of the transistors is essentially unchanged, we notice that any increase of current in Q_1 is balanced by a similar decrease of current in Q_2. As a result, the current I_{EE} does not change, with the further interesting development that neither does the voltage at the coupled emitters, v_e, change. We conclude that for the small-signal difference-input voltage signals, v_e is at a fixed value. Therefore for the small-signal difference mode we refer to v_e as a *virtual ground*. The circuit now reduces to two separate but similar half-circuits. In Fig. 8.6b we show one of the small-signal *difference-mode half-circuits*. Note that the emitter current I_E of the transistor in the half-circuit is one-half of I_{EE}.

Now that we have just a simple CE transistor stage to deal with, we should recall from Table 7.1 that the small-signal voltage gain is given by

$$a_v = \frac{-\beta_0 R_C}{r_\pi}$$

Hence from Fig. 8.6b we obtain the difference-mode voltage gain as

$$a_d = \frac{v_{od}}{v_{id}} = \frac{-\beta_0 R_C}{r_\pi} \tag{8.11}$$

Table 8.1 SUMMARY OF RELATION-SHIP FOR DIFFERENCE- AND COMMON-MODE OPERATION

	Difference mode	Common mode
Input:	$v_{id} = v_{i1} - v_{i2}$	$v_{ic} = \dfrac{v_{i1} + v_{i2}}{2}$
Output:	$v_{od} = v_{o1} - v_{o2}$	$v_{oc} = \dfrac{v_{o1} + v_{o2}}{2}$
Gain:	$a_d = \dfrac{v_{od}}{v_{id}}$	$a_c = \dfrac{v_{oc}}{v_{ic}}$

Input:	$v_{i1} = v_{ic} + \dfrac{v_{id}}{2}$
	$v_{i2} = v_{ic} - \dfrac{v_{id}}{2}$
Output:	$v_{o1} = a_c v_{ic} + \dfrac{a_d v_{id}}{2}$
	$v_{o2} = a_c v_{ic} - \dfrac{a_d v_{id}}{2}$

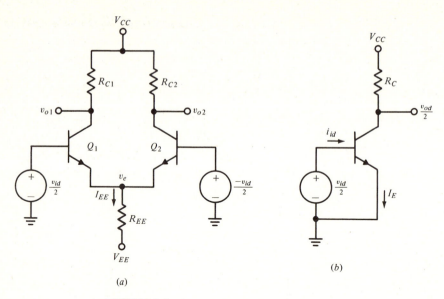

FIGURE 8.6
(a) The difference-mode (full) circuit; (b) the difference-mode half-circuit.

The small-signal input resistance for the CE stage is, again, from Table 7.1,

$$r_i = r_\pi$$

However, notice from Fig. 8.6b that this input resistance is due to $v_{id}/2$ only. Hence the full difference input resistance, due to v_{id}, is twice r_i; then

$$r_{id} = \frac{v_{id}}{i_{id}} = 2r_\pi \qquad (8.12)$$

In a similar fashion we should realize that the magnitude of the output signal in the half-circuit is just one-half the difference output of the complete circuit of Fig. 8.6a.

In summary, the small-signal differential voltage gain of the differential or balanced amplifier is *equal* to that of a simple CE stage biased at the same operating conditions as one transistor of the differential pair. The differential input resistance for the balanced amplifier is *twice* that for the CE stage. The differential output signal voltage is also *twice* that of the half-circuit.

EXAMPLE 8.1   We continue our numerical example by calculating the small-signal differential voltage gain and input resistance for the differential amplifier shown in Fig. 8.4.

SOLUTION   First we determine r_π for the transistor at an operating current of 0.1 mA:

$$r_\pi = \frac{(50)(26 \text{ mV})}{0.1 \text{ mA}} = 13 \text{ k}\Omega$$

We compute the differential voltage gain from Eq. (8.11) as

$$a_d = \frac{-(50)(75)}{13} = -288$$

We use Eq. (8.12) to calculate the difference-mode input resistance:

$$r_{id} = (2)(13) = 26 \text{ k}\Omega$$

With $v_{id} = 1$ mV, note that $v_{id}/2 = 0.5$ mV. Then in Fig. 8.6b, with $a_d = -288$, $v_{od}/2 = -144$ mV, but $v_{od} = -288$ mV. That is, in Fig. 8.6a, $|v_{o1} - v_{o2}| = 288$ mV.

////

Common mode   For the common-mode operation of the differential amplifier, we again redraw the circuit of Fig. 8.5 but this time include the common-mode input signal while setting the difference-mode signal voltages to zero. This is shown in Fig. 8.7a. Here note that the two input signals are of the same polarity as well as the same magnitude.

We can simplify the analysis if we modify our circuit to that of Fig. 8.7b. We have divided R_{EE} into two parallel resistors, each of value $2R_{EE}$, so there is no change in the operation of the circuit. However, now in Fig. 8.7b, with equal voltage signals at the base of Q_1 and Q_2, and with identical transistors, $v_{e1} = v_{e2}$. Hence there is no current in the branch connecting the two emitters; that is, $i_x = 0$. Therefore this branch may be broken with no effect upon the circuit. We then have the *common-mode half-circuit* shown in Fig. 8.7c.

The effect of the resistor $2R_{EE}$ on the circuit in Fig. 8.7c can easily be determined by combining what we have learned in Chap. 7 on the emitter follower and the CE stage. Including a resistor R_E in the emitter circuit increases the input resistance of the CE stage by the factor $(\beta_0 + 1)R_E$. That is,

$$r_i = r_\pi + (\beta_0 + 1)R_E$$

Hence from Fig. 8.7c, we have for the common-mode input resistance,

$$r_{ic} = \frac{v_{ic}}{i_{ic}} = r_\pi + (\beta_0 + 1)2R_{EE} \qquad (8.13)$$

Then recall that the voltage gain of a CE stage is, from Eq. (7.20b),

$$a_v = \frac{-\beta_0 R_C}{r_i}$$

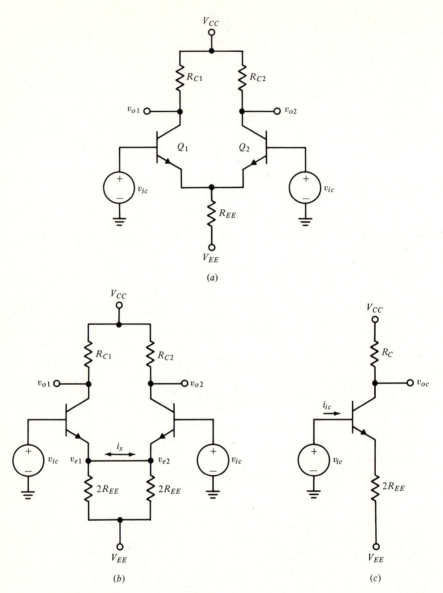

FIGURE 8.7
(a) The common-mode (full) circuit; (b) an equivalent common-mode circuit;
(c) the common-mode half-circuit.

Now in Fig. 8.7c, the common-mode voltage gain is given as

$$a_c = \frac{v_{oc}}{v_{ic}} = \frac{-\beta_0 R_C}{r_{ic}}$$

$$= \frac{-\beta_0 R_C}{r_\pi + (\beta_0 + 1)2R_{EE}} \tag{8.14}$$

Notice in Eqs. (8.13) and (8.14) that as we increase the value of R_{EE} we increase the common-mode input resistance and decrease the common-mode voltage gain.

EXAMPLE 8.2   We continue to use the differential-amplifier circuit shown in Fig. 8.4, this time to determine the small-signal common-mode voltage gain and input resistance.

SOLUTION   From the data we have already calculated, $r_\pi = 13$ kΩ. The common-mode voltage gain is given from Eq. (8.14) as

$$a_c = \frac{-(50)(75)}{13 + (51)(2)(70)} = -0.52$$

We use Eq. (8.13) to find the common-mode input resistance

$$r_{ic} = 13 + (51)(2)(70) = 7150 \text{ kΩ} = 7.15 \text{ MΩ}$$

From this example we note with $v_{ic} = 1$ mV, $v_{oc} = -0.52$ mV. That is, in Fig. 8.7a, $|(v_{o1} + v_{o2})/2| = 0.52$ mV.   ////

We have found that the voltage to ground at the output of the differential-amplifier pair is not only proportional to the difference voltage signal at the input but is also affected by the average voltage of the input signals. Hence we find that we want to keep this common-mode signal as small as possible compared to the difference signal. One figure of merit used for the differential amplifier is the *common-mode rejection ratio* (CMRR), which is given as

$$\text{CMRR} = \left| \frac{a_d}{a_c} \right| \tag{8.15}$$

For our circuit, in Fig. 8.5, this is

$$\text{CMRR} = \frac{288}{0.52} = 554$$

In practice the CMRR is usually expressed in *decibels*, where we express the *voltage ratio* as

$$\text{CMRR (dB)} = 20 \log \left| \frac{a_d}{a_c} \right| \qquad (8.16)$$

Again, for our example this is

$$\text{CMRR} = 20 \log 554 = 54.9 \, \text{dB}$$

We may also express the voltage gain of the differential amplifier in decibels; then for the difference-mode voltage gain

$$a_d \, \text{(dB)} = 20 \log \left| \frac{v_{od}}{v_{id}} \right| \qquad (8.17)$$

$$= 20 \log 288 = 49.2 \, \text{dB}$$

For the common-mode voltage gain

$$a_c \, \text{(dB)} = 20 \log \left| \frac{v_{oc}}{v_{ic}} \right| \qquad (8.18)$$

$$= 20 \log 0.52 = -5.7 \, \text{dB}$$

Now we can also change Eq. (8.16) to express the CMRR as the difference of the two voltage gains, expressed in decibels; namely,

$$\text{CMRR (dB)} = 20 \log |a_d| - 20 \log |a_c| \qquad (8.19)$$

Hence from Eqs. (8.17) and (8.18)

$$\text{CMRR} = 49.2 - (-5.7) = 54.9 \, \text{dB}$$

A summary of the small-signal characteristics of the differential-amplifier pair is listed in Table 8.2. In Demonstration D8.1 we experimentally determine the small-signal characteristics of a differential-amplifier pair similar to the circuit we have analyzed in this section.

**Table 8.2   SUMMARY OF SMALL-SIGNAL CHARAC-
TERISTICS OF DIFFERENTIAL
AMPLIFIER**

	Difference mode	Common mode
Voltage gain:	$a_d = \dfrac{-\beta_0 R_C}{r_\pi}$	$a_c = \dfrac{-\beta_0 R_C}{r_\pi + (\beta_0 + 1)2R_{EE}}$
Input resistance:	$r_{id} = 2r_\pi$	$r_{ic} = r_\pi + (\beta_0 + 1)2R_{EE}$
Common-mode rejection ratio:	$\text{CMRR (dB)} = 20 \log \left\| \dfrac{a_d}{a_c} \right\|$	

EXERCISES

E8.3 Use the half-circuits to calculate r_{id}, r_{ic}, a_d, and a_c for the differential-amplifier circuit of Fig. 8.4, given the following circuit and bias data.

$$V_{CC} = +12 \text{ V} \qquad\qquad V_{EE} = -6 \text{ V}$$

$$I_{C1} = I_{C2} = 0.15 \text{ mA}$$

$$V_{C1} = V_{C2} = 6 \text{ V}$$

$$\beta_F = \beta_0 = 60 \qquad\qquad V_{BE(on)} = 0.7 \text{ V}$$

E8.4 Express the voltage gain of Exercise E8.3 in decibels; also determine the CMRR in decibels.

8.4 CASCADING DIFFERENTIAL PAIRS

A common method of obtaining a greater voltage amplification for the differential amplifier is similar to what we found with the single CE stage in Chap. 7. That is, one gain stage is followed by another. An example of a cascade of two difference amplifiers is shown in Fig. 8.8.

Notice in Fig. 8.8 that the differential output of the input amplifier pair becomes the differential input for the second stage of amplification—the collector of Q_1 is connected to the base of Q_3 and similarly for Q_2 and Q_4. As with a cascade of two CE stages we will have a phase reversal of the signal through Q_1 and Q_2, but this will be canceled by a similar phase reversal through Q_3 and Q_4. Hence the output signal voltage v_{o3} referred to v_{o4} will be in phase with the input signal voltage v_{i1} referred to v_{i2}. Again similar to the cascade of CE stages, we will find that the voltage gain of the complete amplifier is the product of the voltage gain of each differential pair. Moreover, with the common-mode voltage gain of each stage less than unity, we will find the overall CMRR improved over that of just one stage.

We will use the same voltage supplies and operating data for Q_1 and Q_2 that we had before in Fig. 8.4. That is,

$$V_{CC} = +15 \text{ V} \qquad\qquad V_{EE} = -15 \text{ V}$$

$$I_{C1} = I_{C2} = 0.1 \text{ mA}$$

$$V_{CB1} = V_{CB2} = \tfrac{1}{2}V_{CC}$$

$$\beta_F = \beta_0 = 50 \qquad\qquad V_{BE(on)} = 0.7 \text{ V}$$

For Q_3 and Q_4 we have

$$I_{C3} = I_{C4} = 1 \text{ mA}$$

$$V_{CB3} = V_{CB4} = \tfrac{1}{2}(V_{CC} - V_{B3})$$

$$\beta_F = \beta_0 = 50 \qquad\qquad V_{BE(on)} = 0.7 \text{ V}$$

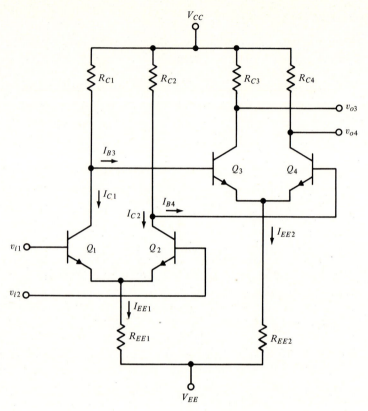

FIGURE 8.8
Cascade of two differential pairs.

8.4.1 DC Conditions

Since $V_{B3} = V_{C1} = \frac{1}{2}V_{CC} = 7.5$ V, we have when $V_{B1} = V_{B2} = 0$ V

$$V_{CB3} = \frac{1}{2}(15 - 7.5) = 3.75 \text{ V}$$

hence

$$V_{C3} = V_{B3} + V_{CB3} = 7.5 + 3.75 = 11.25 \text{ V}$$

Also

$$I_{B1} = I_{B2} = \frac{I_{C1}}{\beta_F} = \frac{100 \ \mu A}{50} = 2 \ \mu A$$

and

$$I_{B3} = I_{B4} = \frac{I_{C3}}{\beta_F} = \frac{1.0 \text{ mA}}{50} = 0.02 \text{ mA}$$

Now I_{B3} must flow through R_{C1}, then

$$R_{C1} = R_{C2} = \frac{V_{CC} - V_{C1}}{I_{C1} + I_{B3}} \tag{8.20}$$

$$= \frac{15 - 7.5}{0.1 + .02} = 62.5 \text{ k}\Omega$$

and

$$R_{C3} = R_{C4} = \frac{V_{CC} - V_{C3}}{I_{C3}} \tag{8.21}$$

$$= \frac{15 - 11.25}{1.0} = 3.75 \text{ k}\Omega$$

We calculate values for I_{EE1} and I_{EE2} as

$$I_{EE1} = I_{E1} + I_{E2} = 0.102 + 0.102 = 0.204 \text{ mA}$$

and

$$I_{EE2} = I_{E3} + I_{E4} = 1.02 + 1.02 = 2.04 \text{ mA}$$

With the input common-mode voltage set to zero, we have

$$R_{EE1} = \frac{V_{E1} - V_{EE}}{I_{EE1}} = \frac{V_{B1} - V_{BE(on)} - V_{EE}}{I_{EE1}} \tag{8.22}$$

$$= \frac{0 - 0.7 - (-15)}{0.204} = 70 \text{ k}\Omega$$

$$R_{EE2} = \frac{V_{E3} - V_{EE}}{I_{EE2}} = \frac{V_{B3} - V_{BE(on)} - V_{EE}}{I_{EE2}} \tag{8.23}$$

$$= \frac{7.5 - 0.7 - (-15)}{2.04} = 10.7 \text{ k}\Omega$$

EXERCISE

E8.5 Calculate the resistance values in Fig. 8.8 but with $V_{CC} = +6$ V and $V_{EE} = -6$ V. All other operating data are as in Sec. 8.4.

8.4.2 Small-signal Characteristics

In this section we will derive equations for the same small-signal parameters as for the single differential pair. We first consider the difference-mode components and then the common-mode components. In each case we make use of the respective half-circuits.

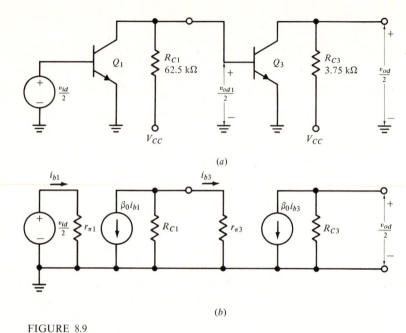

FIGURE 8.9
(a) The difference-mode half-circuit for Fig. 8.8; (b) the small-signal circuit model.

Difference mode To obtain the difference-mode half-circuit we divide each differential-amplifier pair at the points of symmetry and ground (for ac purposes) the center points. We then have the circuit shown in Fig. 8.9a. The small-signal circuit model for this circuit is shown in Fig. 8.9b.

The small-signal input resistance of the first half-circuit is just $r_{\pi 1}$. Therefore the difference-mode input resistance is

$$r_{id} = 2r_{\pi 1} \qquad (8.24)$$

For $I_{C1} = 0.1$ mA, $r_{\pi 1} = 13$ kΩ; hence

$$r_{id} = (2)(13) = 26 \text{ k}\Omega$$

To determine the difference-mode voltage gain for the first stage, a_{d1}, we must account for the loading at the collector of Q_1 due to Q_3. Hence

$$a_{d1} = \frac{v_{od1}}{v_{id}} = \frac{-\beta_0 R_{L1}}{r_{\pi 1}} \qquad (8.25a)$$

where

$$R_{L1} = R_{C1} \| r_{\pi 3}$$

For $I_{C3} = 1$ mA, $r_{\pi 3} = 1.3$ kΩ; then with $R_{C1} = 62.5$ kΩ

$$R_{L1} = \frac{(62.5)(1.3)}{62.5 + 1.3} = 1.27 \text{ k}\Omega$$

therefore

$$a_{d1} = \frac{-(50)(1.27)}{13} = -4.9$$

Since the load at the collector of Q_3 is only R_{C3}, the difference-mode voltage gain of the second stage is given by

$$a_{d3} = \frac{v_{od}}{v_{od1}} = \frac{-\beta_0 R_{C3}}{r_{\pi 3}} \tag{8.25b}$$

$$= \frac{-(50)(3.75)}{1.3} = -144$$

Thus the overall difference-mode voltage gain is

$$a_d = \frac{v_{od}}{v_{id}} = (a_{d1})(a_{d3}) \tag{8.25c}$$

$$= (-4.9)(-144) = 706$$

where from Fig. 8.8, $v_{od} = v_{o3} - v_{o4}$ and $v_{id} = v_{i1} - v_{i2}$.

Common mode To aid in the determination of the common-mode characteristics we show the common-mode half-circuit in Fig. 8.10a and the small-signal circuit model in Fig. 8.10b.

The small-signal common-mode input resistance is given by

$$r_{ic} = r_{\pi 1} + (\beta_0 + 1)2R_{EE1} \tag{8.26}$$

$$= 13 + (51)(2)(70) = 7150 \text{ k}\Omega = 7.15 \text{ M}\Omega$$

The common-mode voltage gain of Q_1 is calculated from

$$a_{c1} = \frac{v_{oc1}}{v_{ic}} = \frac{-\beta_0 R_{L1}}{r_{\pi 1} + (\beta_0 + 1)2R_{EE1}} \tag{8.27a}$$

where

$$R_{L1} = R_{C1} \| [r_{\pi 3} + (\beta_0 + 1)2R_{EE2}]$$

hence

$$R_{L1} = \frac{62.5[1.3 + (51)(2)(10.7)]}{62.5 + 1.3 + (51)(2)(10.7)} = 59 \text{ k}\Omega$$

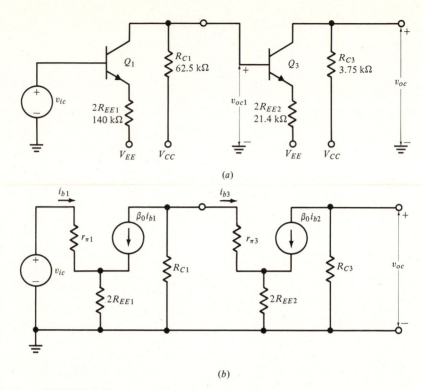

FIGURE 8.10
(a) The common-mode half-circuit for Fig. 8.8; (b) the small-signal circuit model.

Therefore

$$a_{c1} = \frac{-(50)(59)}{13 + (51)(2)(70)} = -0.41$$

For the second stage, a_{c3} is given by

$$a_{c3} = \frac{v_{oc}}{v_{oc1}} = \frac{-\beta_0 R_{C3}}{r_{\pi 3} + (\beta_0 + 1)2R_{EE2}} \qquad (8.27b)$$

$$= \frac{-(50)(3.75)}{1.3 + (51)(2)(10.7)} = -0.172$$

Hence the overall common-mode voltage gain is

$$a_c = (a_{c1})(a_{c3}) \qquad (8.27c)$$

$$= (-0.41)(-0.172) = 0.071$$

The CMRR for our circuit of Fig. 8.8 is

$$\text{CMRR (dB)} = 20 \log \left| \frac{a_d}{a_c} \right| \tag{8.28}$$

$$= 20 \log \left[\frac{706}{0.071} \right] \approx 80 \text{ dB}$$

This is in the range of the CMRR of commercial IC operational amplifiers, which is from 70 to 100 dB.

The characteristics of a modified version of the circuit in Fig. 8.8 are experimentally determined in Demonstration D8.2.

EXERCISE

E8.6 Calculate r_{id}, r_{ic}, a_d, and a_c for the circuit shown in Fig. 8.8 but with $V_{CC} = +6$ V and $V_{EE} = -6$ V. Also determine the CMRR for the circuit. The operating-point data are as in Sec. 8.4.

8.5 CONVERSION TO SINGLE-ENDED

So far our analysis of difference amplifiers has been concerned with obtaining a differential output voltage from a differential input voltage. Such amplifiers are available as commercial ICs, but more common is the type of amplifier we showed in Fig. 8.2a. This is an amplifier or gain block with a single-ended output from a differential input. Additionally, the general requirement is that the single-ended output have a quiescent voltage level of 0 V. Then with the input voltage level also at 0 V we can readily cascade these gain blocks for even more amplification.

One way to restore a single-ended output to 0 V is shown in Fig. 8.11a, where we "float" a battery whose voltage is E volts. For the values we have used in Sec. 8.3 this would require a 7.5-V battery. However, a floating 7.5-V battery in an IC package is not a very practical solution. Other solutions more consistent with IC fabrication techniques are given in the next two sections.

8.5.1 Level Shifting with a Resistor and Current Source

Another solution to the dc offset problem is shown in Fig. 8.11b. Here a resistor R_X is used with an ideal current source I_X to produce a voltage drop $I_X R_X$, so that for $V_0 = 0$ V, $I_X R_X = V_{C2}$. However, we note that the current through R_{C2} will now be the sum of I_X and I_{C2}. Therefore if we want to maintain V_{C2} at a specified

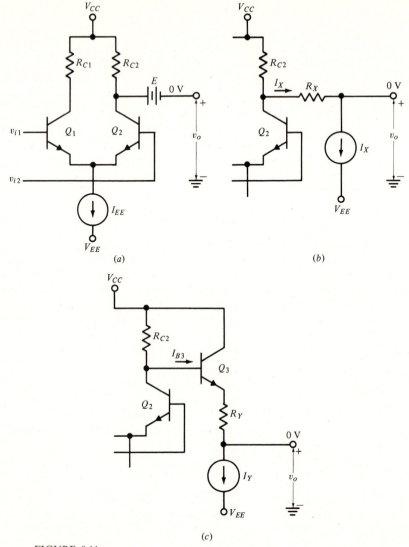

FIGURE 8.11
Voltage-level-shifting circuits, (a) with a "floating battery," (b) with a resistor and
current source, (c) with an emitter follower.

level, we will have to change R_{C2}. If we choose $I_X = 0.1$ mA, and from our
previous example

$$V_{CC} = 15 \text{ V}$$

$$V_{C2} = 7.5 \text{ V}$$

$$I_{C2} = 0.1 \text{ mA}$$

then

$$R_X = \frac{7.5}{0.1} = 75 \text{ k}\Omega$$

but now

$$R_{C2} = \frac{7.5}{0.2} = 37.5 \text{ k}\Omega$$

Now recall the differential voltage gain to the collector of Q_2, and hence the output, is given by

$$a_d = \frac{-\beta_0 R_{C2}}{r_{\pi 2}}$$

Therefore compared to our circuit of Fig. 8.4, where R_{C2} was 75 kΩ, the voltage gain is now reduced by a factor of 2. Also notice that the output resistance of the circuit in Fig. 8.11b is the sum of R_X and R_{C2}. That is,

$$r_o = 75 + 37.5 = 112.5 \text{ k}\Omega$$

This is a large value for the output resistance, and any subsequent loading of the circuit will surely reduce the output signal voltage. Hence the voltage gain to the output is reduced even further.

8.5.2 Level Shifting with an Emitter Follower

A common practice of level shifting found in commercial ICs is shown in Fig. 8.11c. The circuit makes use of an emitter follower Q_3 between the collector of Q_2 and the dropping resistor R_Y.

We again specify $V_{C2} = 7.5$ V and $I_Y = 0.1$ mA. Then the voltage at the emitter of Q_3 is one base-emitter voltage drop below the voltage at the collector of Q_2. That is,

$$V_{E3} = 7.5 - 0.7 = 6.8 \text{ V}$$

Then for 0 V at the output,

$$R_Y = \frac{6.8 - 0}{0.1} = 68 \text{ k}\Omega$$

The current through R_{C2} is now the sum of I_{C2} and I_{B3}, where

$$I_{B3} = \frac{I_Y}{\beta_F + 1} = \frac{0.1}{51} = 0.002 \text{ mA}$$

Therefore

$$R_{C2} = \frac{V_{CC} - V_{C2}}{I_{RC2}} = \frac{15 - 7.5}{0.102} = 73.3 \text{ k}\Omega$$

Compared to our original circuit of Fig. 8.4 the reduction in voltage gain is now very slight. However, before making any final conclusions we need to determine the output resistance for our latest circuit.

From Fig. 8.11c we see that the output resistance is made up of R_Y in series with r_{o3}, the output resistance of the emitter follower Q_3. From Eq. (7.40c) we calculate r_{o3} as

$$r_{o3} = \frac{R_{C2} + r_{\pi 3}}{\beta_0 + 1} = \frac{73.3 + 13}{51} = 1.7 \text{ k}\Omega$$

Hence the output resistance of our new circuit is

$$r_o = 68 + 1.7 = 69.7 \text{ k}\Omega$$

This is smaller than the output resistance of the circuit without the emitter follower but is still too large a value when we consider any practical loading at the output of the circuit. The solution, as one suspects, is to add another emitter follower after R_Y. However, we will leave this to a later section, where we describe the complete circuit of a commercial IC operational amplifier.

In this section we have made use of an ideal dc current source. In the next section we see how a close approximation to an ideal current source is readily obtained in ICs.

EXERCISE

E8.7 Determine the value of the circuit elements of Fig. 8.11a, b, and c by using

$$V_{CC} = +15 \text{ V} \qquad\qquad V_{EE} = -15 \text{ V}$$
$$I_{C1} = I_{C2} = 1 \text{ mA} \qquad\qquad V_{C1} = V_{C2} = 11.25 \text{ V}$$
$$I_X = I_Y = 1 \text{ mA}$$
$$\beta_F = 50 \qquad\qquad V_{BE(\text{on})} = 0.7 \text{ V}$$

8.6   CURRENT SOURCES IN ICs

Beside their use in level shifting we will find a current source useful to replace the emitter resistor R_{EE} in the differential-amplifier pair. In this connection, we have already noted that one figure of merit is the CMRR. We can increase the CMRR if we decrease the value of the common-mode voltage gain. We refer to Eq. (8.14) and note one method of doing just that is to increase the value of R_{EE}. However, for a given value of emitter current for the amplifier pair, increasing the value of R_{EE} can only lead to a large value for V_{EE}. The choice of the magnitude for V_{EE} is generally not unrestricted. Also, increasing V_{EE} increases the power dissipation of the circuit.

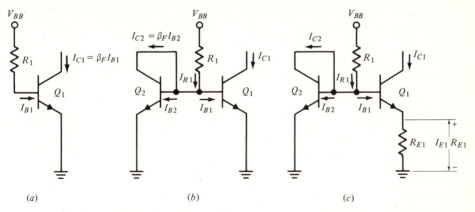

FIGURE 8.12
Current sources in ICs. (*a*) Simple transistor current source; (*b*) diode-biasing scheme; (*c*) improved diode-biasing scheme.

What if we replace R_{EE} with an ideal current source? Then we have an infinite resistance for R_{EE}; hence the common-mode voltage gain goes to zero and the CMRR approaches infinity. However, the output resistance of a practical current source is limited, and therefore there is a finite limit to the value we can obtain for the CMRR.

The collector characteristic of an ideal transistor, operated in the active region, is simply a dc current generator. Therefore to obtain a current source let us look at the circuit of Fig. 8.12*a*. With the transistor in the active region, the collector current is $\beta_F I_{B1}$, where

$$I_{B1} = \frac{V_{BB} - V_{BE(\text{on})}}{R_1}$$

While the base current is reasonably constant with temperature, the value of the collector current is also a function of β_F, and this can vary over a wide range. On a production run of IC wafers, β_F can vary from 50 to 500 from wafer to wafer. Over a temperature range of 50°C, β_F will also change approximately 25 percent. Therefore the circuit of Fig. 8.12*a* is unacceptable as a current source.

8.6.1   Diode-biasing Scheme

An improved scheme to obtain a current source in an IC is shown in Fig. 8.12*b*. Here, two identical transistors in close proximity on the chip are used. Since Q_2 is connected as a diode, this method is sometimes called *diode biasing*.

With transistors Q_1 and Q_2 identical, we can show that the collector currents are also identical if both transistors are operated in the active region with equal V_{BE}. Notice with the collector of Q_2 connected to its base, $V_{CE2} = V_{BE2}$, and for small collector currents Q_2 must be in the active region. Then, from our basic study of transistors (in Sec. 1.4.3), we know

$$I_{C1} = \alpha_{F1} I_{E1} = \alpha_{F1} I_{ES1} e^{V_{BE1}/V_T} \qquad (8.29a)$$

and

$$I_{C2} = \alpha_{F2} I_{E2} = \alpha_{F2} I_{ES2} e^{V_{BE2}/V_T} \qquad (8.29b)$$

Since Q_1 and Q_2 are identical and at the same temperature,

$$\alpha_{F1} = \alpha_{F2} \qquad I_{ES1} = I_{ES2} \qquad V_{BE1} = V_{BE2}$$

Hence

$$I_{C1} = I_{C2}$$

From applying KCL to the circuit of Fig. 8.12b,

$$I_{R1} = I_{B1} + I_{B2} + I_{C2} \qquad (8.30a)$$

and for $\beta_F \gg 1$ we can neglect the contribution of I_{B1} and I_{B2}. Then†

$$I_{R1} \approx I_{C2} = I_{C1} \qquad (8.30b)$$

Now by KVL, we also know

$$I_{R1} = \frac{V_{BB} - V_{BE(on)}}{R_1} \qquad (8.31)$$

Therefore the value of I_{R1}, and hence I_{C1}, is determined by V_{BB}, $V_{BE(on)}$, and R_1, is practically independent of β_F, and is relatively constant with temperature. As an example: for $I_{C1} = 0.1$ mA, with $V_{BB} = 15$ V and $V_{BE(on)} = 0.7$ V,

$$R_1 = \frac{15 - 0.7}{0.1} = 143 \text{ k}\Omega$$

8.6.2 Improved Diode-biasing Scheme

The value of the resistor R_1, just calculated, takes up a significant amount of chip area in the IC if made as a diffused resistor. One way to reduce the value of R_1 and still be able to make a low current source is shown in Fig. 8.12c. Here transistor Q_1 has an additional resistor R_{E1} added in the emitter lead. To explain this circuit let us design for the same I_{C1} that we had in the previous section but now limit R_1 to one-tenth the value used in Fig. 8.12b, and hence to one-tenth the area. This

---

† In most practical situations we may assume $I_{C1} = I_{R1}$. We can then say that the current I_{C1} is the mirror (image) of the current I_{R1}. Hence the circuit of Fig. 8.12b is sometimes referred to as a *current-mirror* circuit.

requires that $R_1 = 14.3 \text{ k}\Omega$. For this condition, with $\beta_F \gg 1$ we calculate a value for $I_{R1} \approx I_{C2}$; namely,

$$I_{C2} = \frac{V_{BB} - V_{BE(\text{on})}}{R_1} \tag{8.32}$$

$$= \frac{15 - 0.7}{14.3} = 1.0 \text{ mA}$$

Now recall that a 60-mV decrease in the voltage across a diode decreases the current through the diode by a factor of 10. Hence with $I_{C1} = 0.1 \text{ mA}$ and $I_{C2} = 1.0 \text{ mA}$, we require that

$$V_{BE1} = V_{BE2} - 60 \text{ mV}$$

Since by KVL in Fig. 8.12c,

$$V_{BE2} = V_{BE1} + I_{E1} R_{E1} \tag{8.33}$$

and for $\beta_F \gg 1$, $I_{E1} = I_{C1}$; then

$$R_{E1} = \frac{V_{BE2} - V_{BE1}}{I_{C1}} \tag{8.34}$$

$$= \frac{60 \text{ mV}}{0.1 \text{ mA}} = 600 \text{ }\Omega$$

A disadvantage of this circuit, compared with that in Fig. 8.12b, is that the power dissipated in the base-biasing circuit is 10 times larger. However, in many circuits this power dissipation is still rather small and not necessarily a limiting factor.

An advantage of the circuit shown in Fig. 8.12c is that the inclusion of resistor R_{E1} increases the output resistance at the collector of transistor Q_1. In a practical transistor circuit, the collector output resistance is increased by a factor of about β_F, the current gain of the transistor (see Prob. P8.26). Hence we practically have almost an ideal current source. As we have already seen this will help improve the CMRR.

The advantage of using a transistor current source in biasing a differential-amplifier pair is experimentally determined in Demonstration D8.3. The common-mode properties of this circuit can then be compared with the resistor-biased differential amplifier of Demonstration D8.1.

EXERCISE

E8.8 In Fig. 8.12c, determine R_{E1} for $I_{R1} = 1 \text{ mA}$ and $I_{C1} = 10 \text{ }\mu\text{A}$.

8.6.3   Current Source with Dissimilar Transistors

The circuit of Fig. 8.12b can be used to solve the problem of the preceding section, provided the transistors Q_1 and Q_2 are identical in all respects except for different areas on the chip. If the effective emitter area of Q_1 is one-tenth that of Q_2, then $I_{ES1} = 0.1 I_{ES2}$. Hence for $V_{BE1} = V_{BE2}$ and $\beta_F \gg 1$,

$$I_{C1} \approx I_{E1} = 0.1 I_{E2} \approx 0.1 I_{C2}$$

Then for $I_{C2} = 1$ mA, $I_{C1} = 0.1$ mA. The value of R_1 is still calculated as in Eq. (8.31). That is, $R_1 = 14.3$ kΩ.

One further point should be mentioned. Several transistors can have their bases connected to the base of the diode-connected transistor. In this manner we can establish multiple constant-current sources by using just one bias circuit, consisting of transistor Q_2 and resistor R_1.

8.7   DESIGN EXAMPLES OF COMPLETE DIFFERENTIAL AMPLIFIERS

Since the differential amplifier is the basic circuit of one of the most popular forms of linear ICs, the operational amplifier, in the next two sections we will discuss the design of two commercial IC op amps. The first reviews all that we have discussed so far in a design that incorporates many of the features of the MC1530 introduced by Motorola Semiconductor. We have elected to simplify the output stage in order to avoid undue complications.

The second example is the type 741 operational amplifier introduced by Fairchild Semiconductor. It contains several new features that we have not yet covered; these are the use of both *n-p-n* and *p-n-p* transistors as active collector load resistors as well as *p-n-p* transistors used for level shifting and in a complementary emitter-follower output stage. The use of *p-n-p* transistors makes possible a significant improvement in many performance parameters over an all–*n-p-n* structure.

8.7.1   Design of All–*n-p-n* Operational Amplifier

In Fig. 8.13 we show a circuit diagram very similar to that for the MC1530 except for the modified output stage. This simple circuit exhibits many of the essential features common to most IC operational-amplifier designs. First, notice the amplifier is completely dc-coupled; no coupling or bypass capacitors are used. The circuit consists essentially of three basic stages: a differential input stage consisting of Q_1 and Q_2, a differential intermediate stage made up of Q_3 and Q_4, and a single-ended output stage incorporating two emitter followers for dc level shift and low output resistance (Q_5 and Q_6). Transistors Q_7 and Q_8 are used as constant-current sources.

The input differential-amplifier pair in Fig. 8.13 is similar to the differential-input–differential-output amplifier of Fig. 8.4. The only change is the use of a transistor current source (made up of Q_7, R_7, R_8, R_9, D_1, and D_2) in place of R_{EE}. The effect of the emitter resistor R_7 is to increase the collector output resistance of Q_7.

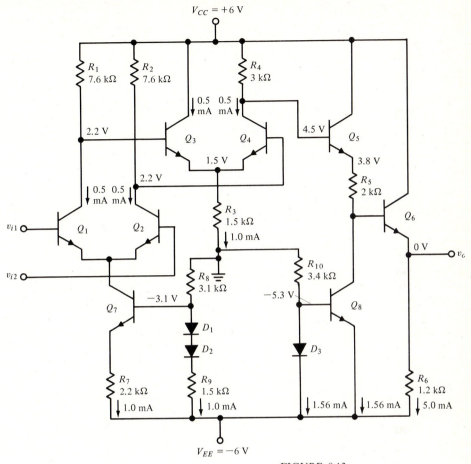

FIGURE 8.13
All–n-p-n transistor operational amplifier.

The intermediate stage of the circuit in Fig. 8.13 is a differential pair biased with resistor R_3. A transistor current source is not necessary in this stage since sufficient common-mode rejection has been established in the input stage. Only a single-ended output is taken from the intermediate stage; thus, only one-half the full differential voltage gain of the stage is utilized.

The dc level shift after the intermediate stage is used to establish the output operating level at 0 V and is achieved by Q_5 and R_5 along with the current source from Q_8. The function of these components is identical to the network shown in Fig. 8.11c. The current source operates in a similar manner to the circuit shown in Fig. 8.12b.

A low output resistance for the operational amplifier is obtained from a single emitter-follower stage: transistor Q_6 and its bias resistor R_6.

8.7.2 DC Conditions

Let us now see how the resistor values in Fig. 8.13 were obtained from the dc design. We will make the following assumptions:

$$V_{CC} = +6 \text{ V} \qquad\qquad V_{EE} = -6 \text{ V}$$

$$\beta_F = \infty \qquad V_{D(\text{on})} = V_{BE(\text{on})} = 0.7 \text{ V}$$

The operating currents for the transistors have been selected as

$$I_{C1} = I_{C2} = 0.5 \text{ mA}$$

$$I_{C3} = I_{C4} = 0.5 \text{ mA}$$

$$I_{C5} = 1.56 \text{ mA}$$

$$I_{C6} = 5.0 \text{ mA}$$

Diodes D_1 and D_2 are diode-connected transistors (collector connected to base) that are identical to Q_7. Similarly for D_3 and Q_8.

The value for R_7, in the emitter of the current source Q_7, is selected on the basis of a high CMRR for the input differential pair. The chosen value is 2.2 kΩ. We first solve values for R_8 and R_9.

Since the current in the current source Q_7 is the sum of the emitter currents for Q_1 and Q_2; that is,

$$I_{C7} = I_{E1} + I_{E2}$$

then

$$I_{C7} = 1 \text{ mA}$$

Moreover for the current source we also require $I_{D1} = I_{D2} = 1$ mA. Hence by KVL,

$$I_{D1} = I_{D2} = I_{R8} = I_{R9} = \frac{0 - 2V_{D(\text{on})} - V_{EE}}{R_8 + R_9}$$

therefore

$$R_8 + R_9 = \frac{0 - 1.4 - (-6)}{1.0} = 4.6 \text{ k}\Omega$$

Also by KVL,

$$I_{E7}R_7 + V_{BE\,7(\text{on})} + I_{R8}R_8 = -V_{EE}$$

With $I_{E7} = 1$ mA and $I_{R8} = 1$ mA, we have

$$1 \text{ mA } (R_7 + R_8) = (6 - 0.7) \text{ V}$$

Since $R_7 = 2.2$ kΩ, then

$$R_8 = 3.1 \text{ k}\Omega \qquad \text{and} \qquad R_9 = 1.5 \text{ k}\Omega$$

The value of R_1 and R_2 is selected on the basis of *input common-mode voltage swing*. This can be explained as follows. One limit of the input common-mode voltage is when this voltage becomes too low; that is, it goes too far in the negative direction. Then with the emitter voltage of Q_1 and Q_2 (i.e., the collector of Q_7) following the input voltage, transistor Q_7 will go into saturation when the base-collector junction of Q_7 becomes forward-biased. Our constant current source and CMRR will then suffer. Since the voltage at the base of Q_7 is fixed by the voltage drop across R_8 to -3.1 V, for $V_{CE(sat)} = 0.2$ V, the collector voltage of Q_7 cannot go below -3.6 V. This means the input common-mode voltage cannot go below -2.9 V, one $V_{BE(on)}$ greater than -3.6 V.

For the maximum positive-going common-mode input signal we are limited by transistors Q_1 and Q_2 going into saturation. The value for R_1 and R_2 was selected to be 7.6 kΩ; thus with $I_{C1} = I_{C2} = 0.5$ mA the operating collector voltage level of Q_1 and Q_2 is about 2.2 V. The input transistors will saturate when $V_{BC} = 0.5$ V; hence the upper limit for the common-mode input voltage is $+2.7$ V. For a symmetrical input waveform, the maximum input common-mode voltage swing is therefore ± 2.7 V.

Now the bases of transistors Q_3 and Q_4 are connected directly to the collectors of Q_1 and Q_2. Hence the voltage at the emitter of the intermediate differential pair is

$$V_{E3} = V_{E4} = V_{C1} - V_{BE(on)}$$

$$= 2.2 - 0.7 = 1.5 \text{ V}$$

Then with $I_{R3} = I_{E3} + I_{E4} = 1.0$ mA,

$$R_3 = \frac{V_{E3}}{I_{R3}} = \frac{1.5}{1.0} = 1.5 \text{ k}\Omega$$

Turning now to the resistor R_4, we choose a value for this resistor so that the common-mode voltage gain of the intermediate stage cannot be greater than 1. Since from Eq. (8.14)

$$a_c \approx -\frac{R_C}{2R_{EE}}$$

we select

$$R_4 = 2R_3$$

hence

$$R_4 = 3 \text{ k}\Omega$$

The operating voltage level at the base of Q_5 is then

$$V_{B5} = V_{CC} - I_{C4}R_4$$

$$= 6 - (0.5)(3) = 4.5 \text{ V}$$

For the current source Q_8, we require

$$I_{C8} = I_{D3} = 1.56 \text{ mA}$$

therefore, using KVL,

$$R_{10} = \frac{6 - 0.7}{1.56} = 3.4 \text{ k}\Omega$$

With a quiescent output voltage level of 0 V, the voltage at the base of Q_6 is 0.7 V. We have already determined that the voltage at the base of Q_5 is 4.5 V. Hence with $I_{C5} = I_{C8} = 1.56 \text{ mA}$,

$$R_5 = \frac{4.5 - 0.7 - 0.7}{1.56} = 2.0 \text{ k}\Omega$$

We specified that $I_{C6} = 5 \text{ mA}$; therefore with 0 V at the output and $V_{EE} = -6 \text{ V}$, $R_6 = 1.2 \text{ k}\Omega$.

This completes the dc design. With this information there are two parameters for the operational amplifier we can compute at this time: the power dissipation for the amplifier and the maximum output voltage swing.

Power dissipation The power dissipation for the circuit in Fig. 8.13 is easily calculated from

$$\begin{aligned}
\text{PD} &= V_{CC}(I_{C1} + I_{C2} + I_{C3} + I_{C4} + I_{C5} + I_{C6}) \\
&\quad + V_{EE}(I_{E7} + I_{D1} + I_{D3} + I_{E8} + I_{E6}) \\
&= 6(0.5 + 0.5 + 0.5 + 0.5 + 1.56 + 5.0) + 6(1.0 + 1.0 + 1.56 + 1.56 + 5.0) \\
&= 112 \text{ mW}
\end{aligned}$$

Output voltage swing With no load, the maximum output voltage swing for this circuit is controlled by the intermediate amplifier stage. The highest voltage that the collector of transistor Q_4 can swing up to is $+6$ V; this is when Q_4 is cut off. The quiescent voltage level for the collector of Q_4 is 4.5 V; therefore there is a 1.5-V positive swing. The attenuation of this signal voltage through the level-shifting network is small because of the high resistance of the current source Q_8, and we are considering no load at the output.

The other limit for the intermediate differential amplifier is when transistor Q_3 is cut off. Now all the current in resistor R_3 is through Q_4. Assuming there is no overdrive to the base of Q_4 (that is, I_{R3} is fixed at 1 mA), the lowest voltage that the collector of Q_4 can swing down to is

$$\begin{aligned}
V_{C4} &= V_{CC} - I_{R3}R_4 \\
&= 6 - (1.0)(3) = 3 \text{ V}
\end{aligned}$$

The negative voltage swing at the collector is therefore from 4.5 to 3 V, that is, 1.5 V.

The signal voltage from the input differential pair could be sufficient to cause transistor Q_4 to saturate. In which case the voltage at the collector of Q_4 is given more precisely by

$$V_{C4} = V_{CC} - I_{C(\text{sat})} R_4$$

where

$$I_{C(\text{sat})} = \frac{V_{CC} - V_{CE(\text{sat})}}{R_3 + R_4} = \frac{6 - 0.2}{1.5 + 3.0} = 1.29 \text{ mA}$$

then

$$V_{C4} = 6 - (1.29)(3) = 2.1 \text{ V}$$

Now the negative voltage swing is -2.4 V.

The output voltage swing is normally listed for a symmetrical voltage change. For the circuit in Fig. 8.13, the maximum output voltage swing is therefore ± 1.5 V. A greater output voltage swing is possible, but then the positive side of the waveform would be *clipped* at 1.5 V, leading to distortion of the output signal.

8.7.3 Small-signal Characteristics

The analysis for the two differential-amplifier stages in Fig. 8.13 follows exactly that for the two cascaded differential pairs of Fig. 8.8, described in Sec. 8.4.3. We now assume $\beta_0 = 100$.

Differential input resistance   The difference-mode input resistance for Q_1 and Q_2 is calculated as

$$r_{id} = 2r_{\pi 1} = \frac{2\beta_0 V_T}{I_{c1}}$$

then

$$r_{id} = \frac{(2)(100)(26 \text{ mV})}{0.5 \text{ mA}} = 10.4 \text{ k}\Omega$$

Differential voltage gain   The load at the collector of Q_1 is due to R_1 and the input resistance of Q_3. Hence

$$R_{L1} = R_1 \| r_{\pi 3}$$

where

$$r_{\pi 3} = \frac{\beta_0 V_T}{I_{C3}} = \frac{(100)(26 \text{ mV})}{0.5 \text{ mA}} = 5.2 \text{ k}\Omega$$

Therefore

$$R_{L1} = 7.6 \text{ k}\Omega \| 5.2 \text{ k}\Omega = 3.1 \text{ k}\Omega$$

Thus

$$|a_{d1}| = \frac{\beta_0 R_{L1}}{r_{\pi 1}} = \frac{(100)(3.1)}{5.2} = 60$$

The load at the collector of Q_4 must be determined as follows. The input resistance of Q_5 is

$$r_{i5} = r_{\pi 5} + (\beta_0 + 1)(R_5 + r_{i6})$$

where the input resistance of Q_6 is

$$r_{i6} = r_{\pi 6} + (\beta_0 + 1)R_6$$

Then

$$r_{\pi 5} = \frac{\beta_0 V_T}{I_{C5}} = \frac{(100)(26 \text{ mV})}{1.56 \text{ mA}} = 1.67 \text{ k}\Omega$$

and

$$r_{\pi 6} = \frac{\beta_0 V_T}{I_{C6}} = \frac{(100)(26 \text{ mV})}{5 \text{ mA}} = 0.52 \text{ k}\Omega$$

Now

$$r_{i6} = 0.52 + (101)(1.2) = 121.7 \text{ k}\Omega$$

and

$$r_{i5} = 1.67 + (101)(123.7) = 12,500 \text{ k}\Omega$$

In the collector circuit of Q_4, with 12,500 k$\Omega \gg$ 3 kΩ, we can ignore the effect of r_{i5} in parallel with R_4; hence $R_{L4} = 3$ kΩ, and the magnitude of the gain for the intermediate differential-amplifier stage, a_{d2}, is given by

$$|a_{d2}| = \frac{\beta_0 R_{L4}}{r_{\pi 4}}$$

where

$$r_{\pi 4} = \frac{\beta_0 V_T}{I_{C4}} = \frac{(100)(26 \text{ mV})}{0.5 \text{ mA}} = 5.2 \text{ k}\Omega$$

Thus

$$|a_{d2}| = \frac{(100)(3)}{5.2} = 58$$

To determine the differential voltage gain to the base of Q_5, we note that in going single-ended we are only using one-half the full differential gain available from Q_3 and Q_4. Hence

$$\frac{v_{b5}}{v_{id}} = a_{d1}\frac{a_{d2}}{2}$$

Also notice that for an input signal at the base of Q_1 there is a phase inversion to the collector of Q_1; however the signal at the base of Q_5 is in phase with the collector of Q_1. For a signal at the base of Q_2 there is a phase reversal to the collector of Q_2, followed by a further phase reversal to the base of Q_5. Hence v_{i2} is termed the noninverting input and v_{i1} is the inverting input. Now

$$\frac{v_{b5}}{v_{id}} = -(60)\frac{58}{2} = -1740$$

where

$$v_{id} = v_{i1} - v_{i2}$$

The rest of the differential-gain calculation is an exercise in the use of emitter followers. First we note that

$$\frac{v_{b6}}{v_{b5}} = \frac{(\beta_0 + 1)r_{i6}}{r_{i5}}$$

and

$$\frac{v_o}{v_{b6}} = \frac{(\beta_0 + 1)R_6}{r_{i6}}$$

Hence

$$\frac{v_o}{v_{b5}} = \frac{(\beta_0 + 1)^2 R_6}{r_{i5}} = \frac{(101)^2(1.2)}{12,500} = 0.98$$

Therefore the differential voltage gain for the complete amplifier circuit of Fig. 8.13 is

$$a_d = \frac{v_o}{v_{id}} = -(1740)(0.98) = -1700$$

Output resistance We determine the output resistance of the operational amplifier by the repeated application of the output resistance of an emitter follower from Eq. (7.40c). Namely,

$$r_o = R_6 \| r_{o6}$$

where

$$r_{o6} = \frac{R_{s6} + r_{\pi 6}}{\beta_0 + 1}$$

and R_{s6} is the small-signal source resistance to the base of Q_6:

$$R_{s6} = R_5 + r_{o5}$$

and

$$r_{o5} = \frac{R_4 + r_{\pi 5}}{\beta_0 + 1}$$

with $r_{\pi 5} = 1.67 \text{ k}\Omega$,

$$r_{o5} = \frac{3 + 1.67}{101} = 0.046 \text{ k}\Omega$$

Then

$$R_{s6} = 2.0 + 0.046 = 2.046 \text{ k}\Omega$$

and

$$r_{o6} = \frac{2.046 + 0.52}{101} = 0.025 \text{ k}\Omega$$

Finally

$$r_o = \frac{(1.2)(0.025)}{1.2 + 0.025} = 0.024 \text{ k}\Omega = 24 \text{ }\Omega$$

Common mode   The common-mode input resistance for a differential-amplifier pair is, from Eq. (8.13),

$$r_{ic} = r_{\pi 1} + (\beta_0 + 1)2R_{EE}$$

The input amplifier pair in Fig. 8.13 uses a transistor current source in place of R_{EE}. If transistor Q_7 was an ideal current generator with an output resistance approaching infinity, we would decide that the common-mode input resistance of the amplifier also approaches infinity. However, in practice the output resistance at the collector of Q_7, while very high, is not infinite. Typically, the output resistance of a transistor current source may range from 10 kΩ to 10 MΩ or more. Hence the common-mode input resistance is generally > 1.0 MΩ.

The common-mode voltage gain is, from Eq. (8.14),

$$a_{c1} = \frac{-\beta_0 R_C}{r_{ic}}$$

For our example, with $R_C = 7.6$ kΩ and r_{ic} taken as 10 MΩ,

$$a_{c1} = \frac{-(100)(7.6)}{10,000} = -0.08$$

Finally the CMRR for the circuit of Fig. 8.13 is given from Eq. (8.28):

$$\text{CMRR} = 20 \log \left| \frac{a_d}{a_c} \right| = 20 \log \left[\frac{1700}{(0.08)(0.98)} \right] = 87 \text{ dB}$$

Demonstration   An all–*n-p-n* operational amplifier similar to what we have analyzed in this section is the subject of Demonstration D8.4. Hence each item of our analysis is made available for experimental verification, and this is unreservedly encouraged.

8.7.4   MC1530/MC1531 Operational Amplifier

As we have previously mentioned, the design example we have just completed is very similar to the MC1530. In this section we will make a comparison between some of the characteristics contained in the data sheet for this operational amplifier (see Fig. 8.14) and the results obtained in the previous section. Also notice from the data-circuit schematic that the MC1531 is similar to the MC1530 except it has a Darlington circuit at each input. Consequently we would expect a higher input resistance for the MC1531. The operating temperature range of the MC1530 and 1531 is from -55 to $+125°C$. The MC1430 and 1431 are intended for the more restricted temperature range of 0 to $+75°C$.

Input bias current   Since when using operational amplifiers provision must be made for the base current of the input transistors, this is an important parameter for the input circuit. The input bias current is given as the average value for the base current of the two input transistors $[I_{IB} = (I_{B1} + I_{B2})/2]$. From the dc conditions of Sec. 8.7.2 we have that $I_{C1} = I_{C2} = 0.5$ mA. The maximum input bias current for the MC1530 is given as 10 μA or $I_{B1} = I_{B2} = 10$ μA. The minimum β_F for the input transistors is therefore 50; typically the value is ≈ 150. As we would expect, the input bias current is about a factor of 100 less for the MC1531.

Input offset voltage   In our analysis of the differential-amplifier pair we have assumed transistor parameters and resistor values were exactly as designed. Hence with zero difference voltage at the input we have 0 V at the output. However, in practice there are departures from design values for the circuit elements. Therefore for 0 V at the output, the input difference voltage is not exactly zero. This is the input offset voltage, the difference voltage required at the input for the output voltage level to be zero. It is small, as shown on the data sheet, and is typically 1.0 mV for the MC1530 and 3.0 mV for the MC1531.

Input offset current   A result of the small dissimilarities in the circuit is that we have to introduce an input offset current. This is the difference in base current of the two input transistors with the output level at 0 V. Again it is typically very small: 0.2 μA for the MC1530 and 0.003 μA for the MC1531.

Input impedance   The typical input impedance at 30 Hz for the MC1530 is 20 kΩ. This is about twice as high as our result, because the low-frequency current gain (β_F) of the input transistors in the MC1530 is closer to 200 than the 100 we have

ELECTRICAL CHARACTERISTICS (V_{CC} = +6.0 Vdc, V_{EE} = –6.0 Vdc, T_A = +25°C unless otherwise noted)

Characteristic	Symbol*	MC1530 Min	Typ	Max	MC1430 Min	Typ	Max	Unit
Input Bias Current	I_{IB}	–	3.0	10	–	5.0	15	µAdc
Input Offset Current	I_{IO}	–	0.2	2.0	–	0.4	4.0	µAdc
Input Offset Voltage $\quad T_A = +25°C$	V_{IO}	–	1.0	5.0	–	2.0	10	mVdc
$\qquad T_A = T_{low}$ ①		–	–	6.0	–	–	11	
$\qquad T_A = T_{high}$ ①		–	–	6.0	–	–	12	
Single-Ended Input Impedance (Open-Loop, f = 30 Hz)	z_{is}	10	20	–	5.0	15	–	kΩ
Common-Mode Input Voltage Swing	V_{ICR}	± 2.0	± 2.7	–	± 2.0	± 2.5	–	Vpk
Equivalent Input Noise Voltage (Open-Loop, R_s = 50 ohms, BW = 5.0 MHz)	e_N	–	10	–	–	10	–	µV(rms)
Common-Mode Rejection Ratio (f = 100 Hz)	CMRR	70	75	–	65	75	–	dB
Open-Loop Voltage Gain	A_{vol}							V/V
$\quad T_A = +25°C$		–	–	–	3000	5000	–	
$\quad T_A = T_{low}$ to T_{high}		4500	5000	12,500	–	–	–	
Bandwidth (Open-Loop, –3.0 dB, no roll-off capacitance)	BW	1.0	2.0	–	1.0	2.0	–	MHz
Output Impedance (f = 100 Hz)	z_o	–	25	50	–	25	50	ohms
Output Voltage Swing (R_L = 1.0 k ohms)	V_O	± 4.5	± 5.2	–	± 4.0	± 5.0	–	Vpk
Power Supply Sensitivity ($R_S \leqslant$ 10 k Ω)	PSRR	–	100	–	–	100	–	µV/V
Power Supply Current	I_D^+, I_D^-	–	9.2	12.5	–	9.2	12.5	mAdc
DC Quiescent Power Dissipation (V_O = 0)	P_D	–	110	150	–	110	150	mW

ELECTRICAL CHARACTERISTICS (V_{CC} = +6.0 Vdc, V_{EE} = –6.0 Vdc, T_A = +25°C unless otherwise noted)

Characteristic	Symbol*	MC1531 Min	Typ	Max	MC1431 Min	Typ	Max	Unit
Input Bias Current	I_{IB}	–	0.025	0.150	–	0.1	0.3	µAdc
Input Offset Current	I_{IO}	–	0.003	0.025	–	0.01	0.1	µAdc
Input Offset Voltage $\quad T_A = +25°C$	V_{io}	–	3.0	10	–	5.0	15	mVdc
$\qquad T_A = T_{low}$ ①		–	–	–	–	–	18	
$\qquad T_A = T_{high}$ ①		–	–	–	–	–	16.5	
Single-Ended Input Impedance (Open-Loop, f = 30 Hz)	z_{is}	1000	2000	–	300	600	–	kΩ
Common-Mode Input Voltage Swing	V_{ICR}	± 2.0	± 2.4	–	± 2.0	± 2.2	–	Vpk
Equivalent Input Noise Voltage (Open-Loop, R_s = 50 ohms, BW = 5.0 MHz)	e_N	–	20	–	–	20	–	µV(rms)
Common-Mode Rejection Ratio (f = 100 Hz)	CMRR	65	65	–	60	75	–	dB
Open-Loop Voltage Gain	A_{vol}							V/V
$\quad T_A = +25°C$		–	–	–	1500	3500	–	
$\quad T_A = T_{low}$ to T_{high}		2500	3500	7000	–	–	–	
Bandwidth (Open-Loop, –3.0 dB, no roll-off capacitance)	BW	–	0.4	–	–	0.4	–	MHz
Output Impedance (f = 30 Hz)	z_o	–	25	50	–	25	50	ohms
Output Voltage Swing (R_L = 1.0 k ohms)	V_O	± 4.5	± 5.2	–	± 4.0	± 5.0	–	Vpk
Power Supply Sensitivity ($R_S \leqslant$ 10 k Ω)	PSRR	–	100	–	–	100	–	µV/V
Power Supply Current	I_D^+, I_D^-	–	9.2	12.5	–	9.2	12.5	mAdc
DC Quiescent Power Dissipation (V_O = 0)	P_D	–	110	150	–	110	150	mW

CIRCUIT SCHEMATICS

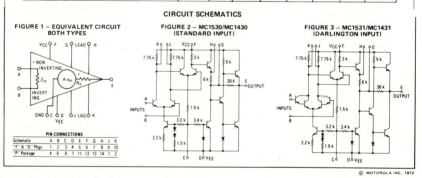

FIGURE 1 – EQUIVALENT CIRCUIT BOTH TYPES

FIGURE 2 – MC1530/MC1430 (STANDARD INPUT)

FIGURE 3 – MC1531/MC1431 (DARLINGTON INPUT)

© MOTOROLA INC., 1972

FIGURE 8.14
Data sheet for MC 1530/MC 1531 operational amplifier. (*Motorola Inc.*)

chosen. The input resistance of the MC1531 is about 100 times that of the MC1530. This indicates that at the lower operating currents of the input transistors in the MC1531 the β_F for these transistors is about 100.

Common-mode input voltage swing With the same voltage applied to both inputs of the amplifier, the maximum for the MC1530 is ± 2.7 V and ± 2.4 V for the MC1531. The ± 2.7-V figure is identical with that obtained in Sec. 8.7.2.

Common-mode rejection ratio The CMRR for the MC1530 is typically 75 dB, and for the MC1531 it is 65 dB. Present-day IC operational amplifiers generally have a CMRR greater than 80 dB.

Open-loop voltage gain This characteristic we have been defining as the difference-mode voltage gain. Typically for the MC1530 this is 5000 or 74 dB, and for the MC1531, 3500 or 71 dB. The more complex output circuit of these amplifiers results in a larger voltage gain than for the circuit of Fig. 8.13.

Output impedance The typical value for both operational amplifiers of 25 Ω compares favorably with our circuit output resistance of 24 Ω.

Output voltage swing The superior output circuit of the 1530-type operational amplifiers results in a typical output voltage swing of ± 5.2 V. Notice this voltage swing is almost from V_{CC} to V_{EE}, which in this case is $+6$ V and -6 V, respectively.

Conclusion Amplifiers like the MC1530, using only *n-p-n* transistors, are useful in many applications. Moreover, in ICs, *n-p-n* transistors have superior frequency characteristics over *p-n-p* transistors that can easily be made in the same chip. Therefore these amplifiers provide greater *bandwidth* than those employing *n-p-n* and *p-n-p* transistors in the same IC. However, a greater voltage gain can be achieved with a mix of *n-p-n* and *p-n-p* transistors in the same circuit, even though the current gain, β_F and β_0, is generally much lower for the *p-n-p* than for the *n-p-n* transistor. Hence circuits using both polarity transistors are extensively used in most low-frequency (less than 1 MHz) applications. We will discuss a representative one of these in the next section.

8.8 DESIGN FEATURES OF THE TYPE 741 OPERATIONAL AMPLIFIER

A photomicrograph of the widely used 741 op amp is shown in Fig. 8.15a. The chip size is 56×56 mils. The circuit schematic taken directly from the photomicrograph is shown in Fig. 8.15b. The circuit differs somewhat from the 741 circuit diagram that is usually shown in manufacturers' data sheets.

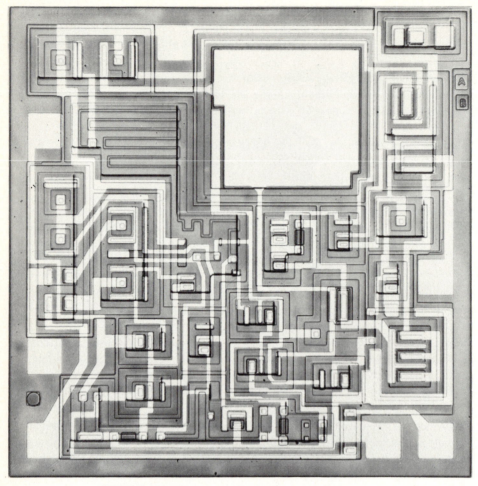

FIGURE 8.15
(*a*) Photomicrograph of the type 741 operational amplifier. (*Fairchild Semiconductor.*)
(*b*) Circuit schematic taken directly from photomicrograph.

In our analysis of the 741 we will discuss only a few selected features relating to the dc and ac design. We will consider the amplifier as being made up of three stages, as shown in Fig. 8.16*a*. Here much of the complexity of Fig. 8.15*b* has been reduced by introducing current sources; i.e., the transistor current source Q_6 in Fig. 8.15*b* has been replaced in Fig. 8.16*a* by the current source I_{KS6}. The resistance r_{o6} in parallel with the current generator is due to the finite collector output resistance of transistor Q_6.

Notice in Fig. 8.15*b* the two collectors for transistor Q_{13}. In describing logic gate circuits, we noted in Sec. 3.4 the use of a multiple-emitter transistor in a TTL gate. Indeed we have a dual-emitter transistor in the 741, shown as Q_{21} in

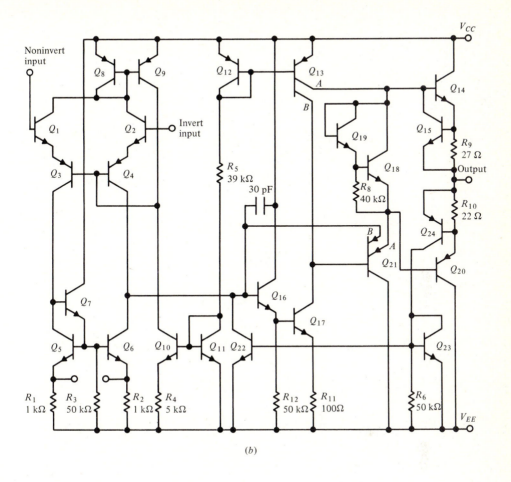

(b)

Fig. 8.15b. The dual-collector transistor can be considered as two separate transistors but with common emitters and bases; the saving is in chip area. A point to mention is that with the split collector it is not necessary for the current in each collector to be identical. Hence in Fig. 8.16a we show the two current sources I_{KS13A} and I_{KS13B} in place of the current-source transistor Q_{13}.

As seen in Fig. 8.16a the input stage consists of Q_1, Q_2, Q_3, Q_4, and the associated biasing circuit. The collector load resistor of transistor Q_4 is the output resistance of Q_6, namely, r_{o6}. Hence Q_6 is termed the *active load* for Q_4.

The second or intermediate stage consists of the emitter follower Q_{16} driving an amplifying transistor Q_{17}, followed by another emitter follower Q_{21}. The high input resistance of Q_{16} helps prevent loading the output of the input stage. Similarly Q_{21} acts as a buffer between the intermediate gain stage and the output stage. The collector of Q_{13B} is the active load for Q_{17}.

The output stage is the *complementary* emitter-follower circuit made up of the n-p-n transistor Q_{14} and the p-n-p transistor Q_{20}. The current source from

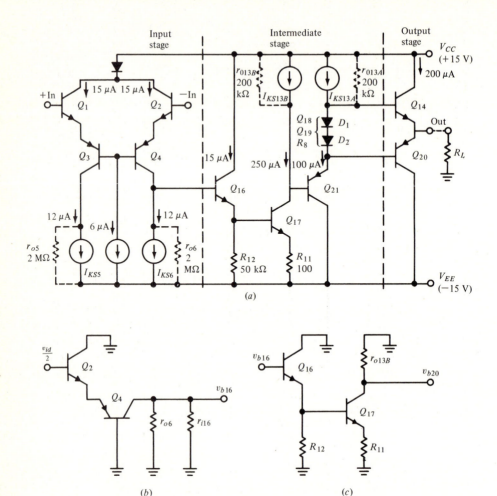

FIGURE 8.16
(a) Simplified circuit diagram of the 741 operational amplifier; (b) small-signal difference-mode half-circuit for the input stage; (c) small-signal circuit for the second stage.

Q_{13A} biases diodes D_1 and D_2 so that for zero differential signal at the input, the dc current in the collectors of both Q_{14} and Q_{20} is about 200 μA. This means that very little power is drawn from the power supplies when the differential input and single-ended output signals are at 0 V.

For full load conditions ($R_L = 2$ kΩ) the overall voltage gain v_o/v_{id} for the 741 is typically 200,000. Thus for $v_{id} = 50$ μV the output signal voltage is $+10$ V. Under this condition, $i_o = 5$ mA, and the collector current of Q_{14} has increased from 200 μA to 5 mA while the collector current of Q_{20} decreased to about 2 μA. Conversely, for $v_{id} = -50$ μV, Q_{20} conducts 5 mA while Q_{14} is practically cut off to 2 μA.

The operation of the complementary emitter follower can be contrasted to the output circuit discussed in Sec. 7.5. With a simple emitter-follower circuit using an *n-p-n* transistor, the output signal on the negative excursion is limited by the quiescent current in the output transistor. With a quiescent current of 200 μA and a load resistance of 2 kΩ, an output signal greater than 0.8 V peak to peak must include some distortion. In the output circuit of the 741, with all components thermally coupled in the same chip and with the accurate matching of voltage drops possible in ICs, we find negligible distortion even with 20 V peak to peak in a load of 2 kΩ.

The dc bias circuit for the input stage is an ingenious arrangement that gives the amplifier a good common-mode characteristic, i.e., both a high CMRR and a high common-mode input resistance. Moreover, the breakdown voltage of the emitter-base diode of the *p-n-p* transistors Q_3 and Q_4 is substantially greater than that of the *n-p-n* transistors such as Q_1 and Q_2. Hence the range of input voltage (both common mode and differential) is greatly improved over the use of an all–*n-p-n* input stage as in Fig. 8.13. The input voltage range for the 741 is typically ± 13 V.

We will not undertake an analysis of the dc bias conditions and common-mode capabilities of the first stage of the 741 but simply use the bias currents given in Fig. 8.16a. We will note, however, that the current sources, transistors Q_5 and Q_6, provide the main portion of the load for transistors Q_3 and Q_4. The 6-μA current source connected to the bases of Q_3 and Q_4 is the result of the two current sources Q_9 and Q_{10} in Fig. 8.15b.

Let us now calculate some of the typical small-signal characteristics of the 741 and compare them with the values on the data sheet shown in Fig. 8.17. We will assume the following transistor parameters:

For all the *n-p-n* transistors:

$$\beta_F = \beta_0 = 200$$

For the *p-n-p* transistors of the input stage, Q_3 and Q_4:

$$\beta_F = \beta_0 = 4$$

For the emitter-follower *p-n-p* transistors Q_{20} and Q_{21}:

$$\beta_F = \beta_0 = 200$$

Unless otherwise indicated in Fig. 8.16a, we assume the collector output resistance of all transistors to be infinite.

8.8.1 Small-signal Characteristics

We will first solve for the small-signal differential input resistance and voltage gain of just the input stage.

FAIRCHILD LINEAR INTEGRATED CIRCUITS • μA741

741

ELECTRICAL CHARACTERISTICS (V_S = ± 15 V, T_A = 25°C unless otherwise specified)

PARAMETERS (see definitions)	CONDITIONS		MIN.	TYP.	MAX.	UNITS
Input Offset Voltage	$R_S \leqslant$ 10 kΩ			1.0	5.0	mV
Input Offset Current				20	200	nA
Input Bias Current				80	500	nA
Input Resistance			0.3	2.0		MΩ
Input Capacitance				1.4		pF
Offset Voltage Adjustment Range				±15		mV
Input Voltage Range			±12	±13		V
Common Mode Rejection Ratio	$R_S \leqslant$ 10 kΩ		70	90		dB
Supply Voltage Rejection Ratio	$R_S \leqslant$ 10 kΩ			30	150	μV/V
Large Signal Voltage Gain	$R_L \geqslant$ 2 kΩ, V_{OUT} = ±10 V		20,000	200,000		
Output Voltage Swing	$R_L \geqslant$ 10 kΩ		±12	±14		V
	$R_L \geqslant$ 2 kΩ		±10	±13		V
Output Resistance				75		Ω
Output Short Circuit Current				25		mA
Supply Current				1.7	2.8	mA
Power Consumption				50	85	mW
Transient Response (Unity Gain)	Risetime	V_{IN} = 20 mV, R_L = 2 kΩ, $C_L \leqslant$ 100 pF		0.3		μs
	Overshoot			5.0		%
Slew Rate	$R_L \geqslant$ 2 kΩ			0.5		V/μs

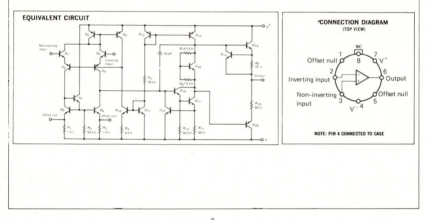

EQUIVALENT CIRCUIT

CONNECTION DIAGRAM
(TOP VIEW)

Offset null   1   NC   7 V^+
              2   8   6
Inverting input   Output
Non-inverting   3   4   5   Offset null
input V^-

NOTE: PIN 4 CONNECTED TO CASE

3

FIGURE 8.17
Data sheet for 741 operational amplifier. (*Fairchild Semiconductor.*)

Input stage   By grounding the points of symmetry in the input stage of Fig. 8.16*a*, we have the small-signal difference-mode half-circuit shown in Fig. 8.16*b*. This circuit is a CC (emitter-follower) stage Q_2, driving a CB stage Q_4. The small-signal input resistance of this half-circuit is then

$$r_{i2} = r_{\pi 2} + (\beta_{02} + 1)r_{i4} \qquad (8.35)$$

where

$$r_{i4} = \frac{r_{\pi 4}}{\beta_{04} + 1}$$

The difference-mode input resistance is twice the input resistance of the half-circuit. Then

$$r_{id} = 2\left(\frac{\beta_{02} V_T}{I_{C2}} + \frac{\beta_{02} + 1}{\beta_{04} + 1}\frac{\beta_{04} V_T}{I_{C4}}\right) \qquad (8.36)$$

With $I_{C2} = 15\ \mu\text{A}$ and $I_{C4} = 12\ \mu\text{A}$,

$$r_{id} = 2\left[\frac{(200)(26\ \text{mV})}{15\ \mu\text{A}} + \frac{201}{5}\frac{(4)(26\ \text{mV})}{12\ \mu\text{A}}\right] = 1.38\ \text{M}\Omega$$

The data sheet gives a typical input resistance of 2.0 MΩ and a minimum value of 0.3 MΩ. From the data sheet we also learn that the typical input bias current is 80 nA. With $I_{C1} = I_{C2} = 15\ \mu\text{A}$ in Fig. 8.16a, and with $\beta_F = 200$, we would calculate the input bias current to be 75 nA.

In order to determine the small-signal differential voltage gain of the input stage, we note that the load at the collector of Q_4 is due to the parallel combination of the output resistance of Q_6 and the input resistance of Q_{16}. Hence

$$R_{L4} = r_{o6}\|r_{i16} \qquad (8.37)$$

Also the output signal voltage v_{b16} is the product of the signal current i_{c4} and R_{L4}. That is,

$$v_{b16} = i_{c4} R_{L4} \qquad (8.38)$$

where

$$i_{c4} = \alpha_{04} i_{e4} \qquad (8.39a)$$

but

$$i_{e4} = (\beta_{02} + 1)i_{b2} \qquad (8.39b)$$

and

$$i_{b2} = \frac{v_{id}}{2r_{i2}} \qquad (8.39c)$$

Therefore, substituting these last three equations in Eq. (8.38), we have

$$\frac{v_{b16}}{v_{id}} = \frac{\alpha_{04}(\beta_{02} + 1)R_{L4}}{2r_{i2}}$$

$$= \left(\frac{\beta_{04}}{\beta_{04} + 1}\right)\left[\frac{(\beta_{02} + 1)R_{L4}}{r_{id}}\right] \qquad (8.40)$$

In the next paragraph we shall determine that $r_{i16} = 4.88$ MΩ; then with $r_{o6} = 2$ MΩ, from Eq. (8.37) we have $R_{L4} \approx 1.42$ MΩ. Hence by using numerical values in Eq. (8.40) we have that the differential voltage gain of the input stage is

$$\frac{v_{b16}}{v_{id}} = \left(\frac{4}{5}\right)\left[\frac{(201)(1.42)}{1.38}\right] = 166$$

Second stage   For the second stage we calculate only the small-signal voltage gain from the base of Q_{16} to the base of Q_{20}. A small-signal representation of the connection of Q_{16} and Q_{17} is drawn in Fig. 8.16c. First we note, with R_{12} removed from the circuit, that due to the current gain of the emitter follower (Q_{16}) the base current to Q_{17} is

$$i_{b17} = (\beta_{016} + 1)i_{b16}$$

then

$$i_{c17} \approx (\beta_{017})(\beta_{016})i_{b16}$$

However, including R_{12}, there is a loss of base current to Q_{17} through this resistor; then the composite current gain β_0^* of Q_{16} and Q_{17} is

$$\beta_0^* = \beta_{016} \frac{R_{12}}{R_{12} + r_{i17}} \beta_{017} \qquad (8.41)$$

Neglecting any signal attenuation through the emitter follower (Q_{21}), the voltage gain of the second stage is

$$\frac{v_{b20}}{v_{b16}} = -\beta_0^* \frac{r_{o13B}}{r_{i16}} \qquad (8.42)$$

From the collector current for Q_{16} and Q_{17}, given in Fig. 8.16a, we have the following small-signal parameters for transistor Q_{16} and Q_{17}.

$$r_{\pi16} = \frac{\beta_{016} V_T}{I_{C16}} = \frac{(200)(26 \text{ mV})}{15 \text{ } \mu\text{A}} = 350 \text{ k}\Omega$$

$$r_{\pi17} = \frac{\beta_{017} V_T}{I_{C17}} = \frac{(200)(26 \text{ mV})}{250 \text{ } \mu\text{A}} = 21 \text{ k}\Omega$$

Now the input resistance to Q_{17} is given by

$$r_{i17} = r_{\pi17} + (\beta_{017} + 1)R_{11}$$
$$= 21 + (201)(0.1) = 41 \text{ k}\Omega$$

Then it follows that the input resistance to Q_{16} is

$$r_{i16} = r_{\pi16} + (\beta_{016} + 1)(R_{12}\|r_{i17})$$
$$= 350 + (201)(50\|41) = 4.88 \text{ M}\Omega$$

which was the value used in the gain calculation of the first stage.
We can now compute the composite current gain, from Eq. (8.41), as

$$\beta_0^* = 200 \frac{50}{50 + 41} 200 = 22,000$$

For the voltage gain of the second stage, including the numerical values in Eq. (8.42), we have

$$\frac{v_{b20}}{v_{b16}} = -22,000 \frac{0.2}{4.88} = -902$$

If we neglect the small attenuation from the output complementary emitter follower, the overall differential voltage gain a_d is

$$a_d = \left(\frac{v_{b16}}{v_{id}}\right)\left(\frac{v_{b20}}{v_{b16}}\right) = (166)(-902) = -150,000$$

This value can be compared with the typical value of 200,000 shown on the data sheet. Note that the input to Q_2, in Fig. 8.16a, is the inverting input for this amplifier.

Output stage   For the output stage we will determine the small-signal output resistance and the maximum output voltage swing.

From the output circuit in Fig. 8.16a we note that quiescently both Q_{14} and Q_{20} are biased in the active region with $I_{C14} \approx I_{C20} \approx 200$ μA. Now due to the current source from Q_{13A} the diodes D_1 and D_2 are also conducting, with $I_{D1} = I_{D2} = 100$ μA. For the small-signal characteristics we will consider that the base of Q_{14} is directly connected to the base of Q_{20}. In this case, the n-p-n transistor Q_{14} is in parallel with the p-n-p transistor Q_{20}. Furthermore the source of the signal for both Q_{14} and Q_{20} is yet another emitter follower, Q_{21}. Now since $\beta_{014} = \beta_{020}$, we can make use of the equation for the small-signal output resistance of an emitter follower in the following:

$$r_o = \frac{(r_{o21}\|r_{o13A}) + (r_{\pi14}\|r_{\pi20})}{\beta_{014} + 1} \tag{8.43}$$

where

$$r_{o21} = \frac{r_{o13B} + r_{\pi21}}{\beta_{021} + 1} \tag{8.44}$$

Calculating values for $r_{\pi14}$, $r_{\pi20}$, and $r_{\pi21}$ from the currents indicated in Fig. 8.16a, we have

$$r_{\pi14} = \frac{\beta_{014} V_T}{I_{C14}} = \frac{(200)(26 \text{ mV})}{0.2 \text{ mA}} = 26 \text{ k}\Omega$$

$$r_{\pi20} = \frac{\beta_{020} V_T}{I_{C20}} = \frac{(200)(26 \text{ mV})}{0.2 \text{ mA}} = 26 \text{ k}\Omega$$

$$r_{\pi21} = \frac{\beta_{021} V_T}{I_{C21}} = \frac{(200)(26 \text{ mV})}{0.1 \text{ mA}} = 52 \text{ k}\Omega$$

Substituting for $r_{\pi21}$ in Eq. (8.44),

$$r_{o21} = \frac{200 + 52}{201} = 1.25 \text{ k}\Omega$$

Since $r_{o13A} = 200$ kΩ, we will neglect the effect of r_{o13A} in parallel with r_{o21}. Then from Eq. (8.43) we can compute the output resistance of the circuit in Fig. 8.16a as

$$r_o = \frac{1.25 + (26\|26)}{201} = 71 \text{ }\Omega$$

Refer now to the complete circuit of the 741, in Fig. 8.15b, where we note that there is a 27-Ω resistor in series with the output from Q_{14} and a 22-Ω resistor in series with the output from Q_{20}. We will discuss the usefulness of Q_{15} and Q_{24} later; for the small-signal output they are nonconducting. Including the two series resistors we have that the small-signal output resistance of the 741 is

$$r_o = (142 + 27)\|(142 + 22) = 83 \ \Omega$$

This value is slightly higher than the typical value of 75 Ω given on the data sheet.

To determine the maximum output voltage swing we note in Fig. 8.16a that with sufficient input drive, transistor Q_{17} will saturate in one direction and cut off in the other. With the transistor saturated, and no load or a small load current at the output, the output voltage is about $2V_{BE(\text{on})}$, due to Q_{21} and Q_{20}, above the voltage at the collector of Q_{17}, which is then approximately V_{EE}. That is,

$$V_{o-} \approx -15 + 1.4 = -13.6 \ \text{V}$$

With Q_{17} cut off, in the limit Q_{21} also cuts off; then the output voltage, with Q_{14} conducting only a small load current, is just one $V_{BE(\text{on})}$ below V_{CC}. That is,

$$V_{o+} \approx +15 - 0.7 = +14.3 \ \text{V}$$

With increasing load currents the output voltage will decrease due to the voltage drop across the base-biasing resistor of Q_{14}, namely, r_{o13A}.

The maximum symmetrical output voltage swing, for the no-load condition, is therefore ± 13.6 V. With $R_L \geq 10$ kΩ the data sheet gives a value of ± 14 V. Since at low operating currents the $V_{BE(\text{on})}$ of the output transistors is less than 0.7 V, this explains the discrepancy.

Output short-circuit current   Another output characteristic given on the data sheet is the output short-circuit current. Notice from Fig. 8.16a that with the output of the amplifier accidentally connected to ground it is possible for large collector currents to flow in Q_{14} and Q_{20}, probably causing catastrophic failure of these transistors. For example, with the output at ground the base voltage of Q_{14} is ≈ 0.7 V. The base current for this transistor can then go as high as

$$I_{B14} = \frac{V_{CC} - V_{B14}}{r_{o13A}} + I_{K13A} = \frac{15 - 0.7}{200} + 0.1 = 0.17 \ \text{mA}$$

Now β_F of Q_{14} may be as high as 500, and the collector current is then

$$I_{C14} = (500)(0.17) = 85 \ \text{mA}$$

With 15 V between the collector and emitter, the power dissipated in this transistor is

$$\text{PD} = (15)(85) = 1,275 \ \text{mW}$$

This is not a healthy state of affairs for this transistor.

Return now to the inclusion of the 27-Ω resistor (R_9) in the output of the complete circuit shown in Fig. 8.15b. With a voltage drop across $R_9 \geq 0.7$ V, transistor Q_{15} will turn on, and current drive to the base of Q_{14} will be diverted to Q_{15}. The maximum current through Q_{15} is limited by the current source transistor Q_{13A} to about 100 μA. Actually Q_{15} will start to turn on when the voltage drop across R_9 is about 0.6 V. Now we see that on a positive voltage excursion at the output, the output current is limited to

$$I_{o+}R_9 = 0.6 \text{ V}$$

therefore

$$I_{o+} = \frac{0.6 \text{ V}}{27 \text{ }\Omega} = 22 \text{ mA}$$

To limit the current on a negative output voltage excusion, the circuit is a little more complicated. When transistor Q_{24} turns on, this turns on Q_{22}, which then diverts current from the base of Q_{16}. Hence the base current to Q_{17} is limited, as is, subsequently, the base current to Q_{21A} and Q_{20}. To turn on Q_{24} we have

$$I_{o-}R_{22} = 0.6 \text{ V}$$

therefore

$$I_{o-} = \frac{0.6 \text{ V}}{22 \text{ }\Omega} = 27 \text{ mA}$$

We can now say that the output stage is short-circuit protected. The current through the output transistors is limited to a safe value. The output current figures we have computed compare favorably with the output short-circuit current given on the data sheet, which is 25 mA typical.

Power dissipation   We can calculate the power consumption of this amplifier from the currents indicated in Fig. 8.16a, where all the sources of current are included except that due to transistors Q_9 and Q_{10}, and transistors Q_{11} and Q_{12}. From Fig. 8.15b we note

$$I_{E11} \approx I_{E12} \approx \frac{V_{CC} - V_{EE}}{R_5} = \frac{15 - (-15)}{39} = 0.77 \text{ mA}$$

$$I_{E9} \approx I_{E8} \approx I_{C1} + I_{C2} = 15 + 15 = 30 \text{ }\mu\text{A}$$

$$I_{E10} \approx I_{E9} + I_{B3} + I_{B4} = 30 + 6 = 36 \text{ }\mu\text{A}$$

Hence the power dissipation for the amplifier is approximately

$$\begin{aligned} \text{PD} &= (V_{CC} - V_{EE})(I_{E5} + I_{E6} + I_{E10} + I_{E11} + I_{E16} \\ &\quad + I_{E17} + I_{C21} + I_{C20}) \\ &= (30)(12 + 12 + 36 + 770 + 15 + 250 + 100 + 200) \\ &\approx 42 \text{ mW} \end{aligned}$$

From the data sheet the power dissipation is typically 50 mW.

Slew rate   We should explain the last item on the data sheet: the slew rate. The slew rate indicates the maximum rate of change of the output voltage under large-signal conditions. This is generally measured using the maximum output voltage swing. Since the rate of change of voltage across a capacitor is limited by the charging current, and some capacitance is always present in a circuit, albeit sometimes not by design, the finite currents in a circuit must lead to a limit on the rate of change of the signal voltage—the slew rate. For the 741, with a load resistance of 2 kΩ, the typical slew rate is given as 0.5 V/μs. That is, the output voltage cannot change from -10 to $+10$ V in less than 40 μs.

Frequency compensation   Finally in our discussion of the type 741 op amp we should make mention of the one capacitor that appears in the circuit diagram of Fig. 8.15b. This is a 30-pF metal-oxide capacitor. The problems of incorporating a capacitor in an IC are illustrated in the photomicrograph of the chip (Fig. 8.15a). Here the 30-pF capacitor is the large grey square area at the top of the chip. This component takes up the largest chip area, by far, of any other component in the circuit. The purpose of the capacitor is to provide frequency compensation to the amplifier. That is, the bandwidth of the amplifier is controlled to prevent undesired oscillations at the output when the op amp is used in negative feedback circuits, such as we shall discuss in Chap. 9.

8.9   COMPARISON OF CIRCUITS

In Table 8.3 we show a comparison of some of the more important characteristics of the two general-purpose IC op amps we have described in this chapter. We also show the characteristics of two other popular IC operational amplifiers.

   The operating characteristics of the popular type 741 op amp are superior to the older MC1530 in almost every respect. Indeed the 741 is generally regarded as a second generation in the development of IC op amps. One characteristic of the

Table 8.3   COMPARISON OF TYPICAL ELECTRICAL CHARACTERISTICS FOR GENERAL-PURPOSE IC OPERATIONAL AMPLIFIERS

$T_A = 25°C$	MC1530	741	LM108	740
Input offset voltage	1.0 mV	1.0 mV	0.7 mV	10 mV
Input offset current	0.2 μA	20 nA	0.05 nA	0.04 nA
Input bias current	3.0 μA	80 nA	0.8 nA	0.1 nA
Input resistance	20 kΩ	2.0 MΩ	70 MΩ	10^6 MΩ
Voltage gain	5,000	200,000	300,000	10^6
CMRR	75 dB	90 dB	100 dB	80 dB
Output voltage swing	±5.2 V(1 kΩ)	±14 V(10 kΩ)	±14 V(10 kΩ)	±14 V(10 kΩ)
Output resistance	25 Ω	75 Ω	$\cdots$	75 Ω
Supply volts	±6 V	±15 V	±15 V	±15 V
Power dissipation	110 mW	50 mW	4.5 mW	126 mW
Unity gain bandwidth	1.75 MHz	1.2 MHz	1.0 MHz	3.0 MHz
Slew rate at unity gain	1 V/μs	0.5 V/μs	0.25 V/μs	6 V/μs
Frequency-compensated	No	Yes	No	Yes

1530 better than the 741 has already been mentioned. This is the unity gain bandwidth. With the voltage gain of the amplifier reduced to unity by the use of negative feedback, to be discussed in the next chapter, the all–*n-p-n* transistor circuit of the 1530 provides a little better frequency performance than the 741 circuit.

We shall also see in the next chapter that unrestricted negative feedback around an amplifier can lead to oscillations unless the frequency response of the amplifier is controlled. In some op amps this frequency compensation is included in the chip, but in others this must be provided, external to the chip, by the user of the device. The 741 has the frequency compensation in the chip; the MC1530 does not.

The two other devices listed in Table 8.3 indicate two different approaches to improving the input characteristics of IC op amps. Ideally we would like the input resistance to be infinite and the input bias current to be zero; then the op amp would have no loading effect on the input signal source. The LM108 uses the so-called super-beta transistors in the input differential-amplifier pair. Super-beta transistors have typical current gains of 5000 at a collector current of 4 μA. In this way the input bias current is typically about 0.8 nA, a factor of 100 less than for the 741.

Another approach is to include junction field-effect transistors (JFETs) in the monolithic IC chip in place of the two input bipolar transistors. This was first done commercially in the 740. The gate circuit of the JFET draws essentially no current, and the input bias current is just due to the leakage current between the drain and gate of the JFET. Typically this is about 0.1 nA for the 740.

However, notice in Table 8.3 that the JFET circuit also has the highest input offset voltage. The lowest offset voltage is obtained with the super-beta transistor input of the LM108, but this device also has the lowest slew rate, which is indicative of low operating currents in the input stage of the amplifier. Notice also that the LM108 has the lowest power dissipation. Generally the higher the power dissipation the greater is the slew rate and also the unity gain bandwidth.

The voltage gain of the three later types of op amps is considered adequate for most general purposes.

8.10 SUMMARY

In this chapter we have progressed from a simple differential-amplifier pair to a complete operational amplifier. However, our discussion of op amps has centered only on one or two general-purpose devices. We have made no mention of many other types of operational amplifiers, all available as ICs. These would include the following op amps:

Low noise
Extended bandwidth
High slew rate
High voltage
Micropower

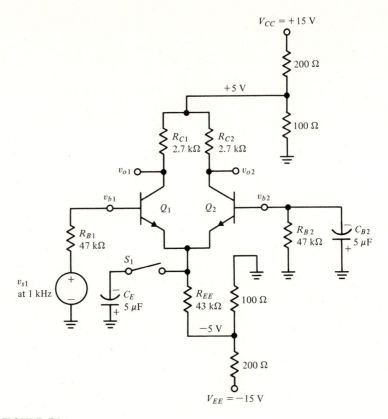

FIGURE D8.1

Q_1 and Q_2 should be a matched pair, such as found in the monolithic transistor array type CA3146E.

The titles are sufficiently suggestive of their use. Also, all these circuits can be represented by the same basic arrangement that we have used in this chapter. That is, an input differential-amplifier pair is followed by an intermediate gain stage that drives some form of power output stage. This same idea also applies to two other IC devices, which while not operational amplifiers do use similar circuit techniques. They are the differential voltage comparator and the sense amplifier which is used in computer magnetic core memories.

Finally there is yet another whole field of linear ICs in so-called consumer applications. These include various amplifiers for use in the entertainment area, like in TV sets, FM radios, AM radios, and even the simple phonograph. There is also a need for linear ICs in the sophisticated electronics associated with the modern automobile. All these circuits make use of the basic ideas we have presented in this chapter.

DEMONSTRATIONS

D8.1   Differential-amplifier pair   With S_1 open, the circuit in Fig. D8.1 is a differential-amplifier pair.

(a) Compute and then measure the dc operating points of Q_1 and Q_2. What would be the reasons for the discrepancies, if any are found? With a curve tracer or β_0 tester, measure the β_0 at the operating point calculated.

(b) Apply a small signal v_{s1} so that $v_{o1} \approx 75$ mV peak to peak and measure v_{s1}, v_{b1}, v_{o1}, and v_{o2}. Now close S_1, readjust v_{s1}, and remeasure the above variables. Compare these measured results to those computed from using the simple r_π and β_0 model and explain any differences.

(c) Disconnect C_{B2} and connect the bottom of R_{B2} to v_{s1} so that you can measure the common-mode properties. With S_1 open, adjust v_{s1} so that the transistors Q_1 and Q_2 are not driven into cutoff or saturation. A signal of 5 V peak to peak should be adequate to obtain sufficient data from which you can measure v_{o1}, v_{o2}, v_{b1}, v_{b2}, and v_s so as to determine the common-mode gain and input resistance. Close S_1 and reduce v_s so that v_{o1} is approximately 75 mV and remeasure the variables. Compare the calculated results from your measurements with those using the r_π and β_0 model.

(d) With a differential input oscilloscope obtain the trace of v_{od} versus v_{id}. Determine the breakpoints in the transfer characteristic and compare them with the computed values.

D8.2   Cascade of differential pairs   The circuit in Fig. D8.2a is a differential-input–differential output amplifier. In this case the outputs v_{o3} and v_{o4} are not at ground potential.

(a) Adjust the offset null so that $v_{o4} - v_{o3}$ is zero when $v_{s1} = v_{s2} = 0$. Set $v_{s2} = 0$ and adjust v_{s1} so that v_{o4} is about 5 V peak to peak. Measure v_{s1}, v_{b1}, v_{c1}, and v_{o4}. Determine the differential gains of each stage and input resistance and compare these to results calculated from the r_π and β_0 model.

(b) The circuit in Fig. D8.2a is used in the triangle of Fig. D8.2b. Zener diode D_1 and potentiometer R_v are added. R_v is adjusted so that v_o is zero with v_{s1} and v_{s2} set to zero. Measure the single-ended voltage gain $v_o/(v_{s1} - v_{s2})$. Shut off V_{CC} and V_{EE}, and with a dc ohmmeter (of proper polarity!) measure R_V. From the results of part (a) and this measurement, calculate the single-ended voltage gain.

D8.3   Transistor constant-current source

(a) The circuit in Fig. D8.3 is similar to that in Fig. D8.1 except that the lower portion replaces the resistor R_{EE}. Determine I_{C3} with S_2 open and then closed. Compare to calculated values. Explain any differences.

(b) Repeat parts (b), (c), and (d) of Demonstration D8.1 (ignoring any use of S_1 and C_E) with both S_2 open and closed. Compare your results to those in Demonstration D8.1.

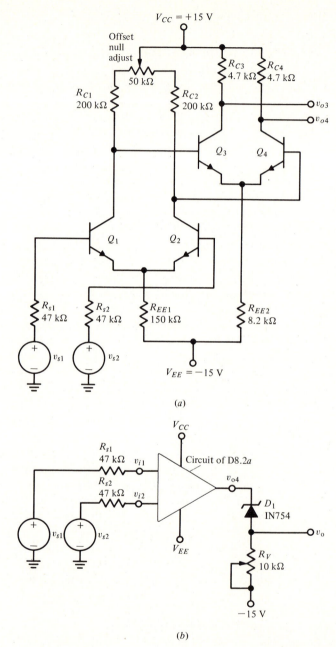

(a)

(b)

FIGURE D8.2

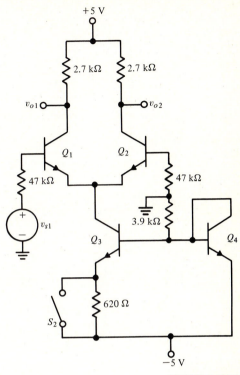

FIGURE D8.3

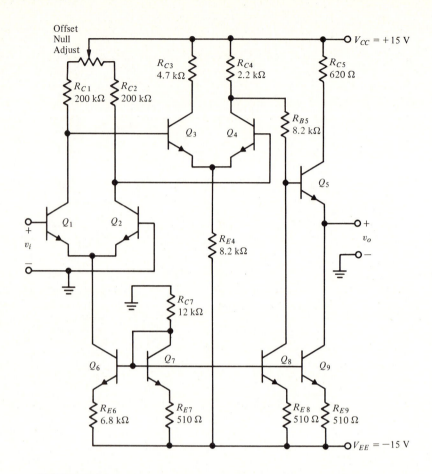

FIGURE D8.4
Differential amplifier for demonstration. Q_1, Q_2, Q_6, and Q_8^* are in one CA3146E; Q_3, Q_4, Q_5, and Q_9^* are in another. (* substrates tied to emitter.)

D8.4 Complete op amp The circuit in Fig. D8.4 can be made using two CA3146Es (monolithic transistor arrays). After nulling the output with the circuit similar to that used in Demonstration D8.2, determine r_{id}, a_d, and a_c. To determine r_o set the input voltage v_s to zero and force a 1-mA peak-to-peak current into the output. Use a 10-V peak-to-peak voltage source and 10-kΩ resistor to approximate this ac current source. Measure v_o (at the emitter of Q_5) and calculate r_o. From circuit parameters and typical values from the CA3146E data sheet, compute the values of r_{id}, a_d, and r_o.

PROBLEMS

P8.1   (a) In Fig. 8.1b, $R_s = 1$ MΩ, $R_L = 1$ kΩ, and $a_v = 100$. Compute v_o/v_s for the following values of r_i and r_o.
    (1) $r_i = \infty$; $r_o = 0$ Ω
    (2) $r_i = \infty$; $r_o = 1$ kΩ
    (3) $r_i = 1$ MΩ; $r_o = 1$ kΩ
    (b) Repeat part (a), but with $R_s = 9$ MΩ, $R_L = 9$ kΩ, and $a_v = 900$.

P8.2   In Fig. 8.2a, with $a_d = 1000$, find v_o for the following input values:

	v_{i1}	v_{i2}
(a)	8 mV	−2 mV
(b)	−28 mV	22 mV
(c)	−428 mV	−422 mV
(d)	4.428 V	4.422 V

P8.3   In Fig. 8.2b, $r_i = 70$ kΩ and $a_d = 1000$. A voltage source v_{i1} is connected to the amplifier inverting input through $R_{s1} = 10$ kΩ. Similarly v_{i2} is connected to the noninverting input through $R_{s2} = 20$ kΩ. Find v_o for the input values of Prob. P8.2.

P8.4   For the strain-gage circuit shown in Fig. 8.3a, assume the nominal resistance value $R_a = R_b = R_c = R_d = 2.5$ kΩ and the bridge supply voltage is 5 V. For peak load strain, the fractional increase on R_a is $(1.6)(10^{-2})$. For this condition, determine:
(a) The Thévenin equivalent circuit at each amplifier input.
(b) The difference voltage at the amplifier input.

P8.5   For the differential-amplifier pair of Fig. 8.4,

$$V_{CC} = +10 \text{ V} \qquad R_{C1} = R_{C2} = 5.2 \text{ k}\Omega \qquad \beta_0 = 100$$

$$V_{EE} = -10 \text{ V} \qquad R_{EE} = 9.3 \text{ k}\Omega \qquad V_{BE(on)} = 0.7 \text{ V}$$

(a) With $V_{I1} = V_{I2} = 0$ V, calculate I_{C1}, I_{C2}, V_{CE1}, and V_{CE2}.
(b) With $V_{I1} = V_{I2} = 5$ V, calculate I_{C1}, I_{C2}, V_{CE1}, and V_{CE2}.

P8.6   (a) For the conditions of Prob. P8.5a determine the small-signal differential voltage gain a_d.
(b) Repeat part (a) for the conditions of Prob. P8.5b.

P8.7   For the differential-amplifier pair of Fig. 8.5,

$$V_{CC} = +15 \text{ V} \qquad R_{C1} = R_{C2} = 200 \text{ k}\Omega \qquad \beta_0 = 70$$

$$V_{EE} = -15 \text{ V} \qquad R_{EE} = 200 \text{ k}\Omega \qquad V_{BE(on)} = 0.6 \text{ V}$$

Use half-circuits to calculate:
(a) The differential-mode input resistance r_{id} and voltage gain a_d.
(b) The common-mode input resistance r_{ic} and voltage gain a_c.

P8.8   For the input differential-amplifier pair of Fig. 8.8,

$$V_{CC} = +12 \text{ V} \qquad R_{C1} = R_{C2} = 120 \text{ k}\Omega \qquad \beta_0 = 50$$

$$V_{EE} = -12 \text{ V} \qquad R_{EE1} = 110 \text{ k}\Omega \qquad V_{BE(on)} = 0.6 \text{ V}$$

Use the input voltage values of Prob. P8.2 and calculate the output voltages v_{o1} and v_{o2}. Ignore the load of the second stage in Fig. 8.8.

P8.9   For the differential amplifier shown in Fig. P8.9, assume $\beta_0 = 100$ and
$V_{BE(\text{on})} = 0.7$ V.
(*a*) Use half-circuits to determine r_{id} and a_d.
(*b*) Use half-circuits to determine r_{ic} and a_c.

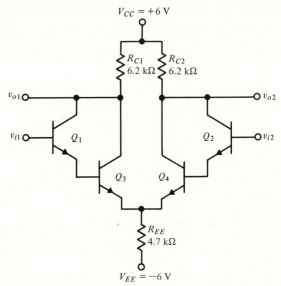

FIGURE P8.9

P8.10   Repeat Prob. P8.9*a* and *b* for the circuit shown in Fig. P8.10.

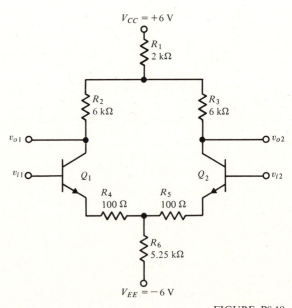

FIGURE P8.10

P8.11 For the differential-amplifier pair of Fig. 8.5,

$$V_{CC} = +15 \text{ V} \qquad I_{C1} = I_{C2} = 100 \ \mu\text{A} \qquad \beta_0 = 100$$

$$V_{EE} = -15 \text{ V} \qquad V_{CE1} = V_{CE2} = 8.2 \text{ V} \qquad V_{BE(on)} = 0.7 \text{ V}$$

(*a*) Calculate values for R_{EE}, R_{C1}, and R_{C2}.
(*b*) Show the differential-mode half-circuits and solve for r_{id} and a_d.
(*c*) Show the common-mode half-circuits and solve for r_{ic} and a_c.

P8.12 In Fig. 8.6, with

$$V_{CC} = +6 \text{ V} \qquad I_{C1} = I_{C2} \qquad \beta_0 = 50$$

$$V_{EE} = -6 \text{ V} \qquad R_{C1} = R_{C2} \qquad V_{BE(on)} = 0.7 \text{ V}$$

(*a*) Solve for R_{EE}, R_{C1}, and R_{C2} under the following conditions:
  (1) $V_{C1} = V_{C2} = \frac{1}{2}V_{CC}$
  (2) $R_{EE} + R_{C1} + R_{C2} = 100 \text{ k}\Omega$
(*b*) With $V_{B2} = 0$ V and a signal v_s applied to base of Q_1:
  (1) Determine v_{o1}/v_s and v_{o2}/v_s.
  (2) Determine the input resistance r_i seen by v_s.

P8.13 In Fig. 8.6 the resistor R_{EE} is replaced by a current source I_{EE}.

$$V_{CC} = +15 \text{ V} \qquad I_{C1} = I_{C2} = 100 \ \mu\text{A} \qquad \beta_0 = 50$$

$$V_{EE} = -15 \text{ V} \qquad V_{CE1} = V_{CE2} = 8 \text{ V} \qquad V_{BE(on)} = 0.7 \text{ V}$$

(*a*) Calculate values for I_{EE}, R_{C1}, and R_{C2}.
(*b*) Solve for the differential voltage gain a_d and input resistance r_{id}.
(*c*) Solve for the common-mode voltage gain a_c and input resistance r_{ic}.

P8.14 For this problem replace the resistor R_{EE} in Fig. 8.6 with a current source $I_{EE} = 2$ mA.

$$V_{CC} = +6 \text{ V} \qquad R_{C1} = R_{C2} = 2 \text{ k}\Omega \qquad \beta_0 = 100$$

$$V_{EE} = -6 \text{ V} \qquad \qquad V_{BE(on)} = 0.7 \text{ V}$$

With $V_{B2} = 0$ V and a signal v_s applied to the base of Q_1:
(*a*) What is the input resistance r_i seen by v_s?
(*b*) What is the small-signal voltage gain v_{o2}/v_s?
(*c*) What is the small-signal current gain i_{o2}/i_s?
(*d*) How does the addition of a resistor R_B in series with the base of each transistor modify the quantity computed in parts (*a*), (*b*) and (*c*)?

P8.15 In Fig. 8.6 replace the resistor R_{EE} with a current source $I_{EE} = 2$ mA. Also include resistor R_{B1} and R_{B2} in series with the signal source to the base of Q_1 and Q_2, respectively.

$$V_{CC} = +10 \text{ V} \qquad R_{C1} = R_{C2} = 5 \text{ k}\Omega \qquad \beta_0 = 70$$

$$V_{EE} = -10 \text{ V} \qquad R_{B1} = R_{B2} = 600 \ \Omega \qquad V_{BE(on)} = 0.7 \text{ V}$$

(*a*) Determine the operating point I_C, V_{CE} for each transistor.
(*b*) Determine the differential-mode input resistance and voltage gain.
(*c*) Determine the common-mode input resistance and voltage gain.

P8.16 In Fig. 8.6, replace the resistor R_{EE} with a current source I_{EE}. Also ground the base of Q_2 and let the signal applied to the base of Q_1 be called v_s.

(a) Derive an expression for i_{c1} as a function of v_s. Assume identical transistors, that never saturate, with a transfer characteristic that can be approximated by

$$i_c = I_K e^{V_{BE}/V_T} \qquad \text{where } I_K \text{ is a constant}$$

(b) Plot this expression over the range $v_s = \pm 150$ mV.

P8.17 In the cascade of two differential pairs in Fig. 8.8, the two resistors R_{EE1} and R_{EE2} are replaced by current sources I_{EE1} and I_{EE2}

$$V_{CC} = +6 \text{ V} \qquad I_{EE1} = I_{EE2} = 2 \text{ mA} \qquad \beta_0 = 50$$

$$V_{EE} = -6 \text{ V} \qquad R_{C1} = R_{C2} = 4 \text{ k}\Omega \qquad V_{BE(on)} = 0.7 \text{ V}$$

$$R_{C3} = R_{C4} = 2 \text{ k}\Omega$$

With $V_{B1} = V_{B2} = 0$ V:

(a) Solve for I_{C1}, I_{C2}, V_{CE1}, V_{CE2}, I_{C3}, I_{C4}, V_{CE3}, and V_{CE4}.

(b) Show the differential-mode half-circuit and small-signal circuit model.

(c) Calculate the differential-mode input resistance and voltage gain.

P8.18 In Fig. 8.8, replace the two resistors R_{EE1} and R_{EE2} with two current sources, $I_{EE1} = 0.204$ mA, $I_{EE2} = 2.55$ mA.

$$R_{C1} = R_{C2} = 24 \text{ k}\Omega \qquad \beta_0 = 50$$

$$R_{C3} = R_{C4} = 1.2 \text{ k}\Omega \qquad V_{BE(on)} = 0.7 \text{ V}$$

(a) For the first stage (loaded with the second stage), find r_{id1} and a_{d1}, where $a_{d1} = v_{od1}/v_{id1}$.

(b) Repeat part (a) for the second stage.

P8.19 In Fig. 8.8,

$$V_{CC} = +15 \text{ V} \qquad R_{C1} = R_{C2} = 200 \text{ k}\Omega \qquad \beta_0 = 50$$

$$V_{EE} = -15 \text{ V} \qquad R_{C3} = R_{C4} = 4.7 \text{ k}\Omega \qquad V_{BE(on)} = 0.7 \text{ V}$$

$$R_{EE1} = 150 \text{ k}\Omega$$

$$R_{EE2} = 8.2 \text{ k}\Omega$$

With $V_{B1} = V_{B2} = 0$ V:

(a) Solve for I_{C1}, I_{C2}, V_{CE1}, V_{CE2}, I_{C3}, I_{C4}, V_{CE3}, and V_{CE4}.

(b) Determine the differential- and common-mode voltage gain a_d and a_c.

(c) What is the CMRR for this circuit?

*P8.20 In Fig. 8.8,

$$V_{CC} = +15 \text{ V} \qquad I_{C1} = I_{C2} = 100 \text{ } \mu\text{A} \qquad \beta_0 = 50$$

$$V_{EE} = -15 \text{ V} \qquad I_{C3} = I_{C4} = 500 \text{ } \mu\text{A} \qquad V_{BE(on)} = 0.7 \text{ V}$$

$$V_{C1} = V_{C2} = \tfrac{1}{2}V_{CC}$$

$$V_{CB3} = V_{CB4} = \tfrac{1}{2}(V_{CC} - V_{C1})$$

With $V_{B1} = V_{B2} = 0$ V:

(a) Calculate values for R_{C1}, R_{C2}, R_{C3}, R_{C4}, R_{EE1}, and R_{EE2}.

(b) Calculate r_{id}, a_d, r_{ic}, and a_c.

(c) With $V_{B2} = 0$ V, what is the maximum voltage at the base of Q_1 that will cause:

(1) Q_4 to be at the edge of saturation $[V_{CE(sat)} = 0.2$ V]?

(2) Q_4 to be at the edge of cut off?

P8.21 In Fig. P8.21,

$$V_{CC} = 10 \text{ V} \qquad R_{C2} = 5 \text{ k}\Omega \qquad \beta_0 = 50$$

$$I_{EE} = 1 \text{ mA} \qquad R_{E3} = 0 \qquad V_{BE(on)} = 0.7 \text{ V}$$

$$I_{E3} = 10 \text{ mA}$$

(a) Determine the dc output voltage for $v_s = 0$.

(b) Determine the small-signal voltage gain v_o/v_s.

(c) Determine the small-signal input resistance at v_s.

(d) Determine the small-signal output resistance at v_o.

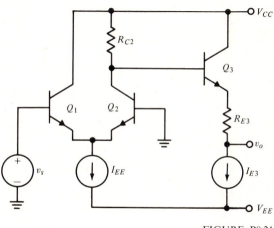

FIGURE P8.21

P8.22 In Fig. P8.21,

$$V_{CC} = 6 \text{ V} \qquad R_{C2} = 30 \text{ k}\Omega \qquad \beta_0 = 100$$

$$I_{EE} = 100 \text{ } \mu\text{A} \qquad V_{BE(on)} = 0.7 \text{ V}$$

$$I_{E3} = 10 \text{ mA}$$

(a) Calculate a value for R_{E3} so that the output voltage $V_o = 0$ V.

(b) Calculate the output resistance at v_o.

(c) Calculate the voltage gain v_o/v_s into a 1-kΩ load.

P8.23 For the amplifier in Fig. P8.23, for $v_{i1} = v_{i2} = 0$ V, $v_o = 0$ V, $\beta_0 = 50$, $V_{BE(on)} = 0.7$ V:

(a) Calculate I_{C1}, I_{C2}, I_{C3}, I_{C4}, V_{CE1}, V_{CE2}, V_{CE3}, V_{CE4}, and I_X.

(b) Calculate v_o/v_{id} and r_{id}.

(c) Calculate v_o/v_{id} if the resistance of the signal source $R_{s1} = R_{s2} = 7.5$ kΩ and a load $R_L = 7.5$ kΩ is added to the output.

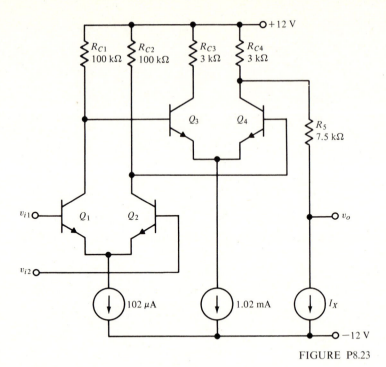

FIGURE P8.23

P8.24 For the amplifier in Fig. P8.24, $I_{C1} = I_{C2} = 10\ \mu A$, $I_{C3} = I_{C4} = 100\ \mu A$, $I_{C5} = 1$ mA, $\beta_0 = 50$, $V_{BE(on)} = 0.7$ V, $V_{D(on)} = 0.7$ V. With $V_{B1} = V_{B2} = 0$ V, $V_{C1} = V_{C2} = \frac{1}{2}V_{CC}$, and $V_{CB4} = \frac{1}{2}(V_{CC} - V_{C1})$:
   (a) Solve for I_1, I_2, R_1, R_2, R_3, R_4, and n (the number of diodes that gives the lowest dc offset voltage at the output).
   (b) Solve for r_{id}, r_{ic}, and r_o.
   (c) Solve for a_d and a_c.

P8.25 For the diode-biasing scheme shown in Fig. 8.12b:
   (a) Write an expression for the *ratio* of the current in R_1 to the collector current in the two identical transistors Q_1 and Q_2. (Do not assume $\beta \gg 1$.)
   (b) Now with $\beta \gg 1$ calculate a value for R_1 if

$$V_{BB} = 15\ V \qquad I_{C1} = 0.1\ mA \qquad V_{BE(on)} = 0.7\ V$$

*P8.26 For the improved biasing scheme shown in Fig. 8.12c:
   (a) Use an appropriate small-signal transistor model to calculate the resistance R seen looking in at the collector of the current-source transistor Q_1.
   (b) Calculate values for R_1 and R_{E1} so that $I_{C1} = 10\ \mu A$ under the condition that $V_{BB} = 6$ V, $V_{BE(on)} = 0.7$ V, and $V_{RE1} = 120$ mV.

P8.27 For the differential amplifier in Fig. P8.27, $I_{E1} = I_{E2} = 150\ \mu A$. For the transistors, $\beta_0 = 60$, $V_{BE(on)} = 0.7$ V, and $r_{oc} = 50$ kΩ.
   (a) Determine the value of R_3.
   (b) Determine r_{id} and a_d.
   (c) Determine r_{ic} and a_c.

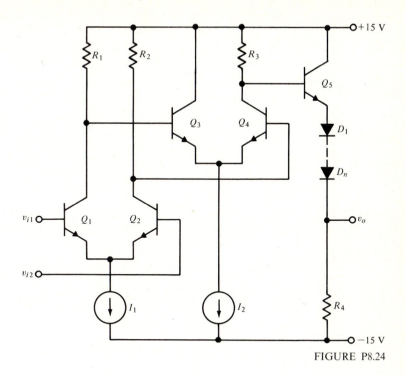

FIGURE P8.24

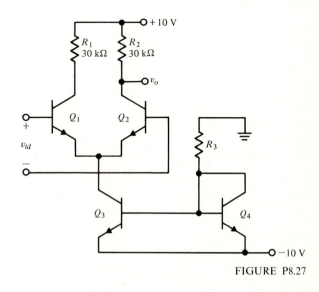

FIGURE P8.27

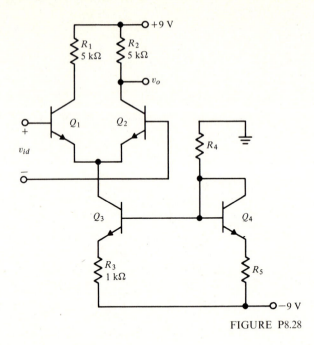

FIGURE P8.28

P8.28 For the differential amplifier in Fig. P8.28, $I_{C1} = I_{C2} = 1$ mA. For the transistor, $\beta_0 = 50$, $V_{BE(\text{on})} = 0.7$ V, and $r_{oc} = 50$ kΩ.
(a) Determine values for R_4 and R_5 if the total drain from the negative supply is 3 mA.
(b) Determine r_{id} and a_d.
(c) Determine r_{ic} and a_c.

P8.29 For the operational amplifier in Fig. P8.29,

$$\beta_0 = 200 \qquad\qquad V_{BE(\text{on})} = 0.7 \text{ V}$$

$$I_{C6} = I_{C7} = 0.1 I_{C8}$$

(a) Calculate a value for R_5 so that $v_o = 0$ V for $v_{i1} = v_{i2} = 0$ V.
(b) Determine the differential voltage gain $a_d = v_o/(v_{i1} - v_{i2})$.

P8.30 For the amplifier circuit in Fig. P8.30,

$$\beta_0 = 50 \qquad\qquad V_{BE(\text{on})} = 0.7 \text{ V}$$

$$I_{C1} = I_{C2} = 100 \text{ μA} \qquad\qquad I_5 = 10 \text{ mA}$$

$$I_{C3} = I_{C4} = 1 \text{ mA} \qquad\qquad I_6 = 10.2 \text{ mA}$$

Solve for
(a) The resistor values R_1, R_2, R_3, R_4, R_5, R_6 under the conditions that $V_{R1} = V_{R2} = 12$ V, $V_{R5} = 5.1$ V, and the quiescent output voltage is 0 V.
(b) The differential input resistance r_{id} and voltage gain a_d.
(c) The small-signal output resistance of the amplifier.

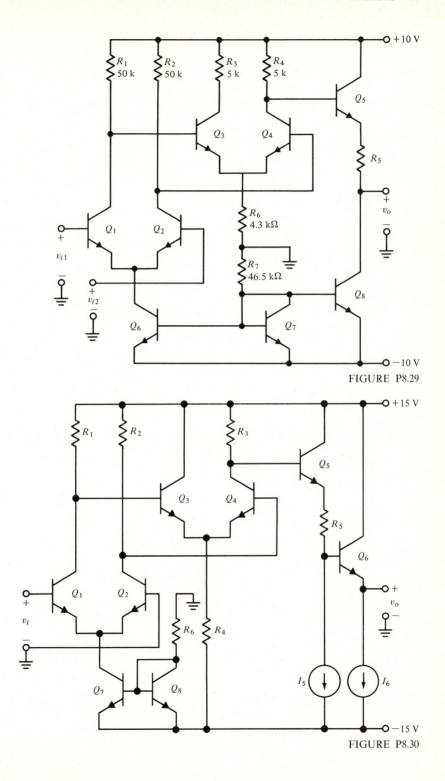

FIGURE P8.29

FIGURE P8.30

P8.31 For the amplifier circuit in Fig. P8.31,

$$\beta_0 = 250 \qquad V_{BE(\text{on})} = 0.7 \text{ V}$$

(a) Solve for the quiescent operating point (I_C and V_{CE}) for all transistors ($Q_1, Q_2, Q_3, Q_4,$ and Q_5).
(b) Calculate the differential voltage gain v_o/v_i.
(c) With $v_i = 0$ V, what is V_o?
(d) For the condition of part (c), what is the input bias current, $(I_{B1} + I_{B2})/2$?

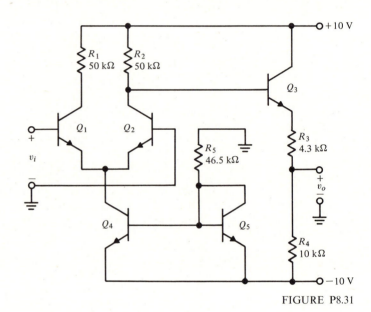

FIGURE P8.31

P8.32 For the circuit in Fig. D8.4, assume

$$\beta_F \to \infty \qquad V_{BE(\text{on})} = 0.7 \text{ V}$$

$$\beta_0 = 100$$

(a) Calculate the operating point (I_C, V_{CE}) for each of the transistors.
(b) Find the dc output voltage with both inputs at 0 V.
(c) Compute $r_{id}, a_v,$ and r_o.

P8.33 The amplifier in Fig. P8.33 is used in an electronic voltmeter.
(a) Calculate values for R_5 and R_6 such that $I_{C1} = I_{C2} = 50 \ \mu\text{A}$ and $I_{C3} = I_{C4} = 500 \ \mu\text{A}$. The input is grounded.
(b) Calculate the input resistance of the voltmeter, that is, r_{id}.
(c) Calculate the input voltage sensitivity for full-scale deflection of the meter.

$$\text{at } I_C = 50 \ \mu\text{A}; V_{BE(\text{on})} = 0.6 \text{ V and } \beta_0 = 150$$

$$\text{at } I_C = 500 \ \mu\text{A}; V_{BE(\text{on})} = 0.65 \text{ V and } \beta_0 = 300$$

P8.34 For the amplifier circuit in Fig. P8.34,

$$\beta_0 = 100 \qquad V_{BE(\text{on})} = 0.7 \text{ V} \qquad V_{D(\text{on})} = 0.7 \text{ V}$$

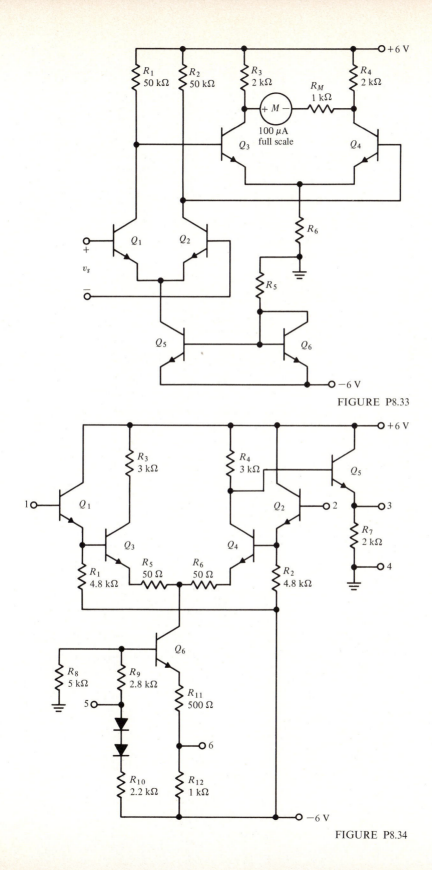

FIGURE P8.33

FIGURE P8.34

    (*a*) With pins 5 and 6 open, calculate:
        (1) The quiescent operating voltage at pin 3.
        (2) The input resistance between pins 1 and 2.
        (3) The voltage gain $[v_o(3 - 4)/v_i(1 - 2)]$.
        (4) The output resistance between pin 3 and 4.
        (5) The maximum symmetrical output voltage swing at pin 3.
    (*b*) Repeat the calculation for pins 5 and 6 connected to -6 V.

*P8.35 For the amplifier circuit of Fig. P8.35,

$$\beta_0 = 20 \qquad V_{BE(on)} = 0.7 \text{ V}$$

    (*a*) Determine the operating points (I_C and V_{CE}) of both Q_1 and Q_2.
    (*b*) Determine the voltage gain of the circuit v_o/v_i.

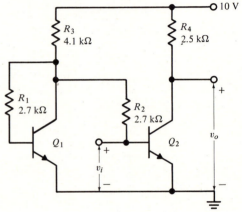

FIGURE P8.35

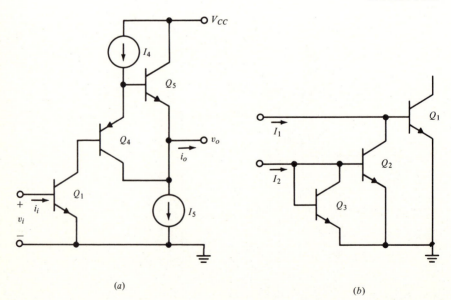

      (*a*)   (*b*)

FIGURE P8.36

*P8.36 (a) Most IC op amps require a dual power supply, typically ± 15 V. A novel IC amplifier that uses only one supply is shown in Fig. P8.36a. This circuit describes a current-mode inverting-type amplifier with a high input resistance and low output resistance. For the circuit of Fig. P8.36a, with a finite load resistor what is the current gain (i_o/i_i) in terms of β_{01}, β_{04}, and β_{05}?

(b) A general-purpose op amp requires both an inverting and noninverting input. For the current-mode amplifier the noninverting input is provided by the addition of Q_2 and Q_3 to the input, as shown in Fig. P8.36b. This is the so-called *current mirror* circuit. For this circuit show that $I_1 = I_2$, if we assume $\beta_F \to \infty$.

9

RESISTIVE FEEDBACK AND FREQUENCY COMPENSATION

In our description of linear ICs we have progressed from the analysis of a single transistor stage operating in the linear active region, through a cascade of such stages, to practical examples of IC operational amplifiers. We have chosen to use the operational amplifier as the basis of describing linear ICs because of its versatility and wide application in processing linear and/or analog signals. In this chapter we will consider the operational amplifier as a basic gain block, characterized by a voltage transfer characteristic (the voltage gain a_v) and the small-signal input and output resistance (r_i and r_o). By the use of a few simple components externally connected to the gain block, we are able to modify the basic small-signal characteristics to meet the particular requirement we may have. This involves the technique known as *negative feedback*.

Indeed, the versatility of operational amplifiers is a direct result of the use of negative feedback. With sufficient feedback the *closed-loop* characteristics of the amplifier (that is, with the feedback connected) become a function of the feedback elements alone. In a typical feedback-amplifier design, the feedback elements are just two resistors, and the precision of the closed-loop voltage gain is set by the value of these two resistors, practically independent of the *open-loop* amplifier characteristic (i.e., with the feedback disconnected). Thus almost any degree of

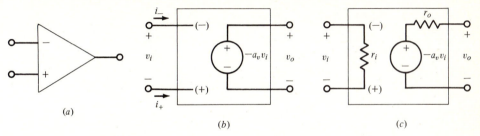

FIGURE 9.1
(a) Block-diagram symbol for operational amplifier; (b) small-signal circuit model of the ideal operational amplifier; (c) small-signal circuit model of a practical operational amplifier.

amplification, providing it is less than the open-loop gain, can be obtained with relative ease and precision. In addition, with negative feedback we are also able to change the input and output characteristics of the operational amplifier. Should we require a large input resistance, much larger than that of the operational amplifier alone, we may use negative feedback. Similarly we may use negative feedback to obtain an output resistance much smaller than that of the operational amplifier alone.

Following a description of the *ideal operational amplifier*, we will consider some simple applications of negative feedback with ideal operational amplifiers. Then follows a more generalized approach to the feedback technique, by which we can more readily learn of the results of negative feedback in practical amplifier design. Finally, because we do wish to consider the practical use of IC operational amplifiers, we will introduce the subject of *frequency response* of these amplifiers and the need of *frequency compensation* with the use of negative feedback.

9.1 IDEAL OPERATIONAL AMPLIFIER

To introduce the subject of the use of operational amplifiers as voltage amplifiers, we make use of an ideal operational amplifier. The symbol used for the operational amplifier is shown in Fig. 9.1a, and the basic elements of the ideal operational amplifier modeled in Fig. 9.1b are described below.

9.1.1 Gain

The primary function of the amplifier is to amplify. Generally the more gain the amplifier has the better it is for our purpose. Excessive gain can always be reduced by external circuitry, so we will assume the amplifier gain to be infinite. That is,

$$a_v \to \infty$$

consequently, for a finite output signal v_o,

$$v_i \to 0$$

9.1.2 Input Resistance

We will assume that the driving source will not have to supply any current to the input terminals of the operational amplifier. That is, the input resistance to the operational amplifier is infinite. Therefore

$$r_i = \infty$$

and for the input currents,

$$i_- = i_+ = 0$$

9.1.3 Output Resistance

For the output signal voltage to be unaffected by any load resistance, we require that the output resistance be zero. Then

$$r_o = 0$$

9.1.4 Input Offset Voltage

This is the name given to the dc differential voltage that must be applied at the input of the amplifier for the output to be at 0 V. For our ideal amplifier we assume that this is zero. That is, the output level of the amplifier will be at 0 V with zero signal appearing across the two inputs.

9.1.5 Response Time

In the ideal amplifier the output must instantaneously respond to a time-varying input signal applied at the input; thus the response time of the amplifier is assumed to be zero. This implies that the frequency response of the amplifier is flat from dc to infinity.

It should be mentioned that the cost of an operational amplifier with these characteristics would also reach to infinity. Fortunately many practical applications can be handled by IC op amps that cost under $1.00.

The basic elements of a practical operational amplifier are shown in Fig. 9.1c. The meaning of the minus $(-)$ and plus $(+)$ signs appearing at the input of the operational amplifier is that the $(-)$ sign indicates the *inverting input* terminal and the $(+)$ sign the *noninverting input* terminal. A positive-going signal v_i appearing at the $(-)$ input [with reference to the $(+)$ input] would cause the output of the voltage source v_o to have the amplified negative-going signal $-a_v v_i$. Referenced to the $(-)$ input, a positive signal at the $(+)$ input would cause a positive signal at the output of the amplifier.

9.2 BASIC CIRCUIT BLOCKS

In this section we will use the basic Kirchhoff's voltage and current laws to derive an expression for the closed-loop voltage gain of an ideal operational amplifier incorporating negative feedback.

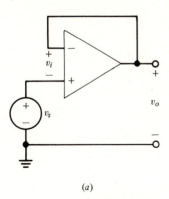

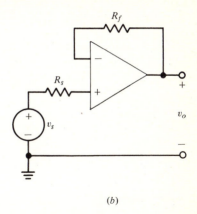

(a)   (b)

FIGURE 9.2
The voltage follower. (a) The basic circuit; (b) the practical circuit, including
bias-current-balancing resistors.

9.2.1 Voltage Follower

A simple example of negative feedback is shown in the *voltage-follower* circuit
shown in Fig. 9.2a. In this circuit the output of the operational amplifier is connected back directly to the inverting input terminal of the amplifier. The input
voltage signal v_s is applied to the noninverting terminal. A positive input signal
will therefore cause a positive output signal, but this output signal also appears at
the inverting input terminal. For the case of the ideal operational amplifier, where
$v_i = 0$, we find the output signal v_o equals the input signal v_s. Thus we have a
circuit with a voltage gain of 1. This circuit is also described as a *unity gain buffer*.
    We can show this algebraically by using KVL in the circuit shown in
Fig. 9.2a. Upon summing the input voltages, we have

$$v_s + v_i = v_o \qquad (9.1)$$

but we have from Fig. 9.1b that

$$v_o = -a_v v_i \qquad (9.2a)$$

or

$$v_i = -\frac{v_o}{a_v} \qquad (9.2b)$$

Now we see that for the ideal amplifier, where $a_v \to \infty$,

$$v_i = 0 \qquad (9.3)$$

Substituting Eq. (9.3) in Eq. (9.1), we have the voltage-follower relation

$$v_s = v_o \qquad (9.4)$$

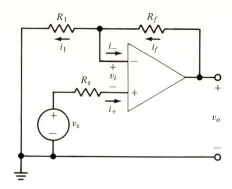

FIGURE 9.3
The noninverting feedback-amplifier circuit.

Unity gain circuits are used as electrical buffers to isolate circuits or devices from one another and prevent undesired interactions. As a voltage-follower power amplifier, this circuit will allow a source with low-current capabilities (i.e., a high resistance source) to drive a low-resistance load with a voltage gain of 1. Moreover, as we shall see later, the input resistance of the voltage-follower circuit is the greatest value that can be obtained from any operational-amplifier circuit using simple resistive feedback.

Since for the ideal operational amplifier $i_- = 0$, no current flows between the output and the inverting input terminal in Fig. 9.2a. We can therefore connect a resistor in this feedback path, as shown in the circuit of Fig. 9.2b. Generally, the value of this resistor will be equal to the source resistance of the input voltage signal. With IC op amps using bipolar transistors, this causes the voltage generated by the base-biasing currents of the input differential-amplifier pair to be balanced out.

9.2.2 Noninverting Amplifier

Another basic circuit block is the *noninverting amplifier*. With this amplifier the output signal is of the same polarity as the input signal, only increased in magnitude; there is no polarity inversion.

As shown in Fig. 9.3 this circuit is similar to the voltage follower with an additional resistor (R_1) connected from the inverting input terminal to ground. Using an ideal operational amplifier, with $v_i = 0$, the voltage to ground at the junction of R_1 and R_f is equal to the input signal voltage v_s. Since no current flows into the input terminals of an ideal amplifier, we then have by simple voltage-divider action of these two resistors, R_1 and R_f,

$$v_s = \frac{R_1}{R_1 + R_f} v_o \qquad (9.5)$$

Or by rearrangement of this equation,

$$\frac{v_o}{v_s} = \frac{R_f + R_1}{R_1} = \frac{R_f}{R_1} + 1 \qquad (9.6)$$

With an operational amplifier whose open-loop voltage gain approaches infinity, we see that the closed-loop voltage gain is solely determined by the value of the two resistors R_1 and R_f. We can make use of Kirchhoff's circuit laws in Fig. 9.3 to verify this statement.

To do this we let the differential input voltage be v_i. The voltage to ground at the inverting input terminal is then

$$v_i + v_s$$

With the input resistance of the operational amplifier equal to infinity (that is, $i_- = 0$), the current in R_1 must be equal to the current in R_f; that is,

$$i_1 = i_f \qquad (9.7a)$$

or

$$\frac{v_i + v_s}{R_1} = \frac{v_o - (v_i + v_s)}{R_f} \qquad (9.7b)$$

But again, as shown in Eqs. (9.2) and (9.3), with $a_v \to \infty$,

$$v_i = 0$$

Substituting for v_i in Eq. (9.7b), we have

$$\frac{v_s}{R_1} = \frac{v_o - v_s}{R_f} \qquad (9.8a)$$

therefore

$$v_s(R_1 + R_f) = v_o R_1$$

and

$$\frac{v_o}{v_s} = \frac{R_f}{R_1} + 1 \qquad (9.8b)$$

In a practical amplifier we would minimize the offset error due to the input bias currents by making

$$R_s = R_1 \| R_f$$

Notice in our analysis that we have essentially summed the currents at the junction of R_1 and R_f. That is, by KCL in Fig. 9.3,

$$i_f - i_1 - i_- = 0$$

Hence this junction is often referred to as the *summing point*. Indeed this simple method of analysis with ideal or near-ideal operational amplifiers is described as the *summing-point method*.

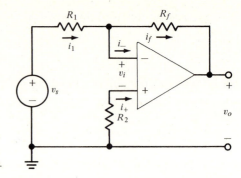

FIGURE 9.4
The inverting feedback-amplifier circuit.

9.2.3 Inverting Amplifier

The third basic gain block is the *inverting-amplifier* circuit, shown in Fig. 9.4. Notice in this circuit we still have the feedback resistor connected between the output and the inverting input terminal, but now the input voltage signal is also connected to the inverting input through the resistor R_1.

Since with an ideal operational amplifier $v_i = 0$, with the noninverting input terminal connected to ground, the voltage to ground at the inverting input terminal is also at 0 V. As a result, the input signal v_s appears across R_1, and the output signal v_o appears across R_f. But since the applied signal is to the inverting input terminal, there is a $180°$ phase inversion between v_o and v_s. Then

$$\frac{v_s}{R_1} = -\frac{v_o}{R_f} \qquad (9.9a)$$

or

$$\frac{v_o}{v_s} = -\frac{R_f}{R_1} \qquad (9.9b)$$

With an ideal operational amplifier, the closed-loop voltage gain is simply the ratio of these two resistors. Again we can make use of Kirchhoff's circuit laws to verify this statement. With the input resistance of the operational amplifier equal to infinity ($i_- = 0$), the same current that flows in R_1 must flow in R_f. That is,

$$i_1 = i_f \qquad (9.10a)$$

or

$$\frac{v_s - v_i}{R_1} = -\frac{v_o - v_i}{R_f} \qquad (9.10b)$$

Again as in Eqs. (9.2) and (9.3), with $a_v \rightarrow \infty$,

$$v_i = 0$$

Substituting for v_i in Eq. (9.10b), we have

$$\frac{v_s}{R_1} = -\frac{v_o}{R_f} \qquad (9.11a)$$

then

$$\frac{v_o}{v_s} = -\frac{R_f}{R_1} \qquad (9.11b)$$

Again, in a practical amplifier we would minimize the offset error due to the input bias currents by connecting a resistor R_2 from the noninverting input terminal to ground. For minimum offset we select

$$R_2 = R_1 \| R_f$$

Any resistance attributable to the signal source could be incorporated into R_1.

We have already noted in this analysis that by using an ideal operational amplifier with $v_i = 0$, the summing point at the junction of R_1 and R_f is at 0 V. Hence in the case of the inverting amplifier, the summing point is referred to as a *virtual ground*. Indeed we will see later that a result of negative feedback in this circuit is to effectively reduce any resistance to ground at the summing point to virtually zero.

In this section we have only been considering the use of operational amplifiers as voltage amplifiers using simple resistive feedback. Historically, operational amplifiers were initially so-called because with the appropriate feedback connections they could perform the analog operations of addition, subtraction, differentiation, and integration. With resistive feedback we can perform the first two operations (see Probs. P9.6 through P9.9). With reactive elements, like capacitors, connected in a feedback network the functions of differentiation and integration can be implemented (see Probs. P9.30 and P9.31). Also, stable oscillators can be made with operational amplifiers by using *positive* feedback to the noninverting input terminal.

In Demonstration D9.1, the ideal operational amplifier is simulated by a type 741, and each of the three feedback amplifiers discussed in this section are implemented with this IC op amp.

EXERCISES

E9.1 By using an ideal operational amplifier and a circuit similar to that shown in Fig. 9.2b with $R_s = 10 \text{ k}\Omega$ and $R_f = 10 \text{ k}\Omega$, what is the closed-loop voltage gain A_v?

E9.2 A circuit similar to that shown in Fig. 9.3 uses an ideal operational amplifier with $R_1 = 1 \text{ k}\Omega$. Calculate the value of R_f for a closed-loop voltage gain A_v of 100.

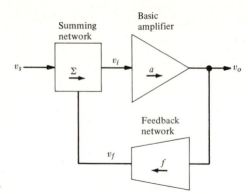

FIGURE 9.5
Block diagram of a feedback system.

E9.3 Again the circuit is similar to Fig. 9.3 with an ideal operational amplifier. To a precision of 1 percent or better, compute the optimal values for R_1 and R_f when $R_s = 1$ kΩ and $A_v = 50$.

E9.4 A feedback-amplifier circuit similar to that shown in Fig. 9.4 uses an ideal operational amplifier. With $R_1 = 100$ Ω and $R_f = 2$ kΩ, what is the closed-loop voltage gain A_v?

9.3 FEEDBACK

We have seen in the previous section that the application of negative feedback to an ideal operational amplifier yields an amplifier circuit with the closed-loop characteristics determined solely by the values of the feedback components. These simple formulations may also be used with practical operational amplifiers, where the characteristics are less than ideal. However, when a more detailed analysis is required, we will find it useful to employ the concepts of a feedback system. This approach does not rely upon the properties of an ideal operational amplifier.

9.3.1 Basic Properties

The general form of a feedback system is illustrated by the block diagram of Fig. 9.5. In this electronic circuit the dimensions of the variables are voltages. But in other applications some of the variables could be current. In a mechanical system, pressure or velocity could be the dimensions of the variables. Irrespective of the physical variables, the basic properties of control by feedback are applied in many different applications and systems.

In the representation of Fig. 9.5 a single feedback loop is assumed. The gain of the basic amplifier is denoted by a and is controlled by transistor parameters that are subject to considerable variation. For our purpose, the feedback network is a precision resistive attenuator, which produces an output v_f that is a replica of its input except reduced in magnitude. The transmission through the feedback network is represented by f, known as the feedback factor. The summing network, as its name implies, sums (adds or subtracts) its two input signals and produces an output v_i, which is the sum (or difference) of the two, that is fed to the input of the basic amplifier. The two signals add for positive feedback. That is, the magnitude of v_i is greater than v_s, the input signal. For negative feedback the signals subtract; then the magnitude of v_i is less than v_s. In order to exchange gain of the basic amplifier for an improvement in the overall amplifier-response characteristic, the feedback must be negative. That is, the feedback signal must be 180° out of phase with the input signal. This phase inversion may be accomplished in either the basic amplifier or the feedback network. As a result of the consequent subtraction by the summing network, the input to, and the output from, the basic amplifier is less than would be the case with no feedback. The overall or closed-loop gain of the feedback system is reduced, but other improvements result particularly with regard to stabilizing performance with respect to component and temperature variations. We also are able to modify selectively the input and output resistance characteristics. We assume that the open-loop gain a is much larger than the closed-loop gain A we require from the feedback system.

We also assume in the block diagram of Fig. 9.5 that there is no effect on the signal from one block by the loading, or connecting, of another. Another simplifying assumption we make is that the transmission in both the basic amplifier and feedback network is considered to be unilateral, that is, in one direction only, as denoted by the arrows within these blocks.

9.3.2 The Feedback Equation

The relationships between the various signals in the feedback system of Fig. 9.5 can be expressed as follows. The signal input to the basic amplifier, v_i, is

$$v_i = v_s + v_f \qquad (9.12a)$$

The output signal of the basic amplifier is

$$v_o = a_v v_i \qquad (9.12b)$$

where a_v is the voltage gain of the basic amplifier. The feedback signal v_f is

$$v_f = f v_o \qquad (9.12c)$$

where f, the feedback factor, is the fraction of the output signal which is fed back. The terminal or closed-loop voltage gain A_v is given by

$$A_v = \frac{v_o}{v_s} \qquad (9.13a)$$

To obtain an expression for A_v in terms of a_v, we first substitute for v_s from Eq. (9.12a) in (9.13a); namely,

$$A_v = \frac{v_o}{v_i - v_f} \qquad (9.13b)$$

Now from Eqs. (9.12b) and (9.12c) we can substitute for v_i and v_f, respectively; then

$$A_v = \frac{v_o}{v_o/a_v - fv_o} \qquad (9.13c)$$

Dividing through by v_o yields

$$A_v = \frac{1}{1/a_v - f} \qquad (9.13d)$$

Now we multiply by a_v to obtain a *generalized form of the feedback equation.* That is,

$$A_v = \frac{a_v}{1 - a_v f} \qquad (9.14a)$$

This equation is also written as

$$A_v = \frac{a_v}{1 + T} \qquad (9.14b)$$

where

$$T = -a_v f \qquad (9.14c)$$

The equations of (9.14) are described as the *basic feedback equations.* We use them along with the following definitions.
Closed-loop gain:

$$A_v = \frac{v_o}{v_s}$$

Open-loop gain:

$$a_v = \frac{v_o}{v_i}$$

Loop gain:

$$T = -\frac{v_f}{v_i}$$

The *loop gain* of the feedback system is the negative of the gain around the feedback loop from the input to the basic amplifier to the output of the feedback network; namely,

$$T = -a_v f = -\left(\frac{v_o}{v_i}\right)\left(\frac{v_f}{v_o}\right) = -\frac{v_f}{v_i} \qquad (9.15)$$

For negative feedback, either a_v or f, but not both, must be negative. Hence T is a positive number. That is,

$$T > 0$$

Then the denominator of Eq. (9.14b) is always positive and greater than 1; consequently $A_v < a_v$.

Conversely, with positive feedback, T is a negative number due to both a_v and f having the same sign. Then

$$T < 0$$

Notice if $T = -1$, the denominator of Eq. (9.14b) goes to zero and the closed-loop gain A_v goes to infinity. This indicates an unstable or oscillatory condition.

For now we shall concentrate only on the case where $T > 0$, that is, the feedback is always negative. Then, under the condition that $T \gg 1$, Eq. (9.14b) reduces to

$$A_v \approx \frac{a_v}{T} = \frac{a_v}{-a_v f} = -\frac{1}{f} \qquad (9.16)$$

For this case the closed-loop gain depends only on the components of the precision feedback network. That is, the situation is similar to that described in Sec. 9.2 using an ideal operational amplifier. Indeed, the large loop gain required to establish the condition of Eq. (9.16) may be achieved through the use of a high-gain basic amplifier, e.g., an IC op amp.

9.3.3 Gain Sensitivity

An important result of negative feedback is the fact that the closed-loop gain A is not as sensitive to parameter changes as is the open-loop gain a. The feedback network of Fig. 9.5 is regarded as a precision network, where the feedback factor f is accurately established and is insensitive to changes in the environment. In contrast, the gain of the basic amplifier is generally not a precise value, being highly dependent on parameter and temperature changes, such as transistor β_0.

Consider a small change in A ($\equiv dA$) resulting from a small change in a ($\equiv da$). Then using the form for A shown in Eq. (9.14a),

$$\frac{dA}{da} = \frac{d}{da}\left(\frac{a}{1 - af}\right)^{\dagger} \qquad (9.17a)$$

$$\frac{dA}{da} = \frac{(1 - af) - a(-f)}{(1 - af)^2} = \frac{1}{(1 - af)^2}$$

---

† This equation is of the form

$$\frac{d}{dx}\left(\frac{x}{v}\right) = \frac{v\dfrac{dx}{dx} - x\dfrac{dv}{dx}}{v^2}$$

that is,

$$\frac{d}{da}\left(\frac{1}{1 - af}\right) = \frac{(1 - af)1 - a(-f)}{(1 - af)^2}$$

By rearrangement

$$dA = \frac{da}{(1 - af)^2} \qquad (9.17b)$$

Expressing the closed-loop change as a fractional change and using Eqs. (9.17b) and (9.14a) directly, we have

$$\frac{dA}{A} = \left[\frac{da}{(1 - af)^2}\right]\left(\frac{1 - af}{a}\right)$$

$$= \frac{da}{a}\left(\frac{1}{1 - af}\right) = \frac{da}{a}\left(\frac{1}{1 + T}\right) \qquad (9.18)$$

The fractional change in the closed-loop gain resulting from the sensitivity of the basic amplifier has been reduced through the application of negative feedback by the factor D_s, where

$$D_s = \frac{1}{1 + T} \qquad (9.19)$$

This factor is often referred to as the *desensitivity factor* of a feedback system.

As a numerical example of the use of these feedback equations, consider the following problem.

EXAMPLE 9.1

(a) Determine the closed-loop gain of a feedback system consisting of a basic inverting amplifier with an open-loop gain of 10,000 and a precision feedback network with a feedback factor of 0.01.

(b) Compute the fractional change in the overall gain for a 50 percent increase in the gain of the basic amplifier.

SOLUTION

(a) Since we are using an inverting amplifier,

$$a = -10^4$$

and for the feedback factor,

$$f = 10^{-2}$$

then from Eq. (9.14b),

$$A = \frac{a}{1 + T}$$

where

$$T = -af = -(-10^4)(10^{-2}) = 100$$

Therefore

$$A = \frac{-10^4}{1 + 100} = -99$$

Note that if we had used the approximation $A \approx -(1/f)$,

$$A = -\frac{1}{10^{-2}} = -100$$

(b) From Eq. (9.18),

$$\frac{dA}{A} = \frac{da}{a}\left(\frac{1}{1 + T}\right)$$

$$= 0.5\left(\frac{1}{1 + 100}\right) = 0.00495$$

That is, the open-loop gain change of 50 percent results in a closed-loop gain change of less than 0.5 percent. ////

The modifying effect of feedback on the parameter variations of an IC op amp is illustrated in Demonstration D9.2.

EXERCISE

E9.5 For a feedback system having a closed-loop gain of 100 and an open-loop gain of 2000, determine the desensitivity factor D_s and compute the fractional change in the closed-loop gain resulting from a 20 percent change in the open-loop gain.

9.3.4 The Noninverting Amplifier

In the next two sections we shall see how the general formulation of feedback described in the previous two sections can be applied with actual circuits. Specifically we will analyze the noninverting and inverting type of feedback amplifier shown in Figs. 9.3 and 9.4, respectively. We will find that the expressions for the closed-loop gain are of the form of Eq. (9.14). In particular we will see how negative feedback modifies the small-signal input and output resistance of the amplifier.

The form of a feedback configuration is generally described in terms of the manner in which the feedback network is connected to the input and output of the basic amplifier. The noninverting amplifier of Fig. 9.3 has been redrawn in Fig. 9.6 to emphasize the nature of the forward and feedback paths and the form of their interconnection. The ground line now appears across the top of the figure, with the basic amplifier and the feedback network shown enclosed within the dashed lines.

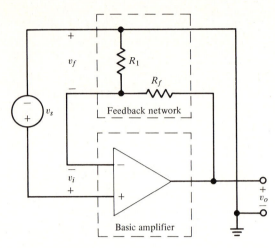

FIGURE 9.6
The noninverting amplifier circuit showing the series-shunt feedback configuration.

From Fig. 9.6 the input to the feedback network is seen to be connected across the output terminals of the basic amplifier. That is, it is connected in *shunt* with the output of the basic amplifier. For such a shunt output connection, the feedback network is regarded as *sampling the output voltage* of the amplifier. At the input to the amplifier the output of the feedback network is connected in *series* with the signal source and the input of the basic amplifier. Hence the feedback signal is *summed with the input voltage* to the amplifier.

The noninverting amplifier of Fig. 9.6 is often referred to as a *series-shunt feedback configuration*. That is, the *input* of the basic amplifier and the feedback signal are in *series*, while the feedback network *shunts* the *output* of the amplifier. Since a voltage signal from the output is fed back in series with the input signal, another name for this particular circuit configuration is *voltage-series feedback*.

If the basic amplifier has the properties of the ideal operational amplifier as shown in Fig. 9.1, it is a straightforward process to characterize the circuit of Fig. 9.6 in the form of the basic feedback equation. Since the ideal operational amplifier has an infinite input resistance and zero output resistance, we have that v_f is the following fraction of v_o:

$$v_f = -\frac{R_1}{R_1 + R_f} v_o \qquad (9.20)$$

and

$$v_o = a_v v_i \qquad (9.21)$$

where a_v is the voltage gain of the operational amplifier.

Now again in Fig. 9.6 we have by KVL,

$$v_s = v_i - v_f \qquad (9.22)$$

then using Eqs. (9.20) and (9.21) to substitute for v_f and v_i, respectively,

$$v_s = \frac{v_o}{a_v} + \frac{R_1}{R_1 + R_f} v_o \qquad (9.23a)$$

that is,

$$v_s = v_o \left(\frac{1}{a_v} + \frac{R_1}{R_1 + R_f} \right) \qquad (9.23b)$$

The closed-loop voltage gain of the amplifier is given by

$$A_v = \frac{v_o}{v_s}$$

that is, from Eq. (9.23b),

$$A_v = \frac{1}{\dfrac{1}{a_v} + \dfrac{R_1}{R_1 + R_f}} \qquad (9.24a)$$

where, multiplying through by a_v, we have

$$A_v = \frac{a_v}{1 + a_v \dfrac{R_1}{R_1 + R_f}} \qquad (9.24b)$$

In terms of the feedback equation,

$$A_v = \frac{a}{1 - af} \qquad (9.25a)$$

where, substituting from Eq. (9.24b),

$$a = a_v \qquad (9.25b)$$

$$f = -\frac{R_1}{R_1 + R_f} \qquad (9.25c)$$

For the ideal operational amplifier, $a_v \to \infty$. In this case,

$$A_v = -\frac{1}{f} = \frac{R_f}{R_1} + 1 \qquad (9.26)$$

which is the same as we had in Eq. (9.8b) when considering the gain of the noninverting amplifier incorporating an ideal operational amplifier. (We can remember to add $+1$ to the ratio of R_f and R_1 in the *noninverting* feedback amplifier if we recall that for this amplifier the input signal is applied to the $(+)$ input.)

Aside from its effect on the gain properties of an amplifier, negative feedback also has an important effect on the input and output resistances. This effect can be

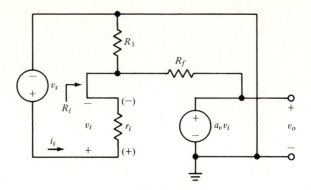

FIGURE 9.7
Small-signal circuit model of the noninverting amplifier, including the input resistance r_i.

demonstrated by considering the noninverting amplifier of Fig. 9.6 with finite resistances included at the input and output of the operational amplifier. To simplify the analysis, the input and output resistance levels will be considered separately.

Finite input resistance   The amplifier circuit of Fig. 9.6 is shown in Fig. 9.7 with a finite input resistance included for the operational amplifier. As before, the output resistance is taken to be zero.

  We wish to determine the input resistance of the overall amplifier. That is, we define the closed-loop input resistance in Fig. 9.7 as

$$R_i = \frac{v_s}{i_s} \qquad (9.27)$$

Now we can express the output signal v_o in terms of the closed-loop voltage gain A_v as

$$v_o = A_v v_s \qquad (9.28)$$

From Fig. 9.7 we also have that

$$v_o = a_v v_i \qquad (9.29)$$

By combining Eqs. (9.28) and (9.29), we may obtain an expression for v_i; namely,

$$v_i = \frac{A_v v_s}{a_v} \qquad (9.30)$$

Now again from Fig. 9.7 we note that

$$i_s = \frac{v_i}{r_i} \qquad (9.31)$$

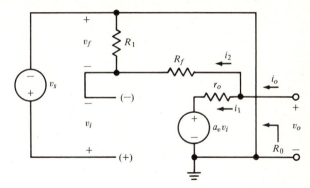

FIGURE 9.8
Small-signal circuit model of the noninverting amplifier, including the output resistance r_o.

Then substituting for v_i from Eq. (9.30), we have

$$i_s = \frac{A_v v_s}{a_v r_i} \qquad (9.32)$$

Now we combine Eq. (9.32) into our definition for R_i in Eq. (9.27); then

$$R_i = \frac{v_s}{i_s} = \frac{a_v r_i}{A_v} \qquad (9.33)$$

But from the generalized form of the feedback equation (9.14b), we have

$$1 + T = \frac{a_v}{A_v} \qquad (9.34)$$

We therefore conclude from Eqs. (9.33) and (9.34) that

$$R_i = (1 + T)r_i \qquad (9.35)$$

That is, due to the *series feedback* at the input, the closed-loop input resistance of the noninverting amplifier is equal to the open-loop input resistance *multiplied* by $1 + T$. This effect can be generalized to the following statement:

Negative feedback connected in *series* leads to an *increase* in the resistance seen by the source voltage by a factor $1 + T$, where T is the loop gain.

Finite output resistance   Shown in Fig. 9.8 is the amplifier of Fig. 9.6 with a finite output resistance r_o included for the operational amplifier. The operational-amplifier input resistance is assumed infinite.

We now wish to determine the effect of the feedback on the overall amplifier output resistance. We refer to Fig. 9.8 and define the closed-loop output resistance, with $v_s = 0$, as

$$R_o = \frac{v_o}{i_o} \qquad (9.36)$$

Also, from Fig. 9.8, with $v_s = 0$, we have that

$$v_i = v_f \qquad (9.37)$$

But by the voltage-divider action of R_1 and R_f,

$$v_f = -\frac{R_1}{R_1 + R_f} v_o \qquad (9.38)$$

By substituting Eq. (9.38) in Eq. (9.37), we can obtain the following expression for $a_v v_i$:

$$a_v v_i = -\frac{a_v R_1}{R_1 + R_f} v_o \qquad (9.39)$$

Again from Fig. 9.8 we note that

$$i_o = i_1 + i_2$$

where

$$i_1 = \frac{v_o - a_v v_i}{r_o} \qquad (9.40a)$$

and

$$i_2 = \frac{v_o}{R_1 + R_f} \qquad (9.40b)$$

Now including Eq. (9.39) in Eq. (9.40a), we have

$$i_1 = \frac{v_o}{r_o} \left(1 + \frac{a_v R_1}{R_1 + R_f} \right) \qquad (9.41a)$$

hence summing i_1 and i_2, from Eqs. (9.41a) and (9.40b),

$$i_o = \frac{v_o}{r_o} \left(1 + \frac{a_v R_1}{R_1 + R_f} + \frac{r_o}{R_1 + R_f} \right) \qquad (9.41b)$$

In general we will find that $R_1 + R_f \gg r_o$. Then from our definition for the output resistance in Eq. (9.36) and by using Eq. (9.41b),

$$R_o = \frac{v_o}{i_o} = \frac{r_o}{1 + a_v \dfrac{R_1}{R_1 + R_f}} \qquad (9.42)$$

Therefore

$$R_o = \frac{r_o}{1 - a_v f} \qquad (9.43)$$

where

$$f = -\frac{R_1}{R_1 + R_f}$$

Or

$$R_o = \frac{r_o}{1 + T} \qquad (9.44)$$

since

$$T = -a_v f$$

Thus due to *shunt feedback* at the output, the closed-loop output resistance of the noninverting amplifier is equal to the open-loop output resistance *divided* by $1 + T$. This result can be generalized to the following statement:

Negative feedback connected in *shunt* leads to a *decrease* in the resistance seen by the source voltage by a factor of $1 + T$, where T is the loop gain.

EXAMPLE 9.2   The operational amplifier shown in Fig. 9.3 can be characterized by the following small-signal parameters:

$$a_v = 1,000 \qquad r_i = 20 \text{ k}\Omega \qquad r_o = 20 \text{ }\Omega$$

With $R_s = R_1 = 100 \text{ }\Omega$, determine the value of R_f for a closed-loop voltage gain of 100. Also calculate the closed-loop input resistance R_i seen by v_s and the output resistance R_o seen by v_o.

SOLUTION   This is a noninverting amplifier with series-shunt feedback. From Eq. (9.25)

$$A_v = \frac{a_v}{1 - a_v f}$$

With $a_v = 10^3$ and $A_v = 10^2$, we compute the loop gain as

$$T = -a_v f = -1 + \frac{10^3}{10^2} = 9$$

Therefore

$$f = \frac{9}{-10^3} = -(9)(10^{-3})$$

but from Eq. (9.25c)

$$f = -\frac{R_1}{R_1 + R_f}$$

Therefore

$$R_f = \frac{0.1 \text{ k}\Omega}{(9)(10^{-3})} - 0.1 \text{ k}\Omega = 11 \text{ k}\Omega$$

Since the value of R_f is so much larger than r_o, we can ignore the effect of output resistance in calculating the closed-loop input resistance. Then from Eq. (9.35),

$$R_i = (1 + T)r_i = (1 + 9)20 \text{ k}\Omega = 200 \text{ k}\Omega$$

The 100 Ω due to R_s is insignificant compared to 200 kΩ.

Again with $R_1 + R_f \gg r_o$, we may use Eq. (9.44) to compute the closed-loop output resistance; namely,

$$R_o = \frac{r_o}{1 + T} = \frac{20\ \Omega}{10} = 2\ \Omega$$

The effective output resistance of this amplifier is only $2\ \Omega$.   ////

We illustrate the effects of negative feedback on the small-signal input and output resistance of an IC op amp in Demonstration D9.3.

EXERCISE

E9.6   For the operational-amplifier configuration shown in Fig. 9.3, the following parameter values apply:

$$a_v = 2000 \qquad r_i = 50\ \mathrm{k\Omega} \qquad r_o = 100\ \Omega$$

With $R_1 = 1\ \mathrm{k\Omega}$ determine the value of R_f so that A_v is 50. For this value of R_f, find the closed-loop input resistance R_i and output resistance R_o.

9.3.5   The Inverting Amplifier

We now wish to analyze the inverting feedback amplifier shown in Fig. 9.4. The circuit has been redrawn in Fig. 9.9 to emphasize the feedback configuration, and again the basic amplifier and the feedback network are shown enclosed within the dashed lines.

Just as for the noninverting amplifier, the input to the feedback network in Fig. 9.9 is connected in *shunt* with the output of the amplifier. The feedback network samples the output voltage. However, the signal fed back to the input is in *shunt* with the signal source and the input of the basic amplifier. The inverting amplifier of Fig. 9.9 is then often referred to as a *shunt-shunt feedback configuration*. Since in this circuit the feedback signal from the output voltage is connected in shunt with the input signal, it follows that this circuit configuration is also known as *voltage-shunt feedback*.

To develop the basic feedback equation from the circuit shown in Fig. 9.9, we first assume the basic amplifier to be an operational amplifier with infinite input resistance and zero output resistance. We then note that

$$i_s = i_f \qquad (9.45a)$$

and

$$v_o = -a_v v_i \qquad (9.45b)$$

Now using KVL we substitute for i_s and i_f in Eq. (9.45a) as

$$\frac{v_s - v_i}{R_1} = -\frac{v_o - v_i}{R_f} \qquad (9.46)$$

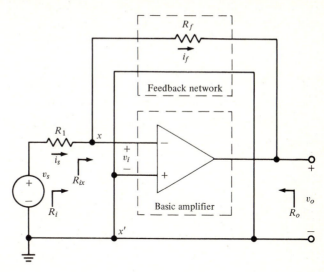

FIGURE 9.9
The inverting amplifier circuit showing the shunt-shunt feedback configuration.

We now use Eq. (9.45b) to eliminate v_i in Eq. (9.46); namely,

$$\frac{v_s + \dfrac{v_o}{a_v}}{R_1} = -\frac{v_o + \dfrac{v_o}{a_v}}{R_f}$$

$$= -\frac{v_o(1 + a_v)}{a_v R_f} \qquad (9.47)$$

As we gather terms in v_o to the right-hand side, we have

$$\frac{v_s}{R_1} = -\frac{v_o}{a_v R_1} - \frac{v_o(1 + a_v)}{a_v R_f}$$

$$= -v_o\left(\frac{1}{a_v R_1} + \frac{1 + a_v}{a_v R_f}\right) \qquad (9.48)$$

Now we can simplify the algebra if we take the reciprical of A_v. That is, from Eq. (9.48),

$$\frac{1}{A_v} = \frac{v_s}{v_o} = -R_1\left(\frac{1}{a_v R_1} + \frac{1 + a_v}{a_v R_f}\right)$$

$$= -\left(\frac{1}{a_v} + \frac{1 + a_v}{a_v}\frac{R_1}{R_f}\right)$$

$$= -\frac{R_f + (1 + a_v)R_1}{a_v R_f} \qquad (9.49)$$

By inverting Eq. (9.49) we have

$$A_v = \frac{v_o}{v_s} = - \frac{a_v R_f}{R_f + (1 + a_v)R_1} \qquad (9.50a)$$

Then with the general case of $a_v \gg 1$,

$$A_v = - \frac{a_v}{1 + a_v \dfrac{R_1}{R_f}} = - \frac{a_v}{1 - \left(-a_v \dfrac{R_1}{R_f}\right)} \qquad (9.50b)$$

In terms of the feedback equation,

$$A_v = \frac{a}{1 - af} \qquad (9.51a)$$

where from Eq. (9.50b),

$$a = -a_v \qquad (9.51b)$$

$$f = \frac{R_1}{R_f} \qquad (9.51c)$$

For the ideal operational amplifier, $a_v \to \infty$. In this case

$$A_v = - \frac{1}{f} = - \frac{R_f}{R_1} \qquad (9.52)$$

This is the same expression as we had in Eq. (9.11b) when considering the voltage gain of an inverting amplifier using ideal operational amplifiers.

Resistance levels The effect of a finite output resistance in the operational amplifier of the shunt-shunt circuit is similar to that for the series-shunt configuration. In both cases the shunt-feedback connection at the output reduces the output resistance by the factor $1 + T$.

Before determining the closed-loop input resistance of the inverting amplifier, we first note the effect of R_f on the input resistance of an ideal operational amplifier. From Fig. 9.9 we note that

$$i_f = - \frac{v_o - v_i}{R_f} \qquad (9.53)$$

To eliminate v_o from this expression, we make use of Eq. (9.45b); we then have

$$i_f = - \frac{-a_v v_i - v_i}{R_f} \qquad (9.54)$$

$$= v_i \frac{1 + a_v}{R_f}$$

We now define the input resistance at the terminal x, x' in Fig. 9.9 as

$$R_{ix} = \frac{v_i}{i_s} \qquad (9.55a)$$

Since for the ideal operational amplifier the input resistance is infinite, then $i_s = i_f$. Therefore

$$R_{ix} = \frac{v_i}{i_f} \qquad (9.55b)$$

Now from Eq. (9.54) we learn the important relationship that

$$R_{ix} = \frac{R_f}{1 + a_v} \qquad (9.56)$$

In this case, the shunt-feedback connection at the input has reduced the effect of the feedback resistance R_f at the input terminal of the operational amplifier by the factor $1 + a_v$. Indeed, it is the general property of all inverting-amplifier circuits that any *impedance* connected between the input and output appears across the input terminals reduced by the factor 1 plus the voltage gain of the amplifier. This is known as the *Miller effect* and is particularly important in the design of high-frequency amplifier circuits.

Now we have previously defined the input resistance of the overall amplifier as

$$R_i = \frac{v_s}{i_s}$$

Therefore, for the circuit shown in Fig. 9.9, where $i_s = i_f$, we have that

$$R_i = R_1 + R_{ix} \qquad (9.57a)$$

For the ideal or near-ideal operational amplifier, with $a_v \to \infty$, we have from Eq. (9.56) that

$$R_{ix} \to 0$$

That is, the resistance to ground from the junction of R_1 and R_f is almost zero. In other words, as we have noted earlier, the summing point is a virtual ground. Hence for all practical purposes

$$R_i = R_1 \qquad (9.57b)$$

If we now consider the operational amplifier to have a finite input resistance, as shown in Fig. 9.10a, we are able to draw an equivalent diagram for the input circuit, as shown in Fig. 9.10b. The output resistance of the operational amplifier is assumed to be zero.

From Fig. 9.10b we see that the closed-loop input resistance of the amplifier is given by

$$R_i = R_1 + R_i' \qquad (9.58a)$$

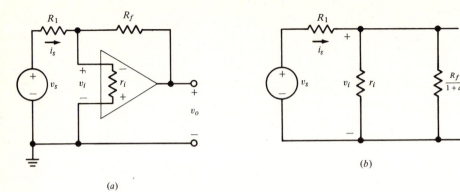

(a)

(b)

FIGURE 9.10

(a) The inverting amplifier circuit, including the input resistance r_i; (b) an equivalent diagram for the input circuit, including the *Miller effect* on the feedback resistance R_f.

where

$$R_i' = r_i \| \frac{R_f}{1 + a_v} \qquad (9.58b)$$

The effect of the finite input resistance of the operational amplifier is to even further assure a virtual ground at the summing point.

EXAMPLE 9.3 The operational amplifier shown in Fig. 9.4 can be characterized by the following small-signal parameters:

$$a_v = 1000 \qquad r_i = 20 \text{ k}\Omega \qquad r_o = 20 \ \Omega$$

With $R_1 = R_2 = 100 \ \Omega$, determine the value of R_f for a closed-loop voltage gain of 100. Also calculate the closed-loop input resistance seen by v_s and the output resistance seen by v_o.

SOLUTION This is an inverting amplifier with shunt-shunt feedback. From the general feedback equation (9.51a),

$$A_v = \frac{a}{1 - af}$$

where $a = -1000$ and $A_v = -100$, therefore

$$af = 1 - \frac{-10^3}{-10^2} = -9$$

and

$$f = \frac{-9}{-10^3} = (9)(10^{-3})$$

but from Eq. (9.51c),

$$f = \frac{R_1}{R_f}$$

Therefore

$$R_f = \frac{0.1 \text{ k}\Omega}{(9)(10^{-3})} = 11.1 \text{ k}\Omega$$

From Eq. (9.58a),

$$R_i = R_1 + R_i'$$

where

$$R_i' \approx 20 \text{ k}\Omega \left\| \frac{11.1 \text{ k}\Omega}{1000} \approx 11.1 \text{ }\Omega \right.$$

then

$$R_i = 100 + 11.1 = 111.1 \text{ }\Omega$$

From Eq. (9.43),

$$R_o = \frac{r_o}{1 - a_v f} = \frac{20}{10} = 2 \text{ }\Omega \qquad ////$$

EXERCISE

E9.7   For the operational-amplifier configuration shown in Fig. 9.4, the following parameter values apply:

$$a_v = 2000 \qquad r_i = 50 \text{ k}\Omega \qquad r_o = 100 \text{ }\Omega$$

With $R_1 = 1 \text{ k}\Omega$, determine the value of R_f so that A_v is -50. For this value of R_f, find the closed-loop input resistance R_i and output resistance R_o.

9.3.6 Summary

In conclusion we have seen that for the cases of series-shunt and shunt-shunt feedback the resulting expression for the closed-loop gain can be arranged in the generalized form of the feedback equation (9.14). It was also noted that for the typical case the closed-loop-gain expressions for the series-shunt and shunt-shunt feedback amplifiers reduce to the simple approximation given for the noninverting amplifier in Eq. (9.8) and the inverting amplifier in Eq. (9.11). Finally we found that the factor $1 + T$ is important not only in the gain equations but also with regard to the closed-loop terminal resistances and amplifier desensitivity.

9.4 FREQUENCY RESPONSE

We have seen that the application of negative feedback to a high-gain amplifier results in a circuit with characteristics dependent almost entirely on the feedback elements. The improvement in gain sensitivity, input resistance, and output resistance is proportional to the amount of feedback, that is, to $1 + T$, which is approximately the loop gain. Thus amplification to any degree of accuracy is possible with sufficient feedback. Large feedback, however, requires that close attention be given to the amplifier open-loop characteristics. Stable amplifiers using negative feedback require that the *magnitude* and *phase* response of the open-loop gain be well defined to frequencies far above the band of interest. A feedback amplifier with a loop gain of 1000 over a 1-kHz *bandwidth*, for example, will usually require that the open-loop characteristics be controlled to over 1 MHz.

9.4.1 Stability

The requirements of a linear amplifier are that it amplify without distortion and without susceptibility to oscillation. The stability or, more properly, the instability problem, can be understood by considering two effects, that of feedback and of frequency.

Feedback effect   From the basic feedback equation of Eq. (9.14*a*),

$$A = \frac{a}{1 - af} = \frac{a}{1 + T}$$

With a or f, but not both, negative, $T > 0$, and we have a stable condition of negative feedback. In the case where both a and f have the same sign, the product is positive, resulting in $T < 0$ and positive feedback. This is a condition of instability, and the possibility of oscillation exists.

Frequency effect   So far in this chapter we have assumed that the gain of the feedback system was constant with frequency. That is, the bandwidth of the operational amplifier is unlimited. We must now examine the more practical case of bandwidth limitation. The open-loop gain of an amplifier, as a function of frequency, can be represented as $a(\omega)$. Here we are using ω, the *radian frequency* equivalent to $2\pi f$, where f is the *cyclic frequency* expressed in *hertz*.

With feedback applied to the amplifier, the closed-loop gain becomes

$$A(\omega) = \frac{a(\omega)}{1 - a(\omega)f} \qquad (9.59)$$

If the product $a(\omega)f$ is negative for all frequencies, then A remains finite and the system is stable. If, however, $a(\omega)f \to 1$, then $A \to \infty$, indicating an unstable or oscillating condition.

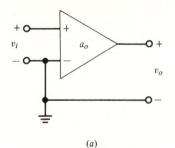 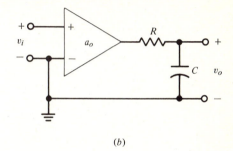

(a) (b)

FIGURE 9.11
Frequency effects in operational amplifiers. (a) Ideal op amp with unlimited frequency response; (b) the frequency response is limited due to the RC integrator.

In order to study the closed-loop response of the feedback amplifier, we introduce the *Bode plot*, which gives a graphic picture of the effect of frequency on the magnitude and phase response of a circuit.

9.4.2 Bode Plots

Consider the ideal amplifier, with unlimited bandwidth, shown in Fig. 9.11a. The small-signal output voltage is given as

$$v_o = a_o v_i$$

where a_o is the voltage gain of the basic amplifier. Now suppose we introduce a simple RC network at the output of the amplifier, as in Fig. 9.11b. We will find that this network will limit the bandwidth of the amplifier. Indeed we already know that at a high enough frequency the capacitor will effectively be a short circuit to ground and then $v_o = 0$. At this frequency the voltage gain of the circuit will be zero. We return to Fig. 9.11b and note that the *impedance* of the series RC network is given by

$$Z = R - jX_c \qquad (9.60)$$

but X_c is a function of the operating frequency, $X_c = 1/\omega C$, so that

$$Z(\omega) = R + \frac{1}{j\omega C} \qquad (9.61)$$

Now due to voltage-divider action of the resistive and reactive elements, the output voltage is given by

$$v_o = \left(\frac{1}{j\omega C}\right) \frac{a_o v_i}{R + \dfrac{1}{j\omega C}}$$

$$= \frac{a_o v_i}{1 + j\omega RC} \qquad (9.62)$$

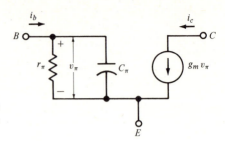

FIGURE 9.12
In this simplified hybrid-π model for the
bipolar transistor, the frequency limitation
is modeled by r_π and C_π.

the terminal voltage gain of the amplifier in Fig. 9.11b is now

$$a_v(\omega) = \frac{v_o}{v_i} = \frac{a_0}{1 + j\dfrac{\omega}{\omega_1}} \qquad (9.63)$$

where

$$\omega_1 = \frac{1}{RC}$$

The complex notation of the voltage gain in Eq. (9.63) may be expressed in terms
of its magnitude and phase angle. The magnitude is given as

$$|a_v(\omega)| = \frac{|a_0|}{\sqrt{1 + \left(\dfrac{\omega}{\omega_1}\right)^2}} \qquad (9.64a)$$

while the phase angle is

$$\phi(\omega) = -\tan^{-1}\frac{\omega}{\omega_1} \qquad (9.64b)$$

We have introduced the arbitrary RC network to show that both the magnitude and phase of the open-loop gain of a practical amplifier are functions of
frequency. In a practical amplifier the frequency-limiting effect is not due to the
loading of an RC network at the output but is primarily due to the frequency
limitations of the transistor. A single-stage CE amplifier shows a similar frequency
dependence, as represented by the circuit in Fig. 9.11b. Indeed it is a simple matter
to utilize Norton's theorem to convert the current-driven input circuit of the
simplified hybrid-π model for the transistor, shown in Fig. 9.12, to the voltage-
driven RC network of Fig. 9.11b.

A circuit which can be represented with a single resistor connected to a
single capacitor, as in Fig. 9.11b, is described as a *one-pole network*. It has just *one
critical frequency*, $\omega_1 = 1/RC$.

Bode plots relate the magnitude and phase response of the gain expression,
with frequency as the independent variable. Usually frequency, in hertz or radians

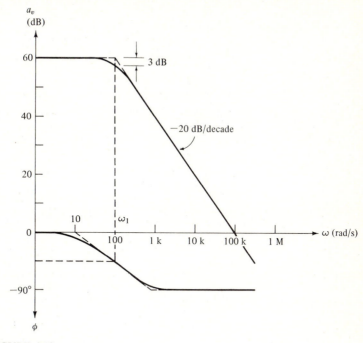

FIGURE 9.13
Bode plot for amplifier of Fig. 9.11*b* with single-pole response (one critical frequency at ω_1).

per second, is plotted as the ordinate on a logarithmic scale, with the logarithm of the magnitude of the gain function as one abscissa and the phase angle on a linear scale as another abscissa. An example is shown in Fig. 9.13. That is,

$$\log |a_v| \text{ versus } \log \omega$$

and

$$\phi \text{ versus } \log \omega$$

One-pole network For the one-pole network we need to obtain values for $a(\omega)$ and $\phi(\omega)$. For $\omega \ll \omega_1$, we have from Eq. (9.64*a*),

$$a_v \approx \frac{a_0}{\sqrt{1}}$$

that is,

$$a_v \approx a_0 \qquad (9.65a)$$

Also with $\omega \ll \omega_1$, we can determine the phase angle from Eq. (9.64*b*); namely,

$$\phi \approx 0° \qquad (9.65b)$$

but notice when

$$\omega = 0.1\omega_1$$

$$\phi = -\tan^{-1} 0.1 \approx -6°$$

These relationships between $a(\omega)$ and $\phi(\omega)$ are shown plotted in Fig. 9.13.
For $\omega = \omega_1$, we compute the magnitude of the gain from Eq. (9.64a) as

$$a_v = \frac{a_0}{\sqrt{2}}$$

therefore

$$a_v = 0.707a_0 \qquad (9.66a)$$

and for the phase angle, again from Eq. (9.64b),

$$\phi = -\tan^{-1} 1 = -45° \qquad (9.66b)$$

Now the gain of an amplifier is normally specified in decibels (dB), where the
voltage gain

$$a_v \text{ (dB)} = 20 \log |a_v| \qquad (9.67)$$

that is, for $a_v = 1000$,

$$a_v = 60 \text{ dB}$$

For the one-pole network we can also express the magnitude expression of
Eq. (9.64a) in logarithms. Then

$$\log |a_v(\omega)| = \log \frac{|a_0|}{\sqrt{1 + \left(\dfrac{\omega}{\omega_1}\right)^2}}$$

$$= \log |a_0| - \tfrac{1}{2} \log \left[1 + \left(\frac{\omega}{\omega_1}\right)^2\right] \qquad (9.68)$$

Using this equation we can determine, at $\omega = \omega_1$, the voltage gain in decibels as

$$a_v = 20(\log |a_0| - \tfrac{1}{2} \log 2)$$

now

$$20(\tfrac{1}{2} \log 2) = 3.01$$

hence

$$a_v = 20 \log |a_0| - 3 \text{ dB} \qquad (9.69)$$

therefore at $\omega = \omega_1$, the magnitude of the gain is reduced by 3 dB. This is illustrated in Fig. 9.13.

For $\omega \gg \omega_1$, from Eq. (9.64a) we have

$$a_v \approx \frac{a_0}{\omega} \qquad (9.70a)$$

hence in this region a_v varies inversely with frequency. The rate is determined by considering a decade of frequency. Let

$$X = a_v(\omega_n) \qquad \text{and} \qquad X' = 20 \log X$$

where X is the magnitude of the gain function at some frequency ω_n, and let

$$Y = a_v(10\omega_n) \qquad \text{and} \qquad Y' = 20 \log Y$$

where Y is the magnitude of the gain at a frequency 10 times ω_n. Then, including Eq. (9.70a), the difference in voltage gain (in dB) over a decade of frequency is given by

$$X' - Y' = 20\left(\log \frac{a_0}{\omega_n} - \log \frac{a_0}{10\omega_n}\right) \qquad (9.71a)$$

$$= 20(\log 10)$$

$$= 20 \text{ dB}$$

That is, the gain at $10\omega_n$ is 20 dB less than it is at ω_n. Therefore, the rate of fall of gain, where $\omega \gg \omega_1$, is -20 dB/decade. The rate can also be expressed as -6 dB/octave. Since an octave is a doubling of the frequency,

$$20(\log 2) = 6 \text{ dB} \qquad (9.71b)$$

From the expression for the phase angle in Eq. (9.64b) we note as $\omega \to \infty$, $\phi \to -90°$, but at $\omega = 10\omega_1$,

$$\phi = -\tan^{-1} 10 \approx -84° \qquad (9.70b)$$

A graph for this one-pole network is shown in Fig. 9.13. The Bode plot makes use of the straight-line segments, the asymptotes, to facilitate the description of the curve. Also note the maximum rolloff is -20 dB/decade and the maximum phase shift is $-90°$.

The 3-dB *bandwidth* of the circuit or amplifier, naturally enough, refers to the frequency at which the gain is down by 3 dB. In this example the 3-dB bandwidth is at ω_1 ($= 100$ rad/s). The *unity gain bandwidth* indicates where the gain of the amplifier is 1 (that is, 0 dB). For the plot shown in Fig. 9.13, this is at 100 krad/s.

Two-pole network We can simulate the effect of cascading transistor amplifiers by cascading two circuits similar to that shown in Fig. 9.11b. We then have the

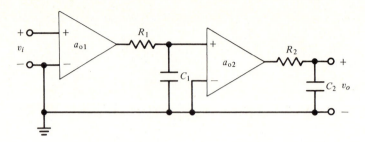

FIGURE 9.14
A two-pole network of two ideal op amps and two RC integrators.

two-pole network shown in Fig. 9.14. The ideal amplifiers provide that there is no interaction between these two circuits; that is, one does not load the other. Similar to Eq. (9.63) we may then write the following transfer function for this network.

$$a_v(\omega) = \frac{a_0}{\left(1 + j\dfrac{\omega}{\omega_1}\right)\left(1 + j\dfrac{\omega}{\omega_2}\right)} \qquad (9.72)$$

where

$$a_0 = a_{01} a_{02}$$

$$\omega_1 = \frac{1}{R_1 C_1}$$

$$\omega_2 = \frac{1}{R_2 C_2}$$

and we assume

$$\omega_2 > \omega_1$$

We may also resolve this number as the magnitude and phase angle of the vector. Then

$$|a_v(\omega)| = \frac{|a_0|}{\sqrt{\left[1 + \left(\dfrac{\omega}{\omega_1}\right)^2\right]\left[1 + \left(\dfrac{\omega}{\omega_2}\right)^2\right]}} \qquad (9.73a)$$

and

$$\phi(\omega) = -\left(\tan^{-1}\frac{\omega}{\omega_1} + \tan^{-1}\frac{\omega}{\omega_2}\right) \qquad (9.73b)$$

A Bode plot for this network is shown in Fig. 9.15. Note here with *two critical frequencies* the maximum rolloff is -40 dB/decade and the maximum phase angle is $-180°$. But the phase angle at ω_1 is $-45°$, and at ω_2 it is $-135°$.

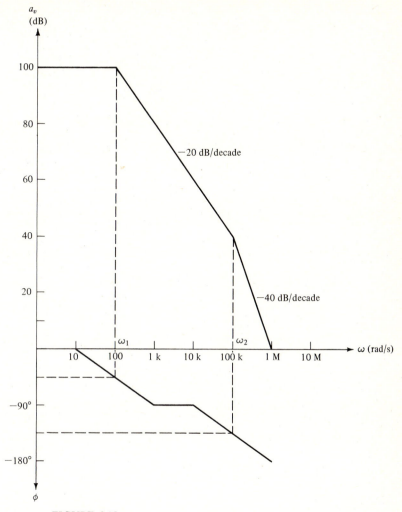

FIGURE 9.15
Bode plot for circuit of Fig. 9.14 with two critical frequencies at ω_1 and ω_2.

General   We can now make some general conclusions regarding Bode plots. For a simple cascade of *RC* networks with no interaction of the critical frequencies (poles), we may conclude that

1   Each pole contributes −20 dB/decade or −6 dB/octave to the rolloff.
2   The maximum phase shift is $(N)(90°)$, where *N* is equal to the number of poles.
3   The phase angle at the first break frequency is −45°. Each subsequent breakpoint contributes 90° to the phase shift.

To further illustrate the above conclusions, we show in Figs. 9.16 and 9.17

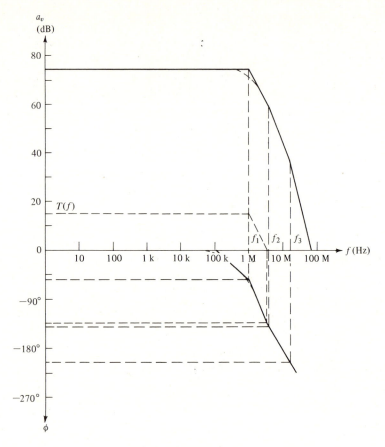

FIGURE 9.16
Bode plot for the type 1530 operational amplifier.

Bode approximation plots for the two *IC* operational amplifiers we considered in Chap. 8, namely, the 1530 and the 741. In each case we also show the frequency response of the open-loop voltage gain. Notice in these plots the frequency is given in hertz.

From Fig. 9.16 we note that the low-frequency open-loop voltage gain of the 1530 is 74 dB; this is equivalent to 5000 or, as it is sometimes written, 5 V/mV. The first break in the frequency response is at 1 MHz. The response then rolls off at a slope of -20 dB/decade to the second break frequency at 6 MHz. A third break frequency is at 20 MHz, with the unity gain bandwidth being at about 85 MHz. As we would expect, the phase angle at 1 MHz, the first breakpoint, is $-57°$ [since $-(\tan^{-1}\frac{1}{1} + \tan^{-1}\frac{6}{1} + \tan^{-1}\frac{20}{1}) = -57°$]. At the second breakpoint the phase angle is $-142°$ and continues to $-205°$ at 20 MHz, the third breakpoint.

For the 741 the plot, in Fig. 9.17, shows the low-frequency voltage gain to be 106 dB ($\equiv 200,000$). The open-loop 3-dB bandwidth is only about 7 Hz, but the

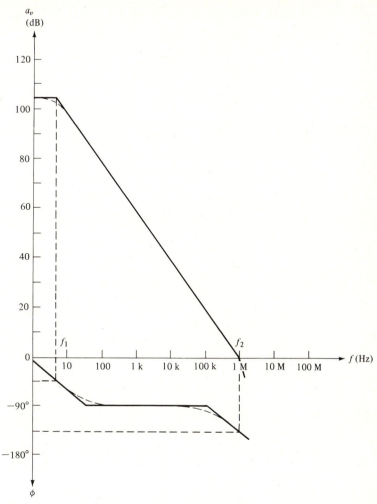

FIGURE 9.17
Bode plot for the type 741 operational amplifier.

high open-loop gain makes this IC useful for frequencies out to the unity gain frequency of 1 MHz. Notice the slope from 10 Hz to 1 MHz is a constant -20 dB/decade. Over much of the useful frequency spectrum the phase angle is $-90°$, changing to $-135°$ at the unity gain frequency.

EXERCISE

E9.8 The open-loop voltage gain of an operational amplifier is 80 dB. The open-loop frequency response shows a first break at 100 kHz, with a -20 dB/decade

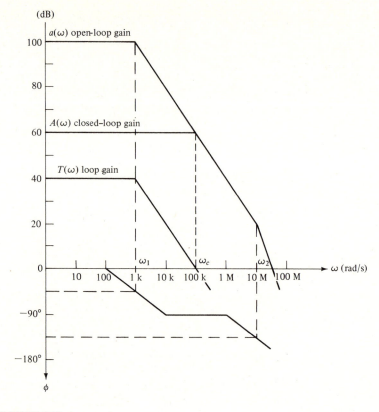

FIGURE 9.18

Bode plot showing relationship of the open-loop gain $a(\omega)$, the loop gain $T(\omega)$, and the closed-loop gain $A(\omega)$.

rolloff to the second break frequency at 10 MHz, where the rolloff becomes -40 dB/decade. Show the magnitude and phase response of the open-loop voltage gain from 1 Hz to 1 GHz.

9.4.3 Operational-amplifier Bode Plots

In a feedback amplifier, with the loop gain $\gg 1$, the closed-loop gain is given as

$$A(\omega) \approx \frac{a(\omega)}{T(\omega)} \qquad (9.74)$$

On a logarithmic scale this is

$$\log|A(\omega)| \approx \log|a(\omega)| - \log|T(\omega)| \qquad (9.75)$$

or

$$\log (\text{closed-loop gain}) \approx \log (\text{open-loop gain}) - \log (\text{loop gain})$$

This is illustrated in the Bode plot of Fig. 9.18. Here we see that the loop gain, being a fraction of the open-loop gain, has a similar characteristic as the open-loop response but is reduced in magnitude. The closed-loop gain (in dB) is shown as the difference between the open-loop gain (in dB) and the loop gain (in dB).

Notice that the approximation given in Eqs. (9.74) and (9.75) is only true with the loop gain $\gg 1$. However, in Fig. 9.18, at ω_c the loop gain $T(\omega_c) = 0$ dB (that is, 1). Therefore, $1 + T$ is equal to 2. Consequently from the basic feedback equation of Eq. (9.14a), $A(\omega_c) = 0.5a(\omega_c)$. In decibels we represent a factor of 2 as 6 dB; that is, at ω_c the closed-loop gain is just 6 dB less than the open-loop gain. In the Bode plot, using straight line asymptotes, we simply take $A(\omega_c) = a(\omega_c)$. For $\omega > \omega_c$, the closed-loop gain follows the open-loop gain, since $T(\omega) < 1$, and again from Eq. (9.14a), $A(\omega) \approx a(\omega)$.

For conditions of stability we have seen it necessary for the product $a(\omega)f$ to be negative at all frequencies. In our examples of negative feedback in voltage amplifiers we have applied the feedback signal (a fraction of the output signal) to the inverting input of the basic amplifier. Hence, at low frequencies, there has been a 180° phase inversion to $a(\omega)f$ through the amplifier, and we have had negative feedback. At higher frequencies we have noted the possibility of the phase angle of $a(\omega) \geq 180°$. This additional 180° phase shift causes $a(\omega)f$ to become positive, resulting in a threat of instability. However, all is not lost; providing the product $a(\omega)f$ is < 1, we can still have a stable system. Here then is the condition for stability: the magnitude of the loop gain must be < 1 when the phase angle is $\geq 180°$.

In general, it is necessary to check the interaction of the closed-loop response with the open-loop response, where $T(\omega) = 1$. If at this frequency the phase angle is less than 180°, the network is stable. That is, in Fig. 9.18, this interaction is at $\omega_c = 100$ krad/s. The phase angle at this frequency is $-90°$, and the amplifier is stable.

However, consider a case where with $T(\omega) = 1$ the phase angle is just $-180°$, and the amplifier is on the edge of oscillation. A *phase margin* is therefore chosen to allow for variation of amplifier and circuit parameters. This is the difference between 180° and the phase angle at the frequency, where the loop gain is equal to 1. The usual practice is to have a minimum phase margin of 45°. The term *gain margin* is also used. This indicates how much less than 1 (0 dB) is the loop gain at the frequency where the phase angle is $-180°$. In Fig. 9.19, at low frequencies the open-loop gain $a(\omega)$ is 100 dB, and the loop gain $T(\omega)$ is 40 dB, resulting in a closed-loop gain $[A(\omega)] = 60$ dB. The phase margin is 90° and the gain margin is 80 dB in this example.

A common problem with the use of IC op amps as feedback amplifiers is given in the following example.

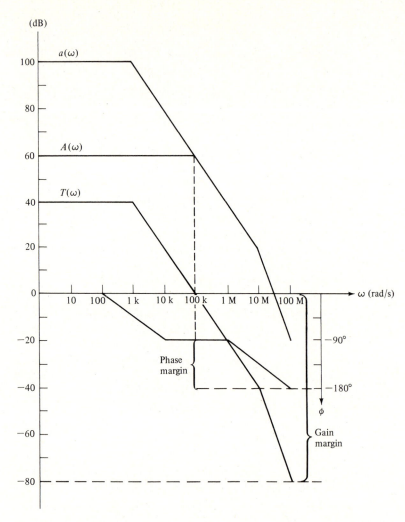

FIGURE 9.19
Bode plot illustrating a phase margin of 90° and a gain margin of 80 dB.

EXAMPLE 9.4 From the frequency response of the MC1530 shown in Fig. 9.16, determine the maximum loop gain permissible for a phase margin $\geq 45°$. For this condition, what is the minimum permissible low-frequency closed-loop gain?

SOLUTION For a 45° phase margin: with $\phi(f_x) = -135°$, $T(f_x) \leq 0$ dB. From Fig. 9.16, $\phi = -135°$ at $f_x = 5$ MHz, now let $T = 0$ dB at $f_x = 5$ MHz. We then draw the graph of $T(f)$. This is a line parallel to the open-loop response with a value of 0 dB at f_x. Hence we find the value $T(0) = 14$ dB. This is equivalent to a loop gain of 5. We determine the minimum low-frequency closed-loop gain from Eq. (9.75) as

$$A(0) = a(0) - T(0)$$
$$= 74 - 14 = 60 \text{ dB} = 1000$$

Notice that if we had made use of the basic feedback equation of Eq. (9.14a), we would obtain for the minimum closed-loop gain,

$$A = \frac{a}{1 + T} = \frac{5000}{1 + 5} = 833$$

Given in decibels, this is equivalent to 58.4 dB. We find from Fig. 9.16 that the intersection of this closed-loop gain with the open-loop response is at a frequency only very slightly greater than f_2, with the phase angle slightly more than $-135°$. Hence using the 1530 in a feedback amplifier, the closed-loop gain must be > 1000 (or for the conservative designer, > 833) if we require the phase margin to be $\geq 45°$.　　　　　////

EXERCISE

E9.9 Determine the minimum permissible closed-loop gain for the operational amplifier described in Exercise E9.8 with a phase margin
  (a) $\geq 45°$; (b) $\geq 90°$
Give your answer as a numeric as well as in decibels.

9.5 FREQUENCY COMPENSATION

A very popular use of IC op amps is in voltage-follower applications. Here the closed-loop voltage gain is unity ($\equiv 0$ dB). The feedback is 100 percent, and the loop gain is approximately equal to the open-loop gain of the amplifier. As we have just seen, the type 1530 would be unstable as a voltage follower unless some other design precautions are taken. These other precautions introduce the subject of *frequency compensation.*

With many operational amplifiers a point in the circuit is brought out to an external pin, where by connecting a resistor and a capacitor, or just a capacitor, the user can change the frequency response of the operational amplifier. The most common form of frequency compensation is *narrowbanding*, whereby the open-loop bandwidth of the amplifier is reduced so that the amplifier can be used with a large amount of feedback, or loop gain, such as the voltage follower we have just cited. Another form of frequency compensation is *broadbanding*. With this technique stability is obtained for the feedback amplifier but with an increased bandwidth as compared with the narrowbanding technique.

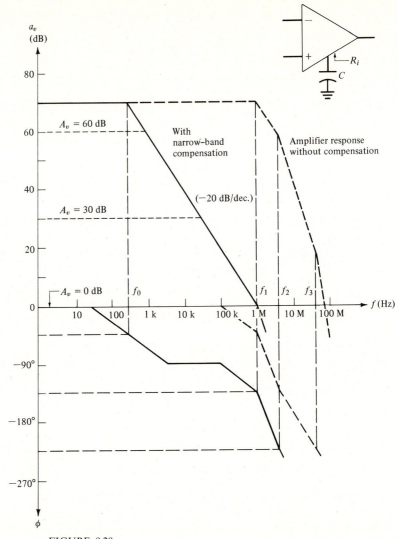

FIGURE 9.20
Narrowband compensation illustrated with the type 702 operational amplifier.

9.5.1 Narrowbanding

The simplest method of narrowbanding a feedback amplifier is to shunt some of the signal to ground with a single relatively large capacitor. That is, an additional pole is introduced into the circuit at a very low frequency so that the rolloff is at -20 dB/decade at least until the unity gain point. Narrowband frequency compensation of the type 702 (an op amp with characteristics similar to the type 1530) is illustrated in Fig. 9.20, where the original open-loop response is plotted as a dashed line. The compensated response is shown as the solid line. The original first pole now becomes the second so that the phase angle at this frequency will now be $-135°$. With an open-loop gain of 1 at this frequency, the amplifier has a phase margin of $45°$ and is stable even as a voltage follower. The closed-loop response for a voltage follower is the $A = 0$-dB curve. Other values of closed-loop gain give the other curves shown dotted.

The narrowbanding method is also referred to as *lag compensation*. The phase of the voltage across the compensating capacitor lags by $90°$ the current through the capacitor. The point where the capacitor is connected is known as the *compensation terminal*. If the internal resistance R_i seen looking into this point in the circuit is specified, the break frequency can be computed as

$$f_0 = \frac{1}{2\pi R_i C} \qquad (9.76)$$

Notice that in Fig. 9.20 the closed-loop 3-dB bandwidths are very much reduced with this frequency-compensation technique. However, by this method, the user is able to have some control over the frequency response of the operational amplifier to suit his particular need for a feedback amplifier.

Other operational amplifiers have the frequency response determined internally in the circuit so that there is only one critical frequency for the open-loop gain greater than 0 dB. That is, the rolloff is at -20 dB/decade through the unity gain point. These are the *internally compensated* operational amplifiers. An example of this type is the 741 op amp. No compensation terminal is provided, and no compensation technique is required. From Fig. 9.17 note that the unity gain frequency of the 741 is at 1 MHz and the phase angle at this frequency is $-135°$. Operated as a voltage follower, the unity gain frequency of the open-loop response also becomes the frequency at which the loop gain is unity. Therefore we have a phase margin of $45°$ or more for all applications of the 741, where the closed-loop voltage gain is equal to or greater than unity.

9.5.2 Broadbanding

Where a compensation terminal is provided and maximum feedback or loop gain is required, the bandwidth may be increased over that of the narrowbanding method by using the natural rolloff of the operational amplifier to provide part of the compensation. This method is shown by the circuit and frequency response of Fig. 9.21. Without compensation the maximum loop gain for the 702 with a $45°$ phase margin is 12 dB.

We can utilize the full 70-dB open-loop voltage gain of the op amp as feedback by using the following broadbanding technique. We modify the original open-loop response to obtain a compensated open-loop response, with the second break frequency at unity gain as shown by the solid line of Fig. 9.21. The series $R_1 C_1$ compensating network gives a rolloff in the loop gain beginning at f_0 and then breaks out (to be a horizontal line) at the first breakpoint of the uncompensated operational amplifier, i.e., at f_1. The amplifier's natural rolloff then pro-

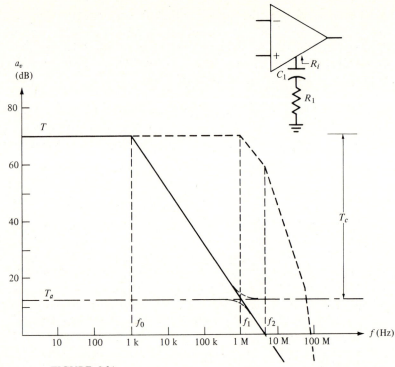

FIGURE 9.21
Broadband compensation illustrated with the type 702 operational amplifier.

vides that the loop gain passes through the unity gain point with adequate phase margin (45°). Notice that this technique gives an open-loop 3-dB bandwidth of 1 kHz in Fig. 9.21 as compared to 300 Hz for the narrowbanding technique used in Fig. 9.20. This broadbanding scheme gives the widest possible bandwidth for any value of closed-loop gain.

The values of the compensation components depend upon the required loop gain and phase margin and the amplifier's internal characteristics. Due to the voltage-divider action of R_i and R_1, the high-frequency gain of the amplifier is reduced by the fraction

$$\frac{R_i}{R_i + R_1}$$

where R_i is the internal resistance of the compensation terminal of the amplifier. Hence the attenuation due to the compensating network is

$$T_c = \frac{R_i + R_1}{R_1} \qquad (9.77)$$

But, at the frequency f_1 the total attenuation from both the compensating network and amplifier rolloff must be equal to the low-frequency loop gain. Using numerics,

$$T = T_a T_c \qquad (9.78a)$$

or in decibels,

$$T(\text{dB}) = T_a \text{ (dB)} + T_c \text{ (dB)} \qquad (9.78b)$$

Therefore, we obtain the required value of the compensating resistor from Eq. (9.77) as

$$R_1 = \frac{R_i}{\dfrac{T}{T_a} - 1} \qquad (9.79)$$

The breakout frequency of the compensating network response should be made equal to the operational-amplifier 3-dB frequency for maximum feedback bandwidth. That is,

$$C_1 = \frac{1}{2\pi f_1 R_1} \qquad (9.80)$$

For the 702 the internal resistance at the compensation terminal is about 3 kΩ. From Fig. 9.21 we have that as a numeric, $T_a = 4 \ (\equiv 12 \text{ dB})$. Operated as a voltage follower we require the loop gain to be equal to the open-loop gain of the amplifier; this is 3200. We obtain a value for the compensating resistor from Eq. (9.79). That is,

$$R_1 = \frac{3 \text{ k}\Omega}{\frac{3200}{4} - 1} \approx 4 \ \Omega$$

The value for the compensating capacitor is, from Eq. (9.80),

$$C_1 = \frac{1}{(2\pi)(10^6)(4)} \approx 0.04 \ \mu\text{F}$$

The use of these compensating techniques to control the frequency response of an IC op amp is further explored in Demonstration D9.4. The ease of using an internally compensated operational amplifier is also demonstrated.

EXERCISE

E9.10   You are required to compensate the op amp described in Exercise E9.8 so that the amplifier has at least 45° of phase margin for all values of closed-loop gain ≥ 1. The source resistance at the compensation terminal is 10 kΩ.

(a) Use narrowband compensation to calculate the minimum value for the compensating capacitor and also to determine the 3-dB bandwidth for a closed-loop low-frequency gain of 20 dB.

(b) Use broadband compensation to calculate values for the compensating resistor and capacitor, and again determine the 3-dB bandwidth of a 20-dB amplifier.

9.6 SUMMARY

The basic small-signal *open-loop* characteristics of an operational amplifier are:

1 The *voltage gain* a_v
2 The *input and output resistances* r_i and r_o

The basic characteristics may be modified by the use of *negative feedback* around the op amp. The resulting *closed-loop* characteristics are identified as A_v, R_i, and R_o.

The *basic feedback equation* is given as

$$A_v = \frac{a_v}{1 - a_v f} = \frac{1}{1 + T}$$

where f = *feedback factor*
    T = *loop gain* $(T = -a_v f)$

For the ideal op amp: $a_v \to \infty$, $r_i \to \infty$, and $r_o \to 0$.
The use of the *summing-point method* with ideal or near-ideal op amps leads to the following expressions for the closed-loop voltage gain:

1 The *voltage-follower*: $A_v = 1$
2 The *noninverting amplifier*: $A_v = (R_f/R_1) + 1$
3 The *inverting amplifier*: $A_v = -R_f/R_1$

Series-shunt feedback is used in the noninverting amplifier. The feedback is connected in series with the input and in shunt with the output. *Shunt-shunt* feedback is used in the inverting amplifier. The feedback is connected in shunt with the input and in shunt with the output.
Negative feedback in *series* with the input (output) leads to an *increase* in the input (output) resistance by the factor $1 + T$. Negative feedback in *shunt* with the input (output) leads to a *decrease* in the input (output) resistance by the factor $1 + T$.
With negative feedback around an op amp we can also perform the functions of:

1 *Addition and subtraction* (with *resistive feedback*)
2 *Differentiation and integration* (with *reactive feedback*)

The *Bode plot* is a graph of the *magnitude* (usually its log) and the *phase angle* (in degrees) of the gain expression versus frequency.
Internally frequency-compensated op amps are unconditionally stable with any amount of loop gain.
Uncompensated op amps may be stabilized by:

1 Narrowbanding (lag compensation)
2 Broadbanding (modified lag compensation)

DEMONSTRATIONS

D9.1   Op amps with resistive feedback   These demonstrations show how transfer characteristics are affected by supply voltages and load and source resistances.

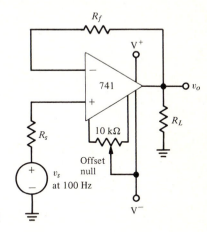

FIGURE D9.1

(*a*) *Voltage follower.* For the circuit in Fig. D9.1, use a 741 op amp and do the operations listed at the conditions below:

	V^+, V^-	R_s	R_f	R_L	Do
1	± 15 V	1 kΩ	1 kΩ	10 kΩ	Determine and plot v_o versus v_s. with v_s ranging from V^- to V^+
2	± 15 V	1 kΩ	1 kΩ	100 Ω	Determine and plot v_o versus v_s. with v_s ranging from V^- to V^+
3	± 5 V	1 kΩ	1 kΩ	10 kΩ	Determine and plot v_o versus v_s. with v_s ranging from V^- to V^+
4	± 5 V	1 kΩ	1 kΩ	100 Ω	Determine and plot v_o versus v_s. with v_s ranging from V^- to V^+
5	± 15 V	1 kΩ	1 kΩ	10 kΩ	Read v_o with $v_s = 0$, then adjust NULL so $v_o = 0$
6	± 15 V	0 kΩ	1 kΩ	10 kΩ	Read v_o with $v_s = 0$
7	± 15 V	2 kΩ	1 kΩ	10 kΩ	Read v_o with $v_s = 0$
8	$+ 15$ V	10 MΩ	10 MΩ	10 kΩ	Read v_o with $v_s = 0$, then adjust NULL so $v_o = 0$
9	± 15 V	1 MΩ	10 MΩ	10 kΩ	Read v_o with $v_s = 0$

From the above data:

(1) Determine the output voltage swing (peak to peak) for ± 5- and ± 15-V supplies and for 10-kΩ and 100-Ω loads.

(2) Determine the short-circuit current that this amplifier can supply.

(3) What is the value of the input offset voltage V_{IO} for this IC?

(4) What is the value of the input offset current I_{IO} for this IC?

(5) What is the value of the input bias current I_{IB} for this IC?

(*b*) Use the same 741 op amp that you have taken data on in part (*a*) to construct a noninverting amplifier, as in Fig. 9.3. Use $R_s = 1$ kΩ to design three different configurations so that A_v is 10, 100, and 1000.

(1) Measure and plot the transfer characteristics, making v_S large enough to get limiting at both polarities.

(2) Measure v_O with $v_S = 0$ in all three configurations. For comparison, use the results of part (*a*) to calculate the output offset voltage.

(*c*) Repeat part (*b*) by using the inverting amplifier of Fig. 9.4. Keep R_1 fixed at 1 kΩ, and design the circuits so that A_v is -1, -10, -100, and -1000.

D9.2   Gain sensitivity   This demonstration shows how the desensitivity factor reduces the effect of open-loop gain on the closed-loop gain.

Use the circuit in Fig. D9.2, where switching the output point to the center of the two 5-kΩ resistors allows us to give the effect of a 50 percent change in a_v (that is $da/a = 0.5$).

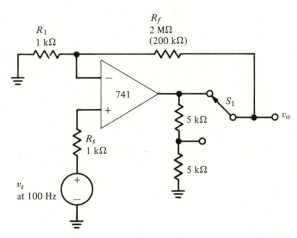

FIGURE D9.2

(*a*) With $R_1 = R_s = 1$ kΩ, $R_f = 2$ MΩ, and $v_S = 0$, NULL the output. Now with S_1 at the top position, increase v_S until v_O is 5 V. Switch S_1 to its bottom position and read the new v_O as $5 - \Delta v_O$ (that is $da/a = \Delta v_O/5$). Determine $1 + T$ from your readings.

(*b*) Attempt to repeat part (*a*) with R_f changed to 200 kΩ. What is your estimate of $1 + T$?

D9.3   Effect of feedback on input and output resistances

(*a*) *Noninverting input resistance.* The circuit in Fig. D9.3*a* allows one to measure the small-signal input resistance R_i seen at the noninverting input. At low frequencies it can be shown that

$$R_i = \frac{v_1}{i_s} = \frac{R_s}{\dfrac{v_{o1}}{v_{o2}} - 1}$$

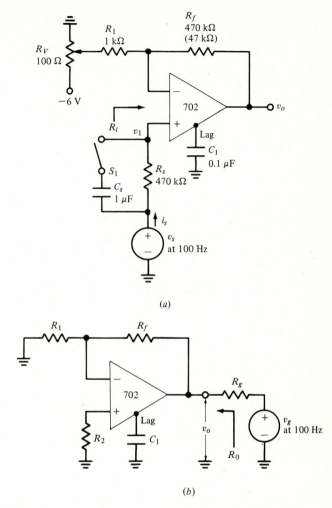

(a)

(b)

FIGURE D9.3

Measuring feedback effects on input and output resistances. (a) Circuit for measuring
input resistance R_{in}; (b) circuit for measuring output resistance R_{out}.

where with constant v_s,

$$v_{o1} = v_o \qquad \text{with } C_s \text{ connected}$$

$$v_{o2} = v_o \qquad \text{with } C_s \text{ disconnected}$$

The variable resistor R_V must be adjusted so that a dc voltage is applied in series
with the inverting input to balance out bias-current effects in R_s. Adjust R_V so that
V_0 is approximately 0 V. Capacitor C_s gives a negligible ac voltage drop at the
signal frequency.

Measure R_i for a 702-type op amp† with R_f set to 470 kΩ and then 47 kΩ. From the value of the open-loop gain a_v, determined as in Demonstration D9.2a, and the closed-loop input resistance R_i, calculate r_{id} and get an estimate for the value of r_{ic}, the common-mode input resistance.

(b) *Output resistance.* The circuit and adjustments made in part (a) can be used to determine the output resistance of a 702. First ground the bottom end of R_s (putting C_s across R_s) and apply the signal generator to the output terminal with a current-limiting resistor R_g, as shown in Fig. D9.3b. For a 702 use $R_g = 10$ kΩ and $v_g \leq 7$ V peak to peak. The closed-loop output resistance is given by the relation

$$R_o = \frac{v_o}{v_g - v_o} R_g$$

(c) *Output impedance.* The circuit in Fig. D9.3b can be used to determine the magnitude of the output impedance of an op amp. Be sure to check that the dc output voltage is ~ 0 V. The signal from the signal generator should be large enough so that v_o can be clearly seen out of the noise but not too large so that the short-circuit-limiting circuitry in the amplifier will come into effect. Thus for a 741-type op amp, the peak current should be under 5 mA. The output impedance with feedback, Z_o, is

$$Z_o = \frac{v_o}{v_g - v_o} R_g$$

which for $v_g \gg v_o$ gives

$$|Z_o| = \left| \frac{v_o}{v_g} \right| R_g$$

Plot $|Z_o|$ versus frequency for A_v of 11, 101, and 1001, and from this give an estimate for the value of $|Z_o|$.

Note: The increase in the magnitude of Z_o at low frequencies for a 741 is due to *thermal feedback* effects within the chip.

D9.4 Frequency effects on open- and closed-loop gains with compensation networks   The open-loop characteristic of an op amp‡ is easily measured by the circuit in Fig. D9.4.

(a) Use a very low frequency (approximately 1 Hz) sawtooth waveform (such as from a VCO as described in Chap. 10) that is amplified to give a 5-V peak-to-peak output for v_s. With an x-y oscilloscope obtain the plot of v_o versus v_x, where, as shown, v_x is $1000v_i$ and the open-loop gain $a_v = v_o/v_i$. From this

---

† We use a 702-type op amp here because of its low input resistance (~ 30 kΩ). This makes it relatively easy to measure R_i.

‡ Type 702 op amp.

plot (1) we can obtain the input offset voltage V_{IO}, (2) we can obtain the magnitude of the open-loop gain at low frequencies, (3) we can see any crossover distortion, and (4) we can determine the output swing capability of the amplifier.

(b) Using a small-signal sine wave as the input v_s (so that $v_o < 8$ V peak to peak for $+12$, -6 V supplies), measure v_x, v_o, and the phase angle between v_x and v_o. Make three measurements per decade (at 1, 2, 5), starting at 20 Hz and going up to as high a frequency as possible permitted by noise and hum limitations. Do this for two compensation circuits $C_1 = 0.1\ \mu F$ and $C_1 = 0\ \mu F$.

Plot log $|a_v|$ and ϕ versus frequency (on a log scale) to obtain a Bode plot.

(c) Use the circuit in Fig. 9.3 (noninverting connection) to plot v_o versus v_s for A_v of 1000, 100, 10, and 1 for both compensation circuits in part (b) above.

(d) In lieu of using a compensated 702 in parts (b) and (c), use a 748 where a single compensation capacitor C_C is set to 1 pF for $A_v = 1000$ or 30 pF for $A_v = 1$.

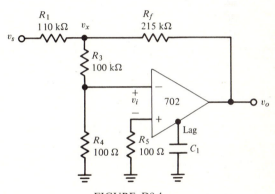

FIGURE D9.4
Circuit for determining open-loop gain.

D9.5 An IC electronic voltmeter ("VTVM") The circuit in Fig. D9.5a uses an IC op amp as a voltage follower to drive a microammeter (50 μA full-scale) and a load resistor so that the microammeter reading is proportional to the input voltage. The dc input voltage V_X is attenuated by a factor of 2 so as to (1) limit the overload current into the IC, (2) minimize the effects of the unknown variable source resistance R_x and the bias current,† and (3) keep the input resistance a constant (20 MΩ). For different full-scale ranges, the range switching in Fig. D9.5b can be used. For zeroing the meter a zero-adjust "pot" can be added, as shown.

Micropower op amps (like the LM4250 that draws only 10 μA from ± 15-V supplies) can be used in this type of application so that battery operation is possible with operating battery life essentially equaling shelf life.

---

† For an LM316A this bias current is only 50 pA, so for $\Delta R_s = \pm 5$ MΩ the uncertainty this generates referred to the noninverting input is only 25 μV.

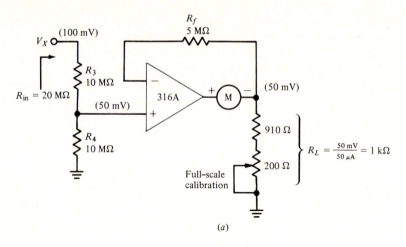

(a)

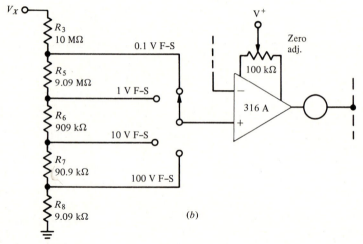

(b)

FIGURE D9.5

(a) Basic electronic dc voltmeter circuit; (b) range switching and zero adjust circuit for part (a).

REFERENCES

The subject matter of this chapter is also found in OLDHAM, W. G., and S. E. SCHWARZ: "An Introduction to Electronics," chap. 12, Holt, New York, 1972. At a more advanced level the subjects of feedback, stability, and frequency compensation are found in chaps. 13, 14, and 15, respectively, of MILLMAN, J., and C. C. HALKIAS: "Integrated Electronics," McGraw-Hill, New York, 1972.

The basic theory of operational amplifiers, including resistive feedback and frequency compensation as well as many applications, is found in the useful little booklet, "Handbook of Operational Amplifiers," published by Burr-Brown Research Corporation, Tucson, 1969. A more advanced reference in design and applications is TOBEY, G. E., et al.: "Operational Amplifiers," McGraw-Hill, New York, 1971.

For more on frequency effects and the compensation of IC op amps, the reader is referred to the manufacturers' data sheets and application notes which are a source of very useful information.

PROBLEMS

P9.1   For the circuit shown in Fig. 9.3, assume an ideal op amp.
(*a*) With $R_1 = 1\ k\Omega$ and $R_f = 10\ k\Omega$, determine the closed-loop voltage gain A_v.
(*b*) Compute an optimal value for R_s.

P9.2   Repeat Prob. P9.1*a* and *b* for the circuit shown in Fig. 9.4 and again assume an ideal op amp.

P9.3   For the circuit shown in Fig. 9.3, assume an ideal op amp and to a precision of 1 percent or better compute the optimal values for R_1 and R_f when $R_s = 1\ k\Omega$ and $A_v = 10$.

P9.4   Repeat Prob. P9.3 for the circuit shown in Fig. 9.4 but with $A_v = -10$ and again assume an ideal op amp.

P9.5   For the circuit shown in Fig. P9.5, assume an ideal op amp.
(*a*) Use Kirchhoff's circuit laws to prove that with $R_{s1} = R_{s2} = R_1 = R_f$,
$$v_o = v_{s1} + v_{s2}.$$
(*b*) What is the input resistance seen by v_{s1} and v_{s2}?

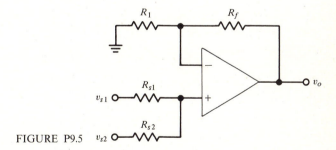

FIGURE P9.5

P9.6   For the circuit shown in Fig. P9.6, assume an ideal op amp.
(*a*) Use Kirchhoff's circuit laws to prove that with $R_{s1} = R_{s2} = R_f$,
$$v_o = -(v_{s1} + v_{s2}).$$
(*b*) What is the input resistance seen by v_{s1} and v_{s2}?

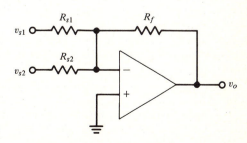

FIGURE P9.6

P9.7 Use an ideal op amp to design a circuit that satisfies the following operation. Let $R_{s1} = 1\,\text{k}\Omega$.
  (a) $v_o = 2.4v_{s1} + 0.6v_{s2}$.
  (b) $v_o = -(2.5v_{s1} + 0.5v_{s2})$.

P9.8 What algebraic equation is described by the operation of the circuit:
  (a) In Fig. P9.8a?
  (b) In Fig. P9.8b?

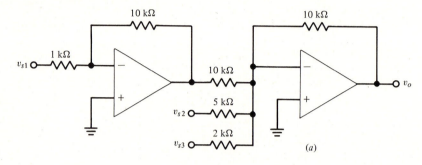

(a)

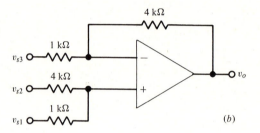

(b)

FIGURE P9.8

P9.9 Use a minimum number of ideal op amps to design a circuit that satisfies the following operation:
  (a) $v_o = 8v_{s1} + 4v_{s2} - 2v_{s3}$ (let $R_{s1} = 1\,\text{k}\Omega$).
  (b) $v_o = 2.61v_{s1} + 17.39v_{s2} - 19v_{s3}$ (let $R_{s3} = 1\,\text{k}\Omega$).

P9.10 For the circuit shown in Fig. 9.3, with $a_v = 1000$:
  (a) Determine A_v, if $R_1 = 1\,\text{k}\Omega$ and $R_f = 24\,\text{k}\Omega$.
  (b) What is the loop gain T?

P9.11 For the circuit shown in Fig. 9.4, with $a_v = 1000$:
  (a) Determine A_v, if $R_1 = 1\,\text{k}\Omega$ and $R_f = 40\,\text{k}\Omega$.
  (b) What is the desensitivity factor D_s?

P9.12 For the circuit shown in Fig. 9.3, with $a_v = 1000$ and $A_v = 10$:
  (a) Determine v_i for $v_o = +10$ V.
  (b) What is the magnitude of the common-mode input signal for $V_o = +10$ V?

P9.13 Repeat Prob. P9.12a and b for the circuit shown in Fig. 9.4 with $a_v = 1000$ and $A_v = -10$.

P9.14 For the circuit shown in Fig. 9.3, with $a_v = 10,000$, $A_v = 100$, $r_i = 100\,\text{k}\Omega$, and $r_o = 100\,\Omega$:
  (a) Calculate the closed-loop input resistance R_i.
  (b) Calculate the closed-loop output resistance R_o.

P9.15   Repeat Prob. P9.14*a* and *b* for the circuit shown in Fig. 9.4 and use the same parameter values as in Prob. P9.14 with $R_1 = 1 \text{ k}\Omega$.

P9.16   Use the circuit diagram of Fig. 9.7 to derive an expression for A_v in terms of a_v, R_1, R_f, and r_i. Check your answer with Eq. (9.24*b*) when $r_i \to \infty$.

P9.17   Use the circuit diagram of Fig. 9.8 to derive an expression for A_v in terms of a_v, R_1, R_f, and r_o. Check your answer with Eq. (9.24*b*) when $r_o = 0$.

P9.18   Use the circuit diagram of Fig. 9.10 to derive an expression for A_v in terms of a_v, R_1, R_f, and r_i. Check that your answer agrees with Eq. (9.50*a*) when $r_i \to \infty$.

P9.19   Use the circuit diagram of Fig. 9.9, with a finite output resistance, to derive an expression for A_v in terms of a_v, R_1, R_f, and r_o. Check that your answer agrees with Eq. (9.50*a*) when $r_o = 0$.

P9.20   The circuit in Fig. P9.20 is used to convert a current i_s into a voltage v_o. Resistors R_1 and R_f are chosen so that $v_o = v_s$:

(*a*) For $a_v \to \infty$, calculate the resistance values for R_1 and R_f and also determine the resistance seen by i_s.

(*b*) Repeat part (*a*) but with $a_v = 10^4$.

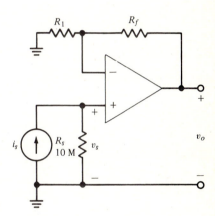

FIGURE P9.20

P9.21   For the circuit in Fig. P9.21, find v_o as a function of i_s for:

(*a*) $a_v = \infty$, $r_i = \infty$.

(*b*) $a_v = 10^4$, $r_i = \infty$.

(*c*) $a_v = 10^4$, $r_i = 10 \text{ k}\Omega$.

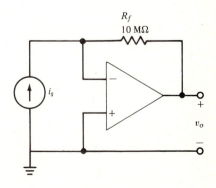

FIGURE P9.21

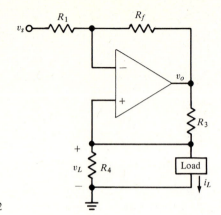

FIGURE P9.22

P9.22 The circuit in Fig. P9.22 is used to convert a voltage v_s into a current i_L. Resistors are chosen so that $R_f/R_1 = R_3/R_4$. Prove that $i_L = -v_s/R_4$. (*Hint:* Solve for i_L as a function of v_s and v_L.)

P9.23 For the improved voltage-to-current converter shown in Fig. P9.23, determine i_L as a function of v_s.

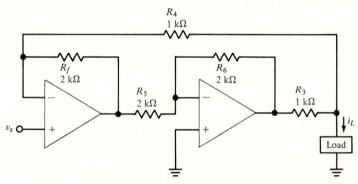

FIGURE P9.23

P9.24 A difference-amplifier circuit is shown in Fig. P9.24. With $R_{f1}/R_1 = R_{f2}/R_2$, prove that $v_o = (R_{f1}/R_1)(v_{s2} - v_{s1})$.

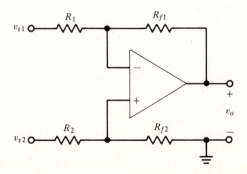

FIGURE P9.24

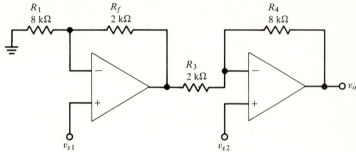

FIGURE P9.25

P9.25 The improved difference circuit of Fig. P9.25 has a high input resistance for both v_{s1} and v_{s2}. Determine v_o as a function of $v_{s2} - v_{s1}$.

P9.26 Use the results of Prob. P9.24 to design an op-amp circuit that provides the gain for the strain-gage amplifier described in Sec. 8.2 and shown in Fig. 8.3a:

(a) With $R_{f1}/R_1 = R_{f2}/R_2$, calculate values for R_1 and R_{f1} so that $A_v = 33.3$.

(b) If we use an LM308A as the op amp, that has a maximum offset voltage of 0.5 mV and a maximum offset current of 1.0 nA, what would be the worst-case offset at the output?

(c) Assuming the offset in part (b) is nulled out, over what temperature range could the circuit operate with less than 1 percent full-scale error if the temperature coefficients of offset voltage and offset current are 5.0 μV/°C and 10 pA/°C, respectively?

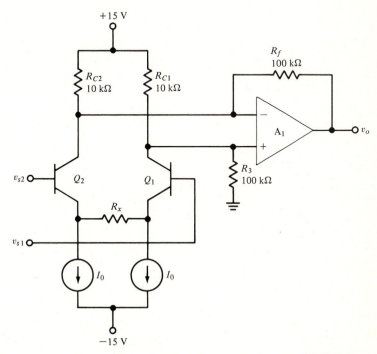

FIGURE P9.27

*P9.27   The circuit shown in Fig. P9.27 is called an *instrumentation amplifier*. It has the virtue that the gain can be programmed simply by changing one resistor, R_x.

(a) Prove that the expression for the voltage gain (with A_1 an ideal op amp) is, for $R_{C1} = R_{C2}$ and $R_3 = R_f$, given by

$$A = \frac{v_o}{v_{s2} - v_{s1}} = \frac{R_f}{\dfrac{R_x}{2} + \dfrac{V_T}{I_0}}$$

(b) Given that $I_0 = 0.26$ mA and $R_f = 100$ kΩ, find the values of R_x that give gains of (1) 1, (2) 10, (3) 1000.

(c) For the three conditions of part (b), find R_i, assuming for Q_1 and Q_2 that $\beta_F = \beta_0 = 200$.

(d) Since R_C does not show in the overall gain expression, what considerations would you use in selecting its value? (*Hint*: Consider dynamic range and offset effects.)

*P9.28   Figure P9.28 shows the use of an op amp in a simple voltage regulator circuit.

(a) Find an expression for V_O as a function of V_Z.

(b) With $V_Z = 5$ V, find values for R_1 and R_f such that V_O can be varied from 5 to 12 V. Limit the maximum current in the feedback network to 0.5 mA.

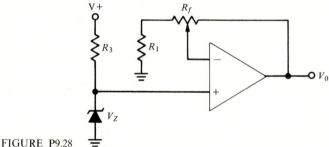

FIGURE P9.28

*P9.29   (a) With a diode connected in the feedback path of an op amp, the output voltage of the amplifier is a logarithmic function of the input voltage. For the circuit in Fig. P9.29 show that at room temperature $v_o = -0.06 \log [v_s/(R_1 I_S)]$, where I_S is the saturation current of the diode and the forward current in the diode is given by $I_D = I_S e^{V_D/V_T}$.

(b) From the information gained in part (a), describe and sketch an op-amp circuit that performs the antilog function, where $v_o = -R_f I_S$ antilog $(v_s/0.06)$.

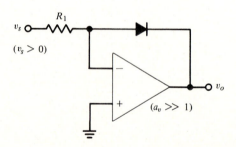

FIGURE P9.29

*P9.30 (a) The combination of two log circuits and one antilog circuit yields a one-quadrant (that is, $v_{s1} > 0$, $v_{s2} > 0$) logarithmic multiplier shown in Fig. P9.30. Assume for this circuit that all diodes are identical, and show that $v_o = -(v_{s1})(v_{s2})/(R_1 I_S)$.

(b) Describe and sketch an op-amp circuit that performs the following function:

$$v_o = 0.06 \log \left(\frac{B}{A}\right) + C$$

where $A > 0$, $B > 0$.

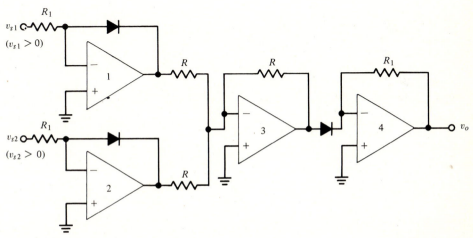

FIGURE P9.30

*P9.31 A common use of op amps is when a voltage is required to be integrated over a period of time. An op-amp integrator is shown in Fig. P9.31. Assume an ideal op amp and prove

$$v_o = -\frac{1}{RC}\int_0^t v_s \, dt$$

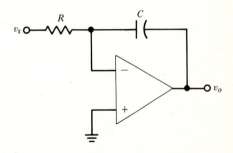

FIGURE P9.31

*P9.32   Combined with an RC circuit the op amp can perform the differentiation function. Show a sketch of an op-amp differentiator and prove that, with an ideal op amp,

$$v_o = -RC \frac{dv_s}{dt}$$

*P9.33   The combination of an op-amp integrator and differentiator shown in Fig. P9.33 is termed an *active filter*. For the component value shown in the figure:
(a) At what input signal frequency is the gain of the filter a maximum?
(b) What is the filter gain at the frequency of part (a)?

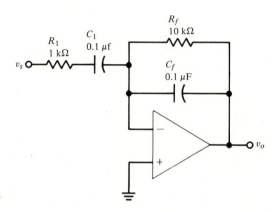

FIGURE P9.33

P9.34   Verify Eqs. (9.73a) and (9.37b) for the two-pole network of Fig. 9.14.
P9.35   An amplifier with a voltage gain of 12 dB is cascaded with a second amplifier that has $A_v = 20$ dB. If the output is now attenuated by a factor of 2, what is the overall voltage gain? Give your answer as a numeric as well as in decibels.
P9.36   The following parameter values are used in the two-pole network of Fig. 9.14.

$$a_{o1} = 20 \qquad a_{o2} = 100$$
$$R_1 = 10 \text{ k}\Omega \qquad R_2 = 1 \text{ k}\Omega$$
$$C_1 = 0.1 \text{ } \mu\text{f} \qquad C_2 = 0.01 \text{ } \mu\text{f}$$

Sketch the frequency and phase response (Bode plot) for the complete network and determine the unity gain frequency and the frequency where the phase angle is 180°.
P9.37   An op amp with $a_v = 66$ dB and one break frequency at 1 MHz is used in a shunt-shunt feedback amplifier with $R_1 = 1$ kΩ and $R_f = 100$ kΩ. Determine the loop gain and the -3-dB frequency for the feedback amplifier.
P9.38   Given an op amp with $a_v = 80$ dB, $f_1 = 10$ kHz, $f_2 = 316$ kHz, and $f_3 = 3.16$ MHz, sketch the Bode plot and:
(a) Show the minimum closed-loop gain A_r for a 0° phase margin.
(b) Repeat part (a) for a 45° phase margin.
P9.39   An op amp with $a_v = 80$ dB, $f_1 = 1$ kHz, and $f_2 = 1$ MHz is used in a series-shunt feedback amplifier with $R_1 = 1$ kΩ.
(a) Determine the minimum A_v for a 45° phase margin.
(b) Compute the minimum value for the feedback resistor R_f.

P9.40 For an op amp with $a_v = 60$ dB, $f_1 = 100$ kHz, and $f_2 = 10$ MHz, what is the -3 dB frequency (f_0) of the open-loop amplifier with simple lag-frequency compensation to ensure a minimum phase margin of $45°$ for all values of $A_v \geq 1$?

P9.41 Broadbanding is applied to the op amp of Prob. P9.40 to extend the bandwidth. If the internal resistance at the compensation terminal is 10 kΩ, determine the values of R_1 and C_1 in the broadbanding network for the same criteria as in Prob. P9.40.

P9.42 Given an op amp with $a_v = 80$ dB, $f_1 = 10$ kHz, $f_2 = 1$ MHz, and $f_3 = 10$ MHz. It is desired to compensate this amplifier so that for $A_v = 20$ dB it will have a $45°$ phase margin. The internal resistance at the compensation terminal is 8 kΩ.

(a) Find the first break frequency of the compensated amplifier (f_0) and the capacitance required by using narrowband compensation.

(b) Find the first break frequency of the compensated amplifier and the values of R_1 and C_1 by using the broadband-compensation technique.

10

APPLICATIONS OF DIGITAL AND LINEAR INTEGRATED CIRCUITS

To facilitate the introduction to the many new concepts and functions that are available as ICs, the subject matter in this text has been classified as being either digital *or* linear. However, in most engineering applications it is extremely rare to find a system that is all digital or all linear. Thus in this chapter we will discuss three examples of small systems, each of which makes use of both digital *and* linear ICs.

The first example is an instrument, the *dual-slope DVM* that we briefly described in the Introduction. Our discussion here will be in greater depth for two reasons: (1) it will provide an excellent review of all the material that we have discussed in the previous nine chapters, and (2) it will give some experience in solving actual control and timing problems. A control and timing block is usually needed whenever complex electronic blocks are interconnected, as in a DVM.

The second item is a simple *digital-to-analog (D-A) converter*. With a D-A converter we can change a digital code into an equivalent analog quantity. This function is commonly required at the output of a digital system, when the digital information is to be shown as a displacement on a CRT, or to turn a positioning motor, or to open a butterfly valve. The particular circuit that we shall study is made by using ICs that we have previously covered. It also has the advantage of using unspecialized parts and could be usefully employed in systems that have modest requirements of accuracy and precision.

The third item is a *phase-locked loop* (PLL) application. This makes use of two new "components," a *phase comparator* and a *voltage-controlled oscillator*, both of which are part of one monolithic chip. By using a PLL, an input signal can be used to *lock* the frequency of a voltage-controlled oscillator to the input signal frequency. Using a PLL, information carried on the incoming signal can also be readily extracted. Such an example is the use of a PLL in the demodulator portion of an FM radio receiver. Another application of PLLs is in a special class of signal generators called *frequency synthesizers*. Here a single precisely fixed frequency source can be used as a basis for generating a large number of frequencies, all of which have the same precision as the fixed source.

10.1 DUAL-SLOPE DIGITAL VOLTMETER

In the Introduction we indicated what were some of the building blocks used in a dual-slope DVM. This type of DVM is based on the principle that, as we saw in Chap. 6, it is relatively easy to make a very precise counter of pulses. Hence, if we were able to generate a train of pulses, where the number of pulses in the train was precisely proportional to an input voltage, we could then readily implement a precise DVM by combining it with a pulse counter.

In the following sections on DVMs, we will discuss the block diagram of an elementary dual-slope DVM and indicate the functions of each different block. Of these functions, three are new, so in the course of our discussion we will cover them in some detail. They are (1) the *ramp generator* (2) the *voltage comparator*, and (3) the *oscillator* (pulse generator) made by using a *monostable multivibrator*. To minimize the electronic switching required in a DVM we will discuss the operating principles of a further simplified dual-slope DVM. Based on the latter we will do a detailed design of the complete instrument, implementing it with as many MSI circuits as possible.

10.1.1 Dual-slope DVM Principles

Let us now see how a dual-slope DVM operates from a block-diagram point of view. In Fig. 10.1a the *electronic* switch S_1, at the input of the DVM, rests at position GND, the ground position. Therefore the input of the *ramp generator* (v_i) is zero. Let us assume that the output of the ramp generator is at 0 V. Momentarily closing the *read* switch S_R causes the following events to happen simultaneously:

1   The *up-counter* is cleared and the *display* reads 000.
2   A signal on the S_1 control line from the *control* causes electronic switch S_1 to move to position X.
3   The gated oscillator is gated on.

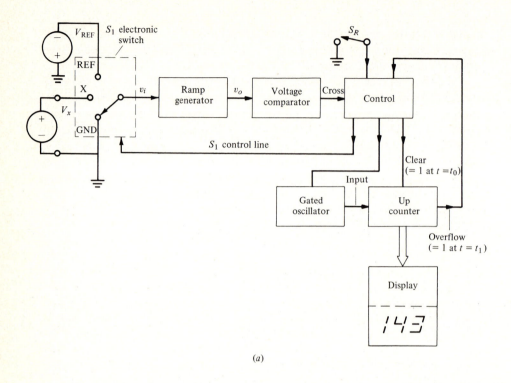

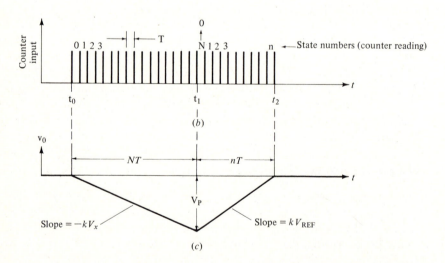

FIGURE 10.1
Dual-slope DVM. (*a*) Block diagram of a DVM; (*b*) pulses at counter input;
(*c*) waveform at output of ramp generator.

For purposes of this analysis let $t = t_0$ at the time the oscillator is turned on. Pulses from the oscillator are applied to the count input of the up-counter. Hence, we start to accumulate the count at t_0, as shown in Fig. 10.1b. While this is occurring, the dc voltage V_x at the input of the DVM causes a *ramp function* to be generated at the output of the ramp generator, shown in Fig. 10.1c in the time interval between t_0 and t_1. The ramp generator produces an output voltage whose slope (dv/dt) is proportional to the instantaneous voltage applied to the input of the DVM. For our purposes we have made $-k$ the constant of proportionality. Therefore for a positive dc voltage V_x, as in our case, the output is a *down-going* ramp.

The ramp continues going down until the counter overflows (goes into its Nth state which contains all zeros†), which produces a HIGH on the OVER-FLOW line at time t_1. By means of the CONTROL, switch S_1 is now set to the REF position so that the input voltage of the ramp generator is now V_{REF}. However, there is no change from the CONTROL to the gated oscillator, so the counter therefore continues counting up from its 000 reading.

For proper operation the REF input causes an upward ramp beginning at time t_1. This up-ramp continues until v_o crosses the axis (that is, $v_o = 0$ V) at time t_2. At this time the output of the *voltage comparator* goes HIGH. This input causes the CONTROL to

1   Turn off the gated oscillator
2   Simultaneously to return switch S_1 to the ground position

The counter has counted n pulses in the time interval $t_2 - t_1$, and this count remains in the counter until the read switch (S_R) is activated again. With a display decoder connected to the counter, the contents of the counter, that is, n, will be visible on the display. For the display to be meaningful we need the count n to be directly proportional to the dc voltage V_x. Let us now prove this is so.

First let us label as V_P the peak value of the ramp voltage v_o at t_1, as shown in Fig. 10.1c. Thus at the end of the down-ramp time we have

$$V_P = -kV_x(t_1 - t_0) \qquad (10.1a)$$

Similarly this is the voltage change during the up-ramp time

$$V_P = -kV_{REF}(t_2 - t_1) \qquad (10.1b)$$

Now the duration between pulses of the oscillator is T (in seconds), so that during the down-ramp time we must have

$$NT = t_1 - t_0 \qquad (10.2a)$$

Similarly for the up-ramp time

$$nT = t_2 - t_1 \qquad (10.2b)$$

---

† This N is the modulo of the counter as discussed in Chap. 6.

On combining Eqs. (10.1a) and (10.1b), we have

$$kV_x(t_1 - t_0) = kV_{REF}(t_2 - t_1) \qquad (10.3a)$$

By using Eqs. (10.2a) and (10.2b) in Eq. (10.3a), we have

$$kV_x NT = kV_{REF} nT \qquad (10.3b)$$

With the value of k and T unchanged during the run-down and run-up time (which is easily obtained in practice), we can cancel these factors in Eq. (10.3b) to obtain

$$V_x N = V_{REF} n \qquad (10.4a)$$

If for example the modulo of the counter, N, is set equal to 1000 and the value of V_{REF} is exactly 1 V, Eq. (10.4a) shows that the unknown voltage V_x is directly and simply related to the final count n stored in the counter by

$$V_x = n \frac{V_{REF}}{N} = n \times 10^{-3} \text{ (volts)} \qquad (10.4b)$$

In words Eq. (10.4b) says the counter reading n is equal to the unknown voltage V_x expressed in millivolts, which is the desired result.

Note that for $N = 1000$, the maximum display value, n_{MD}, that the counter can show is 999. Hence if V_x is 1000 mV, we would be able to read this value only by changing range, i.e., putting in a 10:1 attenuator between the voltage V_x and the DVM input. In order to reduce the number of range changes required in using a DVM, a common choice of N for our example would be 2000. With V_{REF} changed to 2 V, this allows the maximum display value for V_x, V_{xMD}, to be 1999 mV.

The maximum display value is related to the *number* of *digits* of a DVM. If in our example this is chosen as 1999, we would refer to the DVM as a $3\frac{1}{2}$ digit meter. The 3 refers to the last three full digits that can be 0 through 9. Usually the first digit is either 1 or a blank and is by convention valued as $\frac{1}{2}$ of a digit.

EXAMPLE 10.1   Let us determine some of the specifications on the blocks in Fig. 10.1a, given the following set of requirements:

(1) Maximum displayed value of input voltage, V_{xMD}, to be 1.99 V, that is, a $2\frac{1}{2}$ digit meter.

(2) We want a reading 100 ms after the read switch is pushed when a V_x of 1.99 V is applied.

(3) The peak value at the output of the ramp generator, V_P, should be 5 V when 1.99 V is applied.

MODULO OF THE COUNTER AND VALUE OF V_{REF} To make full use of the capabilities of the counter, we set the maximum display value n_{MD} equal to $N - 1$:

$$n_{MD} = N - 1 \qquad (10.5a)$$

or

$$N = n_{MD} + 1 \qquad (10.5b)$$

and thus from requirement 1, $n_{MD} = 199$; hence $N = 200$. Then from Eq. (10.4a) we have the requirement that when the maximum display value is applied,

$$V_{REF} = \frac{N}{n_{MD}} V_{xMD} \qquad (10.6)$$

$$= \frac{(200)(1.99)}{199} = 2 \text{ V}$$

OSCILLATOR PERIOD To obtain the period of the oscillator, we equate the time period $t_2 - t_0$ to the time interval needed to obtain a reading, namely, requirement 2, which is 100 ms. From Fig. 10.1c we see that under maximum display conditions

$$t_2 - t_1 = n_{MD} T$$

also

$$t_1 - t_0 = NT$$

thus

$$t_2 - t_0 = (N + n_{MD})T \qquad (10.7)$$

Hence, using the numeric values obtained above for n_{MD} and N,

$$100 \text{ ms} = (200 + 199)T$$

therefore

$$T = 0.2506 \text{ ms}$$

This corresponds to a frequency

$$f = \frac{1}{T} = 3.99 \text{ kHz}$$

RAMP CONSTANT We next solve for the constant k in the ramp generator. Since the peak value of the ramp output voltage has been set at 5 V, we have for maximum display condition

$$V_P = 5 \text{ V}$$

$$V_{xMD} = 1.99 \text{ V}$$

Now from Eq. (10.1a) we can solve for k to obtain

$$k = \frac{V_P}{V_{xMD}(t_1 - t_0)} \qquad (10.8)$$

From the previous results we have

$$t_1 - t_0 = NT = (200)(0.2506 \text{ ms}) = 50.13 \text{ ms}$$

Substituting this value in Eq. (10.8) gives

$$k = \frac{5}{(1.99)(50.13 \times 10^{-3})} = 50.13 \text{ s}^{-1} \qquad ////$$

We will make use of this constant in the design of the ramp generator.

EXERCISE

E10.1   For a V_{xMD} of 1.999 V that gives a reading after 50 ms with a V_p of 10 V, find the required N, V_{REF}, T, f, and k.

10.1.2   Ramp Generator

The ramp generator is simply implemented by means of the op-amp *integrator* circuit shown in Fig. 10.2a. Assume we have an ideal op amp, then with a virtual ground at v_-, we have $i_x = V_x/R_1$. Also

$$v_o = -v_{C1} \qquad (10.9a)$$

where the capacitor voltage v_{C1} is related to its charge q_1 by

$$v_{C1} = \frac{q_1}{C_1} \qquad (10.9b)$$

Hence the output voltage as a function of time is

$$v_o(t) = \frac{-q_1(t)}{C_1} \qquad (10.9c)$$

Since the capacitor charge is the integral of the current $i_{C1}(t)$ and for $i_- \to 0$, we may write

$$q_1(t) = \int_0^t i_x \, dt + q_1(0) \qquad (10.10a)$$

where $q_1(0)$ is the charge on the capacitor at time $t = 0$. Therefore, from Eq. (10.9c),

$$v_o(t) = -\frac{1}{C_1}\left[\int_0^t i_x \, dt + q_1(0)\right] \qquad (10.10b)$$

$$= -\frac{1}{C_1}\int_0^t i_x \, dt - v_{C1}(0) \qquad (10.10c)$$

where the voltage across the capacitor at time $t = 0$ is

$$v_{C1}(0) = \frac{q_1(0)}{C_1} \qquad (10.10d)$$

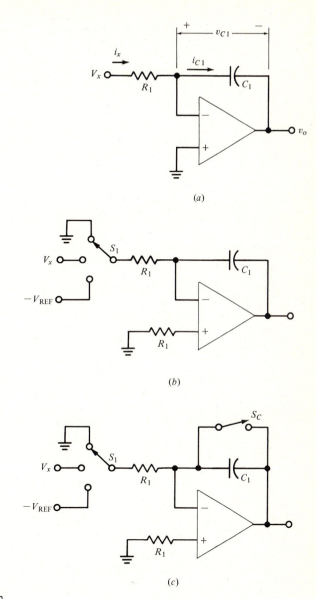

FIGURE 10.2
Ramp generator and input switching. (*a*) Basic ramp generator (integrating circuit);
(*b*) ramp generator with its input switch; (*c*) the switch S_c across C_1 sets the
ramp generator output to zero at $t = t_0$.

Making use of our earlier result that $i_x = V_x/R_1$, we obtain the desired ramp voltage for a dc input voltage V_x:

$$v_o(t) = -\frac{1}{R_1 C_1} \int_0^t V_x \, dt - v_{C1}(0) \quad (10.11a)$$

which on carrying out the integration gives

$$v_o(t) = -\frac{V_x t}{R_1 C_1} - v_{C1}(0) \quad (10.11b)$$

Thus the expression for the ramp constant in Eq. (10.8) is

$$k = \frac{1}{R_1 C_1} \quad (10.12)$$

Returning to Example 10.1, suppose for convenience we use a 1-μF capacitor for C_1. Then with $k = 50.13$ s^{-1}, we calculate the value of R_1 as

$$R_1 = \frac{1}{kC_1} = \frac{1}{(50.13)(1 \times 10^{-6})} = 19,950 \ \Omega$$

Since capacitor C_1 would have in practice a tolerance of about 5 percent, we would then make R_1 out of a 1 percent fixed resistor (that is, 17.8 kΩ) and a 5-kΩ potentiometer that we would adjust to make k the correct value.

Thus the circuit in Fig. 10.2a could be used as the ramp generator in the DVM of Fig. 10.1a. However before we do this, an important detail must be looked into; namely, that in practical IC op amps a finite input (bias) current flows into the op amp. Thus in Fig. 10.2a the effect of the bias current is to charge C_1 even when V_x is zero!

Since the bias current is charging C_1 continuously, the result is at first an increasing error term in the output voltage. Finally the amplifier stabilizes ("saturates") with the output at one end of its range of swing, so that no current flows in C_1.

A partial solution to the bias-current problem is shown in Fig. 10.2b, where we add a resistor R_1 between the noninverting input and ground. If the bias currents were equal (i.e., zero input offset current), they would not affect the capacitor current i_{C1}, and the only effect of the bias current would be to add a small voltage $-I_B R_1$ to v_{C1} in the expression for the output voltage.

However, zero offset current is not attainable, and the result of a finite offset current is to charge (or discharge) C_1 when S_1 is in the ground position. A solution to the offset-current problem is shown in Fig. 10.2c, where we add another electronic switch S_C which shorts out C_1 at time t_0 so that $v_{C1}(0) = 0$. Then the offset current only causes an error during the measuring cycle. For an offset current of 1 nA, a measurement time of 100 ms, and a capacitor value of 1 μF, we obtain an error at the output of

$$\Delta V = \frac{\Delta Q}{C_1} = \frac{I_{IO}(t_2 - t_0)}{C_1} \quad (10.13)$$

$$= \frac{(10^{-9})(10^{-1})}{10^{-6}} = 10^{-4} = 0.1 \text{ mV}$$

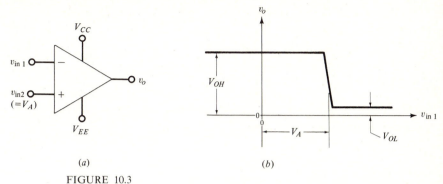

(a)   (b)

FIGURE 10.3
Voltage comparator. (a) Circuit schematic and reference conventions; (b) transfer characteristic.

This value of error in the ramp output voltage can be tolerated in most DVMs since this is a 0.002 percent error when V_P is 5 V. However, we must next consider the error associated with the zero-crossing sensing capability of the voltage comparator which follows the ramp generator in our block diagram of Fig. 10.1. We shall discuss this in the next section.

EXERCISE

E10.2   Referring to Fig. 10.2a for $V_x = 1$ V, $R_1 = 10$ kΩ, $C_1 = 0.5$ μF, and $v_{C1}(0) = -1$ V, find v_o when $t = 15$ ms.

10.1.3   Voltage Comparator

The function of voltage comparison is shown in Fig. 10.3a. This can be done with an IC op amp, with the variable voltage $v_{\text{in, 1}}$ applied to the inverting input and the reference voltage $v_{\text{in, 2}}$ applied to the noninverting input. If we set $v_{\text{in, 2}}$ to the dc value V_A, we have the transfer characteristic shown in Fig. 10.3b. We assume a_v is very large and the offset voltage V_{IO} is small. Then for $v_{\text{in, 1}}$ slightly less than V_A, v_o is at the positive end of the output voltage range. When $v_{\text{in, 1}}$ is slightly greater than V_A, the output is at its most negative value. Thus a differential-input op amp could serve as a voltage comparator, giving

$$v_o = \begin{cases} \text{HIGH} & \text{if } v_{\text{in, 1}} < V_A \\ \text{LOW} & \text{if } v_{\text{in, 1}} > V_A \end{cases}$$

For many applications the performance obtained from a general-purpose op amp as a voltage comparator is not sufficient. Thus special ICs designed to be optimum for this function are called *voltage comparators*. They provide logic level outputs and reasonable fanout compatible with digital ICs. Moreover the response time obtained from these specialized ICs is vastly superior to that obtained if one had used a general-purpose op amp.

Various commercial types of IC voltage comparators are available with response times that range from 12 to 200 ns. Open-loop gains (the slope at $v_{in, 1} = V_A$ in Fig. 10.3b) range from 10^3 to 2×10^5. Maximum offset voltages are from 5 mV to as low as 0.7 mV. Like op amps, voltage comparators can be used open loop or with feedback.

Returning to our DVM with a 5-V ramp, the 5-mV uncertainty due to the offset would cause a 0.1 percent error. However, the offset could be effectively nulled out so that this error could be greatly reduced.

10.1.4   Simplified Dual-slope DVM

DVMs of the type shown in Fig. 10.1a are commonly available. However, to implement the switching functions of S_1, S_R, and S_C (in Fig. 10.2c) adds sufficient complication that a simplified form of dual-slope DVM is often used commercially. This eliminates the need for S_R and S_C and allows one to replace S_1 with a simpler electronic switch. Since we elect to use this scheme in our demonstration model, we now study its operation.

A schematic of the input circuit to the ramp generator for the simplified DVM is shown in Fig. 10.4a. Here current I_x flows continually, and the only current switched is I_{REF}, which is repetitively connected to the integrator only between times t_1 and t_2 as illustrated in Fig. 10.4b.

Assume time t_1 corresponds to the time that the counter is at its terminal count, so that its state number is N where the count in the counter is 0. For a voltage V_x at the input we obtain a current through R_1 of

$$I_x = \frac{V_x}{R_1} \quad (10.14a)$$

The slope of the ramp generator output, from t_0 to t_1, is

$$-\frac{I_x}{C_1} \quad (10.14b)$$

During the time of the up-ramp (t_1 to t_2), S_1 is closed. The slope is then

$$\frac{I_{REF} - I_x}{C_1} \quad (10.14c)$$

When the voltage v_o goes through zero at time t_2, the current I_{REF} is shut off. The content of the counter at this time is n. We now need to be assured that this number n is proportional to the input voltage V_x.

Referring now to Fig. 10.4c, the peak value of v_o due to the down-ramp is given by

$$V_p = -\frac{I_x}{C_1}(t_1 - t_0) \quad (10.15a)$$

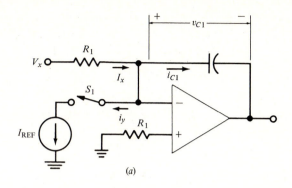

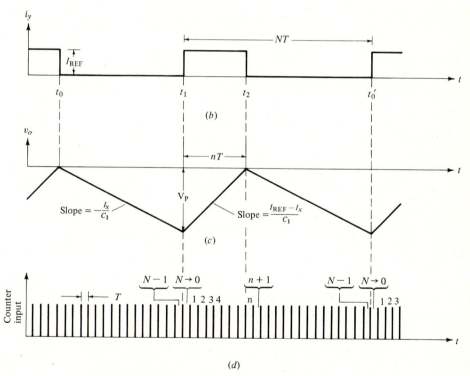

FIGURE 10.4

A simplified dual-slope DVM. (*a*) Block diagram of input switching; (*b*) waveform of current through the switch; (*c*) waveform at output of ramp generator; (*d*) pulse train at counter input.

and this same peak value is obtained during the up-ramp time as

$$V_p = -\frac{I_{REF} - I_x}{C_1}(t_2 - t_1) \quad (10.15b)$$

Moreover from Fig. 10.4d we have

$$t_2 - t_1 = nT \quad (10.15c)$$

and

$$t_1 - t_0 = (N - n)T \quad (10.15d)$$

Combining Eqs. (10.15a) through (10.15d), we have

$$\frac{I_x}{C_1}(N - n)T = \frac{I_{REF} - I_x}{C_1}nT \quad (10.16a)$$

which reduces to

$$\frac{I_x NT}{C_1} = \frac{I_{REF} nT}{C_1} \quad (10.16b)$$

With the average values of T and C_1 the same during the up- and down-ramp, we have

$$I_x N = I_{REF} n \quad (10.16c)$$

Finally, substituting Eq. (10.14a) for I_x, we have the desired result that the voltage V_x is directly proportional to n, the count at time t_2,

$$V_x = \frac{n}{N} I_{REF} R_1 \quad (10.16d)$$

A more complete block diagram of this modified DVM is shown in Fig. 10.5. Significant differences from the block diagram of Fig. 10.1a are:

1 The oscillator is always connected to the counter input, so that the counter is continually receiving pulses.

2 To obtain a fixed display, latches must be added between the counter and the display decoder. The inputs to the latches are briefly enabled by the STROBE line at time t_2 so that they latch onto the count n.

3 A line called OVERLOAD is connected from the counter to the control. This is used to prevent erroneous† readings which can occur when a voltage whose value is larger than V_{xMD}, the maximum display value, is applied to the input. The system in Fig. 10.1 does not have provisions for this.

4 Because the counter is continually receiving pulses, this scheme is most unlikely to give the correct reading within NT seconds after V_x is first applied. Rather the first several displayed readings vary around the final

---

† Voltages greater than the *overload* value can be applied to the input with no damage to the DVM until the *maximum input voltage rating* is exceeded, at which point the instrument's usefulness is imperiled.

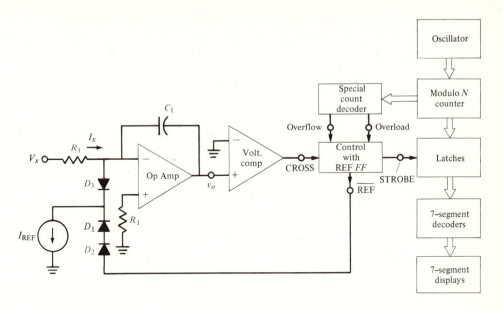

FIGURE 10.5
Complete block diagram of the simplified DVM.

value but soon stabilize at the correct value. In contrast the method shown in Fig. 10.1a does give the correct reading at the end of the initial period of time $(N + n)T$.

10.1.5   Switching the Current Source I_{REF}

The constant current source I_{REF} can be made in a variety of ways, one being the method of Prob. P9.24. The switching of the constant current can readily be done with diode gates made by using three similar silicon diodes, D_1, D_2, and D_3, as shown in Fig. 10.5.

If the anode of D_2 is at ground potential ($\overline{REF}$ = LOW), D_3 will conduct essentially all of I_{REF}. This follows since the inverting input of the op amp is at virtual ground ($v_i = 0$ V), and as can be shown (see Prob. P10.16) that with $\overline{REF} = 0$,

$$I_{D1} = I_{D2} \approx \sqrt{I_{REF} I_S} \qquad (10.17)$$

where I_S is the saturation current in each of the diodes.

Thus if I_{REF} is 10^{-4} A and $I_S = 10^{-14}$ A, when the control output $\overline{REF}$ = LOW, the current through D_1 and D_2 would be about 10^{-9} A (see Fig. 1.4). This gives a 0.001 percent difference between I_{D3} and I_{REF}, which is negligible in our design.

For $\overline{\text{REF}}$ = HIGH and using a standard TTL flip-flop, we typically measure 3.6 V at the anode of D_2. Using $V_{D(\text{on})} = 0.7$ V means that the cathode of D_3 would be at 2.2 V, so that we would obtain a slight reverse leakage through D_3. For most low-power silicon diodes this leakage current would be well under 10^{-9} A, which would also be negligible compared to the usual range of values of I_x.

EXAMPLE 10.2   Let us now use the specification of Example 10.1 in the block diagram of the simplified DVM of Fig. 10.5.

MODULO OF COUNTER AND PERIOD OF OSCILLATOR   The selection of the counter modulo is not as straightforward as it was in Example 10.1. For the circuit in Fig. 10.5 to work, we need

$$N > 2n_{MD}$$

The larger N is, the more rapid is the convergence of the readings to the final value. However, for simplicity we choose

$$N = 2(n_{MD} + 1)$$

thus for $n_{MD} = 199$ we need a counter modulo of 400.

Generally we want the conversion time $t_2 - t_0$ to be a multiple of the power-line frequency so that "hum" will not affect the reading. For a 60-Hz power-line frequency, a selection of $t_2 - t_0$ of 16.67 ms will give us sufficient rejection of any ac signals at the power-line frequency or any of its harmonics. Thus from Fig. 10.4d, where $NT = t_2 - t_0$, we have

$$T = \frac{t_2 - t_0}{N}$$

$$T = \frac{16.67 \text{ ms}}{400} = 41.67 \text{ } \mu s \qquad (10.18)$$

and hence the oscillator frequency f is 24 kHz.

SELECTION OF R_1, C_1, AND I_{REF} For convenience we again choose C_1 to be 1 μF and the absolute value of V_{PMD} as 5 V. To determine the value of I_{REF} we note from Eq. (10.16c) that for maximum display conditions

$$I_{\text{REF}} = \frac{N}{n_{MD}} I_{xMD} \qquad (10.19)$$

Now for the ramp generator in Fig. 10.4a we note that

$$i_{C_1} = C_1 \frac{\Delta V_{C_1}}{\Delta t} \qquad (10.20a)$$

therefore

$$I_{xMD} = C_1 \frac{V_{PMD}}{t_1 - t_0} \qquad (10.20b)$$

But from Eq. (10.15*d*), using maximum display values,

$$t_1 - t_0 = (N - n_{MD})T \qquad\qquad (10.20c)$$

$$= (400 - 199)(41.67 \times 10^{-6}) = 8.38 \text{ ms}$$

Substituting this value into Eq. (10.20*b*), we have

$$I_{xMD} = \frac{(10^{-6})(5)}{8.38 \times 10^{-3}} = 5.97 \times 10^{-4} = 0.597 \text{ mA}$$

and using this in Eq. (10.19), we have

$$I_{REF} = \frac{400}{199} 0.597 = 1.20 \text{ mA}$$

Finally the input resistance R_1 can be calculated as

$$R_1 = \frac{V_{xMD}}{I_{xMD}} = \frac{1.99}{0.597} = 3.33 \text{ k}\Omega \qquad ////$$

10.1.6 Counter Design for DVM

As we want a demonstration DVM based on the design started in Example 10.2, we will now do the detailed design of the digital circuitry. We begin this by designing the modulo 400 counter. This could be made by cascading two decade up-counters and having the terminal count of the second decade drive a modulo 4 up-counter.

Modulo 100 up-counter The block diagram of the decade counter that we will use is shown in Fig. 10.6*a*. These are fully synchronous up-counters† that have the usual clock (C_p) input and the four outputs Q_A, Q_B, Q_C, and Q_D, where Q_A is the LSB and Q_D is the MSB. They also have a master reset input $\overline{MR}$ which when LOW clears all flip-flops to 0. The other terminals provide three additional functions that were not obtained in the decade counter of Chap. 6.

The first of these is that the counter state can be preset to any value by the four parallel inputs $(P_A, P_B, P_C,$ and $P_D)$. The logic values of P_A through P_D are entered into the IC when C_p and $\overline{P_E}$ (parallel enable) are both LOW. Then when C_p goes HIGH, these values will be transferred into the outputs of the flip-flops Q_A through Q_D, respectively.

The second additional item relates to the clock-enabling function that is controlled by two inputs to an AND gate, namely, C_{ET} and C_{EP}. The necessary conditions for counting are $P_E = $ LOW, $C_{ET} = $ HIGH, and $C_{EP} = $ HIGH. The flip-flops in the counter are of the master/slave type, as described in Sec. 5.3.2.

---

† TTL-type 9310. A nearly similar function is performed by TTL-type 74192, which is an up-down decade counter.

(a)

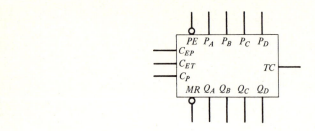

(b)

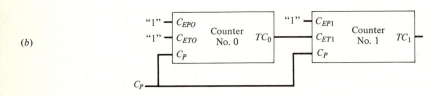

(c)

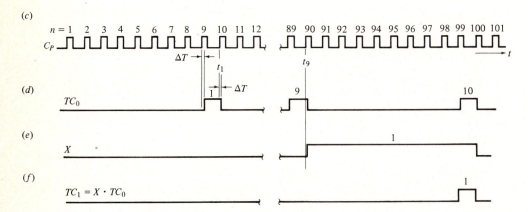

(d)

(e)

(f)

FIGURE 10.6

Modulo 100 counter made from a presettable decade counter. (a) Logic symbol
for synchronous presettable decade up-counter; (b) connections for performing cascading; (c) clock-pulse waveform; (d) terminal count output of counter no. 0; (e) the
waveform for a reference signal X which is HIGH when counter no. 1 is at 1001;
(f) terminal count output of counter no. 1.

Information is entered into the master when C_p is LOW and transferred to the
slave (output) when C_p goes HIGH.

The third feature of this type of counter is the terminal count output TC that
is useful, as we shall soon see, for cascading counters. For a decade counter the
logic expression for the terminal count is

$$TC = C_{ET} \cdot Q_D \cdot \bar{Q}_C \cdot \bar{Q}_B \cdot Q_A \qquad (10.21)$$

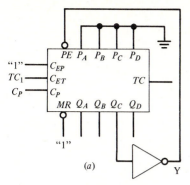
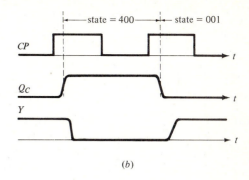

FIGURE 10.7
Modulo 4 counter. (a) Connections for converting a synchronous presettable decade up-counter into a modulo 4 counter; (b) waveforms for the circuit in part (a).

thus the terminal count output is HIGH when the counter state is 9 ($= 1001_2$) and its C_{ET} input is HIGH.

To make a fully synchronous modulo 100 counter we cascade two of the above counter types and arrive at Fig. 10.6b. The least significant decimal (LSD) or "units" counter is called 0 and is to the left as usually drawn. Its terminal count (TC_0) is used as the input signal at the count-enable input C_{ET1} of the "tens" counter, counter 1. All other C_E inputs are tied to HIGH.

The applied clock waveform C_p is indicated in Fig. 10.6c for a series of clock pulses from $n = 1$ to $n = 101$. At $n = 9$ the counter status becomes 9 after a small propagation delay time ΔT. With $C_{ET0} = $ HIGH, then by Eq. (10.21), TC_0 becomes HIGH during the time counter 0 is in state 9, as shown in Fig. 10.6d.

We have already noted that counter 1 can only count up when C_{ET1} is HIGH just before the transition of its C_p from LOW to HIGH. Hence the first clock pulse to affect the state of counter 1 is at $n = 10$ at time $t = t_1$. For reference purposes a waveform X is shown in Fig. 10.6e. X is HIGH whenever the count of counter 1 is 9. Because of the relationship of Eq. (10.21), however, TC_1 will be as shown in Fig. 10.6f. Therefore the first clock pulse to affect the state of counter 2, the "hundreds" counter of modulo 4, is at $n = 100$.

Modulo 4 up-counter   The modulo 4 counter can be made in a variety of ways by using ICs. To get more experience with MSI circuits, we choose to make it by using a modified decade counter of the type used in the modulo 100 part.

With the addition of one inverter we can use the decade counter of Fig. 10.6a to make a different modulo counter (shown in Fig. 10.7a) that sequences from 0000 up through 0100. To verify what is the modulo of the overall counter, we note that the counter counts clock pulses as shown in Fig. 10.7b, where the counter changes state right after the positive-going edge of the clock pulse. Thus with counter state at 399, the next clock pulse advances the counter to 400 and

536

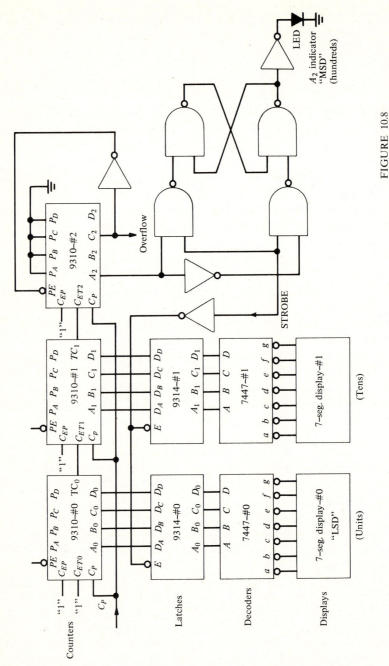

FIGURE 10.8
DVM counter and display circuitry.

Q_C goes HIGH. Then after the short propagation delay time of the inverter, the parallel enable becomes active (Y = LOW) and the parallel inputs (wired to 0000 for this circuit) become the state of the counter when the next clock pulse goes HIGH. The resulting overall counter thus sequences from the states 001 to 400.

In the upper portion of Fig. 10.8 we show the complete counter for N = 400. Its output C_2 serves as the OVERFLOW line. In order to obtain a steady reading for the DVM we need latches that will hold the value of the counter at the time of the zero crossing of the ramp voltage. At this time the STROBE line will briefly go HIGH and *enable* the latches. For each decade counter we need a 4-bit latch which can be obtained in one IC,† as shown in Fig. 10.8. Because the most significant decimal (MSD), which is contained in counter 2, the "hundreds" counter, is only a 0 or a 1, we only need a 1-bit latch for storing this. This is easily made from "discrete" NAND gates (see Fig. 5.9) as indicated in Fig. 10.8. A single buffered LED will serve as our hundreds display character.

The latches that contain the units and tens data drive BCD-to-seven-segment decoder drivers of the type discussed in Sec. 4.5.3. These in turn drive the seven-segment display elements.‡

Now that we have the counter designed, we next look at the CONTROL block of Fig. 10.5.

EXERCISE

E10.3 Using the same techniques as used in reference to Figs. 10.6 to 10.8, construct a counter for the DVM of the type in Fig. 10.1, where the maximum display reading is to be 1999.

10.1.7 Overload Indication

Since the DVM has a maximum displayed value (199 in our example), we need to provide some means to indicate when the input voltage exceeds V_{xMD} (that is, 1.99 V in our example). A simple way to detect this overload condition is to sense if n (the status of the counter at time t_2) is greater than n_{MD}. Thus if it is determined that the counter state reaches n_{MD} before the ramp voltage crosses zero, an overload light is lit.

Another way to indicate that an overload condition exists is to have the display "stick" at the maximum display value when a voltage equal to or greater than V_{xMD} is applied. This can be easily done by having the STROBE line put out

---

† TTL-type 9314 or 7475.

‡ Single-chip LSI circuits (Mostek 5005P or Fairchild 3814) are available that do all the counting, latching, and decoding necessary for a $4\frac{1}{2}$ digit DVM. To minimize output pins on the ICs, they provide seven-segment or BCD outputs that are *multiplexed*.

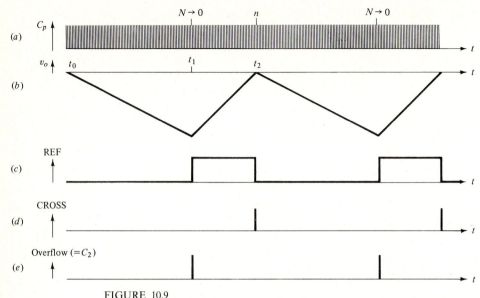

FIGURE 10.9
Waveforms for determining the REF flip-flop design. (a) Clock-pulse waveform; (b) ramp-output waveform; (c) the waveform for the logic variable REF; (d) the waveform for the comparator output CROSS; (e) the waveform used to indicate OVERFLOW

a pulse at the time the counter status is n_{MD} and CROSS is still LOW. After this pulse is obtained on the STROBE line, no further STROBE pulses are to be generated until the counter status goes to 400 again. The details of how this is done will be described in the next section.

10.1.8 Control Logic

The proper operation of most digital systems rests on the control circuitry that senses the status of other blocks and accordingly, at the appropriate time, actuates the flow of information into and out of these other blocks. In the simplified dual-slope DVM this function is contained in the block labeled (appropriately) CONTROL, which is in the center of Fig. 10.5.

To design the control we note from Fig. 10.5 that we require two outputs: (1) $\overline{\text{REF}}$ and (2) STROBE. In Sec. 10.1.5 we pointed out that when $\overline{\text{REF}} = \text{LOW}$, the reference current I_{REF} is permitted to flow through capacitor C_1. When STROBE = HIGH, the latches take on the state of the counter (which is the desired reading n) so that it is visible in the display until the next STROBE pulse arrives.

REF flip-flop design For convenience we again show the waveforms of the DVM clock pulse (C_p in Fig. 10.9a) and the ramp voltage (v_o in Fig. 10.9b). As we have

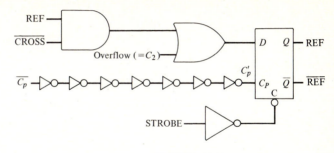

FIGURE 10.10
The control logic in the REF flip-flop.

outlined before, the logic signal REF should be the waveform shown in Fig. 10.9c. The operation of the control block should:

1 Set REF HIGH when the count is 400, which is at time t_1.
2 Set REF LOW whenever the ramp crosses zero, which is at time t_2 when the count is n.
3 Once REF is LOW it must stay LOW until the count is 400 again.

From the third requirement we see that a memory device, that is, a flip-flop, is needed.

For concreteness let us use a D-type flip-flop† that is shown in Fig. 10.10 with its control logic that we shall now design. The true (Q) output of the flip-flop is called REF. The inputs to the control logic which affect D are (1) REF itself, (2) CROSS, the voltage-comparator logic output which is HIGH when the ramp is greater than 0 V (as shown in Fig. 10.9d), and (3) the counter OVERFLOW signal which is still HIGH when the counter enters state 400, as shown in Fig. 10.9e. The signal C_2, from the most significant decade counter (counter 2) is our OVERFLOW signal.

When V_x is less than the overload value (that is, $0 \le V_x < 1.99$ V) the D input of the REF flip-flop has to be programmed so that:

1 D is set HIGH when the counter contains 400 (OVERFLOW = 1). This sets REF to a HIGH state.
2 D stays HIGH as long as CROSS is LOW. This keeps REF in the HIGH state.
3 When CROSS goes HIGH, REF is to go LOW and stay LOW until OVERFLOW becomes HIGH again.

The above requirements can be summarized in the expression

$$D = \text{REF} \cdot \overline{\text{CROSS}} + C_2$$

This function is implemented with an AND and OR gate, as shown in Fig. 10.10.

---

† TTL-type 7474, which is triggered on the positive-going edge of the clock input.

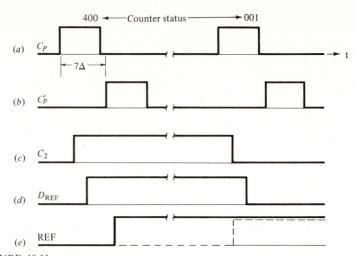

FIGURE 10.11
Detailed timing for the REF flip-flop. (a) The original clock waveform C_p; (b) the delayed clock-pulse waveform C_p'; (c) the waveform for the signal OVER-FLOW ($= C_2$); (d) the waveform for the data signal into the flip-flop, D_{REF}; (e) the waveform for the output of the flip-flop, REF.

The asynchronous CLEAR input to the REF flip-flop is STROBE. When STROBE goes HIGH, output REF goes LOW irrespective of the clocked input D. For a nonoverload value of V_x, CROSS and STROBE have the same waveforms as we shall see later (Fig. 10.12c and d). OVERFLOW ($= C_2$) is used to set REF to HIGH, and then REF is fed back to the AND gate so as to hold REF HIGH as long as CROSS is LOW.

Note that in Fig. 10.10 we show seven inverters between $\overline{C}_p$ and the clock input to the flip-flop, C_p'. In Fig. 10.11 we show these waveforms on a greatly expanded time base: C_p in Fig. 10.11a and C_p' in Fig. 10.11b. The signal C_p' is delayed by 7Δ seconds from $\overline{C}_p$, where Δ is the propagation delay through one inverter. This delay is necessary to ensure that C_2 ($=$ OVERFLOW), shown in Fig. 10.11c, can cause D_{REF} (shown in Fig. 10.11d) to set REF HIGH (as shown in the solid line in Fig. 10.11e) as soon as possible after the counter enters state 400. This is necessary if we are to get a display reading of 00 when V_x is 0 V. If C_p had been used as the clock input to the REF flip-flop, REF would first go HIGH only when the counter was at state 001. This condition is shown as the dotted line in Fig. 10.11e.

STROBE circuitry design The other output that we need from the control block is STROBE which, as we have already specified for a normal reading (that is, $0 \leq V_x < 1.99$ V), would put out a brief pulse to enable the latches when CROSS goes HIGH at $t = t_2$. If the input voltage is overrange (that is, $V_x \geq 1.99$ V), we want a pulse to appear on STROBE during the period of time the counter state is

199. If CROSS goes from LOW to HIGH after this time, the circuit must not permit another STROBE pulse to appear until after the count goes to 400 again.

Let us first synthesize the waveform that is HIGH only when the modulo-400-counter state is 199. Call this logic signal F. One easy way to get F is from the counter by using the three logic signals $(\bar{B}_2, A_2$, and $TC_1)$ shown in Fig. 10.12a. Clearly, if we AND these three signals, we obtain F.

$$F = \bar{B}_2 \cdot A_2 \cdot TC_1 \qquad (10.22a)$$

which can be seen to only be HIGH when the count is 199.

The signal $\bar{B}_2$ can also be used as an enabling signal so that under a normal reading, when

$$0 \leq V_x < 1.99 \text{ V}$$

the CROSS signal generates a STROBE output. CROSS, STROBE, and REF waveforms are shown under this condition in Fig. 10.12c, d, and e, respectively.

Hence we can construct a suitable STROBE signal according to the following expression.

$$\text{STROBE} = \text{CROSS} \cdot \text{REF} \cdot \bar{B}_2 + \text{REF} \cdot F \qquad (10.22b)$$

On substituting Eq. (10.22a) for F, we have

$$\text{STROBE} = \text{CROSS} \cdot \text{REF} \cdot \bar{B}_2 + \text{REF} \cdot \bar{B}_2 \cdot A_2 \cdot TC_1 \qquad (10.22c)$$

Whenever the second term in Eq. (10.22c) has to become 1, REF and $\bar{B}_2$ have the same waveform. This is shown for the OVERLOAD condition in Fig. 10.12f, g, and h, where the crossing if any occurs† when $\bar{B}_2 = 0$ and the strobe is generated in the first portion of the interval when the count is 199. This STROBE sets REF LOW through the asynchronous input CLEAR. Because REF and $\bar{B}_2$ are equal under this condition up to this time, we can further simplify Eq. (10.22c) to

$$\text{STROBE} = \text{CROSS} \cdot \text{REF} \cdot \bar{B}_2 + \text{REF} \cdot A_2 \cdot TC_1 \qquad (10.22d)$$

The circuit for doing this with NAND gates is shown in Fig. 10.15d.

This completes the design of the CONTROL block of the DVM. The only block we still need to design is that labeled oscillator. It can be made from a variety of ICs, one of which we will cover in the next section.

10.1.9 Oscillator (Pulse-generator) Design

A simple way to generate TTL-compatible pulses at a predetermined frequency is to use an IC monostable multivibrator.‡ A block diagram for one type, the 9601, is shown in Fig. 10.13a. In Fig. 10.13b we show the waveforms at the point IN (which is inside the IC and not accessible) and the outputs Q and $\bar{Q}$. When IN first goes HIGH, the output Q goes HIGH after a short propagation delay time t_{pd+}.

---

† The crossing would only occur due to an extraneous noise pulse.
‡ TTL-types 74121 and 9601. These circuits are also called *one-shots* or *single-shots*.

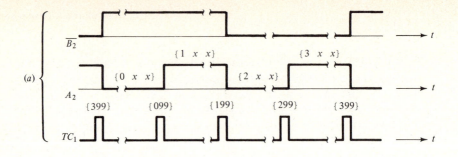

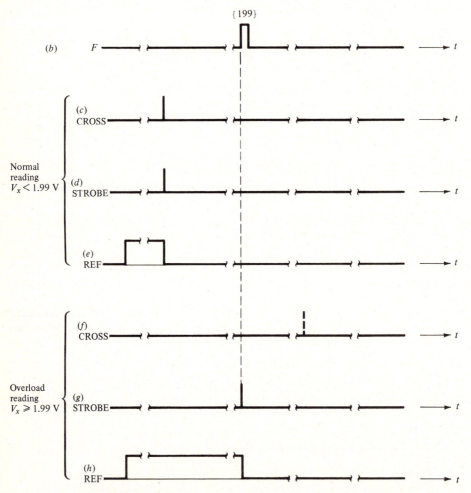

FIGURE 10.12

Waveforms used in determining the STROBE logic signal (count value in { }).
(a) Waveforms of $\bar{B}_2$, A_2, and TC_1; (b) the waveform of the function F; (c) the
waveform of the comparator output CROSS for normal readings; (d) the waveform
of the STROBE signal for normal readings; (e) the waveform of the REF signal
for normal reading; (f) the waveform of the comparator output CROSS for
overload readings; (g) the waveform of the STROBE signal for overload readings;
(h) the waveform of the REF signal for overload readings.

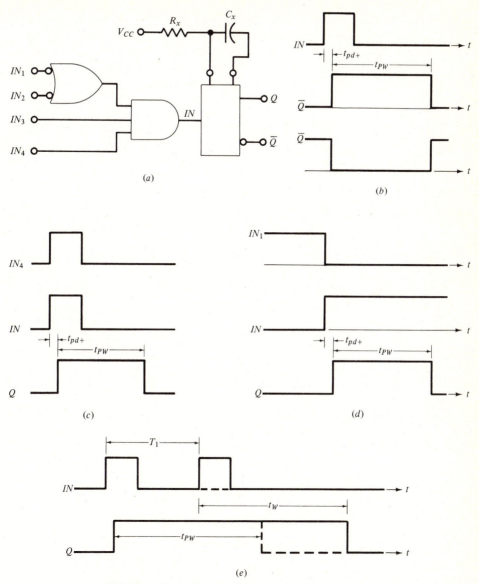

(a)

(b)

(c)

(d)

(e)

FIGURE 10.13
Retriggerable multivibrator (one-shot) characteristics. (a) Block diagram of a one-shot with external timing components R_x and C_x; (b) waveforms for the circuit in part (a); (c) waveforms for the circuit in part (a) when IN_1 = LOW, IN_3 = HIGH; (d) waveforms for the circuit in part (a) when $IN_3 = IN_4 = IN_2$ = HIGH; (e) retriggerable feature shown in solid line holds for $0.3C_x < T_1 < t_{pw}$.

The output Q then stays HIGH for a pulse-width time t_{PW} which is determined by the values R_X and C_X. This relationship for the TTL-type 9601, when $C_X > 1000$ pF, is

$$t_{PW} = 0.32 R_X C_X \left(1 + \frac{0.7}{R_X}\right) \qquad (10.23)$$

with the units as follows: t_{PW} (ns), R_X (kΩ), and C_X (pF). For the TTL-type 74121 the similar relationship is

$$t_{PW} = 0.69 R_X C_X \qquad (10.24)$$

The combinational logic circuit in Fig. 10.13a that generates IN (and is part of the monolithic IC package) gives

$$IN = (\overline{IN_1} + \overline{IN_2}) \cdot IN_3 \cdot IN_4 \qquad (10.25)$$

This feature permits both active HIGH and active LOW inputs.

An example of an active HIGH input triggering the multivibrator is shown in Fig. 10.13c, where the other inputs are such that IN is identical to IN_4. The Q output follows the positive-going (leading) edge of IN_4.

An example of an active LOW input triggering is shown in Fig. 10.13d, where all the other inputs are set HIGH. Thus initially if IN_1 has been HIGH long enough for Q to return to LOW, then after the negative-going (trailing) edge of IN_1, Q will go HIGH and stay HIGH for a time period t_{PW}.

If we use a retriggerable multivibrator like the TTL-type 9601, and if IN consists of two pulses whose leading edges are separated by a time T_1 as in Fig. 10.13e, we obtain the solid line for the output Q. For this to occur, with a TTL-type 9601, T_1 must satisfy the inequality

$$0.3 C_X < T_1 < t_{PW} \qquad (10.26)$$

here C_X is in pF, T_1 is in ns, and t_{PW} is as calculated by Eq. (10.23).

The time t_W that Q is HIGH after the leading edge of the second pulse is labeled in Fig. 10.11e and is

$$t_W = t_{pd+} + t_{PW} \qquad (10.27)$$

If the second pulse did not exist (as shown dotted in the IN waveform) in Fig. 10.13e or if T_1 were too small for inequality (10.26), the Q output would be the dotted line in Fig. 10.13e. This dotted waveform would also exist for a nonretriggerable monostable multivibrator, like the TTL-type 74121, even if T_1 was less than t_{PW}. The retriggerable feature of the 9601 can be inhibited by connecting $\bar{Q}$ to IN_1 or IN_2. The retriggerable type (9601) can produce an output with a duty cycle of 100 percent (that is, Q HIGH continuously), while the nonretriggerable types (74121) are limited to a 90 percent duty cycle.

The potentially 100 percent duty-cycle property is made use of in the clock-pulse generator of Fig. 10.14a. Here a 9601-type retriggerable multivibrator is connected with an inverter between the Q output and IN_4.

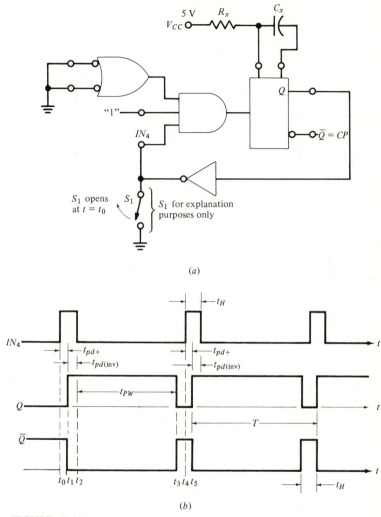

(a)

(b)

FIGURE 10.14
Making a pulse generator from a retriggerable multivibrator. (*a*) Schematic of the pulse generator; (*b*) waveforms at points in the pulse generator.

To simplify the explanation of how the circuit of Fig. 10.14*a* operates, let us assume that a switch S_1 exists that, up to the time $t = t_0$, shorts the output of the inverter to ground. This sets Q LOW. If now we open switch S_1 at $t = t_0$ then, as seen in Fig. 10.14*b*, IN_4 goes HIGH immediately and, after the delay time t_{pd+}, the output Q goes HIGH at $t = t_1$. The IN_4 input then goes LOW after a delay time $t_{pd(inv)}$ which is at $t = t_2$. The Q output stays HIGH during the time t_{PW} (set by R_X and C_X), then drops to LOW at $t = t_3$. After the inverter propagation delay $t_{pd(inv)}$, IN_4 goes back to HIGH, which starts the next pulse on Q at $t = t_5$. This keeps repeating ad infinitum. The switch S_1 is not really necessary for operation since the oscillator will start up when the power supply is turned on.

For a digital system to be least susceptible to noise, we want to make the time of the clock pulse HIGH only as long as required by the switching times of the other components. Thus we could use $\bar{Q}$ as the clock pulse CP. The width of the pulse at $\bar{Q}$ is t_H, which is given by

$$t_H = t_5 - t_3 = t_{pd(inv.)} + t_{pd+} \qquad (10.28)$$

For a TTL inverter (7404-type), $t_{pd(inv)} = 10$ ns, and for the retriggerable multivibrator 9601, $t_{pd+} = 25$ ns. Thus t_H from Eq. (10.28) would typically be 35 ns, which is wide enough for most TTL systems. If we desired a wider pulse we could simply use a larger odd number of inverters between Q and IN_4 in Fig. 10.14a.

For our DVM we required from Eq. (10.18a) that the period of the oscillator was 41.67 μs. From the manufacturer's data-sheet specification we are limited to have R_X greater than 5 kΩ but less than 50 kΩ.

A convenient choice of the capacitor value C_X is a 5000-pF $\pm$ 10 percent value, which allows us to use Eq. (10.23) to determine R_X. Substituting C_X into Eq. (10.23) and solving gives us a nominal value of 25.3 kΩ for R_X. Since the relationship of Eq. (10.23) is only good to about 15 percent accuracy and C_X has a 10 percent tolerance, we should make R_X up from a series combination of a 15-kΩ, 5 percent fixed resistor and a 20 percent tolerance variable resistor (cermet pot) of 20 kΩ. The "pot" would be adjusted in the DVM so that desired value of NT (that is, 16.67 ms) is obtained.

10.1.10 DVM Review

This completes the design of all the blocks required for our simplified DVM. A complete schematic for the whole DVM is shown in Fig. 10.15. The oscillator circuit is slightly different. We feed back to the OR input, and we have added more inverters so as to increase the pulse width. The three additional inverters ensure that CP' is long enough. (See REF flip-flop design.) Also we can now see the clock pulses with a moderate performance oscilloscope.

We also added a voltage follower ahead of the ramp generator. Thus we obtain an input resistance to the DVM that is much higher than the 3.33 kΩ of the ramp generator.

The V_{REF} from which I_{REF} is to be based on can be the regulated 5-V supply used for the TTL logic. An IC power-supply regulator† can be used for this function. The current source I_{REF} is made by using the resistive feedback circuit of Prob. P9.24. Finally we must adjust the input resistor R_1 (the "pot" part is R_{1V}) to calibrate our DVM. We do this by applying a precise dc voltage of a value under 1.99 V, for example, 1.98 V. We should meter this voltage at the time we apply it to our DVM with another instrument whose accuracy is 0.1 percent or better.

---

† National LM309K, Fairchild 7805 or equivalent.

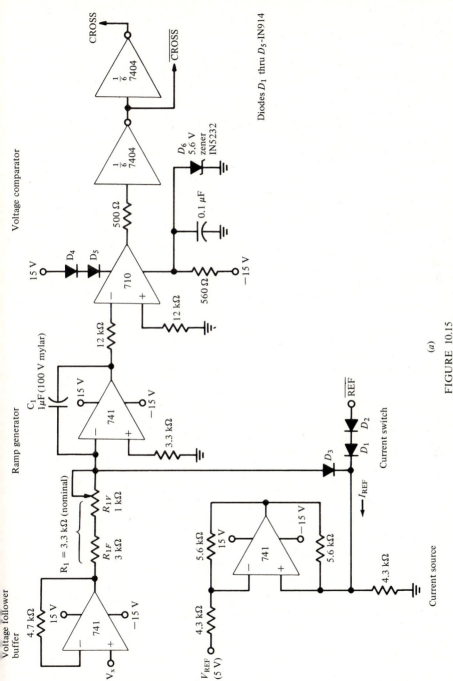

FIGURE 10.15

Complete detailed circuit diagram of the simplified dual-slope DVM. (a) Analog portion of the DVM; (b) oscillator, counter, and display portion of the DVM; (c) control portion of the DVM; (d) strobe portion of the DVM.

(a)

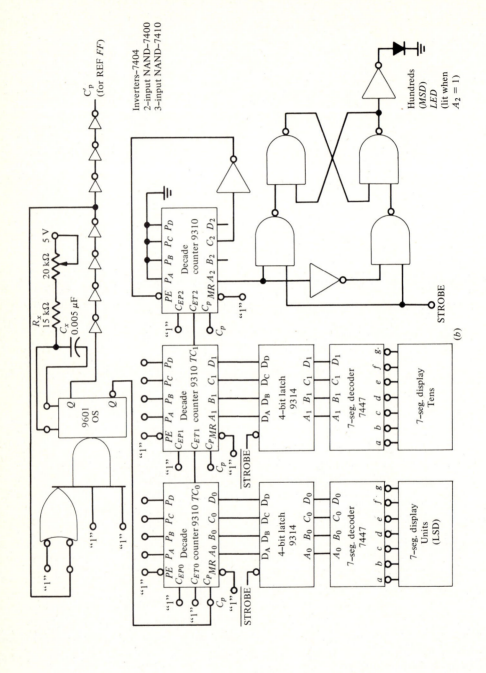

(b)

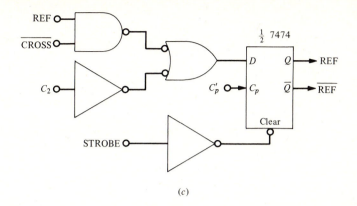

(c)

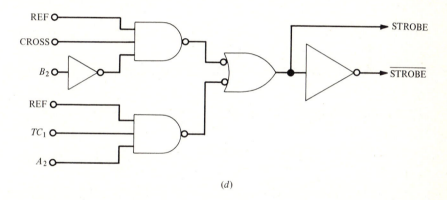

(d)

The DVM that we have designed meets all our original goals. Furthermore this DVM can be the basis of a version with improved performance by just adding more hardware.

As one example we could add more digits in the counter in order to increase the resolution. The errors due to offsets and noise in the analog portion set a limit on how far we can reasonably go in this direction.

Other performance features could be added that would require major changes. For instance our DVM can only measure positive voltages in the range of 0 to 1.99 V. By adding more electronic switching and op amps, we would get automatic polarity, i.e., a reading of the absolute value of the voltage irrespective of polarity, and a polarity indicator.

For voltages greater than the overload value applied to our DVM, we would have to use a resistive attenuator in order to get an "on scale" reading. DVMs are available that automatically switch attenuators and display the result in either a decimal-point change, or the units of the display are indicated.

A multitude of other functions are available in commercial DVMs. Many of these can be used to measure both dc and ac voltages and currents as well as dc resistance; hence they are usually called digital *multimeters*. These instruments have differential inputs and are often *guarded* so that errors caused by common-mode signals are reduced.

Meters intended for use in digital systems have provisions for electronically changing the meter function by external logic signals. These *command* signals plus the meters' BCD outputs are often *optically* coupled to the digital system for the better accuracy given by this electrical *isolation*.

For DVMs that must operate at very high speeds, other techniques are more advantageous than the dual-slope method. Among these techniques are those based on D-A conversion. We will limit our discussion of DVMs to dual-slope types but since the D-A converter is an important building block in many systems, we will discuss one method of D-A conversion in the next section.

10.2   DIGITAL-TO-ANALOG CONVERSION

A common requirement in many digital systems is the need to generate an analog signal that is directly proportional to a digitally coded variable in the system. As an example we may wish to transmit in digital form (possibly via a radio link) the contents of the latch in the DVM of the previous section. At the receiving end we want to convert the digital value into an analog voltage to be used as an oscilloscope input. We want to transmit in digital rather than analog form, since system accuracy is then much less susceptible to being degraded by the noise and nonlinearities in the communication link.

For simplicity let us assume we were only interested in connecting one 4-bit digital word (that is, in 8421 BCD code) into an analog voltage v_o. We want $v_{o(\text{min})}$ to be 0 V when the digital input is 0000_2 and $v_{o(\text{max})}$ to be 9 V with the input as 1001_2. Table 10.1 shows the desired relationship of the converter output v_o as a

Table 10.1   D-A CONVERSION TABLE

n (state no.)	D	C	B	A	v_o (V)
0	0	0	0	0	0
1	0	0	0	1	1
2	0	0	1	0	2
3	0	0	1	1	3
4	0	1	0	0	4
5	0	1	0	1	5
6	0	1	1	0	6
7	0	1	1	1	7
8	1	0	0	0	8
9	1	0	0	1	9

function of the digital inputs D, C, B, and A. Note that in Table 10.1 for the range of voltages given and with the total number of states as N that the step size Δv_o is

$$\Delta v_o = \frac{v_{o(max)} - v_{o(min)}}{N - 1} \qquad (10.29)$$

10.2.1   A Simple Weighted-resistor D-A Converter

One simple way to generate Table 10.1 is by means of the *weighted-resistor* circuit of Fig. 10.16a. Such circuits are called digital-to-analog converters, which is abbreviated to D-A converter or further abbreviated to DAC. The four resistors are switched by the logic signals A, B, C, and D. We assume that whenever one of these logic variables is 1 the associated switch is closed with zero voltage drop across it.

With an ideal op amp for A_1, the output voltage v_o is $-v_{Rf}$. Since v_{Rf} is the product of the feedback resistor value and the current flowing into the summing point, we have

$$v_o = -R_f(i_D + i_C + i_B + i_A) \qquad (10.30a)$$

With the weights given to the input resistors in Fig. 10.16a, we can express Eq. (10.30a) by using the Boolean variables as algebraic variables that take on the numeric values 0 or 1. This gives

$$v_o = -R_f\left(\frac{DV_{REF}}{R_D} + \frac{CV_{REF}}{2R_D} + \frac{BV_{REF}}{4R_D} + \frac{AV_{REF}}{8R_D}\right) \qquad (10.30b)$$

This simplifies to

$$v_o = \frac{-R_f V_{REF}}{8R_D}(8D + 4C + 2B + 1A) \qquad (10.30c)$$

which is the desired function relationship of Table 10.1.

For V_{REF} a positive value and the input logic variables derived from a decade up-counter, we obtain the negative-going staircase shown in Fig. 10.16b.

EXAMPLE 10.3   Let us determine the value of R_f/R_D so that for $V_{REF} = 5$ V we will obtain the 1-V-per-step output shown in Fig. 10.14b. The voltage step can be determined from Eq. (10.30c) or

$$\Delta v_o = \frac{R_f V_{REF}}{8R_D} \qquad (10.30d)$$

Hence

$$\frac{R_f}{R_D} = \frac{8\Delta v_o}{V_{REF}} \qquad (10.30e)$$

$$= \frac{(8)(1)}{5} = 1.6 \qquad ////$$

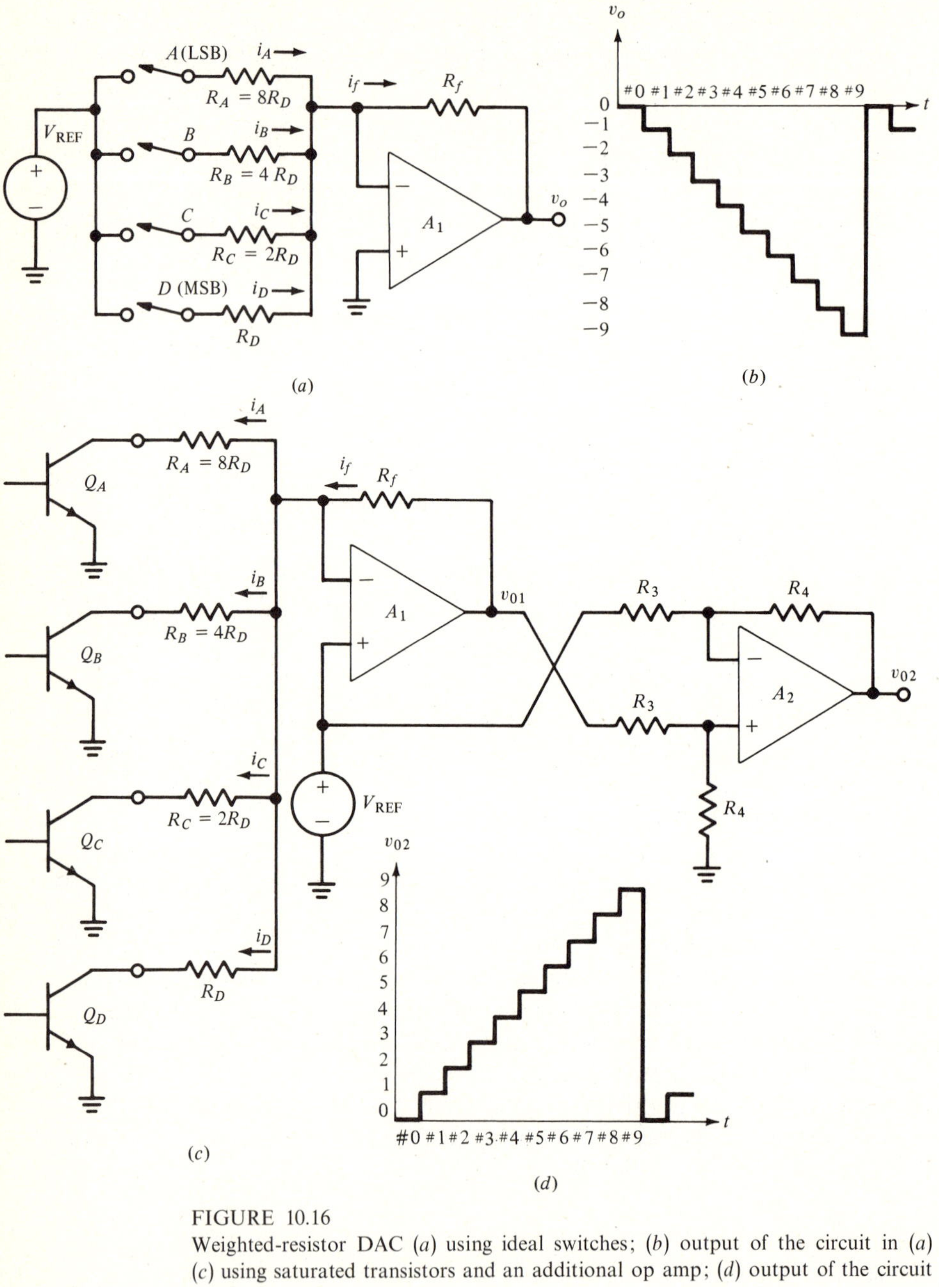

(a)

(b)

(c)

(d)

FIGURE 10.16
Weighted-resistor DAC (a) using ideal switches; (b) output of the circuit in (a)
(c) using saturated transistors and an additional op amp; (d) output of the circuit
in (c).

10.2.2 Transistor-switched Weighted-resistor DAC

The circuit in Fig. 10.16a has the disadvantage that the switches are "floating"; i.e., neither terminal of the switch is grounded. In order to circumvent this, the circuit in Fig. 10.16c is used. Grounded-emitter transistors Q_A through Q_D can now be used as switches, as shown. Transistor Q_D is saturated if $D = 1$; if $D = 0$ it is cut off. Similar conditions hold for Q_C, Q_B, and Q_A. Thus these transistors can be the output transistors of an open-collector hex inverter (TTL-type 7405) where the corresponding inverter inputs are now D, C, B, and A.

Let us for now assume that the $V_{CE(sat)}$ at the inverter output is much smaller than V_{REF}. The voltage at the summing point of an op amp A_1 in Fig. 10.16c is V_{REF}. The current i_f can be determined by the same method as Eq. (10.30a) to be

$$i_f = \frac{V_{REF}}{8R_D} (8D + 4C + 2B + 1A) \quad (10.31a)$$

By KVL the output of A_1 is given by

$$v_{o1} = V_{REF} + \frac{V_{REF} R_f}{8R_D} (8D + 4C + 2B + 1A) \quad (10.31b)$$

To remove the V_{REF} term in Eq. (10.31b), we use the difference amplifier that contains op amp A_2. At its output we obtain our desired result

$$v_{o2} = \left(\frac{R_4}{R_3}\right) \frac{V_{REF} R_f}{8R_D} (8D + 4C + 2B + 1A) \quad (10.31c)$$

With V_{REF} positive we obtain a positive-going staircase, as shown in Fig. 10.14d.

For more than one BCD input, the circuit shown in Fig. P10.22 can be used. For large binary words the method of Fig. 10.16c soon becomes impractical, and the R-2R method discussed in Prob. P10.23 is used.

As a general rule in monolithic ICs we find it easier to switch a precision current source than to switch a voltage source with the same accuracy. (The circuit of Fig. 10.16c, in effect, switches a voltage source.) Current-switching techniques are used in commercial monolithic 10-bit DACs.† Included in these ICs is an accurate current-summing circuit with 10 current sources controlled by digital inputs that are TTL compatible. The exact values of the current in the current sources are set by precision resistors, which may be monolithic diffused, or thin films evaporated on the top surface of the silicon dioxide layer of the IC chip. The voltage reference source is also included in the IC chip. These 10-bit DACs have a linearity of ± 0.05 percent with a 10-V output.

EXERCISE

E10.4 For the circuit in Fig. 10.16c determine the values of R_f/R_D and R_4/R_3 if we drive the inputs from a hexadecimal counter (modulo 16) and want $v_{o2(min)} = 0$ V and $v_{o2(max)} = 7.5$ V. Use $V_{REF} = 5$ V and make the peak value of v_{o1} equal to 10 V.

† For example, see Precision Monolithic monoDAC-02AC, or Analog Devices AD560.

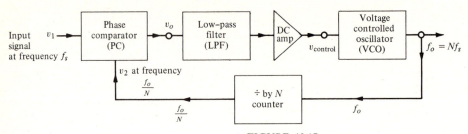

FIGURE 10.17
Block diagram of a digital frequency multiplier.

10.3 PHASE-LOCKED LOOPS

As a final subsystem example involving linear and digital ICs, we will show how to design an oscillator which oscillates at a frequency f_0, where f_0 is an exact integer multiple N of the frequency f_s of an input signal; that is, $f_0 = Nf_s$. When an oscillator is an exact multiple of an input signal, we say that it is *locked* on to the input signal. Locked oscillators are useful in many communication and instrumentation systems.

For example, an aircraft radio transmitter/receiver system may be required to have dozens of different frequencies that can be used as channels. By law these frequencies must be established with a precision of 1 part in 10^4 to 1 part in 10^5. One way of achieving this stability is to use a *quartz crystal* in an oscillator circuit. However cost, size, and weight considerations make it impractical to use a multitude of crystals. By using a digital frequency multiplier, as shown in Fig. 10.17, where the value of N (the modulo of the counter) can be easily changed, it is possible to obtain the desired number of frequencies by using one crystal oscillator.

The circuit in Fig. 10.17 includes some new "components" that we will discuss next. These are the phase comparator (PC) and the voltage-controlled oscillator (VCO). Since we use these components in a feedback *loop*, as in Fig. 10.17, and the feedback is obtained by the comparison of the phases of two signals v_1 and v_2, the whole arrangement is called a phase-locked loop (PLL).

10.3.1 The Phase Comparator†

A common method of making a phase comparator (PC) is to use an electronic switch circuit, as shown in Fig. 10.18a. Here a sinusoidal voltage v_1, as shown in Fig. 10.18b, is applied across resistor R only when switch S_2 is closed. Switch S_2 is closed when a signal v_2 is positive. With the frequency of v_2 equal to that of v_1, switch S_2 is closed and opened in synchronism with v_1. However, as shown in Fig. 10.18c, the time of switch closure and the time v_1 is positive can be different

---

† Other names for this circuit are *phase detector* and *synchronous detector*.

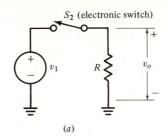

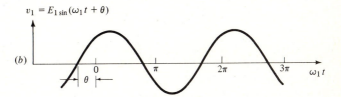

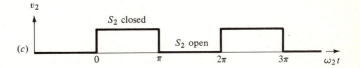

FIGURE 10.18
A switch-type PC and its waveforms. (a) A driven switch circuit; (b) the input voltage waveform; (c) the voltage v_2 controls the closure times of switch S_2 as shown; (d) the instantaneous output voltage of the PC.

because of the relative phase angle θ. With a general phase angle θ the instantaneous output waveform is as shown in Fig. 10.18d. The average value, V_o, of the instantaneous output voltage v_o is

$$V_o = \frac{1}{2\pi} \int_0^\pi E_1 \sin (\omega t + \theta) \, d(\omega t) \qquad (10.32a)$$

$$V_o = \frac{E_1}{2\pi} [-\cos (\omega t + \theta)]_0^\pi = \frac{E_1}{2\pi} [-\cos (\pi + \theta) + \cos \theta] \qquad (10.32b)$$

$$V_o = \frac{E_1}{\pi} \cos \theta \qquad (10.32c)$$

Normally a low-pass filter is located after the PC, so that the high-frequency components are attenuated and the dc component [Eq. (10.32c)] is the dominating portion of the voltage at the filter output. Thus the filtered output of the PC is directly proportional to the cosine of the phase difference between v_1 and v_2. If v_1 was a square wave, a $\cos\theta$ relationship would still hold. Hence as given by Eq. (10.32c) the output of the phase detector is zero when $\theta = 90°$, has its maximum positive value when $\theta = 0°$, and has its minimum value when $\theta = 180°$.

If the frequencies ω_1 and ω_2 are not equal (say, $\omega_2 > \omega_1$), the dc output of the PC would be zero. However, if the low-pass filter in Fig. 10.17 does pass signals at the frequency $\omega_2 - \omega_1$, we will find that the filtered output consists of a sine wave at the *difference* frequency, $\omega_2 - \omega_1$. This *beat* frequency component can still be effective as the control signal on the VCO used in a PLL.

Most IC PCs are made with a differential-amplifier configuration with the signal v_1 applied between the bases of the emitter-coupled pairs (see Fig. P10.28b). The input from the VCO is usually single-ended, and for most IC versions this input should be a symmetrical square wave.

10.3.2 A Voltage-controlled Oscillator

For our purposes, a voltage-controlled oscillator (VCO) should produce at its output a periodic signal (square wave, triangular, or sinusoidal), whose angular frequency ω_0 is linearly related to the control voltage v_c by

$$\omega_0 = \omega_{FR} + K_0 v_c \qquad (10.33)$$

where ω_{FR} is the free-running angular frequency that the VCO has when the control voltage is zero. The constant K_0 is called the VCO sensitivity and has the units of radians per second per volt.

A circuit that produces the relation of Eq. (10.33) is shown in Fig. 10.19a. The circuit functions somewhat like the ramp generator in a dual-slope DVM in that C_X is charged by the current I_A and then discharged for an equal interval of time by the current $-I_A$. In this circuit, a Schmitt trigger (Prob. P5.14) is used to sense when the ramp voltage crosses either one of two boundary values and then switches the current so that the ramp voltage stays within these limits. The transfer characteristic of a Schmitt trigger is shown in Fig. 10.19b.

We began our detailed study of the circuit of Fig. 10.19a by assuming that, at time $t = t_0$, v_X is just slightly above V_{T-} and also that the Schmitt output is LOW, which means Q_3 is cut off. Thus by the extension of KCL, I_A has nowhere to go but into C_X. We assume that the input current of the Schmitt is negligible. Thus v_X increases linearly with a slope I_A/C_X, as shown in Fig. 10.19c. At time $t = t_1$, v_X reaches V_{T+}, the upper trigger level of the Schmitt, and as seen in Fig. 10.17b the output of the Schmitt goes to HIGH, turning Q_3 ON. This allows Q_1 and Q_2 to be turned ON so that the voltage at point Y becomes

$$V_{CE(\text{sat})} + 2V_{D(\text{on})} \qquad (10.34a)$$

As a result of this, for the values given in Fig. 10.19b, D_2 is cut off (see

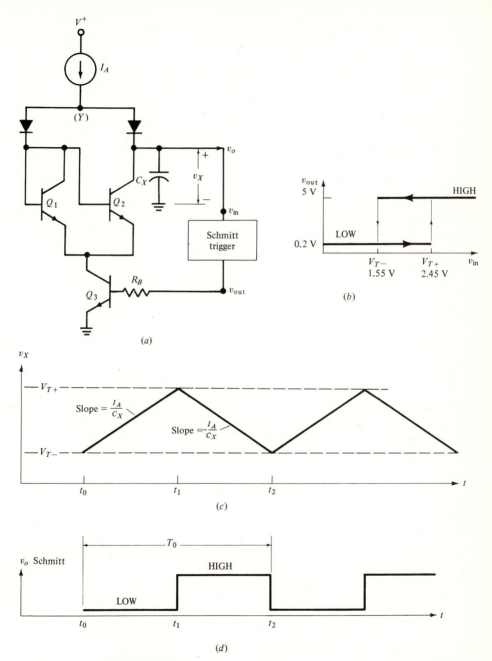

FIGURE 10.19
Voltage-controlled oscillator circuit and waveforms. (a) A form of VCO circuit;
(b) voltage transfer characteristic for the Schmitt trigger; (c) the sawtooth-voltage
waveform is generated across the capacitor; (d) the Schmitt-trigger output voltage.

Prob. P10.29). Assuming β_F of Q_1 and Q_2 to be very large, we can by the same reasoning as applied to current sources in Sec. 8.6.2 show that now

$$i_{C2} \doteq I_A \qquad (10.34b)$$

Hence the capacitor begins to discharge, and its voltage v_X decreases with a slope $-I_A/C_X$, as shown in Fig. 10.19c. The discharging continues until v_X reaches V_{T-}, the lower trigger level. At this time, $t = t_2$, the Schmitt output goes LOW, Q_3 is cut off, and a new charging interval of the next cycle begins.

The period of oscillation T_0 is given by

$$T_0 = t_2 - t_0 = (t_2 - t_1) + (t_1 - t_0) = 2(t_1 - t_0) \qquad (10.35a)$$

The increment of charge put into C_X during the time t_0 to t_1 is

$$\Delta Q = I_A(t_1 - t_0) = C_X \, \Delta v_X \qquad (10.35b)$$

where

$$\Delta v_X = V_{T+} - V_{T-} \qquad (10.35c)$$

Hence

$$t_1 - t_0 = \frac{C_X(V_{T+} - V_{T-})}{I_A} \qquad (10.35d)$$

and

$$T_0 = \frac{2C_X}{I_A}(V_{T+} - V_{T-}) \qquad (10.35e)$$

Since the frequency f_0 is $1/T_0$, we have

$$f_0 = \frac{I_A}{2C_X(V_{T+} - V_{T-})} \qquad (10.35f)$$

For a VCO, the current I_A is made up of two components. The first of these is a dc value I_{A0} that is determined by an external resistor R_X connected between one terminal of the VCO and the power supply V_{CC}. The second component of I_A is to be directly proportional to the control voltage v_c that is applied to change the frequency of the VCO. Expressing this algebraically, we have

$$I_A = I_{A0} + Kv_c \qquad (10.36a)$$

and using

$$\omega_0 = 2\pi f_0 \qquad (10.36b)$$

we have from Eq. (10.35f)

$$\omega_0 = \frac{\pi I_{A0}}{C_X(V_{T+} - V_{T-})} + \frac{K\pi v_c}{C_X(V_{T+} - V_{T-})} \qquad (10.36c)$$

This is the form that we wanted as in Eq. (10.33).

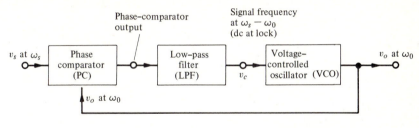

FIGURE 10.20
A simple phase-locked loop.

The constant term in Eq. (10.36c) is the free-running frequency ω_{FR}. We select the external resistor value and the value of C_X so that it is in the center of the range of frequencies that the VCO is to cover. For one commercially available PLL (Signetics NE565) the manufacturer specifies the following relationship between R_X, C_X, and f_{FR}:

$$f_{FR} = \frac{1}{2.7 R_X C_X} \qquad (10.37)$$

This holds for $C_X > 1000$ pF and R_X ranging from 2 to 20 kΩ. This relationship does not hold for small values of C_X, and we find that for very low values of C_X (2.7 pF) and $R_X = 7.5$ kΩ that f_{FR} goes up to about 0.5 MHz. Other types of VCOs in IC form can operate up to 30 MHz.

10.3.3 A Simple Phase-locked Loop

Now that we understand how the individual blocks work, let us put them together in a loop, as shown in Fig. 10.20. Let us further assume that v_s is zero; hence the control voltage is zero and the VCO oscillates at f_{FR}, as shown in Fig. 10.21a. Now we apply a signal v_s of frequency f_s starting at time $t = 0$, as in Fig. 10.21b. We have shown in Fig. 10.21a and b that the initial phase difference θ is 45°. The instantaneous output of the phase comparator is shown in Fig. 10.21c. We further assume that the low-pass filter allows any difference frequency to pass.

The result is that the control voltage drives† the VCO so that the beat frequency is reduced. If f_s is sufficiently close to f_{FR}, the VCO will be "captured" by the input signal so that the VCO will *lock* onto the input signal. This capture process is shown here for the case where the frequencies are equal but the phase difference is not 90°. Hence the VCO speeds up for a few cycles in order to get the 90° phase difference. The VCO output with the loop closed is shown in Fig. 10.21d, where the ×'s are the points for 90° phase difference to v_s. The error is asymptotically approaching zero.

† The explanation of this requires analysis of nonlinear effects, which is outlined in the paper by Grebene in the references at the end of this chapter.

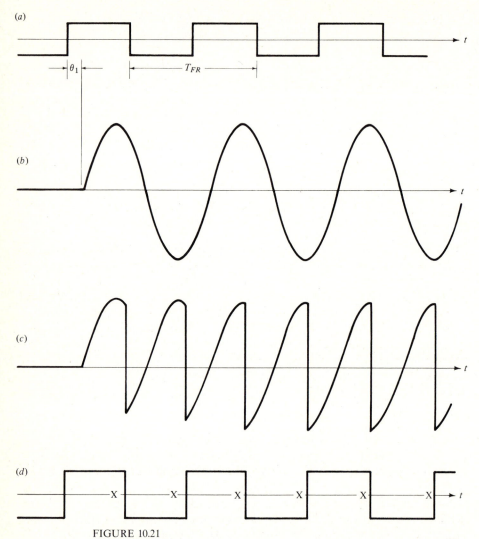

FIGURE 10.21
PLL waveforms for input frequency equal to VCO free-running frequency. (a) VCO output with loop open; (b) v_s input signal to PLL; (c) PC output; (d) VCO output with loop closed.

Once the VCO is locked, the filtered PC output is a dc signal that is just of the correct amount to move the VCO frequency from f_{FR} to f_s. If we then slowly change the input frequency f_s, the VCO will track f_s until the *lock range* is exceeded, at which point the VCO will again free run.

EXAMPLE 10.4 The VCO sensitivity of a PLL $(K_0/2\pi)$ is given as 6600 Hz/V when operating near 10 kHz. If the free-running frequency of the VCO of

Fig. 10.20 is 10.66 kHz and that of v_s is 10 kHz, what must the value of v_c be at lock?

$$f_0 = f_{FR} + \frac{K_0}{2\pi} v_c$$

$$10,000 = 10,660 + (6,600) v_c$$

$$v_c = -\frac{660}{6600} = -0.1 \text{ V} \qquad ////$$

EXERCISE

E10.5 Repeat Example 10.4 to determine the dc control voltage at lock, but use $f_s = 100$ kHz, $K_0/2\pi = 66,000$ Hz/V, and $f_{FR} = 133$ kHz.

In looking at the results obtained from the circuit of Fig. 10.20, one may well question why we go to all this trouble since it appears the same results could be obtained from an amplifier. An example of the usefulness of PPLs is found in practically every TV receiver, where the horizontal and vertical sweep generators have their input signals generated by two PLLs. Here the first benefit of using PLLs is that when no signal is received the picture tube is still scanned since the VCOs are free running. Moreover if the synchronizing signals should be temporarily disrupted by an interfering noise pulse, the VCO output will not be significantly affected due to the smoothing action of the low-pass filter.

10.3.4 A Digital Frequency Multiplier

We now return to the end product of all this discussion, the digital-frequency-multiplier circuit as shown in block-diagram form in Fig. 10.17. We obtain this circuit by simply placing a $\div N$ counter in the feedback path between the VCO output and the lower input to the PC. For example, in Fig. 10.17, f_s is established precisely at 100 kHz by a quartz crystal oscillator. Our problem is to get a 300-kHz signal out of the VCO that will have the stability of the quartz crystal oscillator. To do this we make the VCO free run at about 300 kHz and place a $\div 3$ ($N = 3$) counter in the feedback path. The terminal-count output of the counter, v_2, is near 100 kHz, and this is compared to the 100-kHz signal, v_1. If the VCO frequency is not at 300 kHz, an error signal is generated that brings the VCO to 300 kHz. As in Example 10.4, the magnitude of the dc control signal depends on how far the VCO free-running frequency is from the desired output and also the value of the VCO sensitivity.

EXERCISE

E10.6 With $f_s = 100$ kHz, $K_0/2\pi = 2$ MHz, $f_{FR} = 5$ MHz and $N = 100$ in Fig. 10.17, what is the dc value of the control voltage at lock?

10.4 SUMMARY

\#   Digital and linear ICs are both used in most systems. To utilize them we need (1) to use specialized ICs that interface between the digital and analog circuits or (2) to design special circuits by using a mixture of discrete components and ICs that do this job.

\#   A DVM was designed by using as many ICs as feasible. This design made use of all the major topics studied in the previous chapters. It was also useful in presenting and solving an important problem that exists in any complex system, that is, the control and timing function.

\#   An important building block, the digital-to-analog converter (DAC), was introduced and a simple circuit analyzed.

\#   Phase-locked loops (PLLs) and their internal blocks were analyzed. These blocks and the PLLs constructed from them play an important role in many communication and control systems. We looked into one of these, the digital frequency multiplier.

DEMONSTRATIONS

D10.1   Ramp generator   Construct a ramp generator by using the circuit of Fig. D10.1a. The counter arrangement assures that the duty cycle (τ/T) will be constantly equal to 0.1 irrespective of the frequency of the oscillator. Capacitor C_S blocks the dc voltage so that, as shown in Fig. D10.1b, when v_s is positive, area 1 is equal to area 2 (where v_s is negative). Hence the output ramp waveform is as shown in Fig. 10.1c. Calculate, and then with an oscilloscope, measure the peak-to-peak amplitude of v_o when f_{osc} is 10 kHz. Reduce f_{osc} until the ramp is affected as seen on the CRT. What sets this lower frequency limit?

D10.2   Digital voltmeter   Construct the DVM shown in the schematic of Fig. 10.15a to d. With a known 1.5 V for V_x, adjust the variable part of R_1 so that the display reads 150. Look at the ramp waveform and increase V_x until it is greater than 1.99 V. What happens? Explain why. If no oscilloscope is available, make C_x and C_1 1000 times the values used in Fig. 10.15 so that the ramp period becomes 16 s. Now a regular d'Arsonval voltmeter can be placed across v_o so that we can see v_o as a function of time without an oscilloscope.

D10.3   DAC staircase generator   Construct the circuit in Fig. 10.16c by using 741s for A_1 and A_2. If using 5 percent resistors, make $R_D = 1$ kΩ, $R_C = 2$ kΩ, $R_B = 3.9$ kΩ, and $R_A = 8.2$ kΩ. For $V^+ = 15$ V and $V^- = -15$ V, use $R_f = 0.82$ kΩ, and $R_3 = R_4 = 5.6$ kΩ. If a regulated 5-V supply is available (from the logic portion), use it for V_{REF}. A decade counter (7490 or 9310) driven from the 60-Hz power line, as in Fig. D6.5a, can be used to generate the input signals so that a nine-step staircase can be generated. Check the linearity of

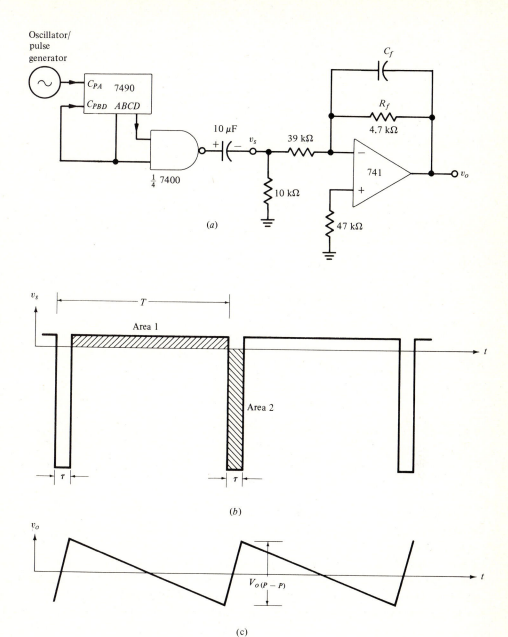

FIGURE D10.1
Ramp generator circuit and waveforms. (*a*) Fixed duty-cycle PC as input to ramp generator; (*b*) waveform used as input to ramp generator; (*c*) output waveform of ramp generator.

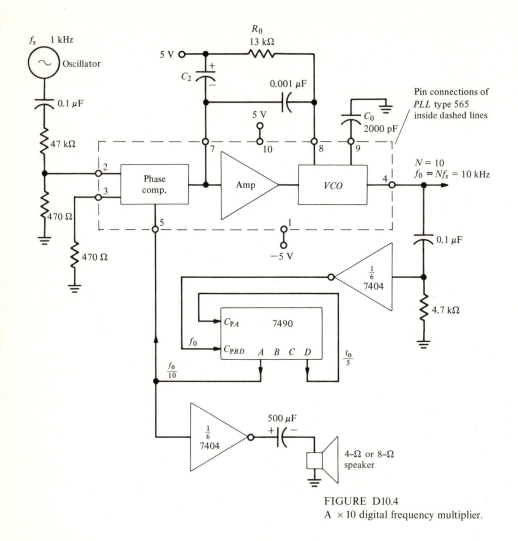

FIGURE D10.4
A $\times 10$ digital frequency multiplier.

your staircase by using an oscilloscope. How could you easily make a staircase going down from the circuit in Fig. 10.16c?

D10.4 PLL frequency multiplier Construct a PLL frequency multiplier by using the circuit of Fig. D10.4. Use a TTL decade counter (7090), where the $\div 5$ is followed by the $\div 2$ so that a square-wave output is obtained for driving the "VCO" input to the PC. Use a 565-type phase-lock loop, setting the free-running frequency at 10 kHz. The filter capacitor C_2 should be set at 5 μF for normal operation, and the input frequency f_s varied to show locking and tracking. A TTL inverter can provide enough drive for a low-impedance loudspeaker so that the effect of capture can be heard as well as seen on an oscilloscope. Change C_2 to 500 μF to show a reduced ability to follow a "moving" f_s.

A CMOS PLL is available† that permits a very wide range of frequencies to be generated by the VCO (from 3 to 999 kHz) without changing the R and C that determine the free-running frequency.

D10.5   Staircase generator for transistor curve tracer   The circuit in Fig. D10.5 can be used as the source of base current for the transistor curve tracer described in Demonstration D1.2. The circuit contains a 4-bit binary counter with feedback so that it is a modulo 11 counter. The weighted-resistor DAC then provides a 10-step staircase by using op amps A_1 and A_2. Op amp A_3 converts the voltage staircase to a current staircase.

By interchanging the left ends of R_4 and R_5, up-going voltage steps can be obtained at the output of A_2 for use in studying n-channel MOSFETs. This also makes the current source suitable for driving p-n-p transistors. The maximum base current of either polarity is 5 mA. Switch S_1 reduces the voltage staircase by a factor of 10 or 100 so that the circuit can be used for lower current applications as well.

REFERENCES

A useful discussion on DVMs and multimeters is given in "Selecting the Right DVM," Application Note 158 from Hewlett-Packard Corporation, Palo Alto, Calif.

For more details on the weighted-resistor DAC, see VANDICK, R. C.: "An Economical D/A or Staircase Generator Using the μA739," Fairchild Semiconductor Application Brief No. 91, February 1969.

A general survey of PLLs and IC versions for them is found in GREBENE, A. B.: The Monolithic Phase-Locked Loop—A Versatile Building Block, *IEEE Spectrum*, vol. 8, no. 3, pp. 38–49, March 1971.

For details on PLLs that have a wide frequency range, see MORGAN, D. C., and G. STEUDEL: The RCA COS/MOS Phase-Locked Loop, A Versatile Building Block for Micro-Power Digital and Analog Applications, RCA Application Note ICAN-6101 in RCA Solid State Data Book SSD-203A (COS/MOS Digital ICs), pp. 319–326, 1973.

PROBLEMS

P10.1   For the DVM circuit in Fig. 10.1, find the ramp generator constant k so the following conditions are met:

$$V_{xMD} = 1.999 \text{ V} \qquad t_1 - t_0 = 80 \text{ ms} \qquad V_P = 8 \text{ V}$$

P10.2   If the ramp generator in Prob. P10.1 is made by using the circuit in Fig. 10.2*a*:
(*a*) Find the product $R_1 C_1$.
(*b*) If $C_1 = 1 \ \mu$F, what value must R_1 be?

P10.3   The circuit in Fig. 10.2*b* initially has S_1 at the GND position and the output is at 0 V. With $V_x = 5$ V and $V_{REF} = 1$ V, the switch is moved to V_x for 2 ms, and then to $-V_{REF}$ for 10 ms. At the end of this time it returns to the GND position.

† RCA CMOS type CD 4046.

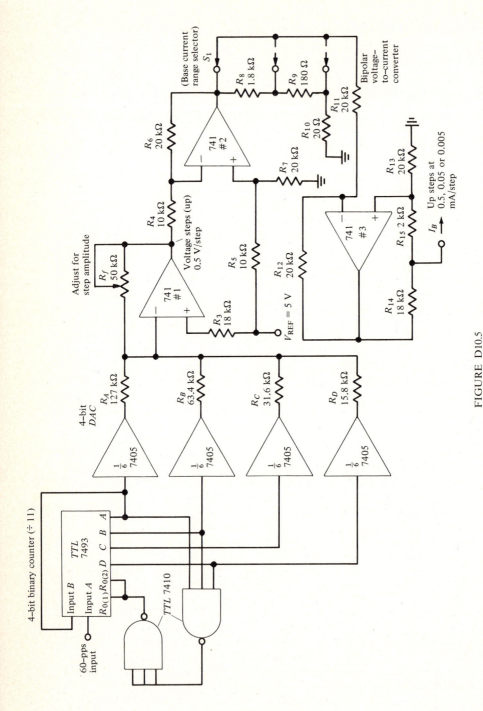

FIGURE D10.5

A 10-step constant current staircase generator using the DAC circuit of Fig. 10.16b. All resistors 1 percent unless noted otherwise. Supply voltage ± 15 V for op amps.

Determine and plot the resulting output voltage for this 12-ms period with $R_1 = 5\ k\Omega$ and $C_1 = 0.5\ \mu F$.

P10.4 (a) If in Prob. P10.3 a resistor $R_f = 400\ k\Omega$ is placed in parallel with C_1 and the input offset current I_{IO} is 0.6 μA, what will the dc voltage be at the output when the switch is at position GND? Assume $V_{IO} = 0$.

(b) If I_{IO} is 0 μA and V_{IO} is 7.5 mV, what would the dc output voltage be?

(c) For $I_{IO} = 0.6$ μA and $V_{IO} = 7.5$ mV, what would the worst-case dc output voltage be?

P10.5 Using the DVM system of Fig. 10.1, if $n_{MD} = 1999$ with $V_{xMD} = 19.99$ V and $V_P = 5$ V:

(a) Find the value of N the modulo of the counter.

(b) If all counters are to be of the decade type how many are required?

(c) If the measurement time $t_2 - t_0$ is to be 5 ms, what is the period of the oscillator?

(d) If the circuit in Fig. 10.2 is used for the ramp generator, what value must $R_1 C_1$ be?

(e) If C_1 is 1 μF, what must R_1 be?

P10.6 For the DVM of Fig. 10.4, where a reading cycle is to be 200 ms and a maximum display value is 1999, find the period and frequency of the oscillator that does this in the simplest fashion.

P10.7 A DVM of the type in Fig. 10.1 is designed with a counter having a modulo of 3000 and a maximum display value of 2999. Design a decoding circuit that can be attached to the counter and control block that will indicate when an overload input voltage is applied.

P10.8 Design a modulo 8 counter by using a 9310-type decade counter of Fig. 10.6a that cycles through state 0, 1, 2, 3, 4, 5, 6, 7.

P10.9 Design a counter that counts modulo 4 through BCD values 6, 7, 8, and 9 by using a 9310 of Fig. 10.6a.

P10.10 Repeat Prob. P10.8, except make a modulo 6 counter starting at 0_{10} and going up to 5_{10}.

P10.11 The TTL-type 9316 is a modulo 16 counter similar in all other respects to the 9310 except that $TC = 1$ when the counter state is 1111. Use this counter to construct a counter that is modulo 8 and counts up from 1000_2 to 1111_2.

P10.12 For the circuit in Fig. P10.12 assume a switch state is 1 when open. Write an expression for the switch states for:

(a) A decade counter;

(b) a hexidecimal counter.

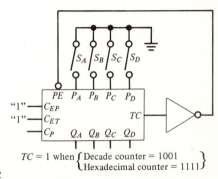

FIGURE P10.12

P10.13 The cascaded up-counter arrangement in Fig. P10.13 has its modulo set by the switches. What is the modulo for the switch setting shown?

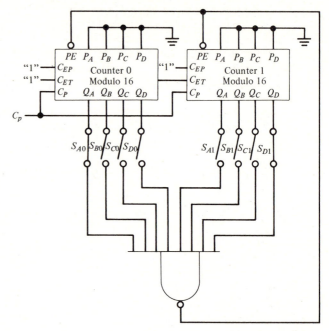

FIGURE P10.13

P10.14 Using only counter 0 in the circuit in Fig. P10.13 and a 4-input NAND gate, set the modulo to be 6. Show the switch position and the waveforms at *A, B, C, D*, and the NAND-gate output.

*P10.15 In a DVM of the type in Fig. 10.5, you want $N = 3000$ and $n_{MD} = 999$. Show the modified and augmented circuit of Fig. 10.15*b, c,* and *d* that will do this.

P10.16 For the diode-gating current switch in Fig. 10.5, show that for all diodes having equal saturation currents I_S and obeying the diode law [Eq. (1.3*a*)] that when the anodes of both D_3 and D_2 are at 0 V that the current I_{D2} is

$$I_{D2} = \sqrt{I_S I_{D3}}$$

*P10.17 In order for a dual-slope DVM to not be affected by power-line voltages, the period of measurement is made equal to an integer multiple of the power-line period. If one desires a DVM that averages to zero both 50- and 60-Hz extraneous input-signal components, what would the shortest reading time be?

P10.18 Using a retriggerable monostable of the type in Fig. 10.13*a*, is it possible to build a pulse generator without any inverters? If so, show the connections and waveforms. Give any disadvantages that this might have compared to the circuits in Figs. 10.14 and 10.15*b*.

P10.19 Use the results of Prob. P10.18 to show how to construct a pulse generator by using two monostable multivibrators, where one *RC* time constant determines the repetition rate and the second the pulse width.

P10.20 Extend the circuit of Fig. 10.16c so that a 6-bit DAC is made by using six binary inputs *A*, *B*, *C*, *D*, *E*, and *F*. Design the circuit so that with V_{REF} of 5 V the maximum inverter current is 10 mA, the maximum value of v_{o1} is 10 V, and the peak value of v_{o2} is 8 V.

P10.21 Design the up-counter for driving the DAC in Prob. P10.20 so as to get a 31-step staircase. Use *D* flip-flops to make the ripple binary counter.

P10.22 Assume one has available two 4-bit DACs, similar to that in Fig. 10.16c, that are to be connected together as two-digit BCD DACs, so that with suitable resistors when the BCD inputs are 1001 the composite output v_{o3} is 9.9 V, etc. Use the circuit in Fig. P10.22, where the outputs v_{o1} and v_{o2} are 10 V when DAC-0 and DAC-1 both have inputs 1001, to find R_1, R_{10}, and R_L. You are to set the resistance seen to ground at the output node at 1 kΩ.

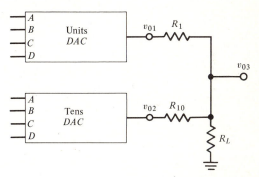

FIGURE P10.22

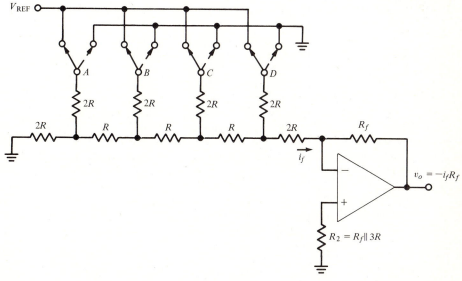

FIGURE P10.23

P10.23 The circuit in Fig. P10.23 is a 4-bit DAC of the R-$2R$ resistor ladder type. Obtain a relationship analogous to that in Eq. (10.30c) for this circuit. (*Hint*: Use superposition to determine the Thévenin equivalent to the left of the rightmost $2R$ resistor in the ladder.)

P10.24 For the circuit in Fig. P10.24 (which is similar to that in Fig. P10.23), determine what v_o is for $DCBA = 1111_2$. If this is full-scale output and R is 50 kΩ, what is the maximum value that the bias current can have for a 0.1 percent error at full scale? Use $V_{\text{REF}} = 5$ V.

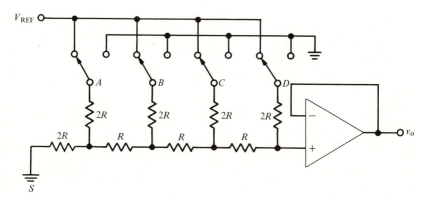

FIGURE P10.24

P10.25 (*a*) Use a hexadecimal counter of the type described in Prob. P10.11 to show how to connect it so that if its outputs drive the inverters in Fig. 10.16c, we would get a 10-step staircase.

(*b*) Cascade two hexadecimal counters with appropriate gating so as to get a 100-step staircase.

P10.26 Two difficulties arise in practice when using open-collector inverters as the switches in the DAC of Fig. 10.16c.

(*a*) The saturated transistors are not perfect switches but for our purposes can be modeled by the circuit in Fig. P10.26. Here $V_{CE(\text{sat, 0})}$ is the saturation voltage under zero current conditions and is about 50 mV. The saturation resistance r_{sat} is about 12 Ω. Assume we originally designed the circuits in Fig. 10.16c so that with perfect switches V_{REF} is 5.00 V and $R_D = 1$ kΩ, $R_C = 2$ kΩ, etc. What should the values of V_{REF} and the four resistors be changed to so that the same currents flow if transistors of the above characteristics were used in place of perfect switches?

(*b*) When an open-collector TTL inverter is cut off and 5 V is applied across the collector to emitter of the output transistor, a small leakage current (typically 5 nA) flows. For worst-case design assume this leakage current I_l becomes 40 μA. In Fig. 10.16c if R_f is 800 Ω and $R_3 = R_4$, what would the typical and worst-case output voltage v_{o2} be if all logic inputs were LOW? What would this be as a percent of full scale if only valid BCD inputs were permitted into the DAC?

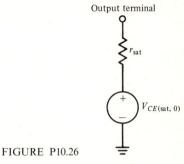

FIGURE P10.26

P10.27   One way to realize the circuit function of Fig. 10.18*a* is to use the ring-modulator circuit in Fig. P10.27*a*. Assume that diodes are all identical and that the bridge-switching inputs Q and $\bar{Q}$, as shown in Fig. P10.27*b* and *c*, are 10-V peak-to-peak levels that are out of phase by 180°.

(*a*) For $R_S = R_L = 100$ kΩ, what is the value of v_0 if $v_s = 0$?

(*b*) For a 0.5-V dc voltage for v_s, what is the waveform of v_o?

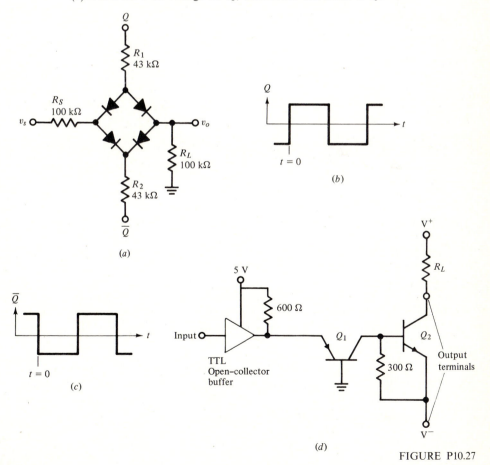

FIGURE P10.27

(c) What is the waveform at v_o if v_s is a sine wave of 0.5-V peak value and is of the same frequency and in phase with Q?

(d) Repeat part (c) if the phase difference is 180°.

(e) Repeat part (c) if the phase difference is 90°.

(f) Using two of the level translator ICs of Fig. P10.27d permits one to shift from TTL levels to obtain the waveforms Q and $\bar{Q}$ in Fig. P10.27b. What values should V^+ and V^- be to get approximately 10-V peak-to-peak levels? If the I_C of Q_2 is to be limited to 50 mA with $V_{CE(\text{sat}, 2)}$ of 0.5 V, what is the minimum value for R_L?

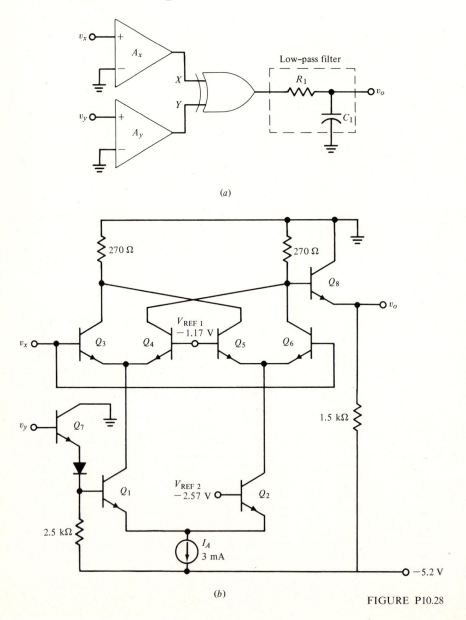

(a)

(b)

FIGURE P10.28

P10.28   (a) The EXCLUSIVE-OR function described in Sec. 4.3.3 is the basis of one type
of PC. The circuit in Fig. P10.28a shows the analog signals v_x and v_y. If these
are not suitable to drive logic gates, this can be easily corrected by adding two
voltage comparators A_x and A_y so that logic signals X and Y are obtained.
The output of the EXCLUSIVE-OR gate has a low-pass filter (R_1 and C_1)
that attenuates the high-frequency components. If v_x and v_y are 180° out of
phase, the output of the EXCLUSIVE-OR gate, v_o, is V_{OH}. Plot the dc output
voltage as a function of the phase difference θ from 0° to 360°.

   *(b) The circuit in Fig. P10.28b is an ECL version for the EXCLUSIVE-OR
function. The two dc reference voltages are generated inside the chip; the input
signals v_x and v_y have been set at ECL levels (-0.75 and -1.66 V). Assume
the current source I_A to be a constant 3 mA and that all transistors
have $V_{BE(on)} = 0.7$ V and $B_F \rightarrow \infty$. Plot the current waveform for transistors
$Q_1, Q_2, \ldots, Q_6$ and the output v_0 for $\theta = 60°$.

P10.29   For the VCO circuit in Fig. 10.19 to operate properly, diode D_2 must not conduct
during the down-ramp time (t_1 to t_2). For the values given for the Schmitt circuit
in Fig. 10.19b, what is the voltage across D_2 when it is closest to being turned ON?
Use $V_{CE(sat)} = 0.2$ V and $V_{BE(on)} = 0.7$ V.

P10.30   (a) Given a VCO whose output radian frequency ω_0 is directly proportional to
the control voltage v_c,

$$\omega_0 = K_D v_c$$

where the constant K_D is set by chip-circuit constants and the external elements R_o and C_o so that

$$K_D = \frac{K_0}{R_o C_o}$$

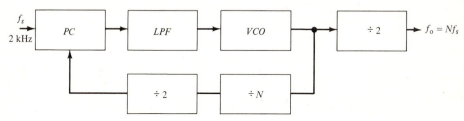

FIGURE P10.30

Assume $K_D = 8.98 \times 10^4$ rad/s/V and that the VCO is driven by a PC of the
type described in Prob. P10.28, which we will now assume has a peak output
voltage of 2.8 V for a 180° phase difference and 0 V for a 0° difference. What
will the dc voltage be on the VCO input if the VCO output is at 10 kHz? If
this was used in a simple PLL, as in Fig. 10.20, what would the phase difference be between the VCO output and the reference signal v_s when the loop is
locked?

   (b) To get the widest control range with this combination of PC and VCO with
the range centered at 10 kHz, what would you change the $R_o C_o$ to in relation
to its value in part (a)?

   (c) Use the results of part (b) to design a frequency multiplier that has N going
from 1 to 9, where f_s is 2 kHz.

(d) For many PCs the input signals have to be of 50 percent duty cycles, i.e., sine or square waves. Thus if we desire to have a × 10 frequency multiplier we use in the feedback loop a decade counter of the type where we first divide by 5 and then divide by 2, as shown in Fig. D10.4. For a general × N multiplier, where the terminal count is not a 50 percent duty-cycle square wave, we can use the technique shown in Fig. P10.30, where two ÷ 2 counters (toggle FF) are added. If we want to repeat part (c) by using this technique, what should the value of $R_o C_o$ be in relation to its value in part (a)?

APPENDIX A

FABRICATION PROCESSES FOR INTEGRATED CIRCUITS

It is not our intention here to give a detailed description of the way ICs are made. Rather we wish to present to the reader sufficient information so that when he uses a packaged IC he will have some concept of what is contained in the package and how it got there.

An IC consists of an arrangement of *components* (principally transistors, diodes, and resistors) contained within a small, thin and *monolithic*† piece of silicon called a *chip*. A single chip may contain thousands of components. The "wiring" together of these components forms the IC. These wiring interconnections take the form of very thin strips of metal (usually aluminum) deposited on the upper surface of the chip. In the fabrication of ICs, the process steps are arranged so that all the components and interconnections are simultaneously made on a wafer. A typical wafer of silicon is 5 cm (2 in.) in diameter and 0.25 mm (0.01 in.) thick. A wafer is cut up *into many chips* after all the processing steps are completed. Since all the processing operates on the upper plane of the wafer, this form of fabrication is known as the *planar* technology.

---

† Monolithic means *single stone* in Greek. Thus, a monolithic IC is one that is contained in one chip.

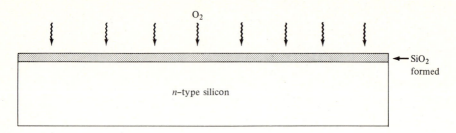

FIGURE A.1
Forming an oxide layer over an *n*-type silicon wafer.

A.1 PLANAR JUNCTION TECHNOLOGY

We begin with a wafer of *n*-type silicon,† as shown in Fig. A.1. The wafer is placed in a high-temperature furnace ($\approx$ 1000°C) with a gaseous oxygen atmosphere. The oxygen and silicon soon interact to form an insulating oxide layer (SiO_2) over the surface of the silicon. The thickness of the oxide layer can be accurately controlled by the time and temperature of the oxidizing cycle. Typically the oxide is about 0.5 μm (0.00002 in.) thick.

Upon treatment of the oxide layer by means of photographic and chemical-etching techniques similar to that used in making PCBs, one selectively removes the oxide in regions where it is desired to add another chemical element that will change the conductivity of the underlying silicon. Figure A.2 shows a cross section through part of a wafer that has had this oxide removed in selected regions and was then placed in a diffusion furnace.

Here a *p*-type impurity is passed over the heated silicon in the gaseous form. After a suitable length of time, the upper portion of the *n* material was converted to *p*-type material. The oxide blocks diffusion at the other areas of the silicon. We are left with pockets of *p*-type material, with the rest of the silicon still being *n*-type. The *p*-type impurity diffuses

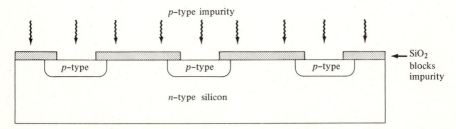

FIGURE A.2
Diffusion of *p*-type impurities proceeds at the areas where the oxide has been removed.

† Silicon is referred to as *n*-type if normal electrical conduction is primarily due to electrons being in the majority (*n*-type since electrons are charged negatively). Conversely, if holes are the majority carriers of current, it is *p*-type (since holes are positively charged).

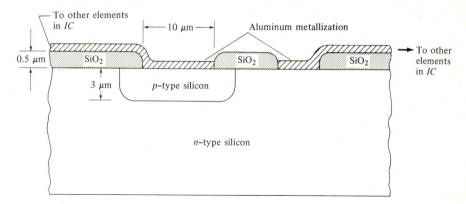

FIGURE A.3
Detailed cross section of a *p-n* junction diode showing the metallic interconnections
to other elements in the IC.

both vertically and laterally under the edge of the oxide, which means that in all cases the
junction (where the *p* changes to *n*) is always underneath the thermally grown oxide.

The oxide acts as an insulating medium over which metallized conducting paths are
placed to act as interconnecting "wire" between the various elements. Figure A.3 shows this
metallizing in cross section. The metallizing makes contact with the *p* and *n* regions, as
shown in the figure, and then travels to other parts of the IC over the silicon dioxide. A
photograph of the top of an LSI chip is shown in Fig. I.6*b*.

A.2   CIRCUIT ELEMENTS

IC circuit elements are created by multiple application of the techniques used to create *p-n*
junctions.

A.2.1   IC Bipolar Transistors

A simplified version of a bipolar† *n-p-n* transistor structure that would be fabricated using
IC technology is shown in Fig. A.4. Here the elements of the transistor are in the upper
portion, as labeled with the *p* (base) region contained completely inside the *n* (collector)
region. The emitter is the *n* + region inside the *p* base. There is a "topside" collector contact
connected to the collector *n* region. In addition, we see a contact to the underlying *p*-type
material, which is called the *isolation region*. In Fig. A.5 we show the conventional circuit
schematic for such a transistor with the elements labeled as in Fig. A.4. In all IC use of
transistors, the isolation diode (which is shown as the diode connecting the collector and
the isolation terminal) is always back-biased and hence has a negligible effect on the circuit.
Thus these diodes are never shown on circuit schematics.

---

† Bipolar junction transistors have also been referred to as junction transistors. Since junction
field-effect transistors also exist, the preferred shortened name of bipolar junction transistor is
bipolar transistor. ICs made using bipolar transistors as the only form of transistor are called
bipolar ICs.

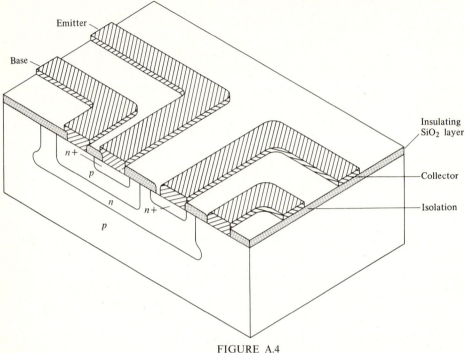

Emitter
Base
Insulating SiO$_2$ layer
$n+$
p
n
$n+$
p
Collector
Isolation

FIGURE A.4
Cross section of an *n-p-n* transistor as made in an IC.

A.2.2 IC Diffused Resistors

Figure A.6 shows a diffused resistor made by utilizing monolithic IC technology, where the two terminals of the resistor are brought out on the top of the silicon dioxide. Again, the underlying junctions must be reverse-biased for normal circuit operation. Diffused resistors are suitable for circuits that can be designed with the sum of all the resistor values being up to 10^6 Ω. To obtain higher resistance elements that occupy less of the expensive silicon "real estate," thin-film resistors are used. These are made from deposited alloys of metals and semiconductors that are placed on top of the silicon dioxide layer shown in Fig. A.7. Again, aluminum metallizing is used to interconnect the resistor to other elements in the circuit.

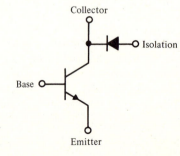

Collector
Isolation
Base
Emitter

FIGURE A.5
Schematic symbol for an *n-p-n* transistor and the associated isolation diode in an IC.

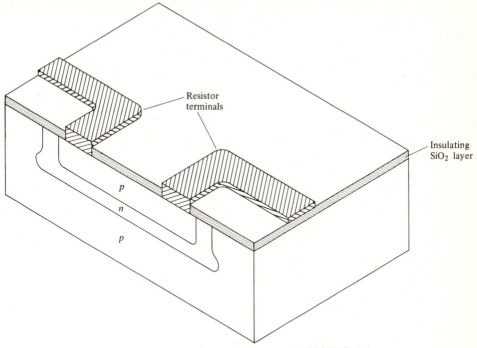

FIGURE A.6
A diffused resistor in an IC structure.

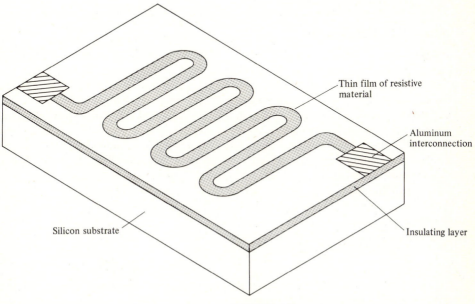

FIGURE A.7
A thin-film resistor over the insulating layer in an IC.

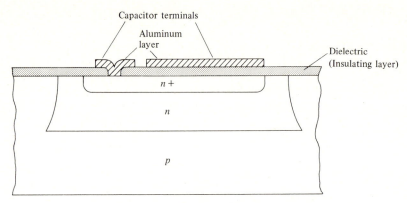

FIGURE A.8
Cross-section view of a MOS capacitor in an IC.

A.2.3 IC Capacitors

Capacitors can be made by using reverse-biased *p-n* junction diodes. However, capacitors made this way have a capacitance that varies with voltage. To eliminate this variation, capacitors are made by using the silicon dioxide layer as a dielectric, as shown in Fig. A.8. Since this structure is a sandwich made with layers of metal, oxide, and semiconductor, the capacitors are called *MOS* capacitors. Here the desired capacitance is between the two external circuit leads and is determined by the area of the metallizing, the thickness, and the dielectric constant of the silicon dioxide. Typically, the capacitance is $160 \, \text{pF/mm}^2$ ($0.1 \, \text{pF/mil}^2$). Hence ICs are usually designed so that the total capacitance is less than 100 pF.

A.2.4 MOS Transistors

Another active element used in ICs is the MOS field-effect transistor† whose cross section is shown in Fig. A.9a. These devices are often concisely referred to as MOSFETs, and circuits made using them are called MOS ICs.

The exact same technology used for making *p-n* junctions is used in forming the *n* regions in the MOSFET. A metallized area identical to that used for the upper plate of the MOS capacitor is placed, as shown in Fig. A.9a, overlapping the two *n* regions. This overlapping structure is called the *gate electrode*. The silicon area underneath it is called the *channel*. If the silicon at the source and drain is *n*-type, as in Fig. A.9a, the MOSFET is called an *n*-channel device. If the source and drain are *p*-type with all the rest of the silicon *n*-type, the MOSFET is called a *p*-channel device.

---

† Another type of field-effect transistor is the junction field-effect transistor (JFET). Junction field-effect devices are advantageously used as components in certain instrumentation and communication applications. In linear ICs, JFETs are used in conjunction with bipolar transistors on the same monolithic chip. We do not discuss JFETs in detail in this text, so that when we refer to a transistor we are always referring to a bipolar (junction) transistor, and when we refer to a field-effect transistor, it is a MOSFET unless noted otherwise.

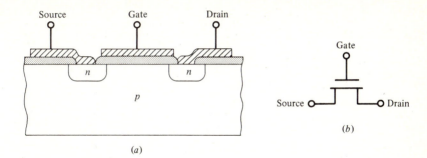

FIGURE A.9
(*a*) Cross-section view of an *n*-channel MOSFET; (*b*) schematic symbol for a MOSFET.

The schematic for this circuit element is shown in Fig. A.9*b*. Because of the simple structure required for the MOSFET, we do not require the isolation diffusion, and hence the area that is required for a given transistor is much smaller than in the bipolar technology. Moreover, a MOSFET by itself can substitute for a diffused resistor that takes up more area. Consequently, it is advantageous to employ MOSFETs in lieu of bipolar transistors in that MOS circuits can be made in one-third to one-fifth the area required for an equivalent bipolar circuit. Moreover, the MOS IC can be made with less process steps. However, the maximum speeds possible with MOSFET circuits are considerably slower than that in bipolar circuits. In particular, for circuits that are required to rapidly drive a large amount of cable and wiring capacity, we find that bipolar circuits give considerably better performance. Hence MOSFET circuits are popular in LSI functions, where high circuit density is possible on a chip with a minimum number of interconnections and maximum circuit speeds are not necessary. A good example of this is a pocket calculator. For very-high-speed circuits that require a large number of external connections (such as the central processor in a scientific computer) with the attendant parasitic wiring capacity, bipolar circuits are advantageous. There are many cases in which both types are adequate, and other factors will enter into deciding which one to choose.

A.3   PACKAGING THE IC CHIP

Figure A.10 shows a technical artist's representation of the most common form of IC package. This is the type shown on the circuit card in Fig. I.6*a*. In Fig. A.10 the bulk of the material is a plastic whose purpose is to protect the inner portion from any adverse atmospheric contaminants, as well as to carry heat away. The plastic also provides mechanical support for the thin external metal *leads* that protrude out of the plastic. These metal leads are inserted into IC sockets if sockets are used. If sockets are not used, then usually these metal leads are soldered into the holes of the PCB.

Inside the package, each metal lead serves as a terminal to which one end of a very fine wire of approximately 12 μm (0.0005 in.) diameter is attached by pressure welding. The other end of the fine wire is also attached to a metallized bonding *pad* on the surface of the silicon *chip* which lies in the center of the package. This silicon chip is soldered to a wider

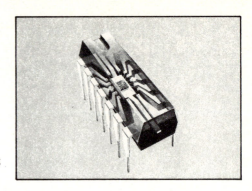

FIGURE A.10
An artist's view of a packaged IC showing
the chip within the package.

version of the metal leads, and this metal lead is used also to spread the heat that is generated in the chip throughout the plastic package.

REFERENCES

For a comprehensive source of information on IC fabrication, see WARNER, R. M., and J. N. FORDEMWALT: (eds.), "Integrated Circuits: Design Principles and Fabrication," chaps. 11–15, McGraw-Hill, New York, 1965.

For a briefer version on IC fabrication, see OLDHAM, W. G., and S. E. SCHWARTZ: "An Introduction to Electronics," chap. 7, Holt, New York, 1972.

In this appendix we will introduce some notation and conventions used in analyzing ICs. We also will illustrate the circuit laws of Kirchhoff. Using these as a basis, we shall prove, with a specific example, the network theorems of Thévenin and Norton. In addition, we will present the superposition theorem, which is useful in linear circuit analysis.

We begin by discussing dc voltage sources. In ICs these can be power supplies or signals that are the output of one IC and the input to another IC.

B.1 VOLTAGE SOURCES: REAL AND IDEAL

Consider a 12-V auto battery, where the symbol for a battery is shown in Fig. B.1a. For many applications this can be considered as a pure or *ideal voltage source*. The symbol for an ideal voltage source is shown in Fig. B.1b. Here the voltage at the terminals (x, x') of the voltage source is a constant V_0 irrespective of the load current I. (The load current is that current supplied by the source voltage.) A VI plot for such an ideal voltage source is shown in Fig. B.1c. Thus a battery which is a real nonideal voltage source (Fig. B.1a) can sometimes be *modeled* by means of an ideal voltage source (the word *ideal* is often dropped) as in Fig. B.1b.

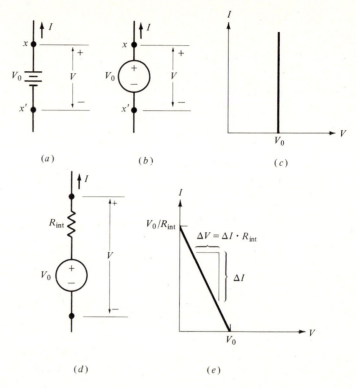

FIGURE B.1
Voltage sources. (*a*) A battery symbol with reference directions; (*b*) circuit symbol for an ideal voltage source; (*c*) VI characteristic of an ideal voltage source; (*d*) a model for a battery with a nonzero internal resistance; (*e*) VI characteristic for the model in part (*d*).

For applications where small changes in the supply voltage need to be considered, the voltage at the terminal of the battery will be affected by the load current due to internal effects inside the battery. For these applications, the simple model of Fig. B.1*b* has been modified by adding an internal resistance R_{int}, as shown in the new *circuit model*, Fig. B.1*d*. The VI characteristic of this model is shown in Fig. B.1*e*.

For operation at low currents (< 10 A) a value of 0.02 Ω would be an appropriate value for R_{int} of a fully charged car battery. From this model, we calculate the short-circuit current as V_0/R_{int}, that is, 600 A. The measured short-circuit current could be quite different, since to accurately model a battery (as almost any other physical entity) over an extremely wide range of current and/or voltage we generally need a more complex model, or we must change the parameters of the model, depending upon the region in which we are operating.

We will use ideal voltage sources as elements in circuit models of diodes and transistors. For dc analysis we will use voltage sources that are fixed. In Chap. 7 we use

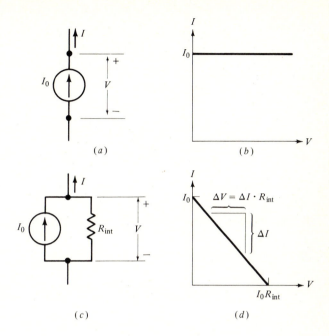

FIGURE B.2
Current sources. (*a*) Symbol for an ideal current source; (*b*) VI characteristic for an ideal current source; (*c*) a model for a nonideal current source; (*d*) VI characteristic for the model in part (*c*).

variable voltage sources, where the magnitude of the voltage source is dependent upon some other variable in the circuit. Such dependent or *controlled* sources are of great value in modeling transistors and ICs.

B.2   CURRENT SOURCES: REAL AND IDEAL

Because most realizations of current sources are made with an electronic circuit, it is not possible to introduce them with a familiar concept like a battery. Nevertheless, current sources are such a useful concept in modeling and are so widely used as circuit "components" in the IC field that we need a good understanding of them.

An ideal constant dc current source would produce a constant current at its terminal regardless of the voltage across the terminals. The symbol for an ideal dc current source is shown in Fig. B.2*a*, and its VI characteristic is shown in Fig. B.2*b*. A nonideal current source can be approximated by the model shown in Fig. B.2*c*, where an ideal current source is paralleled by a resistor R_{int}. The resulting VI characteristic is shown in Fig. B.2*d*. Note that the VI characteristic of Fig. B.2*d* can be made to exactly match that found in Fig. B.1*e* by the appropriate choice of V_0 and I_0 and by using the same R_{int}. Thus we can choose whichever representation is the most useful to our needs.

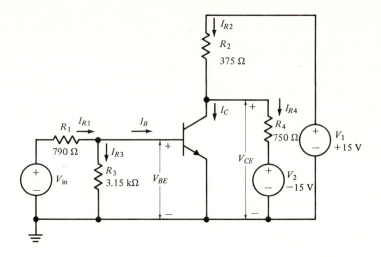

FIGURE B.3
A transistor-inverter circuit with reference polarities for currents and voltages.

Practical electronic current sources also depart from the ideal in that there are limitations to the terminal voltages that can exist. These are set either by the power-supply voltages used in the circuitry or by the breakdown voltage of the transistors used to make the current source.

B.3 KIRCHHOFF'S LAWS

Two circuit laws which are fundamental to the analysis of any electronic circuit are:

1 Kirchhoff's current law (KCL), which is a statement that reflects the conservation of electric charge
2 Kirchhoff's voltage law (KVL), which is a restatement of the conservation of energy

Hence both laws are always valid irrespective of the elements used in the circuit.

B.3.1 Kirchhoff's Current Law

Kirchhoff's current law (KCL) states that the algebraic sum of all currents into (or alternatively, out of) a node point is zero.† In analyzing any node we assign a plus value to those currents with reference-direction arrows pointing toward the node and a negative value to currents whose reference-direction arrows point away from the node.

---

† This law has the useful extension which states that the algebraic sum of all the currents flowing into an enclosed region is zero.

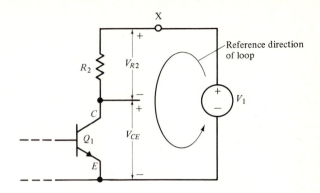

FIGURE B.4
Applying Kirchhoff's voltage law to part of the collector circuit.

For an example, let us apply KCL to the circuit in Fig. B.3, specifically to the node that joins R_1, R_3, and the base of Q_1. Thus we obtain

$$I_{R1} - I_{R3} - I_B = 0 \qquad (B.1a)$$

We can rewrite this in a more convenient form as

$$I_{R1} = I_{R3} + I_B \qquad (B.1b)$$

The choice of reference directions (assumed direction of flow) is important since the proper choice can help in checking the validity of the model chosen in calculating the numerical values of the circuit. For instance, in Fig. B.3, Q_1 is an *n-p-n* transistor, and thus its dc base current I_B is a positive value if Q_1 is conducting current under normal conditions and if the reference direction for I_B is taken as in Fig. B.3. Moreover, the base of Q_1 will, for normal operation, be at a positive potential relative to the emitter (which is grounded), and hence the current I_{R3} will be positive if the reference direction is given as shown in Fig. B.3. If, on calculating I_B and I_{R3}, we obtain negative values for either variable, we will be instantly alerted to the fact that our initial assumptions of the operation of the circuit and the resulting circuit model were both wrong.

B.3.2 Kirchhoff's Voltage Law

Kirchhoff's voltage law (KVL) states that the algebraic sum of the branch voltages going around any closed path (loop) in a circuit is zero. We show a loop in the circuit of Fig. B.4, where we have chosen the reference direction to be counterclockwise. In the same figure we have also assigned reference polarities to the voltages of the three branches that are between the node points C, E, and X. In applying KVL, if the branch voltage increases as one traverses the loop, the branch voltage is entered into the summation as a positive quantity; otherwise it is entered as a negative quantity.

Applying this to Fig. B.4, we have

$$V_1 - V_{R2} - V_{CE} = 0 \qquad (B.2a)$$

This can be rewritten as

$$V_1 = V_{CE} + V_{R2} \qquad (B.2b)$$

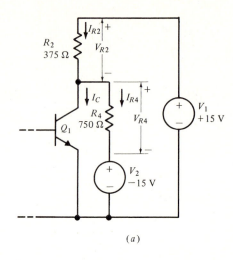

(a)

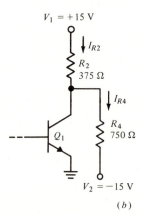

FIGURE B.5
Reference polarities in the collector circuit.
(a) The chosen reference polarities for the
collector circuit shown in detail; (b) conventional electronic schematic convention
for the circuit in part (a).

(b)

To facilitate checking our assumptions, we have chosen reference directions of voltage
on the transistor which will give us positive values for the normal operation of the transistor.
In the next section, we discuss some conventions used in schematics and the labeling of
current directions.

B.3.3 Current-flow Convention for IC Analysis

Since all elements in an IC system except the power supplies (such as batteries) dissipate dc
power, it is best to choose our current directions so that when they are positive the
"battery" is discharging and all other elements are absorbing power. Thus in Fig. B.5a we
have shown a power-supply current leaving the positive terminal of the battery V_1. Since
this current flows through R_2, we have chosen to call the current I_{R2}. Notice in Fig. B.5a
we also show a current I_{R4} with a reference direction into the plus sign of the voltage source
symbol for V_2. However V_2 is a negative supply; hence the plus sign is, in reality, the
negative terminal of V_2. Therefore I_{R4} is leaving the positive terminal of the "battery," that
is, inside of V_2, which is the minus sign of the symbol for the voltage source.

Now power is absorbed by an element when the current flow is into the terminal with the highest potential. Since a resistor is a power-absorbing element, V_{R2} is given the polarity as shown.

The relationship between this voltage and current through the resistor is the familiar Ohm's law:

$$V_{R2} = I_{R2}R_2 \qquad \text{(B.3)}$$

with the polarities as shown in Fig. B.5a.

In Fig. B.5b we show the conventional method of drawing circuit schematics with the highest potential point at the top of the diagram. V_1 and V_2 are the "hot" terminals of two power supplies, while the other terminals are connected to ground. The directions of dc current are all assumed "down hill" in all parts of the schematic. This convention makes it very easy to check dc calculations since rarely will a variable have a negative value.

B.4   AN APPLICATION OF THE CIRCUIT LAWS

Let us apply the laws of Kirchhoff and Ohm to the collector circuit of Fig. B.3. We will solve the currents I_{R2} and I_{R4} under the condition that $I_C = 10$ mA. Applying KCL, we have

$$I_{R2} = I_{R4} + I_C \qquad \text{(B.4)}$$

From KVL, we have

$$V_1 = V_{R2} + V_{R4} + V_2 \qquad \text{(B.5a)}$$

Since we are interested in obtaining values for the currents, we can then rewrite Eq. (B.5a) by using Ohm's law as

$$V_1 = I_{R2}R_2 + I_{R4}R_4 + V_2 \qquad \text{(B.5b)}$$

As we know V_1, V_2, R_2, R_4, and I_C, we have in Eqs. (B.4) and (B.5b) two equations with two unknowns (I_{R2} and I_{R4}) that we can solve. Substituting in the numerical values, we obtain

$$I_{R2} = I_{R4} + 10$$

$$15 = 0.375I_{R2} + 0.750I_{R4} - 15$$

Note that we have used currents in milliamperes and resistances in kilohms which, when multiplied together, gives us the units of volts; that is, $(\text{mA})(\text{k}\Omega) = \text{V}$. This is a convenient choice for numerical work in most ICs.

Solving the above simultaneous equations gives us

$$I_{R4} = 23.3 \text{ mA}$$

$$I_{R2} = 33.3 \text{ mA}$$

A quick check on our algebra and arithmetic can be made by calculating V_{R4} and V_{R2} and verifying that these values with $V_1 = 15$ V and $V_2 = -15$ V satisfy Eq. (B.5a).

$$V_{R4} = I_{R4}R_4 = (23.3)(0.75) = 17.5 \text{ V}$$

$$V_{R2} = I_{R2}R_2 = (33.3)(0.375) = 12.5 \text{ V}$$

$$15 = 12.5 + 17.5 - 15$$

In the above example we had to solve simultaneous equations in order to determine the currents (and voltages) in the circuit. For brevity and clearer insight into circuit operation we would like to avoid doing even this amount of algebra. Fortunately it is possible to solve these circuit problems in a more straightforward manner by the judicious use of two simple circuit theorems.

EXERCISES

EB.1 For the circuit shown in Fig. EB.1, use Kirchhoff's laws to determine the current I_B, given $V_{BE} = 0.7$ V.

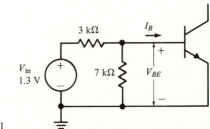

FIGURE EB.1

EB.2 For the circuit shown in Fig. EB.2, use Kirchhoff's laws to determine the voltage V_{CE} when $I_C = 6$ mA.

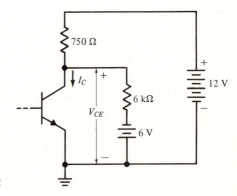

FIGURE EB.2

B.5 THÉVENIN'S THEOREM

Thévenin's theorem simplifies the analysis of one selected portion of a circuit. Consider the input circuit of Fig. B.3, which we select to treat as two parts as shown in Fig. B.6a. The circuit consists of the *source* network (V_{in}, R_1 and R_3) and the *load*, that is, the transistor.

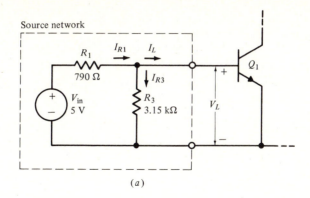

(a)

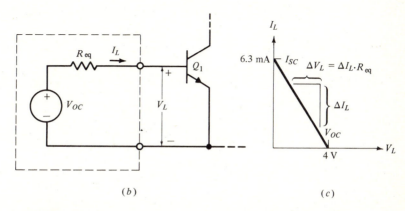

(b)   (c)

FIGURE B.6
The source network in the base circuit. (a) Polarity conventions chosen; (b) the
Thévenin equivalent to the circuit in part (a); (c) the $V_L I_L$ characteristic of the
circuit in part (a).

Thévenin's theorem permits us to replace a linear source network by its Thévenin
equivalent, which is shown in Fig. B.6b. This equivalent is an ideal voltage source V_{OC} in
series with a resistance R_{eq}. V_{OC} is the open-circuit voltage at the load (with the load
removed). R_{eq} is the resistance seen looking toward the source from the load. The resistance
is determined under the condition all independent sources (either voltage or current) are set
to zero. In Fig. B.6a this means V_{in} is replaced by a short circuit. If V_{in} had a significant
internal resistance, this resistance must be left in the branch.

As a "proof" of Thévenin's theorem let us first use KVL and KCL to calculate the
relationship between I_L and V_L for the circuit in Fig. B.6a. By KVL we note

$$V_{in} = R_1(I_L + I_{R3}) + V_L \qquad (B.6a)$$

hence

$$V_L = V_{in} - R_1(I_L + I_{R3}) \qquad (B.6b)$$

From Ohm's law we have

$$I_{R3} = \frac{V_L}{R_3} \qquad \text{(B.6}c\text{)}$$

Using Eq. (B.6c) in Eq. (B.6a) and then rearranging terms, we can write the expression for V_L as a function of V_{in} and I_L:

$$V_L = \frac{V_{in}}{1 + \dfrac{R_1}{R_3}} - \frac{R_1}{1 + \dfrac{R_1}{R_3}} I_L \qquad \text{(B.6}d\text{)}$$

or

$$V_L = \frac{R_3}{R_1 + R_3} V_{in} - \frac{R_1 R_3}{R_1 + R_3} I_L \qquad \text{(B.6}e\text{)}$$

When $V_L = 0$, we have from Eq. (B.6d) the short-circuit current (I_{SC}) at the output of the source network as

$$I_L = I_{SC} = \frac{V_{in}}{R_1} \qquad \text{(B.7}a\text{)}$$

On using the numeric values when $V_{in} = 5$ V,

$$I_L = \frac{5 \text{ V}}{0.79} = 6.3 \text{ mA}$$

Now, when $I_L = 0$, from Eq. (B.6e) the voltage at the output of the open-circuited source network is

$$V_L = V_{OC} = \frac{R_3}{R_1 + R_3} V_{in} \qquad \text{(B.7}b\text{)}$$

Again, using the numeric values,

$$V_{OC} = \frac{3.15}{0.79 + 3.15} 5 = 4 \text{ V}$$

The relationship in Eq. (B.6e) can now be written in terms of three newly defined variables V_{OC}, I_{SC}, and R_{eq} as follows:

$$V_L = V_{OC} - R_{eq} I_L \qquad \text{(B.8)}$$

The relationship between V_L and I_L is graphically illustrated in Fig. B.6c. The intercept on the V_L axis (where I_L is 0) is the open-circuit voltage V_{OC}. On the I_L axis the intercept is at the short-circuit current I_{SC} (that is, where $V_L = 0$). If the load current I_L decreases by an amount ΔI_L, the load voltage increases by an amount ΔV_L, which is equal to $\Delta I_L \cdot R_{eq}$, where R_{eq} is the Thévenin resistance and has the units of ohms. Hence, from Fig. B.6c and Eq. (B.8), the expression for R_{eq} is

$$R_{eq} = \frac{V_{OC}}{I_{SC}}$$

$$R_{eq} = \frac{4.0}{6.3} = 0.63 \text{ k}\Omega \qquad \text{(B.9)}$$

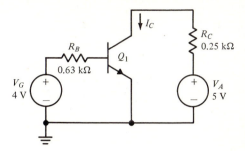

FIGURE B.7
The inverter circuit simplified with Thévenin equivalents for both the base and collector networks.

B.5.1   Thévenin Equivalent

Let us now check the relationship between I_L and V_L by using Thévenin's theorem to directly evaluate V_{OC} and R_{eq}. Thus in Fig. B.6a we can directly solve for V_{OC} which gives the same value as in Eq. (B.7b). That is,

$$V_{OC} = 4 \text{ V}$$

To solve for R_{eq}, we note in Fig. B.6a that with V_{in} replaced by a short circuit, R_{eq} is equal to R_1 in parallel with R_3. That is,

$$R_{eq} = R_1 \| R_3 = \frac{R_1 R_3}{R_1 + R_3}$$

$$= \frac{(0.79)(3.15)}{0.79 + 3.15} = 0.63 \text{ k}\Omega$$

Hence we have the same slope and intercept as in Fig. B.6c, and we have shown that the Thévenin circuit is a valid equivalent for the input circuit of Fig. B.3.

We may similarly use Thévenin's theorem to simplify the output circuit of Fig. B.3, consisting of V_1, V_2, R_2, and R_4. With the transistor in Fig. B.3 removed,

$$V_{CE} = V_{OC} = V_2 + \frac{R_4}{R_2 + R_4}(V_1 - V_2) \tag{B.10}$$

$$= -15 + \frac{0.75}{0.375 + 0.75}[15 - (-15)] = 5 \text{ V}$$

We determine R_{eq} by setting V_1 and V_2 to 0 V; thus

$$R_{eq} = R_2 \| R_4$$

$$= \frac{(0.375)(0.75)}{0.375 + 0.75} = 0.25 \text{ k}\Omega$$

We may now diagram a simplified form of the circuit in Fig. B.3. This is shown in Fig. B.7, where in place of V_{OC} and R_{eq} we use V_G and R_B for the input circuit, and V_A and R_C for the output circuit.

Now to return to our earlier problem in Sec. B.4, where we solved for the currents I_{R2} and I_{R4} of Fig. B.3, under the condition that $I_C = 10$ mA. With reference to Fig. B.7, we may solve for V_{CE} with $I_C = 10$ mA,

$$V_{CE} = V_A - I_C R_C \tag{B.11}$$

$$= 5 - (10)(0.25) = 2.5 \text{ V}$$

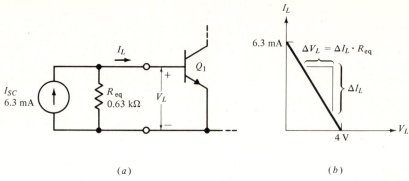

(a)   (b)

FIGURE B.8
Norton equivalents. (a) The Norton equivalent circuit for the base network of
Fig. B.6a; (b) the $V_L I_L$ characteristic of the circuit in part (a).

With $V_{CE} = 2.5$ V we can now solve directly for I_{R2} and I_{R4} in Fig. B.3.

$$I_{R2} = \frac{V_1 - V_{CE}}{R_2} = \frac{15 - 2.5}{0.375} = 33.3 \text{ mA}$$

and

$$I_{R4} = \frac{V_{CE} - V_2}{R_4} = \frac{2.5 + 15}{0.75} = 23.3 \text{ mA}$$

These values are the same as we obtained in Sec. B.4, but the calculations have been
simplified so we can almost do them "in our head."

B.6   NORTON'S THEOREM

For determining the current and voltage of a load, Norton's theorem permits us to replace
a source network by its Norton equivalent. Thus the input circuit shown in Fig. B.6a can be
replaced by the Norton equivalent shown in Fig. B.8a. This is a current source I_{SC}
paralleled by a resistance R_{eq}. The current source I_{SC} is determined by measuring or calculating the current I_L that flows if the load is replaced by a short circuit.

B.6.1   Norton Equivalent

We can rewrite Eq. (B.6b) by expressing I_L as a function of V_L:

$$I_L = \frac{V_{in} - V_L}{R_1} - I_{R3} \qquad \text{(B.12a)}$$

Then, by substituting the numeric values, we have,

$$I_L = \frac{5}{0.79} - \frac{V_L}{0.79} - \frac{V_L}{3.15}$$

$$= 6.33 - 1.58V_L$$

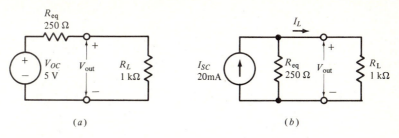

FIGURE B.9
An illustrative example of the use of Norton's theorem. (a) A circuit to be analyzed by Norton's theorem; (b) The Norton equivalent of part (a).

A plot of this relationship is shown in Fig. B.8b with the same value of intercepts (6.3 mA and 4 V) as in Fig. B.6c.

Equation (B.12a) can be rewritten in the following form:

$$I_L = I_{SC} - \frac{V_L}{R_{eq}} \qquad \text{(B.12b)}$$

The Norton equivalent circuit that generates Eq. (B.12b) is shown in Fig. B.8a.

From Fig. B.8a we see that for $I_L = 0$,

$$V_L = V_{OC} = I_{SC} R_{eq} \qquad \text{(B.13)}$$

Notice that the Norton equivalent circuit† consists of a resistor R_{eq} *in parallel with a current source* I_{SC} (where $I_{SC} = V_{OC}/R_{eq}$), whereas the Thévenin equivalent circuit consists of a resistor R_{eq} *in series with a voltage source* V_{OC} (where $V_{OC} = I_{SC} R_{eq}$).

EXAMPLE B.1   Using Norton's theorem consider the circuit of Fig. B.9a, where we desire to use Norton's theorem to find the voltage across R_L (that is, V_{out}). The source network is given as a Thévenin equivalent of 5 V in series with 0.25 kΩ. The Norton equivalent current source is

$$I_{SC} = \frac{V_{OC}}{R_{eq}} = \frac{5}{0.25} = 20 \text{ mA} \qquad \text{(B.14)}$$

This current source is in parallel with R_{eq} in Fig. B.9b. Now to determine the voltage V_{out} we see that the current source is now in parallel with an equivalent resistance R_p, where

$$R_p = R_{eq} \| R_L$$

---

† The value of the reciprocal of the Thévenin resistance is called the *Norton conductance* G_{eq}. That is, $G_{eq} = 1/R_{eq}$, which for our example is 1.58 millimhos.

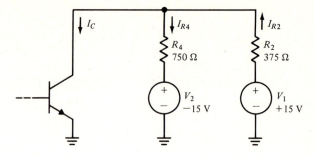

FIGURE B.10
Applying the superposition theorem to the collector circuit of Fig. B.3.

Then for $R_L = 1$ kΩ, we have

$$R_p = \frac{(1)(0.25)}{1 + 0.25} = 0.2 \text{ k}\Omega$$

Hence

$$V_{out} = I_{SC} R_p$$

$$V_{out} = (20)(0.2) = 4 \text{ V}† \qquad (B.15)$$

B.7 SUPERPOSITION THEOREM

The superposition theorem states that for any *linear* network the total response to the applied voltage or current sources is the sum, i.e., superposition, of the effect of each individual current or voltage source. In particular let us again consider the output portion of the circuit of Fig. B.3 as now shown in Fig. B.10. We will again determine the currents I_{R2} and I_{R4} under the condition that $I_C = 10$ mA. However this time we will make use of the superposition theorem.

We first consider the effects of only applying the voltage source V_1 with the other sources shut off; that is, $V_2 = 0$ and $I_C = 0$. For this condition we will designate the current in R_2 as $I_{R2(1)}$. Similarly for V_2 applied while $V_1 = 0$ and $I_C = 0$ gives us a current in R_2 designated as $I_{R2(2)}$.

For both voltage sources set to zero‡ and the current source I_C applied, we have a

---

† This can be done also directly from Fig. B.9a by using the voltage-divider relationship

$$V_{out} = \frac{R_L}{R_{eq} + R_L} V_{oc}$$

$$= \frac{1}{1 + 0.25} 5 = 4 \text{ V}$$

‡ This is an analytical "setting to zero," since if we tried to do this experimentally the transistor would not function properly.

current in R_2 of $I_{R2(3)}$. Thus for all three sources applied we have by superposition the actual current in R_2 as

$$I_{R2} = I_{R2(1)} + I_{R2(2)} + I_{R2(3)} \qquad \text{(B.16)}$$

Similarly for I_{R4} we have, using the same subscript notation,

$$I_{R4} = I_{R4(1)} + I_{R4(2)} + I_{R4(3)} \qquad \text{(B.17)}$$

Return now to our numeric example with $V_1 = 15$ V, $V_2 = -15$ V, and $I_C = 10$ mA. First, with $V_2 = 0$ and $I_C = 0$,

$$I_{R2(1)} = \frac{V_1}{R_2 + R_4} = \frac{15}{0.375 + 0.75} = 13.3 \text{ mA}$$

With $V_1 = 0$ and $I_C = 0$,

$$I_{R2(2)} = \frac{-V_2}{R_2 + R_4} = \frac{-(-15)}{0.375 + 0.75} = 13.3 \text{ mA}$$

With $V_1 = 0$ and $V_2 = 0$,

$$I_{R2(3)} = \frac{R_4}{R_2 + R_4} I_C = \frac{0.75}{0.375 + 0.75} 10 = 6.66 \text{ mA}$$

We can similarly solve the components of I_{R4}.

$$I_{R4(1)} = \frac{V_1}{R_2 + R_4} = 13.3 \text{ mA}$$

$$I_{R4(2)} = \frac{-V_2}{R_2 + R_4} = 13.3 \text{ mA}$$

$$I_{R4(3)} = -\frac{R_2}{R_2 + R_4} I_C = -\frac{0.375}{0.375 + 0.75} 10 = -3.33 \text{ mA}$$

$I_{R2(3)}$ and $I_{R4(3)}$ are determined by the current-divider relationship.† The minus sign in $I_{R4(3)}$ is due to the difference in reference direction. Hence by the superposition theorem of Eqs. (B.16) and (B.17),

$$I_{R2} = 13.3 + 13.3 + 6.6 = 33.3 \text{ mA}$$

$$I_{R4} = 13.3 + 13.3 - 3.3 = 23.3 \text{ mA}$$

These values are the same as obtained in Sec. B.4 or B.5. We will often find the superposition theorem useful for both numerical and conceptual purposes.

---

† This is readily derived from the following:

$$V_{R2(3)} = I_C(R_2 \| R_4) = I_C \frac{R_2 R_4}{R_2 + R_4}$$

$$I_{R2(3)} = \frac{V_{R2(3)}}{R_2} = I_C \frac{R_4}{R_2 + R_4}$$

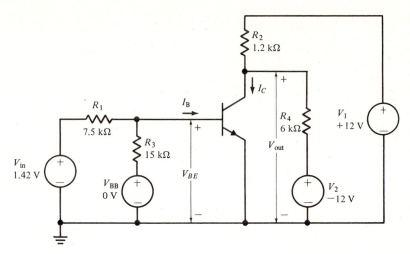

FIGURE EB.3

EXERCISE

EB.3 For the circuit in Fig. EB.3:

(*a*) Find the Thévenin equivalent circuit for the input circuit.

(*b*) For $V_{BE} = 0.7$ V, find I_B by using the results of part (*a*).

(*c*) Find the Thévenin equivalent circuit for the output circuit.

(*d*) For $I_C = 3$ mA and the results of part (*c*), determine the value of V_{out}.

(*e*) (1) Find the Norton equivalent circuit for the input circuit.

(2) Repeat part (*b*) but using the results of part (*e*, 1).

(3) Find the Norton equivalent circuit for the output circuit.

(4) Repeat part (*d*) but using the results of part (*e*, 3).

(*f*) For $V_{in} = 0$ V and $V_{BB} = -1.5$ V, repeat part (*a*).

(*g*) For $V_{in} = 1.42$ V and $V_{BB} = -1.5$ V, determine the Thévenin equivalent circuit for the input by using previous results and the superposition theorem.

Now that you have had the exposure to the preceding circuit theorems, you probably are saying to yourself, "Fine, but how do I decide when I should use one or the other of these theorems?" Unfortunately there is no ready answer to this most important question, and in many cases one method is as good as another and personal preferences will be the deciding factor. However, by considering factors as in the following examples we can often see a clear choice.

First consider how to model the auto battery in Sec. B.1. If we are using it where the terminal current will be about 1 A, we would find the Thévenin equivalent more suitable. This is because the Norton equivalent has a current source of 600 A which mostly goes down the 0.02-Ω internal resistance. Hence numerical accuracy and appreciation of interaction of circuit constants will be lost if a Norton equivalent is used.

Now consider a case where a 1-mA current source is to be used that has a 1-MΩ shunting resistance in parallel with it. This is generated in an electronic circuit that has all supply voltages of 15 V or less. If we tried to use a Thévenin equivalent for this, we would have a 1000-V voltage source to deal with. If we use this large voltage in the analysis, we would have a harder time seeing interactions and limitations of the models as compared to using a Norton equivalent.

The superposition theorem is useful for analyzing circuits that have multiple voltage and/or current sources.

REFERENCES

For additional material on circuit laws and theorems, see OLDHAM, W. G., and S. E. SCHWARZ: "An Introduction to Electronics," chaps. 1–3, Holt, New York, 1972; also SMITH, R. J.: "Circuits, Devices and Systems," 2d ed., chaps. 2 and 8, Wiley, New York, 1971.

APPENDIX C

NUMBER SYSTEMS AND CODES

Data presented to and received from a digital system are generally in decimal form, that is, readily recognizable to the human operator. For example, the DVM described in our Introduction displays the dc voltage as a decimal number in the readout devices. Likewise, time in a digital clock is also displayed as a decimal number. However, we have seen that digital circuits have just two operating states; this is a characteristic of the binary system. The principal objective of this appendix is to develop a procedure for converting binary numbers to decimal numbers (decoding) and vice versa (encoding).

We will begin with a consideration of how numbers are represented in the decimal system, enabling us to identify several basic ideas that are common to all number systems in terms of the use that is most familiar to us.

C.1 THE DECIMAL NUMBER SYSTEM

The number of discrete states to a counting system identifies the *base* of the system. In the decimal system there are 10 states, which we recognize as the integers (whole numbers) 0 through 9. The base of the decimal number system is therefore 10. The base of a number system is sometimes referred to as the *radix* of the system.

When we count by integers in the decimal system, we first accumulate integers 0 through 9 in the units column, then carry 1 to the tens column and reset the units column to 0. We then count 0 through 9 in the units column again, carry another 1 to the tens column (so that the total in the tens column is $1 + 1$, or 2), and so forth. If there are enough objects in the set being counted, we accumulate 10 in the tens column, whereupon we now carry a 1 into the hundreds column and reset the tens column to 0. Notice that the largest digit appearing in the decimal system is 9. From the process just described, it can be seen that the largest digit in a number system is the base of the system minus 1.

For integer numbers the *least significant column* in the decimal system is the units column (10^0). The next least significant is the tens column (10^1), then the hundreds column (10^2), and so on. As a simple example, the decimal integer 3567 consists of 7 units, 6 tens, 5 hundreds, and 3 thousands. Here the 3 in the thousands column represents the *most significant digit*. Now the decimal number 3567 can be written out as a summation of powers of 10, with a multiplier (0 through 9) associated with each power of 10.

EXAMPLE C.1

$$3567_{10}† = 3(10^3) + 5(10^2) + 6(10^1) + 7(10^0)$$

We can perform this operation readily because of our thorough familiarity with the decimal system. However, other number systems are not so familiar, and we need a more systematic approach to determine the multiplier of the base. Let us reconsider the process by which we associate 7 with the units column in the preceding example.

Dividing the number by 10 (the base of the system) yields an integer quotient and a remainder which is the least significant number in the integer. This remainder becomes the multiplier for the units column.

$$\frac{3567}{10} = 356 \ \& \ 7$$

With each successive division of the quotient by the base of the system, a remainder is obtained which is the multiplier for each succeeding power of the base number.

$$\frac{3567}{10} = 356 \ \& \ 7$$

$$\frac{356}{10} = 35 \ \& \ 6$$

$$\frac{35}{10} = 3 \ \& \ 5$$

$$\frac{3}{10} = 0 \ \& \ 3$$

$$3567_{10} = 3(10^3) + 5(10^2) + 6(10^1) + 7(10^0)$$

Remember, the *least* significant number is obtained at the *first* division and the *most* significant number at the *final* division, which is when the quotient is zero.

---

† The subscript 10 indicates the base of the number.

C.2 THE BINARY NUMBER SYSTEM

With this review of the decimal system, we now turn our attention to the binary system. The base of the binary system is 2. Hence there are just two integers, represented by 0 and 1. However the basic rules for writing numbers are the same in the binary system and the decimal system. We count 0 through 1 for each column of the number and carry 1 for each time we pass the base of the system, namely, 2. The similarity to the decimal system is shown in the example, where we also convert a binary number to its decimal equivalent.

EXAMPLE C.2

$$1011_2 = 1(2^3) + 0(2^2) + 1(2^1) + 1(2^0)$$
$$= 1(8) + 0(4) + 1(2) + 1(1)$$
$$= 8 + 0 + 2 + 1$$
$$= 11_{10}$$

C.2.1 Conversion of Decimal Integers

It is also useful to be able to convert a decimal integer to its binary equivalent. A simple method has already been described, that of successive division by the base of the new system. Converting to binary, successive division by 2 will yield remainders which are multipliers for increasing orders of powers of 2.

EXAMPLE C.3 Convert 356_{10} to its binary equivalent.

SOLUTION

$$\frac{356}{2} = 178 \;\&\; 0$$

$$\frac{178}{2} = 89 \;\&\; 0$$

$$\frac{89}{2} = 44 \;\&\; 1$$

$$\frac{44}{2} = 22 \;\&\; 0$$

$$\frac{22}{2} = 11 \;\&\; 0$$

$$\frac{11}{2} = 5 \;\&\; 1$$

$$\frac{5}{2} = 2 \;\&\; 1$$

$$\frac{2}{2} = 1 \;\&\; 0$$

$$\frac{1}{2} = 0 \;\&\; 1$$

$$356_{10} = 1(2^8) + 0(2^7) + 1(2^6) + 1(2^5) + 0(2^4) + 0(2^3) + 1(2^2) \qquad ////$$
$$+ 0(2^1) + 0(2^0)$$
$$= 101100100_2$$

Notice again, the least significant binary digit or bit is obtained at the first division and the most significant bit at the final division. We may also check our result as follows:

$$\text{CHECK} \qquad 101100100_2 = 1(2^8) + 0(2^7) + 1(2^6) + 1(2^5) + 0(2^4) + 0(2^3)$$
$$+ 1(2^2) + 0(2^1) + 0(2^0)$$
$$= 256 + 64 + 32 + 4$$
$$= 356_{10}$$

EXERCISES

EC.1 Convert from the decimal to the binary number system:
(*a*) 60_{10}; (*b*) 3601_{10}.

EC.2 Convert from the binary to the decimal system:
(*a*) 10110011_2; (*b*) 1011001101_2.

C.3 THE OCTAL SYSTEM

With some digital computer systems, the input and output data are organized in the octal system. The data are still processed within the system in binary fashion. An advantage of the octal system is the ease of conversion to and from the binary system.

C.3.1 Decimal Conversion

The base of the octal system is the number 8. Here the eight states are numbered 0 through 7. The largest digit appearing in the octal system is then 7. The rules of conversion between the octal system and the decimal system are the same as those applied in the conversion of binary numbers to and from the decimal system. The base number 8 is used in place of the base 2 of the binary system. An octal number is converted to a decimal number in the following example.

EXAMPLE C.4 Convert 356_8 to its decimal equivalent.

SOLUTION
$$356_8 = 3(8^2) + 5(8^1) + 6(8^0)$$
$$= 3(64) + 5(8) + 6(1)$$
$$= 192 + 40 + 6 = 238_{10} \qquad ////$$

Now for an example of converting a decimal number to the octal number system.

EXAMPLE C.5   Convert 356_{10} to its octal equivalent.

SOLUTION

$$\frac{356}{8} = 44 \ \& \ 4$$

$$\frac{44}{8} = 5 \ \& \ 4$$

$$\frac{5}{8} = 0 \ \& \ 5$$

$$356_{10} = 544_8 \qquad\qquad ////$$

CHECK

$$544_8 = 5(8^2) + 4(8^1) + 4(8^0)$$
$$= 320 + 32 + 4$$
$$= 356_{10}$$

Again, notice in converting to the octal equivalent by successive division, the least significant digit is obtained at the first division.

C.3.2   Binary Conversion

Binary to octal conversion and the reverse is simple. We recognize that 7 is the highest digit that can appear in the octal system and that this is equivalent to 111 in the binary system. In converting binary numbers to the octal system, the technique is to start from the LSB, divide the number into groups of three binary digits each, and convert each group into its octal equivalent.

EXAMPLE C.6   Convert 11110010_2 to the octal system.

SOLUTION

$$11110010_2 = 011 \quad 110 \quad 010_2$$
$$011_2 = 3_8$$
$$110_2 = 6_8$$
$$010_2 = 2_8$$
$$11110010_2 = 362_8 \qquad\qquad ////$$

Converting octal numbers to the binary system is even simpler. Each octal digit is converted to a group of three binary digits.

EXAMPLE C.7   Convert 356_8 to the binary system.

SOLUTION

$$356_8 = 011 \quad 101 \quad 110_2$$

$$= 011101110_2 \qquad ////$$

In summary, it can be stated that the digits in a number only have significance when the base of the number is known. This is identified by the subscript following the number. With the common decimal system, the subscript is generally omitted and taken for granted. However, whichever system is used there are simple techniques for conversion from one system to another. Digital ICs are available which make this conversion, and we will see how this is done in Chap. 4. A comparison of the common number systems is made in Table C.1.

Table C.1   COMMON NUMBER SYSTEM

System: Base:	Decimal 10	Binary 2	Octal 8
	0	0	0
	1	1	1
	2	10	2
	3	11	3
	4	100	4
	5	101	5
	6	110	6
	7	111	7
	8	1000	10
	9	1001	11
	10	1010	12
	11	1011	13
	12	1100	14
	13	1101	15
	14	1110	16
	15	1111	17
	16	10000	20
	17	10001	21
	18	10010	22
	19	10011	23
	20	10100	24
	30	11110	36
	40	101000	50
	50	110010	62
	60	111100	74
	70	1000110	106
	80	1010000	120
	90	1011010	132
	100	1100100	144

EXERCISES

EC.3 Convert from the decimal to the octal number system:
(a) 710_{10}; (b) 351_{10}.

EC.4 Convert from the binary to the octal system:
(a) 10101100_2; (b) 11001101_2

EC.5 Convert the octal numbers 725_8 and 512_8 to
(a) the decimal system; (b) the binary system.

C.4 NUMBER CODES

There are many schemes, or codes, for representing a decimal number in binary form besides the straightforward binary conversion. In this section, we will deal with the most general scheme, the binary-coded decimal (BCD).

We have seen that representing decimal numbers as binary numbers leads to a long string of digits. While easily handled in a digital system, they are difficult for a human being to remember or evaluate in magnitude. Thus with an electronic wristwatch, which records each minute by counting 60 pulses, we have some concept of what 60_{10} means but little idea of what 111100_2 means until it is converted to 60_{10}.

The octal system has fewer digits and can be digested by a computer, but it is still not too readily intelligible to the human being accustomed to the decimal system. For example, we are unaccustomed to an arithmetic system in which 66 and 24 add to give 112, as they do in the octal system.

C.4.1 BCD Code

With the BCD scheme, each decimal digit is coded or represented by four binary digits. The weighting of the binary digits readily yields the decimal equivalent. Various BCD systems have been used, but the most popular is the one with the 8421 weighting. The MSB has a weight of 8, with the LSB having a weight of 1. This code is shown in Table C.2 for each of the 10 digits in the decimal system, along with another weighting scheme (2421).

Each decimal digit forms a 4-bit word in the BCD code. A number containing two decimal digits would then be represented by two 4-bit words. For example,

$$60_{10} = 0110 \quad 0000 \quad \text{(8421 BCD)}$$

since

$$6_{10} = 0110_2$$

and

$$0_{10} = 0000_2$$

EXERCISE

EC.6   Convert the following decimal numbers to 8421 BCD:
(a) 56792_{10}; (b) 7337_{10}.

Table C.2   TWO COMMON BCD CODES

Decimal	8421 BCD				2421 BCD			
	8	4	2	1	2	4	2	1
0	0	0	0	0	0	0	0	0
1	0	0	0	1	0	0	0	1
2	0	0	1	0	0	0	1	0
3	0	0	1	1	0	0	1	1
4	0	1	0	0	0	1	0	0
5	0	1	0	1	1	0	1	1
6	0	1	1	0	1	1	0	0
7	0	1	1	1	1	1	0	1
8	1	0	0	0	1	1	1	0
9	1	0	0	1	1	1	1	1

APPENDIX D

A GLOSSARY OF INTEGRATED CIRCUIT TERMINOLOGY†

ac coupled flip-flop An edge-triggered flip-flop that changes state due to the rise or fall of a clock pulse. It is characterized by having a maximum allowable rise or fall time.

accumulator A register for doing such arithmetic and logic functions as addition and shifting.

active elements Those components in a circuit which have gain, or direct current flow, such as SCRs, transistors, thyristors, or tunnel diodes. They change the basic character of an applied electrical signal by rectification, amplification, switching, etc. (Passive elements have no gain characteristics. Examples: inductors, capacitors, resistors.)

active pull-up Similar to a pullup resistor, except a transistor replaces the resistor connected to the positive supply voltage. This allows low output impedance without high power consumption.

A-D Analog-to-digital.

A-D encoder Analog-to-digital encoder for changing an analog quantity to equivalent digital representation.

adder A switching circuit which combines bits to generate their sum and carry.

address (*Noun*) A code label that identifies a specific location in a computer's memory where certain information is stored. (*Verb*) Selection of stored information for retrieval from a computer's memory.

† Excerpted by permission from "A Glossary of Integrated Circuit and Related Terminology." Compiled and Edited November 1970 by Motorola Semiconductor Products, Inc.

ambient temperature   The environmental temperature in which a device operates, e.g., perhaps $+150°C$ in a rocket, or $-45°C$ in a remote Arctic weather station.

amplifier   A circuit used to increase the power, voltage, and/or current level of a signal.

AND (gate)   If and only if all inputs are true, the output is true.

arithmetic unit   The part of a computer that performs arithmetic operations.

astable   Refers to a device which has two temporary states. The device oscillates between the two states with a period and duty cycle predetermined by time constants. (*See* bistable.)

asynchronous   Nonclocked operation of a switching or logic network wherein the completion of one instruction or operation triggers the next.

asynchronous inputs   The terminals in a flip-flop which affect the output state of the flip-flop independently of the clock. Called set, preset, reset, or clear. Sometimes these are referred to as dc inputs.

bar   A symbolization of the inverse or complement of a logic function. (*See* Q and $\bar{Q}$—read as Q-bar.)

BCD   Binary-coded decimal.

binary   A numbering system using only two symbols: 0 and 1.

binary-coded-decimal representation (BCD)   A system of representing decimal numbers. Each decimal digit is represented by a combination of four binary digits (bits).

binary counter   Flip-flops connected in cascade to perform binary counting. Four flip-flops have a counting capacity of 2^4, or 16; five cascaded flip-flops have a capacity of 2^5, or 32, etc. (Also called *binary scaler*.) (*Also see* counter *and* ring counter.)

binary logic   Digital logic elements operating with two distinct states called true and false or high and low or on and off or 1 or 0. In computers the two states are represented by two different voltage levels.

binary-to-1-of-8 line decoder   This digital function takes three input lines with eight possible states and decodes them internally, enabling only the one output line which corresponds to that particular combination of the three inputs.

bi-quinary counter   A 4-bit counter capable of divide-by-two, divide-by-five, or divide-by-10 functions.

bistable   A logic device that has two stable states. The flip-flop is an example of the bistable element. (A device with two temporary states is an astable device.) The flip-flop can be caused to go to either of the two states by input signals, but remains in that state after the input signals are removed. (*See* astable.)

bit   An acronym for binary digit. In the binary numbering system, each of the two marks 0 and 1 is called a binary digit, or bit. The decimal number 39, converted to a binary number, becomes 100111 and is made up of six 0 or 1 marks, or six bits. A bit is the smallest divisible amount of information that can be stored in a flip-flop.

block diagram   A geometric drawing of the basic units in a circuit, interconnected by straight lines indicating current flow, signal flow, etc.

Boolean algebra   A logical calculus, named for mathematician George Boole, using alphabetic symbols to stand for logical variables and 0 and 1 to represent states. (AND, OR, and NOT are the three basic logic operations in this algebra. NAND and NOR are combinations of the three basic operations.)

buffer A digital circuit element used to increase fanout (the number of outputs a circuit can drive) or to convert input or output levels for signal-level compatibility. Also, any isolating amplifier stage.

buss (line) A connection line for distribution of signals, power supply, or ground.

cascadable counter A logic counting block that has the necessary connections available to allow the operation of more than one counter in series thus increasing the modulus of the counter subsystem. (*See* cascading.)

cascading Placing of two or more circuits in series, the output of one being connected to the input of the next in the series.

central processing unit (CPU) The arithmetic and control portions of a computer.

chip A tiny piece of semiconductor material scribed or etched from a semiconductor slice on which one or more electronic components are formed. (Also called *die*.)

clear To return a memory or storage element to its "standard" state, usually 0. Also called reset.

clock A pulse generator or signal waveform which synchronizes the timing of switching circuits and memory in a digital computer system. It determines the speed of the CPU.

clocked A term that refers to gating that is added to a basic flip-flop. This gating permits the flip-flop to change state only when a change in the clocking input or an enable level of the clocking input happens to be present.

clock input That terminal on a flip-flop whose condition or change of condition controls the admission of data into a flip-flop through the synchronous inputs and thereby controls the output state of the flip-flop. The clock signal performs two functions: (1) it permits data signals to enter the flip-flop; (2) after entry, it directs the flip-flop to change state accordingly.

clock rate The speed (frequency) at which the major portion of a computer or logic element operates, determined by the rate at which words or bits are transferred from one internal element to another. Clock rate is expressed in megahertz (if a parallel-operation machine: words; if a serial operation machine: bits).

CML Current-mode logic, in which transistors operate in the unsaturated mode. This logic has ultra-fast switching speeds and low logic swings. (Also called ECL or MECL.†)

CMOS (Complementary MOS) With extra diffusions, a circuit with both *p*- and *n*-channel FETs on the same MOS wafer. CMOS, complementary metal-oxide-semiconductor logic, is designed to have extremely low power dissipation (essentially zero during stand-by) making it especially useful for remote applications where power is expensive. Other attributes: high noise immunity, high fanout, full power-supply logic swings, and ready acceptance of a wide range of power supplies.

comparator (digital) Used to check to see if both input words are identical, in which case EXCLUSIVE-NOR or coincidence gates can be used. Additional logic may also be added to indicate if one word is greater than or less than the other.

comparator (linear or analog) A device that compares an input voltage with a reference level. The output may indicate greater than, less than, or even equal to.

counter (1) a device capable of changing states in a specified sequence upon receiving appropriate input signals; (2) a circuit which provides an output pulse or other indication after receiving a specified number of input pulses.

CPU *See* central processing unit.

current-mode logic *See* CML.

† Trademark of Motorola Inc.

current sinking logic   A logic family in which fanout is limited by how much current can sink through the output to ground, e.g., DTL and TTL.

current sourcing logic   A logic family in which fanout is limited by how much current the devices can source through the output from the supply to the driven inputs, e.g., RTL.

D-A   Digital-to-analog.

D-A decoder   Digital-to-analog decoder for changing a digital word to an equivalent analog value.

dc amplifier   A direct-coupled multistage amplifier, which is highly desirable for designing integrated circuits because reactive components, such as inductances and high-value capacitances used for coupling, are difficult to fabricate.

dc flip-flop   Clocking is accomplished by an input voltage level rather than by a clock waveform edge. Clocking may be done by very slowly changing waveforms.

DCTL   Direct-coupled transistor logic, the first logic form that was considered for integrated circuits. (RTL is DCTL modified by the use of resistance coupling.)

decade counter   A logic device that has 10 stable states. The device may be cycled through these 10 states with 10 clock or pulse inputs. A decade counter usually counts in a binary sequence from state 0 through 9 and then cycles back to 0. It is sometimes referred to as a divide-by-10 counter.

decoder   A unit that translates a combination of signals into a single signal representing that combination. Often used to extract information from a complex signal.

delay time (pulse or propagation)   The time interval from a point at which the leading edge of the input pulse has risen to 10% of its maximum amplitude to a point at which the leading edge of the output pulse has risen to 10% of its maximum amplitude. (*Note:* Can be measured from 50% to 50% points instead of 10% to 10% points. Can also be measured from one input level to one output level such as 1.5 V.)

demultiplexer   A circuit which directs information from a single input to one of several outputs in a specific sequence determined by the information applied to the control inputs.

D flip-flop   D stands for delay. A D flip-flop is one whose output is a function of the input which appeared one pulse earlier, e.g., if a 1 appeared at the input, the output after the next clock pulse will be a 1.

differential amplifier   A circuit which amplifies the difference of potential between two input signals.

differential comparator   A circuit using differential-amplifier design techniques to compare an input voltage with a reference voltage. When the input voltage is below the reference, the circuit output is in one state; when the input voltage is above the reference, the output is in the opposite state. Commonly used for pulse amplitude detector circuits, A-D conversion, and differential twisted pair line receivers for data transmission in noisy environments.

diffusion   A thermal process by which minute amounts of impurities are deliberately impregnated and distributed into semiconductor material.

digital circuit   A circuit which operates like a switch and can make logical decisions. It is used in computers or similar decision-making equipment. The more common families of digital integrated circuits (called *logic forms*) are RTL, DTL, HTL, ECL, and TTL.

digitize The translation of a quantitative measurement into a coded numerical equivalent.

discrete circuits Circuits built of separate, individually manufactured, tested and assembled electronic components (transistors, diodes, resistors, etc.)

divide-by-16 counter A logic device, containing four flip-flops, that will count from the binary number 0 through 15 and then recycle to 0. All 16 states of the four flip-flops are used. Sometimes referred to as a hexadecimal counter.

dot A dot, or bubble, drawn on the input of a logic symbol to indicate the active signal input is a negative input. The lack of a dot indicates a positive active signal.

dot-AND Externally connecting separate circuits or functions so that the combination of their outputs results in an AND function. The point at which the separate circuits are wired together will be a 1 if all circuits feeding into this point are 1. (Also called *wired-AND*.)

dot-OR Externally connecting separate circuits or functions so that the combination of their outputs results in an OR function. The point at which the separate circuits are wired together will be 1 if any of the circuits feeding into this point are 1. (Also called *wired-OR*.)

driver A device in a logic family that can be controlled with normal logic levels while its output has the capability of sinking or sourcing high current. The output may control a lamp, relay, or a very large fanout of other logic devices. Also a device "driving" a higher output device or transistor by supplying power, voltage, or current to it.

DTL Diode-transistor logic. The logic is done by diodes; the transistors are used as inverting amplifiers.

dual in-line (package) A container for a circuit, a package, having two parallel rows or lines of leads. This facilitates automatic insertion of the integrated circuit package on the assembly line.

dynamic shift register A shift register that stores information using temporary charge-storage techniques. This results in simple register design, thus allowing the production of large, inexpensive shift registers. The major drawback of this technique is that the information is lost if the clock repetition rate is reduced below a minimum value.

ECL Emitter-coupled logic, an unsaturated logic performed by emitter-coupled transistors. It was developed to achieve the ultra-high speeds required by advanced computers, unattainable with saturated logic circuits.

edge-triggered flip-flop Changes state only during clock transition.

element A part of an integrated circuit which contributes directly to its electrical characteristics. An active element exhibits gain, as a transistor; a passive element does not have gain, such as a resistor or capacitor.

emitter-coupled logic *See* ECL.

encoder A unit which changes discrete inputs into coded combinations of outputs.

EXCLUSIVE-NOR (gate) If one and *only* one of the inputs is true the output is *not* true.

EXCLUSIVE-OR (gate) If one and *only* one of the inputs is true the output is true.

expandable gate A logic gate for which the number of inputs can be expanded by the simple addition of an expander block.

expander A logic block which can be easily connected to an expandable gate to increase the number of available logic inputs.

fan-in The number of inputs connected to a gate.

fanout The number of loads connected to a gate.

fanout (maximum) The worst-case number of loads a circuit can drive.

flat pack A description of a slab-shaped (*flat*) integrated circuit package (*pack*).

flip-flop A digital circuit used to store information. The flip-flop has two stable states: 0 or 1. When the flip-flop is set to one of these states, it remains in that state until the application of a control signal causes it to "flip" or "flop" to the other state (0 or 1). Among the many kinds of flip-flops are delay (*D*), gating (*J-K*), set-reset (*R-S*), and toggle flip-flops. (*See* specific kinds: *D*, *J-K*, *R-S-T*, *R-S*, and *T*.)

frequency divider A counter having a gating structure added to it for providing an output pulse after receiving a specified number of input pulses.

frequency synthesizer A frequency-generating circuit that can produce a multitude of output frequencies from a single input frequency.

full adder A logic device which will add two binary bits, taking into consideration the possible carry from a previous addition. Sum and carry outputs are provided.

full subtractor A device which can take the difference of two input bits, also subtract a borrow from those bits, and provide the difference and borrow outputs.

function A word description of what a given circuit does.

gate A digital logic element with usually one output and many inputs, designed so that the output is enabled only when certain input conditions are met. The circuit then "gates," or passes, the desired signal.

gated The action in which data may be either passed or blocked along some path, depending on the level of a control input to the "gating" logic.

gray code A binary code in which each number differs from the sequential numbers in one place—and in that place it differs by one unit.

half-adder A logic element which will add two input bits. It has no provision for adding in the carry from a previous addition. (*See* full adder.)

hardware The physical components of a computer or a system. (*Software* is the term used to describe the programs and instructions for a computer.)

heat sink A mounting base for semiconductor devices, usually metallic, which serves to dissipate, carry away, or radiate into the surrounding atmosphere heat which is generated within a device. The device's package itself often serves as a heat sink for the semiconductor chip, but, for higher power devices, a separate heat sink on which one or more packages are mounted is required to prevent overheating, and subsequent destruction, of the semiconductor junctions.

hexadecimal counter *See* divide-by-16 counter.

hex inverter Six logic inverters in a single package.

HTL High-threshold logic, developed as the economical answer to the need for shielding circuitry from electrical noise in a system.

hybrid circuit Sometimes used to define multichip circuits in which two or more silicon chips are interconnected within one package. More often used to denote circuits made by two different processes (i.e., semiconductor and thick-film or thin-film) or by a combination of integrated structures and discrete components.

IC *See* integrated circuit.

IF amplifier An IF amplifier is an intermediate-frequency amplifier usually used in superheterodyne receivers. The IF amplifier normally operates somewhere between 100 kHz and 100 MHz. Most AM broadcast-band receivers have IF amplifiers at either 262 or 455 kHz. Most FM broadcast-band receivers have IF amplifiers at 10.7 MHz. The IF amplifier is used to provide high gain per stage and selectivity, both of which are more difficult to achieve at higher operating frequencies.

implied-AND or dot-AND or wired-AND   The combined outputs are true if and only if all outputs are true.

implied-OR or wired-OR   The combined outputs are true if one or more of the outputs is true.

inhibit   To prevent an action or acceptance of data by applying an appropriate signal to the appropriate input (generally a logic 0 in positive logic).

input   The signal (current, voltage, or power) fed into a circuit. Also, that point (terminal) at which the incoming signal is applied.

integrated circuit (IC)   An integrated circuit is a tiny electrical device that is built into a single package and performs a complete circuit function. It replaces a given number of transistors, diodes, resistors, and capacitors that would be needed to perform the equivalent function.

inverter   A device that complements a logic variable, i.e., changes the logic levels from high to low or low to high. (Also called a *NOT* circuit.)

J-K flip-flop   A flip-flop having two inputs designated *J* and *K*. At the application of a clock pulse, a 1 on the *J* input and a 0 on the *K* input will set the flip-flop to the 1 state. A 1 on the *K* input and a 0 on the *J* input will reset it to the 0 state. 1s simultaneously on both inputs will cause the flip-flop to change state regardless of the previous state; 0s simultaneously will prevent change.

junction   A boundary between *n*-type and *p*-type semiconductor material.

large-scale integration (LSI)   A large number of interconnected integrated circuits manufactured simultaneously on a single slice of semiconductor material (usually over 100 gates or basic circuits, with at least 500 circuit elements).

latch   A very simple logic storage element. The most basic form is two cross-coupled logic gates which store a pulse presented to one logic input until the other input is pulsed, thus storing the complementary information in the latch.

latch voltage   The effective voltage on an input at which a flip-flop switches to its opposite state.

lead   One of the wires or pins on the outside of a semiconductor device, used to connect the device with a circuit.

linear (or analog) circuit   The linear circuit operates directly on a continuous electrical signal to change the signal's shape, increase its strength, or otherwise modify it for a specific end function. Typical linear circuits are operational amplifiers, differential amplifiers, sense amplifiers, regulators and multipliers, etc. Normally the output is some continuous (linear) function of the input and the device does not operate in saturation.

line driver   An integrated circuit specifically designed to transmit logic information through long lines, normally several feet or more in length.

line receiver   Used in conjunction with a line driver to detect signals at the receiving end of a long line.

logic   A means of solving complex problems through the use of symbols to define basic concepts. (The three basic logic symbols are *AND*, *OR*, and *NOT*. *NAND* and *NOR* are combinations of the logic operations which those three basic symbols represent.)

logical operations   Nonarithmetical operations such as selecting, searching, sorting, matching, comparing, etc.

logic level   One of two possible states, or levels: 0 or 1.

logic diagram   A drawing which represents the logical functions of AND, NAND, etc.

logic swing   The voltage difference between the two levels representing logical 1 and 0.

LSB   Least significant bit; the lowest weighted digit of a binary number.

LSI   *See* large-scale integration.

main frame   The central processing unit of a computer plus the input/output unit and the random-access and read-only memories. The main frame is the computer without peripherals.

majority gate   If more than half the number of inputs are true the output is true, otherwise the output is false.

master/slave flip-flop   A normally synchronous flip-flop which stores information in a master section on the one clock edge or level and transfers it to the slave section on the next clock edge or level.

metal-oxide-semiconductor   A circuit in which the active region is a metal-oxide-semiconductor sandwich. The oxide acts as the dielectric insulator between the metal and the semiconductor. (*See* MOS.)

microcircuit   Another name for integrated circuit.

microelectronics   Refers to circuits built from miniaturized components and includes the broad category of integrated circuits.

micron   One-millionth of a meter.

microsecond   One-millionth of a second.

mil   One-thousandth of an inch.

millisecond   One-thousandth of a second.

minimum toggle frequency   The guaranteed minimum toggle frequency of a flip-flop. The maximum toggle frequency for a typical device is usually about 20% higher.

monolithic   Refers to the single slice of silicon substrate on which an integrated circuit is built; hence, monolithic integrated circuit.

monostable multivibrator   A logic element which has two states, one of which is stable and the other temporary. A monostable may be triggered into the temporary state where it will remain for a predetermined period of time before reverting to the stable state.

MOS   An acronym for metal-oxide-semiconductor. It is one of the solid-state technologies used for the fabricating of large, low-cost memories with very high input impedance. The insulator used is an oxide of the semiconductor substrate material. (*See* metal-oxide-semiconductor.)

MSB   Most significant bit; the highest weighted digit of a binary number.

MSI   Medium-scale integration. (*See* LSI.) Smaller than LSI, but having at least 12 gates or basic circuits with at least 100 circuit elements.

multiplexer (analog or linear)   A device that can be used to select one out of a multiple number of inputs and switch its information to the output. The output voltage follows the input voltage with a small error. Field-effect and MOS devices are frequently used in this application.

multiplexer (digital)   A device that can select one of its multiple inputs and pass that input logic level on to the output. Analog data cannot be handled (*only digital information*). Input channel select information is usually presented to the device in binary weighted form, and decoded internally, selecting the proper input. The device acts as a single-pole, multiposition switch that will pass digital information in only one direction.

multiplexing   The combining of a number of data lines into a single channel. Multiplexing has already reduced the tons of communications wiring normally used in modern jet liners to a single-wire system in the new 747s. (*See* multiplexer.)

multiplier, four-quadrant   The linear, four-quadrant monolithic block that can multiply, double frequency, perform control functions, modulate, demodulate . . . provide the correctly signed product for all four possible combinations of positive and negative values of its two inputs. The versatile four-quadrant multiplier can perform any function that can be expressed mathematically as the product of two quantities, generating such arithmetic functions as products, squares, quotients, and square roots.

multivibrator   A two-state device. (*See* monostable multivibrator.)

MUX   Multiplex.

NAND (gate)   An AND circuit that delivers an inverted output signal.

nanosecond (ns)   One-billionth of a second, or one millimicrosecond.

negative logic   Logic in which the less positive voltage represents the 1 state. (Positive logic is that in which the more positive voltage stands for the 1 state.)

negative voltage regulator   A negative voltage regulator maintains a constant negative output voltage (*minus* with respect to ground), even with wide input voltage changes or variable output loading.

noise   Extraneous signals, any disturbance, which causes interference with the desired signal or operation.

noise immunity   How much insensitivity a logic circuit has to triggering or reaction to undesirable electrical signals or noise.

noise margin   The amount of voltage of extraneous signal which can be tolerated before a circuit's output voltage deviates from the allowable logic voltage levels.

NOR (gate)   An OR circuit that delivers an inverted output signal.

ns   An abbreviation for nanosecond.

n-type   A semiconductor may be one of two types: *n*-type means that an excess of negative charges are present; *p*-type semiconductor has an excess of positive charges.

octal numbering system   One which uses numbers with the radix (base) 8.

offset   In digital circuits, offset is the dc voltage on which a signal is impressed. In a linear amplifier, offset is the input voltage change needed to cause a zero output voltage.

one-shot   A monostable device which when triggered by an external pulse or level will switch into the temporary state, holding the output at a given logic level until a predetermined time delay is up. At this time the output will revert to its stable state.

op amp   *See* operational amplifier.

open-loop gain   The gain of an amplifier with no external feedback.

operational amplifier   An amplifier having high dc stability and high gain, high input impedance and low output impedance. Its operating characteristics are a function of external feedback components. They are used in such analog functions as summing amplifiers, integrators, function generators, gain blocks, etc. They are a common linear IC building block used in many systems applications.

OR (gate)   If one or more inputs is true the output is true.

oxidation   A chemical reaction in which a thin portion of the surface of a silicon wafer or slab is converted to silicon dioxide. In general, the conversion of any element to its oxide.

package   The metal, ceramic, or plastic protective housing (cover) which encloses a semiconductor device or function.

package count   The number of packaged circuits within a subsystem or system.

parallel operation Type of information transfer whereby all digits of a word are handled simultaneously, each bit being transmitted on separate lines in order to speed operation, as opposed to serial operation in which the bits are transmitted one at a time along a single line.

parameter A definable, stated constant, electrical characteristic of a device.

parasitics Stray reactances or elements associated with the desired components diffused into an integrated circuit. Such parasitics may consist of capacitances, resistances, diodes or transistors effectively in series or in shunt with the diffused components. They tend to limit the performance of the desired components in a circuit unless compensated for in device and circuit design.

passive elements Those components in a circuit which have no gain characteristics, such as capacitors, resistors, and inductors. (*See* active elements.)

PCB Printed circuit board.

phase detector A circuit that provides an output signal indicating the relative phase, or timing difference between input signals. It is often used to operate a voltage-controlled oscillator, thus keeping it synchronized with a reference signal. (Also called *phase discriminator*.)

phase-locked loop (PLL) A closed-loop electronic servomechanism whose output will lock onto and track a reference signal. Phase lock is accomplished by comparing the phase of the output signal (or multiple thereof), with the phase of the reference signal. Any phase difference between these two signals is converted to a correction voltage that changes the phase of the output signal to make it track the reference.

picosecond (ps) One-thousandth of a nanosecond.

PLL *See* phase-locked loop.

positive logic Logic in which the more positive voltage stands for the 1 state. The less positive voltage represents the 0 state. (Negative logic is that in which the less positive voltage represents the 1 state.)

positive voltage regulator A positive voltage regulator operates from a plus supply voltage (referenced to ground) and keeps a constant positive output voltage even with wide changes in either input voltage or in output loading.

printed circuit A substrate on which a predetermined pattern of printed wiring and printed elements has been formed.

programmable counter Also called *modulo N counter* as it can be logically programmed to count to any number from 0 to its maximum possible modulus.

propagation delay The time difference between the application of a signal to a particular system or circuit and its appearance at the output of the system or circuit.

p-type A semiconductor may be one of two types: p-type semiconductor has an excess of positive charges; n-type means that an excess of negative charges are present in the semiconductor.

pulldown resistor A resistor connected to a negative voltage or ground.

pullup resistor A resistor connected to a positive supply voltage.

pulse A brief voltage or current surge of measurable duration (called *pulse width* or *pulse length*) and magnitude (called *pulse height* or *pulse amplitude*).

pulse-triggered binary A flip-flop that will change state when a pulse or waveform of short duration is applied to the input.

Q output The reference output of a flip-flop. When this output is 1, the flip-flop is said to be in the 1 state; when it is 0, the output is said to be in the 0 state.

$\overline{Q}$ (Q-bar) output The second output of a flip-flop. It is always opposite in logic level to the Q output.

quad latch A group of four flip-flops, each of which can store a "true" or "false" level, and that are all normally enabled by a single control line. When the flip-flops are all enabled, new data may be stored in each of them.

quiescent dissipation The power dissipated by a circuit when no dynamic signals are applied to the inputs or activating the circuitry.

RAM *See* random access memory.

random access memory (RAM) A memory that has the stored information immediately available when addressed, regardless of the previous memory address location. As the memory words can be selected in any order, there is equal access time to all.

read-only memory (ROM) A memory which permits the reading of a predetermined pattern of 0s and 1s. This predetermined information is stored in the ROM at the time of its manufacture. A ROM is analogous to a dictionary, where a certain address results in predetermined information output. The information is available at any time, but it can not be modified during normal system operation.

readout The manner in which a computer displays the processed information. May be digital, visual display, punched tape, punched cards, automatic typewriter, etc.

register A short-term storage circuit whose capacity is usually one computer word. Certain variations may also include provisions for shifting, calculating, etc. (*See* static shift register, or dynamic shift register.)

regulator A regulator is a device used to maintain a desired output voltage or current constant regardless of normal changes to the input or to the output load.

reliability The measure of a device's ability to function without failure over a period of time.

reset To return a memory or storage element to its "standard" state, usually 0. Also called *clear*.

RF amplifier An RF amplifier is a radio-frequency amplifier normally operating at a higher frequency than an IF or intermediate-frequency amplifier. It is used for signal isolation, to provide gain for the signal so that it is amplified above the noise level, and to provide input selectivity to a receiver. The RF stage is normally the first stage of amplification in a radio receiver.

ring counter A sequential closed loop of flip-flops. A pattern of binary bits is shifted around the loop (ring) by a common clock. There are usually n (equal to the number of flip-flops) unique states or counts. A ring counter is often used to control a sequence of n events.

ripple Serial transmission of data, corresponding to the way a row of dominoes falls after the first domino has been felled.

ROM *See* read-only memory.

R-S flip-flop A flip-flop consisting of two cross-coupled NAND gates having two inputs designated R and S. A 1 on the S input and 0 on the R input will reset (clear) the flip-flop to the 0 state; 1 on the R input and 0 on the S input will set it to 1. It is assumed that 0s will never appear simultaneously at both inputs. If both inputs have 1s, it will stay as it was. 1 is considered nonactivating.

R-S-T flip-flop A flip-flop having three inputs: R, S, and T. It works as the R-S flip-flop except that the T input is used to cause the flip-flop to change states.

RTL Resistor-transistor logic, developed to overcome the problem of variations in the base-emitter voltage of DCTL, the forerunner of RTL. In RTL, the logic is done by resistors, while the transistors are used to amplify and obtain an inverted output from any positive input.

saturated logic   A form of logic in which one output state is the saturation voltage of a transistor, e.g.: resistor-transistor logic (RTL), diode-transistor logic (DTL), and transistor-transistor logic (TTL). (*See* unsaturated logic.)

schematic   A diagram of a circuit in which symbols are used to stand for the circuit's basic components.

Schmitt trigger   A binary circuit that can take a slowly changing input waveform and convert it to an output waveform which has sharp transitions. Normally hysteresis is present between an upper triggering and a lower triggering level.

Schottky barrier   A simple metal semiconductor interface which exhibits nonlinear impedance.

semiconductor   Materials may be classified as one of three types by resistivity: low resistivity materials are called *conductors;* very high resistivity materials are called *insulators* or *dielectrics;* materials with intermediate resistivities are called *semiconductors* —examples are germanium and silicon.

serial operation   Type of information transfer within a digital computer whereby the bits are handled sequentially rather than simultaneously, as in parallel operation. Serial operation is slower than parallel operation, but utilizes less complex circuitry.

set (as opposed to clear)   An input on a flip-flop used to affect the Q output. It is this input through which signals can be entered to get the Q output to go from 0 to 1. It cannot get Q to go to 0.

shift register   A digital storage circuit which uses a chain of flip-flops to shift data from one flip-flop to its adjacent flip-flop on each clock pulse. Data may be shifted several places to the right or to the left depending on additional gating and the number of clock pulses received. Depending on the number of positions shifted, in a right shift the rightmost characters are lost; in a left shift the leftmost characters are lost. (*Also see* dynamic shift register *and* static shift register.)

single-shot multivibrator   A monostable multivibrator which, after being triggered to the quasi-stable state, will "flop" back by itself to the stable state after a certain period of time.

skewing   Refers to time delay or offset between any two signals in relation to each other.

slew rate   The maximum speed with which an output can be driven from one limit of the dynamic range to the other.

slice   A single wafer or slab, one of many cut from a silicon ingot to form the thin substrate on which semiconductors are fabricated. After processing, the slice may contain up to hundreds or thousands of semiconductor dice.

state   The logic 0 or 1 condition of a circuit's output.

static shift register   A shift register that uses logic flip-flops for storage. This technique in integrated form results in larger storage cell size and therefore shorter shift register lengths. Its primary advantage is that information will be retained as long as the power supply is connected to the device. A minimum clock rate is not required; in fact, it can be unclocked.

switch-tail ring counter   A type of ring counter where one stage has its output inverted before providing an input to the next stage. An even number of states equal to $2n$ (n is equal to the number of flip-flops) is normally generated. For example: A modulo 10 counter can be built from five flip-flops. Each flip-flop changes states once every five counts. Decoding of all 10 states is easily accomplished with 10 two-input gates.

synchronous operation   The performance of a switching network, wherein all major circuits in the network are switched silultaneously by a clock pulse generator.

_T_ flip-flop   A flip-flop having only one input. A pulse appearing on the input will cause the flip-flop to change states.

threshold voltage   That input voltage level at which a binary circuit goes from one logic level to the other, e.g., 0 to 1, or 1 to 0.

toggle   To switch back and forth between two states, as in a flip-flop. (Like a toggle switch.)

toggle flip-flop   _See_ flip-flop.

trigger   A timing pulse which initiates logic circuit operations.

truth table   A tabular chart which lists all the possible combinations of the inputs and outputs of a circuit.

TTL (T²L)   Transistor-transistor logic, which was evolved from a desire for more speed than was possible with DTL. DTL was modified by replacement of the diode cluster with a multiple-emitter transistor.

two-phase dynamic   Dynamic logic circuits using two clock signals to control the progress of information through the circuit or logic system.

unsaturated logic   A form of logic with transistors operated outside the saturated region, e.g.: current-mode logic (CML) and emitter-coupled logic (ECL). (_See_ saturated logic.)

up-down counter   A counter which can count in either an ascending or descending order.

voltage comparator   A circuit that compares two analog voltages and generates a logic output when the two voltages are equal or one is greater or lesser than the reference. (_See_ comparator.)

voltage regulator   A circuit which keeps an output voltage at a predetermined value, or which varies the voltage according to a predetermined plan, regardless of normal voltage changes to the input or impedance changes in the output load.

wafer   The thin slice or flat disc of silicon sliced from a silicon ingot, on which integrated circuits and other semiconductor devices are simultaneously fabricated. (_Also see_ slice.)

wired-AND   _See_ dot-AND.

wired-OR   _See_ dot-OR.

SOLUTIONS TO EXERCISES

Chapter 1

E1.1   1.3 V

E1.2   0.7 V

E1.3   (a) 0.25 mA, 0.63 V

       (b) (1) 0.6 mA; (2) 0.25 mA; (3) 0.3 mA

E1.4 $V_G = 0.76$ V, $V_A = 5$ V

E1.5   0.23 V, 0.25 V

E1.6   (a) Normal active, $I_C = 4.9$ mA, $I_E = 5.0$ mA, $V_{BE} \simeq 0.7$ V, $V_{CE} \simeq 1.7$ V

       (b) Inverted active, $I_C = -0.8$ mA, $I_E = -0.01$ mA, $V_{BC} \simeq 0.7$ V

E1.7   Graphical $I_B = 0.8$ mA

       Analytical $I_B = 0.4$ mA

E1.8   2.5 V

E1.9   EOC: $V_G = 0.7$ V, $V_{\text{out}} = 5$ V

       EOS: $V_G = 1.5$ V, $V_{\text{out}} = 0.2$ V

E1.10 3.62 V

E1.11 (a) 1.5 V; (b) 5.5 V

Chapter 2

E2.1 $K = CA + CB + BA$

E2.2 $F = W + X$

E2.3 $F = C(D + M)$

E2.4 See answer to E2.1

E2.5 See answer to E2.2

E2.6 $F = AB + \bar{A}\bar{C}\bar{D} + \bar{A}B\bar{C} + BCD$

or

$F = AB + B\bar{C}\bar{D} + \bar{A}B\bar{C} + BCD$

E2.7 $F = (B + D)(\bar{A} + \bar{B} + \bar{D})(A + \bar{C} + \bar{D})$

E2.8 (a) $V_F = 1.0$ V; $I_{D2} = 4.0$ mA

(b) $V_F = 4.0$ V; $I_{D2} = 1.0$ mA

E2.9 (a) $V_F = 0.2$ V; $I_{D2} = 0.44$ mA

(b) $V_F = 1.8$ V; $I_{D2} = 0.76$ mA

Chapter 3

E3.1 (a) $V_{OL} = 0.2$ V; $V_{OH} = 3.57$ V

(b) $NM_L = 0.5$ V; $NM_H = 1.91$ V

(c) $k = 2.99$

E3.2 (a) Max $I_{C(\text{sat})} = 1.6$ mA

(b) Min $I_{B(\text{sat})} = 0.16$ mA

(c) $N = 6$

E3.3 (a) Logic swing $= 1.78$ V

(b) $NM_L = 0.42$ V; $NM_H = 1.28$ V

(c) Transition width $= 0.08$ V

E3.4 (a) Logic swing $= 0.78$ V

(b) $NM_L = 0.50$ V; $NM_H = 0.23$ V

(c) $k = 5.84$

E3.5 $N = 16$

E3.6 (a) Logic swing $= 14.8$ V

(b) $V_{IL} = 6.7$ V; $V_{OH} = 15$ V

$V_{IH} = 6.7$ V; $V_{OL} = 0.2$ V

(c) $NM_L = 6.5$ V; $NM_H = 8.3$ V

(d) $N = 41$

E3.7 (a) $I_E = 0.39$ mA (into input terminal)

(b) $I_E = -1.03$ mA (out of input terminal)

E3.8 (a) $N_L = 51$; $N_H = 86$

(b) $I_{CCL} = 0.28$ mA; $I_{CCH} = 0.10$ mA; $PD_{\text{av}} = 0.96$ mW

E3.9 $NM_H = 0.22$ V

Chapter 4

E4.1

D_1	D_2	D_3	G_1	G_2
0	0	1	0	0
0	1	1	0	0
1	0	1	0	0
1	1	1	0	0
1	0	0	1	0
0	1	0	0	1
0	0	0_1	1	0
1	1	0	1	0

E4.2 $184.86

E4.3 $A\bar{B} + A + \bar{C}B + D\bar{A}$

E4.4 $(\bar{A} + \bar{C})(B + A)(\bar{D})(\bar{A} + D)$

E4.5 $\bar{A}\bar{C} + \bar{B}\bar{C} + \bar{B}A + B + C$

E4.6

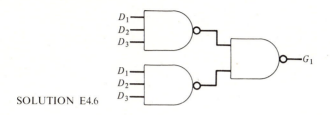

SOLUTION E4.6

E4.7

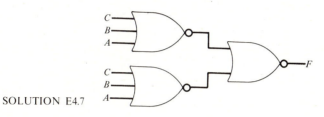

SOLUTION E4.7

E4.8

SOLUTION E4.8

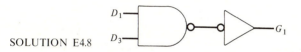

E4.9

SOLUTION E4.9

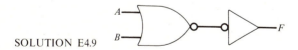

E4.10
C	1	1	1	0	0
A	...	1	0	1	1
B	...	0	1	1	0
S	...	0	0	0	1
K	...	1	1	1	0

E4.11

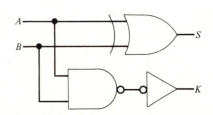

SOLUTION E4.11

E4.12

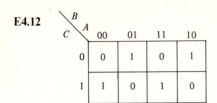

SOLUTION E4.12

E4.13 (a) $\bar{C} + BA$; (b) $A \cdot (\bar{C} + B)$

E4.14 $S_0 = \bar{B}\bar{A}$, $S_1 = \bar{B}A$, $S_2 = BA$, $S_3 = B\bar{A}$; $B = S_3 + S_2$, $A = S_2 + S_1$

E4.15

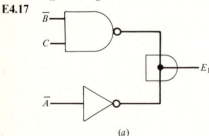

SOLUTION E4.15   (a)   (b)

E4.16 $N_H = 1$; $N_L = 1$

E4.17

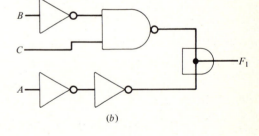

(a)   (b)

SOLUTION E4.17

E4.18

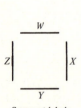

Segment labels

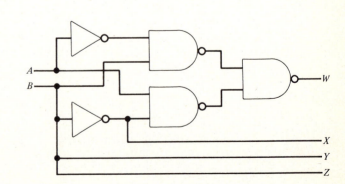

SOLUTION E4.18

Chapter 5

E5.1   (a) $V_{OL} = 0.2$ V; $V_{OH} = 3.18$ V

         (b) $k = 4.34$

E5.2   (a) 9; (b) 36 ns

E5.3   (a)

Q_n	Q_{n+}	S_n	R_n
0	0	Ø	1
0	1	1	0
1	0	0	1
1	1	1	Ø

(b)

n	S	R	Q_n	Q_{n+}
1	0	0	0	?
2	1	0	0	1
3	1	0	1	1
4	0	0	1	?
5	0	1	1	0
6	0	1	0	0
7	1	1	0	0
8	1	1	1	1

E5.4

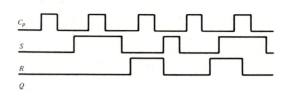

SOLUTION E5.4 Q

E5.5   (a)

Q_n	Q_{n+}	J_n	K_n
0	0	Ø	1
0	1	Ø	0
1	0	0	Ø
1	1	1	Ø

(b)

n	J_n	K_n	Q_n	Q_{n+}
1	0	0	0	1
2	1	0	1	1
3	1	0	0	1
4	0	0	1	0
5	0	1	0	0
6	0	1	1	0
7	1	1	0	0
8	1	1	1	1

E5.6

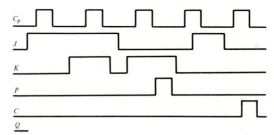

SOLUTION E5.6 Q

E5.7 $J_n = D_n$; $K_n = \bar{D}_n$

Chapter 6

E6.1   (a) $N = 4$; (b) $N = 6$; (c) $N = 8$

E6.2   (a) 256; (b) 4096; (c) 65,536

E6.3

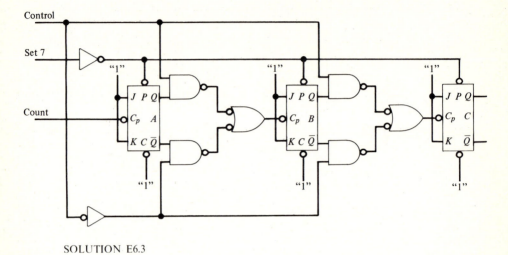

SOLUTION E6.3

E6.4

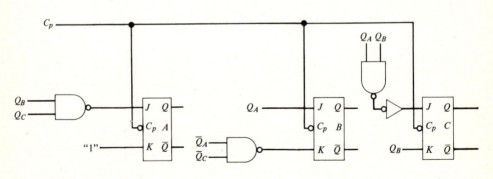

SOLUTION E6.4

E6.5

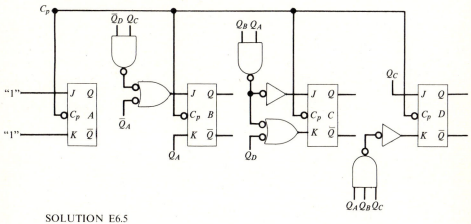

SOLUTION E6.5

E6.6

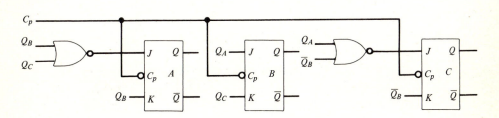

SOLUTION E6.6

E6.7

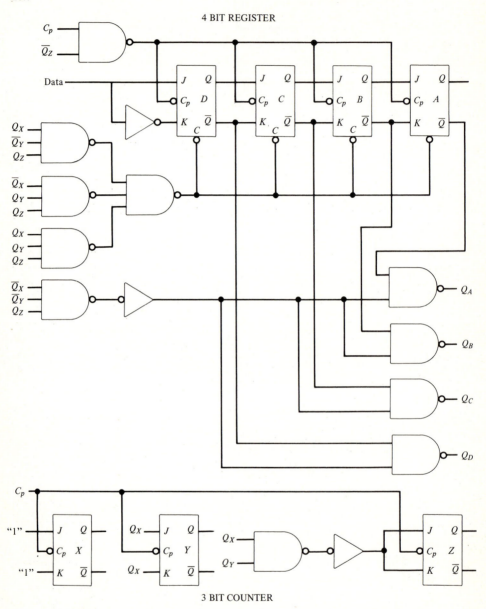

SOLUTION E6.7

E6.8

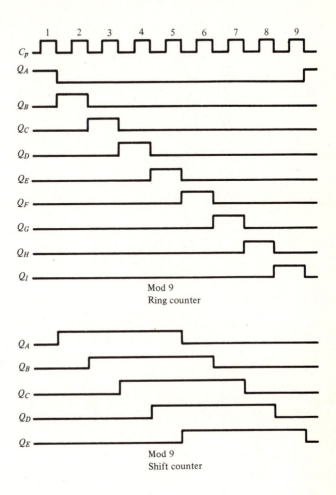

Mod 9
Ring counter

Mod 9
Shift counter

SOLUTION E6.8

Chapter 7

E7.1 At 27°C: $g_m = 0.039$ mho. $r_\pi = 1.3$ kΩ
At 0°C: $g_m = 0.042$ mho, $r_\pi = 1.2$ kΩ
At 70°C: $g_m = 0.031$ mho, $r_\pi = 1.6$ kΩ

E7.2 $I_C = 1.99$ mA; $V_{CE} = 9.03$ V; $r_i = 0.65$ kΩ; $r_o = 3$ kΩ; $a_v = -230$

E7.3 $E_{bat} = 8.33$ V; $a_v = 9410$; $P_0 = 7.09$ μW

E7.4 $r_i = 18.4$ kΩ; $r_o = 51.6$ Ω; $a_v = 0.834$

E7.5 $r_{i3} = 50.2$ kΩ; $r_{o4} = 4.5$ Ω; $P_0 = 231$ μW

E7.6 $r_i = 54.9$ Ω; $r_o = 1.2$ kΩ; $a_v = 21.8$

E7.7

	CE	CC	CB
r_i	1.04 kΩ	82.0 kΩ	13 Ω
r_o	5 kΩ	13 Ω	5 kΩ
a_i	80	81	0.99
a_v	−385	0.99	380

Chapter 8

E8.1 (a) 810; (b) 0.0826

E8.2 $R_{C1} = R_{C2} = 15$ kΩ; $R_{EE} = 13.1$ kΩ

E8.3 $r_{id} = 20.8$ kΩ $a_d = -231$
$r_{ic} = 2.13$ MΩ $a_c = -1.13$

E8.4 $a_d = 47.3$ dB $a_c = 1.1$ dB
CMRR = 46.2 dB

E8.5 $R_{C1} = R_{C2} = 25$ kΩ $R_{EE1} = 26$ kΩ
$R_{C3} = R_{C4} = 1.5$ kΩ $R_{EE2} = 4.1$ kΩ

E8.6 $r_{id} = 26$ kΩ $a_d = 275$
$r_{ic} = 2.66$ MΩ $a_c = 0.08$
CMRR = 71 dB

E8.7 (a) $R_{C2} = 3.75$ kΩ; $E = 11.25$ V
(b) $R_{C2} = 1.88$ kΩ; $R_X = 11.25$ kΩ
(c) $R_{C2} = 3.68$ kΩ; $R_Y = 10.55$ kΩ

E8.8 $R_{E1} = 12$ kΩ

Chapter 9

E9.1 $A_v = 1$

E9.2 $R_f = 99$ kΩ

E9.3 $R_1 = 1.02$ kΩ; $R_f = 50$ kΩ

E9.4 $A_v = -20$

E9.5 $D_s = 0.05$; $dA/A = 1$ percent

E9.6 $R_f = 50.28$ kΩ; $R_i = 2$ MΩ; $R_o = 2.5$ Ω

E9.7 $R_f = 51.28$ kΩ; $R_i = 1.026$ kΩ; $R_o = 2.5$ Ω

E9.8

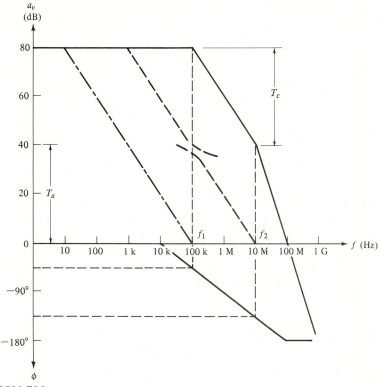

SOLUTION E9.8

E9.9 (a) $A_v = 40$ dB $\equiv 100$; (b) $A_v = 60$ dB $\equiv 1000$

E9.10 (a) $C = 1.6$ μF. For $A_v = 20$ dB, $f_{3\text{dB}} = 10$ kHz.

(b) $R_1 = 100$ Ω; $C_1 = 0.016$ μF. For $A_v = 20$ dB, $f_{3\text{dB}} = 1$ MHz.

Chapter 10

E10.1 $N = 2000$; $V_{\text{REF}} = 2$ V; $T = 12.5$ μs; $f = 80$ kHz; $k = 200$ s^{-1}

E10.2 -2 V

E10.3 $N = 2000$ (cascade three decade counters with PE of MSD being driven from inverter whose input is Q_B)

E10.4 $R_f/R_D = 8/15$; $R_4/R_3 = 1.5$

E10.5 -0.5 V

E10.6 2.5 V

Appendix B

EB.1 0.1 mA

EB.2 6 V

EB.3 (a) 0.947 V; 5 kΩ

(b) 0.049 mA

(c) 8 V; 1 kΩ

    (*d*)   5 V
    (*e*)   (1)   0.189 mA; 5 kΩ
           (2)   0.049 mA
           (3)   8 mA; 1 kΩ
           (4)   5 V
    (*f*)   −0.5 V; 5 kΩ
    (*g*)   0.447 V; 5 kΩ

Appendix C

EC.1   (*a*) 111100_2;   (*b*) 111000010001_2
EC.2   (*a*) 179_{10};   (*b*) 717_{10}
EC.3   (*a*) 1306_8;   (*b*) 537_8
EC.4   (*a*) 254_8;   (*b*) 315_8
EC.5   (*a*) 469_{10}; 111010101_2
        (*b*) 330_{10}; 101001010_2
EC.6   (*a*)   0101   0110   0111   1001   0010   (8421 BCD)
        (*b*)   0111   0011   0011   0111   (8421 BCD)

INDEX

8–Lead metal can
(Top view)

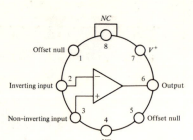

Note: Pin 4 connected to case

8–Lead metal can
(Top view)

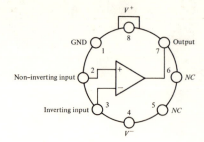

14–Lead dip
(Top view)

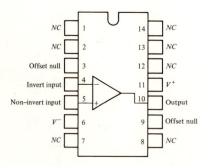

14–Lead dip
(Top view)

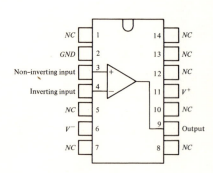

Type 710 voltage comparator

8–Lead minidip
(Top view)

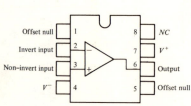

Type 741 operational amplifier

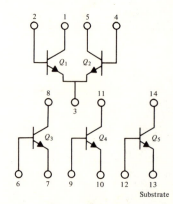

Type 3046 transistor array